T0184873

Texts in Computer Science

Editors
David Gries
Fred B. Schneider

For other titles published in this series, go to
http://www.springer.com/series/3191

Arindama Singh

Elements of Computation Theory

Springer

Arindama Singh
Department of Mathematics
Indian Institute of Technology Madras
Sardar Patel Road
Chennai - 600036
India
asingh@iitm.ac.in

Series Editors
David Gries
Department of Computer Science
Upson Hall
Cornell University
Ithaca, NY 14853-7501, USA

Fred B. Schneider
Department of Computer Science
Upson Hall
Cornell University
Ithaca, NY 14853-7501, USA

ISSN 1868-0941 ISSN 1868-095X (eBook)
ISBN 978-1-4471-6142-4 ISBN 978-1-84882-497-3 (eBook)
DOI 10.1007/978-1-84882-497-3
Springer Dordrecht Heidelberg London New York

British Library Cataloguing in Publication Data
A catalogue record for this book is available from the British Library

Cover design: SPi Publisher Services

Printed on acid-free paper

Springer is part of Springer Science+Business Media (www.springer.com)

Preface

The foundation of computer science is built upon the following questions:

What is an algorithm?
What can be computed and what cannot be computed?
What does it mean for a function to be computable?
How does computational power depend upon programming constructs?
Which algorithms can be considered feasible?

For more than 70 years, computer scientists are searching for answers to such questions. Their ingenious techniques used in answering these questions form the theory of computation.

Theory of computation deals with the most fundamental ideas of computer science in an abstract but easily understood form. The notions and techniques employed are widely spread across various topics and are found in almost every branch of computer science. It has thus become more than a necessity to revisit the foundation, learn the techniques, and apply them with confidence.

Overview and Goals

This book is about this solid, beautiful, and pervasive foundation of computer science. It introduces the fundamental notions, models, techniques, and results that form the basic paradigms of computing. It gives an introduction to the concepts and mathematics that computer scientists of our day use to model, to argue about, and to predict the behavior of algorithms and computation. The topics chosen here have shown remarkable persistence over the years and are very much in current use.

The book realizes the following goals:

- To introduce to the students of computer science and mathematics the elegant and useful models and abstractions that have been created over the years for solving foundational problems about computation
- To help the students develop the ability to form abstract models of their own and to reason about them
- To strengthen the students' capability for carrying out formal and rigorous arguments about algorithms

- To equip the students with the knowledge of the computational procedures that have hunted our predecessors, so that they can identify similar problems and structures whenever they encounter one
- To make the essential elements of the theory of computation accessible to not-so-matured students having not much mathematical background, in a way that is mathematically uncompromising
- To make the students realize that mathematical rigour in arguing about algorithms can be very attractive
- To keep in touch with the foundations as computer science has become a much more matured and established discipline

Organization

Chapter 1 reviews very briefly the mathematical preliminaries such as set theory, relations, graphs, trees, functions, cardinality, Cantor's diagonalization, induction, and pigeon hole principle. The pace is not uniform. The topics supposedly unknown to Juniors are discussed in detail.

The next three chapters talk about regular languages. Chapter 2 introduces the four mechanisms such as regular expressions, regular grammars, deterministic finite automata, and the nondeterministic finite automata for representing languages in their own way. The fact that all these mechanisms represent the same class of languages is shown in Chap. 3. The closure properties of such languages, existence of other languages, other structural properties such as almost periodicity, Myhill–Nerode theorem, and state minimization are discussed in Chap. 4.

Chapters 5 and 6 concern the class of context-free languages. Here we discuss context-free grammars, Pushdown automata, their equivalence, closure properties, and existence of noncontext-free languages. We also discuss parsing, ambiguity, and the two normal forms of Chomsky and Greibach. Deterministic pushdown automata have been introduced, but their equivalence to LR(k) grammars are not proved.

Chapters 7 and 8 discuss the true nature of general algorithms introducing the unrestricted grammars, Turing machines, and their equivalence. We show how to take advantage of modularity of Turing machines for doing some complex jobs. Many possible extensions of Turing machines are tried and shown to be equivalent to the standard ones. Here, we show how Turing machines can be used to compute functions and decide languages. This leads to the acceptance problem and its undecidability.

Chapter 9 discusses the jobs that can be done by algorithms and the jobs that cannot be. We discuss decision problems about regular languages, context-free languages, and computably enumerable languages. The latter class is tackled greedily by the use of Rice's theorem. Other than problems from language theory, we discuss unsolvability of Post's correspondence problem, the validity problem of first order logic, and of Hilbert's tenth problem.

Chapter 10 is a concise account of both space and time complexity. The main techniques of log space reduction, polynomial time reduction, and simulations including Savitch's theorem and tape compression are explained with motivation and rigour. The important notions of NLS-completeness and NP-completeness are explained at length. After proving the Cook–Levin theorem, the modern approach of using gadgets in problem reduction and the three versions of optimization problems are discussed with examples.

Special Features

There are places where the approach has become nonconventional. For example, transducers are in additional problems, nondeterministic automata read only symbols not strings, pushdown automata require both final states and an empty stack for acceptance, normal forms are not used for proving the pumping lemma for context-free languages, Turing machines use tapes extended both ways having an accepting state and a rejecting state, and acceptance problem is dealt with before talking about halting problem. Some of the other features are the following:

- All bold-faced phrases are defined in the context; these are our definitions.
- Each definition is preceded by a motivating dialogue and succeeded by one or more examples.
- Proofs always discuss a *plan of attack* and then proceed in a straightforward and rigorous manner.
- Exercises are spread throughout the text forcing lateral thinking.
- Problems are included at the end of each section for reinforcing the notions learnt so far.
- Each chapter ends with a summary, bibliographical remarks, and additional problems. These problems are the unusual and hard ones; they require the guidance of a teacher or browsing through the cited references.
- An unnumbered chapter titled *Answers/Hints to Selected Problems* contains solutions to more than 500 out of more than 2,000 problems.
- It promotes interactive learning building the confidence of the student.
- It emphasizes the intuitive aspects and their realization with rigorous formalization.

Target Audience

This is a text book primarily meant for a semester course at the Juniors level. In IIT Madras, such a course is offered to undergraduate Engineering students at their fifth semester (third year after schooling). The course is also credited by masters students from various disciplines. Naturally, the additional problems are tried by such masters students and sometimes by unusually bright undergraduates. The book (in notes form) has also been used for a course on Formal Language Theory offered to Masters and research scholars in mathematics.

Notes to the Instructor

The book contains a bit more than that can be worked out (not just covered) in a semester. The primary reason is: these topics form a prerequisite for undertaking any meaningful research in computer science. The secondary reason is the variety of syllabi followed in universities across the globe. Thus, courses on automata theory, formal languages, computability, and complexity can be offered, giving stress on suitable topics and mentioning others. I have taught different courses at different levels from it sticking to the core topics.

The core topics include a quick review of the mathematical preliminaries (Chap. 1), various mechanisms for regular languages (Chap. 2), closure properties and pumping lemma for regular languages (Sects. 4.2 and 4.3), context-free languages (Sects. 5.2–5.4), pushdown automata, pumping lemma and closure properties of context-free languages (Sects. 6.2, 6.4, and 6.5), computably enumerable languages (Chap. 7), a noncomputably enumerable language (Chap. 8), algorithmic solvability (Sects. 9.2–9.4), and computational complexity (Chap. 10). Depending on the stress on certain aspects, some of the proofs from these core topics can be omitted and other topics can be added.

Chennai,

Arindama Singh

January 2009

Acknowledgements

I cheerfully thank

My students for expressing their wish to see my notes in the book form,

IIT Madras for keeping me off from teaching for a semester, for putting a deadline for early publication, and for partial financial support under the Golden Jubilee Book Writing Scheme,

Prof. David Gries and Prof. Fred B. Schneider, series editors for Springer texts in computer science,

Mr. Wayne Wheeler and his editorial team for painstakingly going through the manuscript and suggesting improvements in presentation,

Prof. Chitta Baral of Arizona State University for suggesting to include the chapter on mathematical preliminaries,

Prof. Robert I. Soare of the University of Chicago, *Dr. Abhaya Nayak* of Maquarie University, and *Dr. Sounaka Mishra* of IIT Madras for suggesting improvements,

My family, including my father Bansidhar Singh, mother Ragalata Singh, wife Archana, son Anindya Ambuj, daughter Ananya Asmita, for tolerating my obsession with the book, and

My friends Mr. Biswa R Patnaik (in Canada) and Mr. Sankarsan Mohanty (in Orissa) for their ever inspiring words.

Arindama Singh

Contents

1 Mathematical Preliminaries

1.1 Introduction

An African explorer conversant with the language of the Hottentot tribe asks a native, "How many children do you have?" The tribesman answers, "Many." The determined explorer persists on. He shows his index finger, meaning "one?" Promptly comes the answer, "no." He adds his middle finger, meaning "two"; the answer is "no"; "three?," "no"; "four," "no." Now all the five fingers on the explorer's right hand are straight. Answer comes, "yes." The puzzled explorer experiments with another tribesman. Over the next week, he discovers that they have only three kinds of numbers, one, two, and many.

It is an old story, but perhaps not without morals. The Hottentot tribesman does not have a way of naming the numbers more than two. How does he manage his cattle?

Our mathematical tradition has gone so far and so deep that it is indeed difficult to imagine living without it. In this small chapter, we will discuss a fragment of this tradition so that the rituals of learning the theory of computation can be conducted relatively easily. In the process, we will fix our notation.

1.2 Sets

A **set** is a collection of objects, called its members or **elements.** When writing a set by showing its elements, we write it with two closing curly brackets. The curly brackets serve two purposes: one, it shows the elements inside, and two, it says that the whole thing put together is another object on its own right. Sometimes, we become tired of writing out all the elements, and we put three dots. For example,

{pen, pencil, knife, scissors, paper, chalk, duster, paper weight, . . .}

is the set of names of educational stationery. It is the *extensional* way of representing a set. But what about the expression, "the set of names of educational stationery?" It is a set, nonetheless, and the same set as above. We thus agree to represent a

A. Singh, *Elements of Computation Theory*, Texts in Computer Science,

set by specifying a property that may be satisfied by each of its elements. It is the *intensional* way of representing a set. (Note the spelling of "intensional.") If A is a set and a is an element in A, we write it as $a \in A$. The fact that a is not an element of A is written as $a \notin A$.

When specifying a set by a property, we write it as $\{x : P(x)\}$, meaning that this set has all and only those x as elements which satisfy a certain property $P(\cdot)$. Two sets A, B are said to be the same set, written $A = B$, if each element of A is in B and each element of B is in A. In that case, their defining properties must be logically equivalent. For example, the set $\{2, 4, 6, 8\}$ can be written as $\{x : x = 2 \text{ or } x = 4 \text{ or } x = 6 \text{ or } x = 8\}$. Also, $\{2, 4, 6, 8\} = \{x : 2 \text{ divides } x \text{ and } x \text{ is an integer with } 0 < x < 10\}$. Further, $\{2, 4, 6, 8\} = \{4, 8, 6, 2\}$; the order of the elements when written down explicitly, does not matter, and it is assumed that there are no repetitions of elements in a set.

Two sets A, B are **equal**, written $A = B$, whenever they have precisely the same elements. We say that A is a **subset** of a set B, written $A \subseteq B$, whenever each element of A is in B. Similarly, we say that A is a **proper subset** of B, written $A \subsetneqq B$, whenever $A \subseteq B$ but $A \neq B$. Thus, $A = B$ iff $A \subseteq B$ and $B \subseteq A$. We abbreviate the phrase "if and only if" to "iff."

A mathematical discourse fixes a big set, called the universal set often denoted by U. All other sets considered are subsets of this big set in that particular context. As a convention, this big set is never mentioned, and if strict formal justification is required, then this is brought into picture.

Let A, B be sets and let U be the universal set (in this context, of course). The **union** of A, B is written as $A \cup B = \{x : x \in A \text{ or } x \in B\}$. The **intersection** of A, B is $A \cap B = \{x : x \in A \text{ and } x \in B\}$. The **difference** of A, B is $A - B = \{x : x \in A \text{ but } x \notin B\}$. The **complement** of A is $\overline{A} = U - A = \{x : x \notin A\}$.

We define the **empty set** $\emptyset$ as a set having no elements; $\emptyset = \{x : x \neq x\} = \{\,\}$. For any set A, $\emptyset = A - A$. We find that $A \cup \emptyset = A$ and $A \cap U = A$. Moreover, $\emptyset$ is unique, whatever be the universal set. When two sets A, B have no common elements, we say that they are **disjoint** and write it as $A \cap B = \emptyset$. For example, with the universal set as the set of all natural numbers $\mathbb{N} = \{0, 1, 2, \ldots\}$, A as the set of all prime numbers, and B as the set of all composite numbers, we see that

$$A \cap B = \emptyset, \quad A \cup B = \mathbb{N} - \{0, 1\} = \overline{\{0, 1\}}.$$

The **power set** of A, denoted by 2^A, is the set of all subsets of A. An **ordered pair** of two objects a, b is denoted by (a, b), which can also be written as a set. For example, we may define $(a, b) = \{\{a\}, \{a, b\}\}$. We see that the ordered pair satisfies the following property:

$$(a, b) = (c, d) \text{ iff } a = c \text{ and } b = d.$$

In fact, it is enough for us to remember this property of the ordered pairs. The **Cartesian product** of the sets A and B is $A \times B = \{(x, y) : x \in A \text{ and } y \in B\}$.

The operations of union, intersection, and the (Cartesian) product can be extended further. Suppose $\mathcal{A} = \{A_i : i \in I\}$ is a collection of sets A_i, where I is some set, called an index set here. Then, we define

$$\cup\mathcal{A} = \cup_{i\in I} A_i = \{x : x \text{ is in some } A_i\}.$$

$$\cap\mathcal{A} = \cap_{i\in I} A_i = \{x : x \text{ is in each } A_i\}.$$

For the product, we first define an n-tuple of objects by

$$(a_1, a_2, \ldots, a_n) = ((a_1, a_2, \ldots, a_{n-1}), a_n), \quad \text{when } n > 2.$$

Finally, we write

$$A_1 \times A_2 \times \cdots \times A_n = \{(x_1, x_2, \ldots, x_n) : \text{ each } x_i \in A_i, \text{ for } i = 1, \ldots, n\}.$$

When each $A_i = A$, we write this n-product as A^n. Similarly, arbitrary Cartesian product can be defined though we will use only a finite product such as this. Because of the above property of ordered pairs, and thus of n-tuples, the kth coordinate becomes meaningful. The kth coordinate of the n-tuple $(x_1, x_2, \ldots, x_n)$ is x_k.

Clearly, $A \times \emptyset = \emptyset \times A = \emptyset$. In addition, we have the following identities:

Double Complement: $\overline{\overline{A}} = A$.

De Morgan: $\overline{A \cup B} = \overline{A} \cap \overline{B}$, $\overline{A \cap B} = \overline{A} \cup \overline{B}$.

Commutativity: $A \cup B = B \cup A$, $A \cap B = B \cap A$.

Associativity: $A \cup (B \cup C) = (A \cup B) \cup C$, $A \cap (B \cap C) = (A \cap B) \cap C$.

Distributivity: $A \cup (B \cap C) = (A \cup B) \cap (A \cup C)$, $A \cap (B \cup C) = (A \cap B) \cup (A \cap C)$,

$A \times (B \cup C) = (A \times B) \cup (A \times C)$, $A \times (B \cap C) = (A \times B) \cap (A \times C)$,

$A \times (B - C) = (A \times B) - (A \times C)$.

1.3 Relations and Graphs

We use the relations in an extensional sense. The binary relation of "is a son of" between human beings is thus captured by the set of all ordered pairs of human beings, where the first coordinate of each ordered pair is a son of the second coordinate. A **binary relation** from a set A to a set B is a subset of $A \times B$. If R is such a relation, a typical element in R is an ordered pair (a, b), where $a \in A$ and $b \in B$ are suitable elements. The fact that a and b are related by R is written as $(a, b) \in R$; we also write it as $R(a, b)$ or as aRb.

Any relation $R \subseteq A \times A$ is called a **binary relation on the set** A. Similarly, an n-ary relation on a set A is some subset of A^n. For example, take P as a line and a, b, c as points on P. Write $B(a, b, c)$ for "b is between a and c." Then B is a ternary relation, that is, $B \subseteq P^3$, and $B(a, b, c)$ means the same thing as $(a, b, c) \in B$. **Unary relations** on a set A are simply the subsets of A.

Binary relations on finite sets can conveniently be represented as diagrams. In such a diagram, the elements of the set A are represented as small circles (**points, nodes, or vertices**) on the plane and each ordered pair $(a, b) \in R$ of elements

$a, b \in A$ is represented as an arrow from a to b. We write inside each circle its name. The diagrams are now called **digraphs** or **directed graphs**.

Example 1.1. The digraph for the relation $R = \{(a, a), (a, b), (a, d), (b, c), (b, d), (c, d), (d, d)\}$ on the set $A = \{a, b, c, d\}$ is given in Fig. 1.1. □

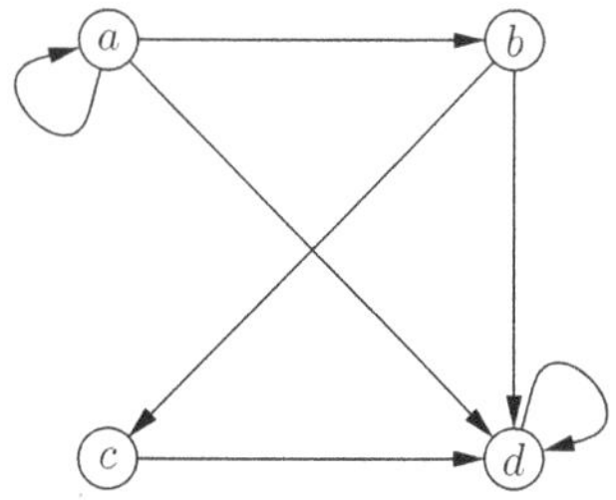

Fig. 1.1. Digraph for R in Example 1.1.

Sometimes we need to give names to the edges (arrows) in a digraph as we give names to roads joining various places in a city. The resulting digraph is called a labeled digraph. Labeled digraphs are objects having three components: a set of vertices V, a set of **edges** E, and an **incidence relation** $I \subseteq E \times V \times V$, which specifies which edge is incident from which vertex to which vertex.

Example 1.2. Figure 1.2 depicts the labeled digraph (V, E, I), where the vertex set $V = \{a, b, c, d\}$, the edge set $E = \{e_1, e_2, e_3, e_4, e_5, e_6, e_7\}$, and the incidence relation $I = \{(e_1, a, a), (e_2, a, b), (e_3, a, d), (e_4, b, c), (e_5, b, d), (e_6, c, d), (e_7, d, d)\}$. □

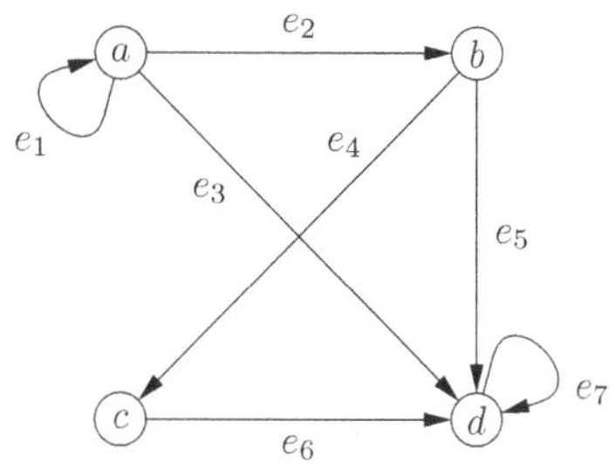

Fig. 1.2. Labeled digraph for Example 1.2.

Sometimes we do not need to have the arrows in a digraph; just the undirected edges suffice. In that case, the digraph is called an **undirected graph** or just a **graph**. Usually, we redefine this new concept. We say that a graph is an object having a set of vertices and a set of edges as components, where each edge is a two-elements set (instead of an ordered pair) of vertices.

Example 1.3. The graph in Fig. 1.3 represents the graph $G = (V, E)$, where $V = \{a, b, c, d\}$ and $E = \{\{a, a\}, \{a, b\}, \{a, d\}, \{b, c\}, \{b, d\}, \{c, d\}, \{d, d\}\}$. □

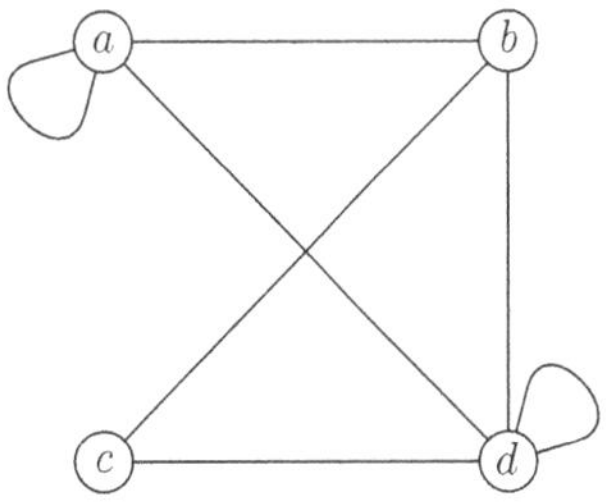

Fig. 1.3. Graph for Example 1.3.

Two vertices in a graph are called **adjacent** if there is an edge between them. A **path** in a graph is a sequence of vertices $v_1, v_2, \ldots, v_n$, such that each v_i is adjacent to v_{i+1}. For example, in Fig. 1.3, the sequence a, a, b, c, d is a path, and so are a, b, d and a, d, c. We say that the path $v_1, v_2, \ldots, v_n$ **connects** the vertices v_1 and v_n. Moreover, the **starting point** of the path $v_1, v_2, \ldots, v_n$ is v_1 and its **end point** is v_n. Similarly, directed paths are defined in digraphs.

A graph is called **connected** if each vertex is connected to each other by some path (not necessarily the same path). For example, the graph of Fig. 1.3 is a connected graph. If you delete the edges $\{a, d\}$, $\{b, c\}$, $\{b, d\}$, then the resulting graph is not connected. Similarly, if you remove the edges $\{a, b\}$, $\{a, d\}$, then the resulting graph is not connected either.

A path $v_1, v_2, \ldots, v_n, v_1$ is called a **cycle** when $n > 2$ and no vertex, other than the starting point and the end point, is repeated. A connected cycleless graph is called a **tree.** By giving directions to the edges in a tree, we obtain a directed tree.

In a directed tree, if there is exactly one vertex towards which no arrow comes but all edges incident with it are directed outward, the vertex is called the **root** of the tree. Similarly, any vertex in a directed tree from which no edge is directed outward is called a **leaf.** It is easy to see that there can be only one edge incident with a leaf and that edge is directed toward the leaf. Because, otherwise, there will be a cycle containing the leaf! A tree having a root is called a **rooted tree.**

Example 1.4. The graph on left side in Fig. 1.4 is a rooted tree with c as its root. It is redrawn on the right in a different way. □

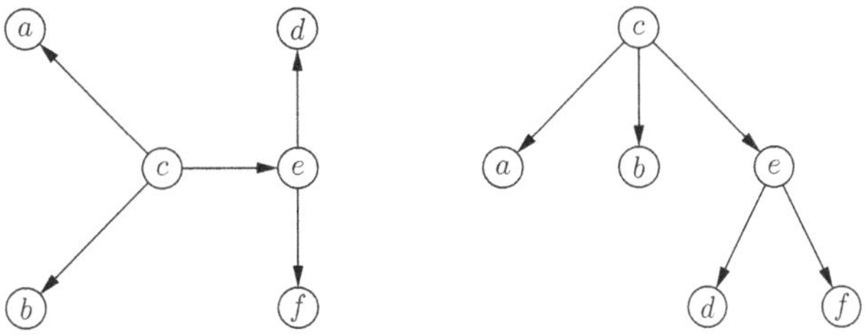

Fig. 1.4. Trees for Example 1.4.

The trees in Fig. 1.4 are redrawn in Fig. 1.5 omitting the directions. Since the direction of edges are always from the root toward any vertex, we simply omit the directions. We will use the word "tree" for rooted trees and draw them without directions. Sometimes, we do not put the small circles around the vertices. The tree on left side in Fig. 1.5 uses this convention of drawing the trees in Fig. 1.4. It is further abbreviated in the right side tree of the same figure.

In Fig. 1.5, all **children** of a vertex are placed below it and are joined to the **parent** vertex by an undirected edge. We say that the root c has depth 0; the children of the root are the vertices of depth 1 in the tree; the depth 2 vertices are the vertices that have an edge from the vertices of depth 1; these are children of the children of the root, and so on. In a tree, the **depth** is well defined; it shows the distance of a vertex from the root. The depth of a tree is also called its **height**, and trees in computer science grow downward!

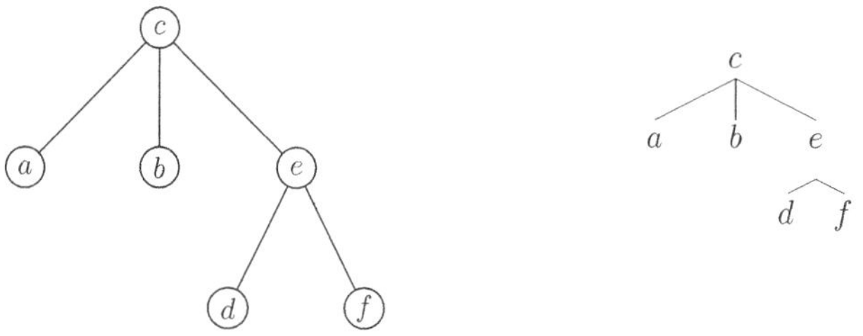

Fig. 1.5. Trees for Example 1.4 redrawn.

Leaves of the tree in Fig. 1.5 are the vertices $a, b, d,$ and f. The nonleaf vertices are called the **branch nodes** (also, branch points or branch vertices).

With this short diversion on representing relations as graphs, we turn towards various kinds of properties that a relation might satisfy. Suppose R is a binary relation on a set A. The most common properties associated with R are

Reflexivity: for each $x \in A$, xRx.
Symmetry: for each pair of elements $x, y \in A$, if xRy, then yRx.
Antisymmetry: for each pair of elements $x, y \in A$, if xRy and yRx, then $x = y$.
Transitivity: for each triple of elements $x, y, z \in A$, if xRy and yRz, then xRz.

A binary relation can be both symmetric and antisymmetric, or even neither. For example, on the set $A = \{a, b\}$, the relation $R = \{(a, a), (b, b)\}$ is reflexive, symmetric, transitive, and antisymmetric. On the other hand, the relation $S = \{(a, b), (a, c), (c, a)\}$ on A is neither reflexive, nor symmetric, nor antisymmetric, nor transitive.

Given a binary relation R on a set A, we can extend it by including some more ordered pairs of elements of A so that the resulting relation is both reflexive and transitive. Such a minimal extension is called the **reflexive and transitive closure** of R.

Example 1.5. What is the reflexive and transitive closure of $R = \{(a, b), (b, c), (c, b), (c, d)\}$ on the set $A = \{a, b, c, d\}$?

Solution. Include the pairs $(a, a), (b, b), (c, c), (d, d)$ to make it reflexive. You have

$$R_1 = \{(a, a), (a, b), (b, b), (b, c), (c, b), (c, c), (c, d), (d, d)\}.$$

Since $(a, b), (b, c) \in R_1$, include (a, c). Proceeding similarly, You arrive at

$$R_2 = \{(a, a), (a, b), (a, c), (b, b), (b, c), (b, d), (c, b), (c, c), (c, d), (d, d)\}.$$

Since $(a, b), (b, d) \in R_2$, include (a, d) to obtain

$$R_3 = \{(a, a), (a, b), (a, c), (a, d), (b, b), (b, c), (b, d), (c, b), (c, c), (c, d), (d, d)\}.$$

You see that R_3 is already reflexive and transitive; there is nothing more required to be included. That is, the reflexive and transitive closure of R is R_3. □

In Example 1.5, interpret $(x, y) \in R$ to mean there is a communication link from city x to city y, possibly, a one-way communication link. Then the reflexive and transitive closure of R describes how messages can be transmitted from one city to another either directly or via as many intermediate cities as possible. On the set of human beings, if parenthood is the relation R, then its reflexive and transitive closure gives the ancestor–descendant relationship, allowing one to be his/her own ancestor.

The **inverse** of a binary relation R from A to B, denoted by R^{-1}, is a relation from B to A such that for each $x \in A, y \in B$, $xR^{-1}y$ iff yRx. The inverse is also well defined when R is a binary relation on a set A. For example, on the set of human beings, the inverse of the relation of "teacher of" is the relation of "student of." What is the inverse of "father of?"

Suppose R is a binary relation on a set A. We say that R is a **partial order** on A if it is reflexive, antisymmetric, and transitive. The relation of $\leq$ on the set of all integers is a partial order. For another example, let S be a collection of sets. Define the relation R on S by "for each pair of sets $A, B \in S$, ARB iff $A \subseteq B$." Clearly, R is a partial order on S.

A relation R is called an **equivalence relation** on A if it is reflexive, symmetric, and transitive. The equality relation is obviously an equivalence relation, but there can be others.

For example, in your university, two students are related by R if they are living in the same dormitory. It is easy to see that R is an equivalence relation. You observe that the students are now divided into as many classes as the number of dormitories.

Intuitively, the elements related by an equivalence relation share a common property. The property divides the underlying set into many subsets, where elements of any of these typical subsets are related to each other. Moreover, elements of one subset are never related to elements of another subset. Let R be an equivalence relation on a set A. For each $x \in A$, we define the **equivalence class** of x, denoted as $[x]$, by

$$[x] = \{y \in A : xRy\}.$$

Besides x, each element $y \in [x]$ is a **representative** of the equivalence class $[x]$. We define a **partition** of A as any collection $\mathcal{A}$ of disjoint nonempty subsets of A whose union is A. The equivalence classes form the set, so to speak. We will see that its converse is also true.

Theorem 1.1. *Let R be a binary relation on a set A. If R is an equivalence relation, then the equivalence classes of R form a partition of A. Conversely, if $\mathcal{A}$ is a partition of A, then there is an equivalence relation R such that the elements of $\mathcal{A}$ are precisely the equivalence classes of R.*

Proof. Suppose R is an equivalence relation on a set A. Let $\mathcal{A} = \{[x] : x \in A\}$, the collection of all equivalence classes of elements of A. To see that $\mathcal{A}$ is a partition of A, we must show two things.

(a) Each element of A is in some equivalence class.
(b) Any two distinct equivalence classes are disjoint.

The condition (a) is obvious as each $x \in [x]$. For (b), let $[x]$ and $[y]$ be distinct equivalence classes. Suppose, on the contrary, that there is $z \in [x] \cap [y]$. Now, xRz and yRz. By the symmetry and transitivity of R, we find that xRy. In that case, for each $w \in [x]$, we have wRy as xRw and yRx implies that yRw. This gives $w \in [y]$. That is, $[x] \subseteq [y]$. For each $v \in [y]$, we have similarly vRx. This gives $v \in [x]$. That is, $[y] \subseteq [x]$. We arrive at $[x] = [y]$, contradicting the fact that $[x]$ and $[y]$ are distinct.

Conversely, suppose $\mathcal{A}$ is a partition of A. Define the binary relation R on A by

For each pair of elements $x, y \in A$, xRy iff both x and y are elements of the same subset B in $\mathcal{A}$.

That means if $a, b \in B_1 \in \mathcal{A}$, then aRb, but if $a \in B_1 \in \mathcal{A}$ and $c \in B_2 \in \mathcal{A}$, then a and c are not related by R (assuming that $B_1 \neq B_2$). Clearly, R is an equivalence relation. To complete the proof, we show that

(i) Each equivalence class of R is an element of $\mathcal{A}$.
(ii) Each element of $\mathcal{A}$ is an equivalence class of R.

For (i), let $[x]$ be an equivalence class of R for some $x \in A$. This x is in some $B \in \mathcal{A}$. Now $y \in [x]$ iff xRy iff $y \in B$, by the very definition of R. That is, $[x] = B$.

Similarly for (ii), let $B \in \mathcal{A}$. Take $x \in B$. Now $y \in B$ iff yRx iff $y \in [x]$. This shows that $B = [x]$. □

1.4 Functions and Counting

Intensionally, a function is a map that associates an element of a set to another, possibly in a different set. For example, the square map associates a number to its square. If the underlying sets are $\{1, 2, 3, 4, 5\}$ and $\{1, 2, \ldots, 50\}$, then the square map associates 1 to 1, 2 to 4, 3 to 9, 4 to 16, and 5 to 25. Extensionally, we would say that the graph of the map (not the same graph of the last section, but the graph as you have plotted on a graph sheet in your school days) is composed of the points $(1, 1)$, $(2, 4)$, $(3, 9)$, $(4, 16)$, and $(5, 25)$. We take up the extensional meaning and define a function as a special kind of a relation.

Let A, B be two sets. A **partial function** f from the set A to the set B is a relation from A to B satisfying

for each $x \in A$, if xfy and xfz, then $y = z$.

This conforms to our intensional idea of a map, as no element x can be associated with different elements by f. It is a partial function, as it is not required that each element in A has to be taken to some element of B by f. We use the more suggestive notation for a partial function by writing $f : A \to B$. When xfy, we write $y = f(x)$. The set A is called the **domain** of f and the set B is called the **co-domain** of f.

A partial function $f : A \to B$ is called a **total function** if for each $x \in A$, there is some $y \in B$ such that $y = f(x)$, that is, when f takes each element of A to some (hence a unique, corresponding to that element of A) element of B. Note that it does not say that all elements of A should be mapped to the same element of B. Following tradition, we use the word **function** for a total function and use the adjective "partial" for a partial function. To emphasize, a partial function is not necessarily strictly partial.

The **range** of a partial or a total function is the subset of B, which are attained by f, that is, the set $\{y \in B : y = f(x) \text{ for some } x \in A\}$. If $D \subseteq A$, we write the range of f as $f(D) = \{f(x) \in B : x \in D\}$.

Partial functions can be composed by following the internal arrows in succession, as it is said. For example, if $f : A \to B$, $g : B \to C$ are two maps, and $a \in A$, then we can get $f(a)$ in B, and then go to $g(f(a))$ in C by following the maps. This can be done provided $f(a)$ is defined, and also if $g(f(a))$ is defined. When both f, g are total functions, this is obviously possible. The composition map is written as $g \circ f$. Notice the reverse notation; it helps in evaluation, that is, when it is defined for an element $a \in A$, we have $(g \circ f)(a) = g(f(a))$. The composition map $g \circ f : A \to C$.

The **inverse** of a partial function $f : A \to B$ is well defined; it is the relation f^{-1} defined from B to A. But f^{-1} is not necessarily a partial function. For example, the square map on $\{1, -1\}$ is a partial function (in fact, total) whose inverse is the relation $\{(1, 1), (1, -1)\}$. This inverse, the square-root relation is not a partial function as 1 is taken to two distinct elements 1 and -1. This happens because the square map is not one to one, it is many to one. A partial function $f : A \to B$ is called **one–one** if

for each $x, y \in A$, if $x \neq y$, then $f(x) \neq f(y)$.

Equivalently, for each $x, y \in A$, if $f(x) = f(y)$, then $x = y$. It is easy to see that the inverse of a one–one partial function is again a one–one partial function. But inverse of a total function (even if one–one) need not be a total function. Because, there might be elements in its co-domain that are not attained by the map. We call a partial function $f : A \to B$ an **onto** partial function (or say that f is a partial function from A onto B) if

for each $y \in B$, there exists an $x \in A$ such that $y = f(x)$.

Equivalently, for an onto function, the range of f coincides with the co-domain of f. It is easy to see that a one–one total function from A onto B has an inverse, which is also a one–one total function from B onto A. A one–one total function is also called an **injection** and an onto total function is called a **surjection.** A **bijection** is a one–one onto total function. Two sets A and B are said to be in **one-to-one correspondence** if there is a bijection from one to the other.

Suppose $f : A \to B$, $C \subseteq A$, and $D \subseteq B$. The **image** of C under f is the set $\{f(x) : x \in C\}$, and is denote by $f(C)$. Similarly, the **inverse image** of D under f is the set $\{x : f(x) \in D\}$ and is denoted by $f^{-1}(D)$. The notation $f^{-1}(D)$ should not mislead you to think of f^{-1} as a partial function; as you have seen, it need not be so. f^{-1} is a total function only when f is a bijection. In such a case, $f \circ f^{-1}$ is the identity map on B and $f^{-1} \circ f$ is the identity map on A. The **identity map** takes each element to itself.

An arbitrary Cartesian product of sets can be defined using functions. An n-tuple of elements from a set A can be given as a function $f : \{1, 2, \ldots, n\} \to A$. Here, we simply rewrite the kth coordinate in the n-tuple as $f(k)$. We use this observation for an extension of the product. Suppose $\mathcal{A} = \{A_i : i \in I\}$ is a collection of sets A_i, where I is an index set. Then the product is defined as

$$\times\mathcal{A} = \times_{i\in I} A_i = \text{ the set of all functions } f : I \to \cup\mathcal{A} \text{ with } f(i) \in A_i.$$

Recall that the Hottentot tribesman knew the meaning of one-to-one correspondence; he could say that he had as many children as the fingers on the right hand of the African explorer. The tribesman did not know the name of any number beyond two. He could count but could not name the number he has counted. Probably, he had a bag full of pebbles as many as the sheep he owned. This is how he used to keep track of the sheep in his possession. The idea behind counting the elements of a set is the same as that of the tribesman.

We say that two sets A and B have the **same cardinality** when there is a bijection between them. We write cardinality of a set A as $|A|$. Cardinality of a set intuitively captures the idea of the number of elements in a set. Notice that we have not defined what $|A|$ is; we have only defined $|A| = |B|$.

To make the comparison of cardinalities easier, we say $|A| \leq |B|$ if there is an injection from A to B (a one–one total function from A to B). We say that $|A| \geq |B|$ if $|B| \leq |A|$. *Cantor–Schröder–Bernstein Theorem* says that if $|A| \leq |B|$ and $|B| \leq |A|$, then $|A| = |B|$. (A proof is outlined in Problem 1.29.) Further, we write $|A| < |B|$ when $|A| \leq |B|$ but $|A| \neq |B|$; similarly, $|A| > |B|$ when $|A| \geq |B|$ but $|A| \neq |B|$.

Since the empty set $\emptyset$ has no elements, we define $0 = |\emptyset|$. We go a bit further and define $1 = |\{0\}|$. And then define inductively $n + 1 = |\{0, 1, \ldots, n\}|$. These are our **natural numbers**, elements of the set $\mathbb{N} = \{0, 1, 2, \ldots\}$. Then the operations of $+$ and $\times$ are defined on $\mathbb{N}$ in the usual way. Notice that $+$ is a function that maps a pair of natural numbers to a natural number, and so is $\times$. Of course, we simplify notation by writing mn for $m \times n$.

Once the natural numbers are defined, we define the **integers** by extending it to $\mathbb{Z} = \mathbb{N} \cup \{-1, -2, \ldots\} = \mathbb{N} \cup \{-n : n \in \mathbb{N}\}$, with the convention that $-0 = 0$. Notice that $-n$ is just a symbol, where we put a minus sign preceding a natural number. The set of **positive integers** is defined as $\mathbb{Z}_+ = \mathbb{N} - \{0\} = \mathbb{Z} - \{-n : n \in \mathbb{N}\}$. Then the operations $+$ and $\times$ are extended to $\mathbb{Z}$ in the usual way, $n + (-n) = 0$, etc. This extension allows us to solve simple equations such as $x + 5 = 2$ in $\mathbb{Z}$.

However, equations such as $5x = 2$ cannot be solved in $\mathbb{Z}$. We thus extend $\mathbb{Z}$ by including symbols of the form m/n. We arrive at the set of **rational numbers** $\mathbb{Q} = \{p/q : p \in \mathbb{Z}, q \in \mathbb{Z}_+\}$, with the usual conventions like $(p/q) \times q = p$ and $p/q = r/s$ when $ps = rq$, etc. We then see that each rational number can be represented as a decimal number like $m.n_1n_2n_3\cdots$ with or without the minus sign, where each n_i is one of the digits $0, 1, \ldots, 9$ and $m \in \mathbb{N}$. Such decimals representing rational numbers satisfy a nice property:

> Beyond some finite number of digits after the decimal point, a finite sequence of digits keep recurring.

Further, we see that these recurring decimals uniquely represent rational numbers with one exception: a decimal number with recurring 0's can also be written as another decimal with recurring 9's. We agree to use the latter and discard the former if uniqueness is at vogue. For example, the decimal 0.5 is written as $0.499999\cdots$. This guarantees a unique decimal representation of each number in $\mathbb{Q}$. Also, this convention allows us to consider only the recurring infinite decimals instead of bothering about terminating decimals.

We then extend our numbers to the **real numbers**. The set of real numbers, $\mathbb{R}$, is the set of all (infinite) decimals. The nonrecurring decimals are called the **irrational numbers**, they form the set $\mathbb{R}-\mathbb{Q}$. This extension now allows us talking about square roots of numbers. For example, $\sqrt{2} \in \mathbb{R}$, but $\sqrt{2} \notin \mathbb{Q}$. However, we find that it is not enough for solving polynomial equations, for example, there is no real number x satisfying the equation $x^2 + 1 = 0$.

For solving polynomial equations, we would need the complex numbers. We define the set of complex numbers as $\mathbb{C} = \{x + \imath y : x, y \in \mathbb{R}\}$, where $\imath$ is taken as $\sqrt{-1}$. Notice that $\imath$ is again a symbol which is used as $\sqrt{-1}$. The operations of $+$, $\times$, taking roots, etc. are extended in the usual way to $\mathbb{C}$. It can be shown that our quest for solving polynomial equations stop with $\mathbb{C}$. The *Fundamental Theorem of Algebra* states that each polynomial of degree n with complex coefficients has exactly n complex roots.

Besides, there are complex numbers that are not roots of any polynomial equation. Moreover, we require to distinguish between surds like $\sqrt{2}$ and numbers like π. For this purpose, we restrict our polynomials to have rational coefficients. We define an **algebraic number** as a complex number, which is a solution of a polynomial equation, where the polynomials have rational coefficients. Other complex numbers are called **transcendental numbers**. In fact, there are more transcendental numbers than the algebraic numbers (Problem 1.28), though we know a very few of them.

Further, there is a natural partial order on $\mathbb{N}$, the $\leq$ relation. It so happens that this relation can be extended to $\mathbb{Z}$, $\mathbb{Q}$, and $\mathbb{R}$. However, it stops there; it cannot be extended to $\mathbb{C}$. This does not mean that there cannot be any partial order on $\mathbb{C}$. For example, define $\leq$ on $\mathbb{C}$ by $a + \imath b \leq c + \imath d$ iff $a < c$, or ($a = c$ and $b \leq c$), taking the $\leq$ on $\mathbb{R}$ as the basis. You can verify that this defines a partial order on $\mathbb{C}$. But this is not an extension of the $\leq$ relation on $\mathbb{R}$. Because our definition of the relation $<$ says that $0 < \imath$, we should have $0 \times 0 < \imath \times \imath = -1$, which is not true.

Observe that by this process of extension, we have *constructed* some infinite sets, the sets whose cardinalities cannot be written as natural numbers. Infinite sets can be defined without using numbers. We say that a set is infinite or has infinite cardinality if it has the same cardinality as one of its proper subsets. And a finite set is a set which is not infinite. Naturally, a finite set has greater cardinality than any of its proper subsets. For example, $\mathbb{N}$ is an infinite set as the function $f : \mathbb{N} \to 2\mathbb{N}$ defined by $f(n) = 2n$ is a bijection, where $2\mathbb{N}$ denotes the set of all even natural numbers.

It can further be shown that a set is finite iff either it is $\emptyset$ or it is in one-to-one correspondence with a set of the form $\{0, 1, \ldots, n\}$ for some natural number n. We take it as our definition of a **finite set**. Using this, we would define an **infinite set** as one which is not finite. Cardinalities of finite sets are now well defined: $|\emptyset| = 0$, and a set that is in one-to-one correspondence with $\{0, 1, \ldots, n\}$ has cardinality $n + 1$. Can we similarly define the cardinalities of infinite sets?

Well, let us denote the cardinality of $\mathbb{N}$ as $\aleph_0$; read it as aleph-null. Can we say that all infinite sets have cardinality $\aleph_0$? Again, let us increase our vocabulary. We call a set **denumerable** (also called enumerable) if it is in one-to-one correspondence with $\mathbb{N}$, that is, having cardinality as $\aleph_0$. The one-to-one correspondence with $\mathbb{N}$ gives an enumeration of the set: if $f : \mathbb{N} \to A$ is the bijection, then the elements of the denumerable set A can be written as $f(0), f(1), f(2), \ldots$. The following statement should then be obvious.

Theorem 1.2. *Each infinite subset of a denumerable set is denumerable.*

Proof. Let A be an infinite subset of a denumerable set B. You then have a bijection $f : \mathbb{N} \to B$. That is, elements of B are in the list: $f(0), f(1), f(2), \ldots$. All elements of A appear in this list exactly once. Define a function $g : \mathbb{N} \to A$ by induction:

Take $g(0)$ as the first element in the list, which is also in A.
Take $g(k + 1)$ as the first element in the list occurring after $g(k)$, which is in A.

This g is a bijection since A is infinite. □

Further, we say a set to be **countable** if it is either finite or denumerable. Since each number $n < \aleph_0$, (Why?) it follows from Theorem 1.2 that a set A is countable iff $|A| \leq \aleph_0$ iff A is in one-to-one correspondence with a subset of $\mathbb{N}$ iff there is a one–one function from A to $\mathbb{N}$ iff A is a subset of a countable set. Moreover, in all these iff statements, $\mathbb{N}$ can be replaced by any other countable set. Further, A is denumerable iff A is infinite and countable. A set that is not countable is called **uncountable**.

Theorem 1.3. *$\mathbb{Z}$ and $\mathbb{Q}$ are denumerable; thus countable.*

Proof. For the denumerability of $\mathbb{Z}$, we put all even numbers in one-to-one correspondence with all natural numbers, and then put all odd natural numbers in one-to-one correspondence with the negative integers.

To put it formally, observe that each natural number is in one of the forms $2n$ or $2n + 1$. Define a function $f : \mathbb{N} \to \mathbb{Z}$ by $f(2n) = n$ and $f(2n + 1) = -(n + 1)$. To visualize, $f(0) = 0, f(1) = -1, f(2) = 1, f(3) = -2, \ldots$. It is easy to see that f is a bijection. Therefore, $\mathbb{Z}$ is denumerable.

For the denumerability of $\mathbb{Q}$, let $\mathbb{Q}_P$ denote the set of all symbols of the form p/q, where $p, q \in \mathbb{Z}_+$. Also, denote the set of positive rational numbers by $\mathbb{Q}_+$. When we look at these symbols as rational numbers, we find many repetitions. For example, corresponding to the single element 1 in $\mathbb{Q}_+$, there are infinitely many elements $1/1, 2/2, 3/3, \ldots$ in $\mathbb{Q}_P$. We construct a one–one function from the set $\mathbb{Q}_P$ to $\mathbb{Z}_+$.

The elements of $\mathbb{Q}_P$ can be written in a two-dimensional array as shown below.

$$
\begin{array}{ccccccccccccc}
\frac{1}{1} & \rightarrow & \frac{1}{2} & & \frac{1}{3} & \rightarrow & \frac{1}{4} & & \frac{1}{5} & \rightarrow & \frac{1}{6} & & \cdots \\
& \swarrow & & \nearrow & & \swarrow & & \nearrow & & \swarrow & & & \\
\frac{2}{1} & & \frac{2}{2} & & \frac{2}{3} & & \frac{2}{4} & & \frac{2}{5} & & \frac{2}{6} & & \cdots \\
\downarrow & \nearrow & & \swarrow & & \nearrow & & \swarrow & & & & & \\
\frac{3}{1} & & \frac{3}{2} & & \frac{3}{3} & & \frac{3}{4} & & \frac{3}{5} & & \frac{3}{6} & & \cdots \\
& \swarrow & & \nearrow & & \swarrow & & & & & & & \\
\frac{4}{1} & & \frac{4}{2} & & \frac{4}{3} & & \frac{4}{4} & & \frac{4}{5} & & \frac{4}{6} & & \cdots \\
\downarrow & \nearrow & & \swarrow & & & & & & & & & \\
\frac{5}{1} & & \frac{5}{2} & & \frac{5}{3} & & \frac{5}{4} & & \frac{5}{5} & & \frac{5}{6} & & \cdots \\
& \swarrow & & & & & & & & & & & \\
\vdots & & \vdots & & \vdots & & \vdots & & \vdots & & \vdots & & \ddots
\end{array}
$$

In the first row are written all the numbers of the form $1/m$, varying m over $\mathbb{Z}_+$; in the second row, all the numbers of the form $2/m$; etc. Any number $p/q \in \mathbb{Q}_P$ is the qth element in the pth row. Thus the array exhausts all numbers in $\mathbb{Q}_P$.

Now, start from $\frac{1}{1}$ and follow the arrows to get an enumeration of numbers in the array. This means that the (enumerating) function $f : \mathbb{Q}_P \to \mathbb{Z}_+$ is defined by

$$f\left(\frac{1}{1}\right) = 1,\ f\left(\frac{1}{2}\right) = 2,\ f\left(\frac{2}{1}\right) = 3,\ f\left(\frac{3}{2}\right) = 4,\ f\left(\frac{2}{2}\right) = 5,\ \cdots$$

We see that f is a one–one function.

Let $\mathbb{Q}_A = \mathbb{Q}_P \cup \{0\} \cup \{-p/q : p/q \in \mathbb{Q}_P\}$. This set contains $\mathbb{Q}$ in the same way as $\mathbb{Q}_P$ contains $\mathbb{Q}_+$. Extend $f : \mathbb{Q}_P \to \mathbb{Z}_+$ to the function $f : \mathbb{Q}_A \to \mathbb{Z}$ by taking $f(0) = 0$ and $f(-p/q) = -f(p/q)$ for $p/q \in \mathbb{Q}_P$. This extended f is also a one–one function. (Show it.) We thus have $|\mathbb{Q}_A| \le |\mathbb{Z}|$. Since $\mathbb{Q}_A$ is infinite, it is denumerable. Since $\mathbb{Q}$ is an infinite subset of $\mathbb{Q}_A$, by Theorem 1.2, it is also denumerable. Finally, denumerability implies countability. □

The method of proof in Theorem 1.3 proves that $\mathbb{Z}_+ \times \mathbb{Z}_+$ is denumerable. All you have to do is keep the ordered pair (m, n) in place of m/n in the array. For another alternative proof of this fact, you can search for a one–one map from $\mathbb{Z}_+ \times \mathbb{Z}_+$ to $\mathbb{Z}_+$. One such is defined by $f(m, n) = 2^m 3^n$. Just for curiosity, try to prove that the function $g : \mathbb{Z}_+ \times \mathbb{Z}_+ \to \mathbb{Z}_+$ given by $g(m, n) = 2^m(2n - 1)$ is a bijection.

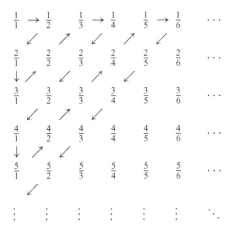

It is then clear that (Cartesian) product of two countable sets is countable. You can further extend to any finite number of products, as $A \times B \times C$ is simply $(A \times B) \times C$, etc. The proof method also shows that a *countable union of countable sets*

is countable. Keep on the first row, the first countable set, on the second row, the second countable set, and so on, and then proceed as in the above proof! Similarly, a denumerable union of finite sets is denumerable. However, a denumerable product of denumerable sets is not countable. Check whether you can prove it following the proof of Theorem 1.4 below!

Sometimes a result can be too counter intuitive; Theorem 1.3 is one such. Unlike $\mathbb{N}$, if you choose any two numbers from $\mathbb{Q}$, you can always get another number (in fact, infinitely many numbers) between them. But this does not qualify $\mathbb{Q}$ to have more elements than $\mathbb{N}$. What about $\mathbb{R}$, the set of real numbers?

Theorem 1.4. $\mathbb{R}$ *is uncountable.*

Proof. Let $J = \{x \in \mathbb{R} : 0 < x < 1\}$ be the open interval with 0 and 1 as its end points; the end points are not in J. We first show that J is uncountable.

We use the famous *diagonalization method* of Georg Cantor. Suppose, on the contrary, that J is countable. Then, we have a bijection $g : \mathbb{N} \to J$. The elements of J can now be listed as $g(0), g(1), g(2), \ldots$. But each number in J is a nonterminating decimal number. Write all these numbers in J as in the following:

$$
\begin{aligned}
g(0) &= 0.a_{11}\ a_{12}\ a_{13}\ a_{14}\ \cdots \\
g(1) &= 0.a_{21}\ a_{22}\ a_{23}\ a_{24}\ \cdots \\
g(2) &= 0.a_{31}\ a_{32}\ a_{33}\ a_{34}\ \cdots \\
g(3) &= 0.a_{41}\ a_{42}\ a_{43}\ a_{44}\ \cdots \\
&\ \ \vdots \\
g(n-1) &= 0.a_{n1}\ a_{n2}\ a_{n3}\ a_{n4}\ \cdots\ a_{nn}\ \cdots \\
&\ \ \vdots
\end{aligned}
$$

where each a_{ij} is one of the digits $0, 1, \ldots, 9$. Using this array of decimals, construct a real number $d = 0.d_1\ d_2\ d_3\ d_4\ \cdots$, where for each $i \in \mathbb{Z}_+$,

d_i equals 0 when $a_{ii} = 9$, otherwise, d_i equals $a_{ii} + 1$.

This number d is called the diagonal number. It differs from each number in the above list. For example, $d \neq g(0)$ as $d_1 \neq a_{11}$, and $d \neq g(n-1)$ as $d_n \neq a_{nn}$. But this d is in J, contradicting the fact that the list contains each and every number in J. Therefore, J is uncountable.

Since $J \subseteq \mathbb{R}$, uncountability of $\mathbb{R}$ is proved. Of course, a stronger fact holds: $|J| = |\mathbb{R}|$. To see this, define a function $f : J \to \mathbb{R}$ by $f(x) = (x - 1/2)/(x - x^2)$. Verify that f is a bijection. □

Recall that we have agreed to write sets by specifying a defining property of the form $\{x : P(x)\}$. Existence of an uncountable set such as $\mathbb{R}$ dispenses the wrong belief that every set can be expressed by a property and each property can give rise to a set.

To see this, suppose you want to express properties in English. (In fact, any other language will do.) Each such property is a finite sequence of symbols from the Roman alphabet. For any fixed n, there are clearly a finite number of properties having n occurrences of symbols. Hence, there are only a countable number of properties. But there are uncountable number of sets, for example, sets of the type $\{r\}$, where r is a real number. Hence, there are sets that do not correspond to any property. For the converse, see the following example.

Example 1.6. Consider the property of "$A \notin A$." This property is perhaps meaningful when A is a set. Let S be the set of all sets A such that $A \notin A$. Now, is $S \in S$ or $S \notin S$?

Solution. If $S \in S$, then by the very definition of S, we see that $S \notin S$. Conversely, if $S \notin S$, then S satisfies the defining property of S, and thus $S \in S$. Therefore, $S \in S$ iff $S \notin S$. □

The contradiction in Example 1.6 shows that there is no set corresponding to the property that $x \notin x$. See *Russells' paradox* if you are intrigued.

This is the reason why axiomatic set theory restricts the definition of new sets as subsets of old sets. In Example 1.6, if you take the big set as the set of all sets, then S could be a subset of that. In fact, axiomatic set theory does a clever thing so that existence of such a big set can never be justified. Moreover, it prevents constructing a set that may also be a member of itself. For information on axiomatic set theory, you may search for set theories of Zermelo–Fraenkel, or of Gödel–Berneys–Von Neuman, or of Scott–Potter.

In the proof of Theorem 1.4, we have constructed a set by changing the diagonal elements of the array of numbers listed as $g(0), g(1), \ldots$. Below we give a very general result that cardinality of any set must be strictly less than the cardinality of its power set, which was first proved by Georg Cantor using (and inventing) the diagonalization method.

Theorem 1.5 (Cantor). *No function from a set to its power set can be onto. Therefore,* $|A| < |2^A|$.

Proof. Let $f : A \to 2^A$ be any function. Let $x \in A$. Then $f(x) \subseteq A$. Thus, x is either in $f(x)$ or it is not. Define a subset B of A by $B = \{x \in A : x \notin f(x)\}$. We show that there is no $y \in A$ such that $B = f(y)$.

On the contrary, suppose there exists a $y \in A$ such that $B = f(y)$. Is $y \in B$? If $y \in B$, then as per the definition of B, $y \notin f(y)$. That is, $y \notin B$. On the other hand, if $y \notin B$, then $y \notin f(y)$. Again, because of the definition of B, $y \in B$. We thus arrive at the contradiction that $y \in B$ iff $y \notin B$. Hence, there is no $y \in A$ such that $B = f(y)$. That is, f is not an onto map.

Finally, take $g : A \to 2^A$ defined by $g(x) = \{x\}$, for each $x \in A$. This map is one–one. Hence $|A| \le |2^A|$. But $|A| \ne |2^A|$, as there is no function from A onto 2^A. Therefore, $|A| < |2^A|$. □

Now it is obvious that the power set of a denumerable set is uncountable. You can derive the uncountability of the open interval J defined in the proof of Theorem 1.4

from Cantor's theorem by using binary decimals instead of the usual decimals. This representation will first prove the fact that $|J| = |2^{\mathbb{N}}|$. In the last paragraph of the proof of Theorem 1.4, we have shown that $|J| = |\mathbb{R}|$. Hence, $\mathbb{R} = |2^{\mathbb{N}}|$. Cantor conjectured that each infinite set in between $\mathbb{N}$ and $\mathbb{R}$ must be in one-to-one correspondence with one of $\mathbb{N}$ or $\mathbb{R}$, now known as the *Continuum Hypothesis*. Because of this reason, we denote the cardinality of $2^{\mathbb{N}}$ as $\aleph_1$. Then, the continuum hypothesis asserts that any subset of $\mathbb{R}$ is either finite or has cardinality $\aleph_0$ or $\aleph_1$. There are interesting results about the continuum hypothesis, but you should be able to look for them on your own.

Notice that we have only defined the cardinalities of finite sets. For infinite sets, we know how to compare the cardinalities. Moreover, for notational convenience, we write the cardinality of a denumerable set as $\aleph_0$. Cardinality of the power set of a denumerable set is written as $\aleph_1$. We may thus extend this notation further by taking cardinality of the power set of the power set of a denumerable set as $\aleph_2$, etc., but we do not have the need for it right now. The countability results discussed so far can be summarized as:

$\mathbb{Z}$, $\mathbb{Q}$, and the set of algebraic numbers are countable.
$\mathbb{R}$, $\mathbb{C}$, $\mathbb{R} - \mathbb{Q}$, and the set of transcendental numbers are uncountable.
An infinite subset of any denumerable set is denumerable.
Subsets of countable sets are countable.
Denumerable union of finite sets is denumerable.
Denumerable union of denumerable sets is denumerable.
Countable union of countable sets is countable.
Finite product of denumerable sets is denumerable
Finite product of countable sets is countable.
Countable product of countable sets need not be countable.
Power set of a denumerable set is uncountable.

1.5 Proof Techniques

In all the theorems except Theorem 1.1, we have used the technique of *proof by contradiction*. It says that statement S is considered proved when from the assumption that S is not true follows a contradiction. To spell it out explicitly, suppose we have a set of premises Ω. We want to prove that if all the statements in Ω are true, then the statement S must be true. The method of proof by contradiction starts by assuming the falsity of S along with the truth of every statement in Ω. It then derives a contradiction. If the premises in Ω are $S_1, S_2, \ldots, S_n$, then the method can be schematically written as

Required: $S_1, S_2, \ldots, S_n$. Therefore, S.
We prove: $S_1, S_2, \ldots, S_n$ and *not* S. Therefore, a contradiction.

Keeping the premises in Ω in the background, the method may be summarized as

not S implies a contradiction. Therefore, S.

It works because, when *not* S implies a contradiction, *not* S must be false. Therefore, S must be true. Conversely, when S is true, *not* S is false, and then it must imply a contradiction.

The method of proof by contradiction appears in many disguises. Calling the above as the first form, the second form of the method is

S_1 and *not* S_2 implies a contradiction. Therefore, "if S_1, then S_2."

This is justified due to the simple reason that by asserting the falsity of "if S_1, then S_2," we assert the truth of S_1 and the falsity of S_2.

The third form of proof by contradiction is proving the *contraposition* of a statement. It says that for proving "if S_1, then S_2," it is sufficient to prove its contraposition, which is "if S_2 is false, then S_1 is false." In fact, a statement and its contraposition are logically equivalent. Why is it another form of the "proof by contradiction?" Suppose you have already proved the contrapositive statement "if S_2 is false, then S_1 is false". Then, *not* S_2 and S_1 together give the contradiction that S_1 is true as well as false. And then the second form above takes care. The principle of proving the contraposition can be summarized as

If *not* S_2, then *not* S_1. Therefore, if S_1, then S_2.

The contrapositive of a statement is not the same as its converse. The converse of "if S_1 then S_2" is "if S_2 then S_1," which is equivalent to "if *not* S_1 then *not* S_2".

The fourth form does not bring a contradiction from the assumption that S is false. Rather it derives the truth of S from the falsity of S. Then, it asserts that the proof of S is complete. It may be summarized as

not S implies S. Therefore S.

Justification of this follows from the first form itself. Assume *not* S. Since you have proved *not* S implies S, you also have S. That is, by assuming *not* S you have got the contradiction: S and *not* S.

The fifth form is the so-called *argument by cases.* It says that to prove a statement S, pick up any other statement P. Assume P to deduce S. Next, assume that P is false and also deduce S. Then, you have proved S. It may be summarized as

P implies S. *not* P implies S. Therefore, S.

Why does it work? Since, you have proved P implies S, its contraposition holds. That is, you have *not* S implies *not* P. But you have already proved that *not* P implies S. Thus, you have proved *not* S implies S. Now the fourth form takes care.

All the while you are using the law of *double negation,* that is, *not not* S is equivalent to S. Along with it comes the law of *excluded middle* that one of S or *not* S must be true. For example, in the argument by cases, you use the fact that one of P

or *not* P must hold. There have been many objections to the law of excluded middle. One of the most befitting example is by J. B. Bishop. It is as follows.

Example 1.7. Show that there are irrational numbers x, y such that x^y is rational.

Solution. Gelfand–Schneider theorem states that if $\alpha \notin \{0, 1\}$ is an algebraic number and β is an irrational number, then α^β is a transcendental number. $\sqrt{2}$ is algebraic as it is a solution of the equation $x^2 = 2$. We also know that $\sqrt{2}$ is irrational. Therefore, $\sqrt{2}^{\sqrt{2}}$ is transcendental, and hence, irrational. Now, with $x = \sqrt{2}^{\sqrt{2}}$ and $y = \sqrt{2}$, you see that

$$x^y = (\sqrt{2}^{\sqrt{2}})^{\sqrt{2}} = \sqrt{2}^{(\sqrt{2}\times\sqrt{2})} = (\sqrt{2})^2 = 2,$$

a rational number. Hence we have an affirmative answer to the above question. You may also use the fact that e and $\ln 2$ are irrational but $e^{\ln 2} = 2$, a rational number. However, look at another proof given below that uses the argument by cases.

The alternative proof: Either $\sqrt{2}^{\sqrt{2}}$ is rational or irrational. If it is rational, we take $x = y = \sqrt{2}$. If it is irrational, take $x = \sqrt{2}^{\sqrt{2}}$ and $y = \sqrt{2}$. Then, $x^y = 2$, a rational number. With argument by cases, the proof is complete. □

The alternative proof in the solution of Example 1.7 does not give us a pair x, y of irrational numbers satisfying the requirements. However, it proves the existence of such irrational numbers. In the mainstream mathematics, this is a well appreciated proof. The question there is not only about accepting the law of excluded middle, but also about appreciating the nature of existence in mathematics.

When we say that there exists an object with such and such property, what we understand is: it is not the case that the property is false for every object in the domain of discourse. It may or may not be always possible to construct that object with exactitude. See the following example.

Example 1.8. Show that there exists a real number x satisfying $x^{x^5} = 5$.

Solution. Let $f : \mathbb{R} \to \mathbb{R}$ be given by $f(x) = x^{x^5}$. We see that $f(1) = 1$ and $f(2) = 2^{32}$. Also, f is a continuous function. Since $f(1) < 5 < f(2)$, by the intermediate value theorem, there exists an a with $1 < a < 2$ such that $f(a) = 5$. □

The a is not obtained exactly, but we know that there is at least one such point between 1 and 2. Of course, there is a better way of getting such an a. For example, $a = 5^{1/5}$ does the job! However, even if we could not have got this simple a, the solution in Example 1.8 is still valid.

Example 1.9. Show that there exists a program that reports correctly whether tomorrow by this time, I will be alive or not.

Solution. Consider the following two programs:

Program-1: Print "I will be alive tomorrow by this time."
Program-2: Print "I will not be alive tomorrow by this time."

Either I will be alive tomorrow by this time or not. That is, either Program-1 correctly reports the fact or Program-2 correctly reports the fact. Hence, we have a program that does the job, but we do not know which one. □

What about *Program-3*: Wait till tomorrow. See what happens; then report accordingly? This, of course, does the job correctly. But this is constructive, whereas the existence in the solution of Example 1.9 is not. Looked in a different way, the statement in Example 1.9 is ambiguous.

One meaning of it has been exploited in the example. The other meaning asks for a program that "justifiably predicts" whether I will be alive till tomorrow or not. The solution there does not answer this nor does Program-3. Also, none of Program-1 and Program-2 work correctly in all cases. Program-1 can be wrong if I really die today, and Program-2 is wrong when I do live up to day after tomorrow.

Nonetheless, the method combined with diagonalization can be used for showing nonexistence of programs. Choose any programing language in which you can write a program that would compute functions with domain $\mathbb{N}$ and co-domain $\{0, 1\}$. Computing a function here means that if $f : \mathbb{N} \to \{0, 1\}$ is a given function, then you can possibly have a program P_f in your language that takes input as any $n \in \mathbb{N}$ and outputs $f(n)$. This program P_f computes the function f.

Example 1.10. Prove that there exists a function $g : \mathbb{N} \to \{0, 1\}$ that cannot be computed by any program in whatever language you choose.

Solution. Choose your language, say, C. Since the C-programs can be enumerated in alphabetical order, they form a countable set. The set of all C-programs that compute functions is a subset of the set of all C-programs, and hence, is countable. Enumerate the programs that compute functions in alphabetical order. Call them $C_0, C_1, C_2, \ldots$. Each C_j takes a number $n \in \mathbb{N}$ as an input, and outputs either 0 or 1. Define a function $f : \mathbb{N} \to \{0, 1\}$ by

$$f(n) = 0, \text{ if } C_n \text{ outputs 1 on input } n, \text{ and } f(n) = 1 \text{ if } C_n \text{ outputs 0 on input } n.$$

Now, if there exists a C-program that computes f, then it must be one of $C_0, C_1, C_2, \ldots$. Suppose it is C_m. But on input m, C_m outputs a different value than f. So, it does not compute f. Hence no C-program can compute this f. □

A nonconstructive version of the solution to Example 1.10 uses the fact that there are uncountable number of functions from $\mathbb{N}$ to $\{0, 1\}$, whereas there are only a countable number of C-programs. It is because the set of all such functions is in one-to-one correspondence with the power set of $\mathbb{N}$. See Problem 1.16.

In the above solution, we have used a form of proof by contradiction. All the forms of proof by contradiction are propositional in nature, that is, they simply play with the simple propositional connectives like "and," "or," "not," etc. One more proof method that uses the propositional connectives is the so-called proof employing a *conditional hypothesis.* It is summarized as

Assume S_1. Prove S_2. Thereby you have proved: if S_1, then S_2.

As P *or* Q is logically equivalent to (*not* P) *implies* Q, the method of conditional hypothesis can be used to prove such a disjunction. That is, a typical proof of P *or* Q starts with assuming *not* P and concluding Q.

Sometimes the proofs using conditional hypothesis can be confusing and misleading if it is combined with other spurious elements. See the following example.

Example 1.11 (Fallacious Proof?). Show that there is no life on earth.

Solution. Let S be the statement: If S is true, then there is no life on earth.

We first show that S is true. As S is in the form "if S_1, then S_2" to show it, we assume S_1 and prove S_2. Here is such a proof:

Proof of "S is true" begins.

Assume S_1, that is, S is true. As S is true, we have

If S is true, then there is no life on earth.

Owing to our assumption that S is true, we see that

There is no life on earth.

This is S_2. So we have proved "if S_1, then S_2", that is, If S is true, then there is no life on earth. That is, S is true.

Proof of "S is true" ends.

As S is true, we have

If S is true, then there is no life on earth.

The truth of the last statement and the fact that S is true imply

There is no life on earth. □

Certainly, there is something wrong. You see that there is nothing wrong with the proof using the technique of conditional hypothesis. It is wrong to denote the statement

If S is true, then there is no life on earth.

by S. To further understand what is going on, see the following commercial of this book using *three seductive questions.*

Example 1.12 (Trap). This is a conversation between you and me.

I : I'll ask you three questions. Would you like to answer each with "Yes" or "No?"

You : Yes.

I : That's my first question. Will you answer the same to the third as to the second?

You : No.

I : Will you promise me that you will read only this book on Theory of computation and no other book on the topic throughout your life?

Now you are trapped. Since you have answered "No" to the second question, you cannot answer "No" to the third question. Had you answered "Yes" to the second question, then also you had to answer "Yes" to the third. □

Of course, had you chosen to answer "No" to my first question, you would not have been trapped. But as it is, why does it happen? The reason is the same as in Example 1.11, a spurious self-reference. When you give notation to a statement, it should have both the possibilities of being true or false. The notation itself cannot impose a truth condition. In Example 1.11, the notation S violates this, as S cannot be false there. If S is false, then the "if... then..." statement that it stands for becomes true, which is untenable. The same way, you are trapped in Example 1.12.

Along with the propositional methods, we had also used the diagonalization technique of Cantor. There are two more general proof methods we will use in this book. The first is the principle of *mathematical induction.* It is, in fact, a deductive procedure. It has two versions: one is called the strong induction and the other is called induction, without any adjective.

Writing $P(n)$ for a property of natural numbers, the two forms of the principle can be stated as

Strong Induction: $P(0)$. If $P(k)$ holds for each $k < n$, then $P(n)$.
Therefore, for each $n \in \mathbb{N}$, $P(n)$.

Induction: $P(0)$. If $P(n)$ then $P(n+1)$.
Therefore, for each $n \in \mathbb{N}$, $P(n)$.

Verification of $P(0)$ is called the *basis step* of induction, and the other "if ... then" statement is called the *induction step.* In case of strong induction, the fact that all of $P(0), P(1), \ldots P(n-1)$ hold is called the *induction hypothesis*; and in case of induction, the induction hypothesis is "$P(n)$ holds." Both the principles are equivalent, and one is chosen over the other for convenience.

In the case of strong induction, the induction step involves assuming $P(k)$ for each $k < n$ and then deriving $P(n)$. While the induction step in the other case consists of deriving $P(n+1)$ from the single assumption $P(n)$. Thus it is safer to start with the strong induction when we do not know which one of them will really succeed.

The principle is also used to prove a property that might hold for all natural numbers greater than a fixed m. This is a generalization of the above. The formulation of the principle now looks like:

Strong Induction: $P(m)$. If $P(k)$ holds for each k with $m \le k < n$, then $P(n)$.
Therefore, for each natural number $n \ge m$, $P(n)$.

Induction: $P(m)$. For each $n \ge m$, if $P(n)$ then $P(n+1)$.
Therefore, for each natural number $n \ge m$, $P(n)$.

As earlier, verification of $P(m)$ is the basis step of induction, and the other "if ... then" statement is the induction step. In case of strong induction, the fact "$P(k)$ holds for each k with $m \le k < n$" is the induction hypothesis; and in case of induction, the induction hypothesis is "$P(n)$ holds."

Not only on $\mathbb{N}$, but wherever we see the structure of $\mathbb{N}$, we can use this principle. For example, it can be used on any set via the well ordering principle, which states that every set can be well ordered; see Problem 1.10. However, we will not require this general kind of induction.

Example 1.13 (Hilbert's Hotel). Hilbert has a hotel having rooms as many as numbers in $\mathbb{Z}_+$. Show that he has rooms for any number of persons arriving in groups, where a group might contain infinite number of persons.

Solution. Naturally, we take the infinite involved in the story as $\aleph_0$. If only one such group asks for rooms, Hilbert just assigns one to each. Suppose he has accommodated n number of such groups. The $(n + 1)$th group arrives. Then, he asks the incumbents to move to other rooms by the formula:

Person in Room-n moves to Room-$2n$.

Now, all odd numbered rooms are free. And the persons in the just arrived group get accommodated there. □

If a group of persons contains $\aleph_1$ or more people, then certainly Hilbert fails to meet the demands. Notice that in the solution to Hilbert's hotel, I have used induction. You can have a shorter solution mentioning the fact that a finite union of countable sets is countable.

We had defined graphs with finite sets of vertices implicitly. But a graph can also have an infinite set of vertices. For example, in $\mathbb{Z}_+$, take the relation of "divides," that is, $R = \{(n, mn) : m, n \in \mathbb{Z}_+\}$. In the graph of this relation, there will be edges from 1 to every positive integer, 2 to every even number, 3 to each multiple of 3, etc. Similarly, trees on infinite sets are defined.

A rooted tree in which each vertex has at most k number of children, for some $k \in \mathbb{N}$, is called a **finitely generated tree.** A **branch** in a rooted tree is a path from the root to a leaf. An infinite branch is then a sequence of vertices $v_0, v_1, v_2, \ldots$ such that v_0 is the root and v_{i+1} is a child of v_i, for each $i \in \mathbb{N}$. We demonstrate the use of induction in an infinite tree.

Example 1.14 (König's Lemma). Show that each finitely generated infinite tree has an infinite branch.

Solution. Let T be a finitely generated infinite tree. We show by induction that we have a branch $v_0, v_1, \ldots, v_n$, with v_n as the root of an infinite subtree of T for each $n \in \mathbb{N}$. Notice that such a branch cannot be finite.

In the basis step, We choose v_0, the root of the tree T. As T is a subtree of itself, for $n = 0$, the statement holds.

For the induction step, suppose we have already a sequence $v_0, v_1, \ldots, v_n$ of vertices such that v_n is the root of an infinite subtree of T. Suppose the vertex v_n has m children, where $m \leq k$; this k is fixed for the tree. Consider the subtree T' of T having v_n as the root. If we remove v_n from T', we get m subtrees with the children of v_n as the roots of the subtrees. At least one of these subtrees has infinite number of vertices, otherwise, T' will become finite. Take one such subtree which is infinite.

Choose its root as v_{n+1}. Now, we get a sequence $v_0, v_1, \ldots, v_n, v_{n+1}$ such that v_{n+1} is the root of an infinite subtree of T. Here ends the induction step. □

However, there is a danger in misusing the principle of induction. See the following example.

Example 1.15 (Fallacious Induction). In a certain tribe, each boy loves a girl. Show that each boy loves the same girl.

Solution. In the basis step, consider any single boy. Clearly, the statement holds for him. For the induction step, assume that if you take any group of n boys, you find that they love the same girl. Now, take any group of $n+1$ boys. Call them $b_1, b_2, \ldots, b_{n+1}$. Form two groups of n boys each. The first group has the boys $b_1, b_2, \ldots, b_n$ and the second group has $b_2, b_3, \ldots, b_{n+1}$. Now, by the induction hypothesis, all of $b_1, b_2, \ldots, b_n$ love the same girl and all of $b_2, b_3, \ldots, b_{n+1}$ love the same girl. As $b_2, b_3, \ldots, b_n$ are common to both the groups, we see that all of b_1, b_2, b_n, b_{n+1} love the same girl. □

For the argument of the induction step to hold, the set $\{b_2, b_3, \ldots, b_n\}$ must be nonempty. That means, the basis step is not $n = 1$ but $n = 2$. You will see plenty of induction proofs later. Combining the principle of induction and proof by contradiction, we get *Fermat's principle of finite descent.* It is stated as

If $P(n+1)$, then $P(n)$. But *not* $P(0)$. Therefore, *not* $P(n)$, for each $n \in \mathbb{N}$.

Example 1.16 (Surprise Quiz). Your teacher (not of this course!) declares in the class on a Friday that he will be conducting a surprise quiz some time on the next week. When returning to the dormitory, your friend says – "so nice of him; he will not be able to conduct the quiz this time." He thus argues, "You see, he cannot afford to keep the quiz on Friday, for, in that case, he does not conduct the quiz till Thursday. Certainly then, we infer the quiz to be on Friday and it would not be a surprise quiz. Now agreed that he has to conduct the quiz on or before Thursday, can he afford not to conduct the quiz till Wednesday? No, for then, we infer that only on Thursday he conducts the quiz. Continuing three more steps, you see that he cannot even conduct the quiz on Monday." □

This is an application of Fermat's principle of finite descent. Of course, there has been some rhetoric involved, coining on the ambiguous meaning of the surprise in the quiz. But there is nothing wrong in your friend's argument! We will not have occasions for the use of the principle in this book. But another consequence of induction will be used at many places. It is the *Pigeon hole principle.* It states that if there are n pigeons and $m < n$ pigeon holes to accommodate all the pigeons in the pigeon holes there must be at least two pigeons in same pigeon hole. A formal version goes as follows

Let A, B be two finite sets. If $|A| > |B|$ and $f : A \to B$ is a total function, then f cannot be one–one.

Even finiteness of the sets can be dropped, but that would require transfinite induction to justify the principle. Try proving this principle by using induction on $|A|$. We see an application of the principle.

Example 1.17. Show that if seven points are chosen randomly from inside a circle of radius 1, then there are at least two points whose distance is less than 1.

Solution. Take a circle of radius 1 and divide it into six equal parts by drawing six radii. Each of the six sectors of the circle is bounded by two radii and an arc. If seven points are chosen at random from inside the circle, then by Pigeon hole principle, at least two of them are from the same sector; which sector, we do not know. Now, the distance between those two points is less than 1. □

You have already seen how induction could be used for defining certain objects. For example, we have defined the cardinalities of finite sets as $|\emptyset| = 0$, $|\{0\}| = 1$, $|\{0, 1, 2, \ldots, n\}| = n + 1$. This definition uses induction. Another common example of *definition by induction,* sometimes called a *recursive definition*, is of the factorial function. It is defined by $0! = 1$, $(n + 1)! = n!\,(n + 1)$, for each $n > 0$. The Fibonacci sequence is defined recursively by $f_0 = 1$, $f_1 = 1$, $f_{n+1} = f_n + f_{n-1}$, for each $n > 1$. The construction of a suitable branch in König's lemma is by induction.

Sometimes a definition by induction does not use any number. For example, in defining an arithmetic expression involving the variables x, y, and the only operation as $+$, you would declare that each of x, y is an arithmetic expression. This is the basis step in the definition. Next, you will declare that if E_1, E_2 are expressions, then $(E_1 + E_2)$ is an expression, and nothing else is an expression. In such a case, suppose you want to prove that in every expression there is an equal number of left and right parentheses. How do you proceed?

Obviously, the proof is by induction on the number of left parentheses, or on the number of right parentheses, or on the number of $+$ signs. But this has to be identified. Suppose we pick up the later. In the basis step, if there is no $+$ sign in an expression, then there are no parentheses. Hence, the number of left and right parentheses are equal, equal to 0.

Assume the induction hypothesis that if an expression has less than n number of $+$ signs, then the number of left and right parentheses are equal. We plan to use strong induction to be on the safe side. Suppose E is an expression having n number of $+$ signs,. Then, $E = (E_1 + E_2)$ for some expressions E_1 and E_2. Now, both of E_1, E_2 satisfy the induction hypothesis. Thus they have equal number of left and right parentheses. Then so does $(E_1 + E_2)$.

In the above proof, we can avoid the parameter n, which we had chosen as the number of $+$ signs. Here, we verify that the statement holds in the basis case of the inductive definition of expressions. Next, in the inductive step, we see if E_1, E_2 are expressions satisfying the conclusion that the left and right parentheses are equal in number, then so does the new expression $(E_1 + E_2)$. There ends the proof.

Such a use of induction without identifying a parameter is named as the principle of *structural induction.* To keep the matter straight, we will rather identify a suggestive integer parameter than using this principle.

1.6 Summary and Problems

As you have observed, the mathematical preliminaries are not at all tough. All that we require is a working knowledge of set theory, the concept of cardinality, trees, induction, and the pigeon hole principle.

A good reference on Set Theory including cardinality that covers all the topics discussed here is [50]. For induction, see [102]. For a reference on formal derivations and their applications to discrete mathematics, see [48]. These books also contain a lot of exercises. For an interesting history of numbers and number systems, see [35]. The story of the Hottentot tribes is from this book; of course I have modified it a bit to provide motivation for counting.

Unlike other chapters, I have neither included exercises nor problems in each section; probably you do not require them. If you are really interested, here are some.

Problems for Chapter 1

1.1. Prove all the laws about set operations listed at the end of Sect. 1.2

1.2. What is wrong in the following fallacious proof of the statement that each symmetric and transitive binary relation on a nonempty set must also be reflexive:

Suppose aRb. By symmetry, bRa. By transitivity, aRa?

1.3. Label the edges of the graph in Fig. 1.3 as $e_1, \ldots, e_7$ and then write the corresponding labeled graph as a triple, now with an incidence relation.

1.4. If $|A| = n$, then how many binary relations on A are there? Among them, how many are reflexive? How many of them are symmetric? How many of them are both reflexive and symmetric?

1.5. Show that the inverse of an equivalence relation on a set is also an equivalence relation. What relation is there between the equivalence classes of the relation and those of its inverse?

1.6. Let A be a finite set. Let the binary relation R on 2^A be defined by xRy if there is a bijection between x and y. Construct a function $f : 2^A \to \mathbb{N}$ such that for any $x, y \in 2^A$, $f(x) = f(y)$ iff xRy.

1.7. Let R be an equivalence relation on any nonempty set A. Find a set B and a function $f : A \to B$ such that for any $x, y \in A$, $f(x) = f(y)$ iff xRy.

1.8. Let R, R' be two equivalence relations on a set A. Let P and P' be the partitions of A consisting of equivalence classes of R and R', respectively. Show that $R \subseteq R'$ iff P is *finer than* P', that is, each x in P is a subset of some y in P'.

1.9. Let A be a nonempty set. Suppose $\mathcal{A}$ denotes the empty collection of subsets of A. What are the sets $\cup\mathcal{A}$ and $\cap\mathcal{A}$?

1.10. An order relation R on a set A is a binary relation satisfying the following properties:

Comparability: For any $x, y \in A$, if $x \neq y$, then either xRy or yRx.
Irreflexivity: For each $x \in A$, it is not the case that xRx.
Transitivity: For any $x, y, z \in A$, if xRy and yRz, then xRz.

An element $a \in B \subseteq A$ is a smallest element of B (with respect to the order R) if for each $x \in B$, $x \neq a$, you have $a < x$. A set A with an order relation R is said to be well ordered if each nonempty subset of A has a smallest element.

The *Well Ordering Principle* states that given any set, there exists a binary relation with respect to which the set A is well ordered.

The *Axiom of Choice* states that given a collection $\mathcal{A}$ of disjoint nonempty sets, there exists a set C having exactly one element in common with each element of $\mathcal{A}$.

Prove that the axiom of choice and the well ordering principle are equivalent.
[Hint: Zermelo proved it in 1904; and it startled the mathematical world.]

1.11. Let $f : A \to B$ be a (total) function. Suppose $C, C' \subseteq A$ and $D, D' \subseteq B$. Recall that $f(C) = \{f(x) : x \in C\}$ and $f^{-1}(D) = \{x \in A : f(x) \in D\}$. Show that
(a) If $C \subseteq C'$, then $f(C) \subseteq f(C')$.
(b) If $D \subseteq D'$, then $f^{-1}(D) \subseteq f^{-1}(D')$.
(c) $C \subseteq f^{-1}(f(C))$. Equality holds if f is injective.
(d) $f(f^{-1}(D)) \subseteq D$. Equality holds if f is surjective.
(e) $f(C \cup C') = f(C) \cup f(C')$.
(f) $f(C \cap C') \subseteq f(C) \cap f(C')$. Give an example, where equality fails.
(g) $f(C) - f(C') \subseteq f(C - C')$. Give an example, where equality fails.
(h) $f^{-1}(D \cup D') = f^{-1}(D) \cup f^{-1}(D')$.
(i) $f^{-1}(D \cap D') = f^{-1}(D) \cap f^{-1}(D')$.
(j) $f^{-1}(D - D') = f^{-1}(D) - f^{-1}(D')$.
(k) Statements (e), (f), (h), and (i) hold for arbitrary unions and intersections.

1.12. Suppose $f : A \to B$ and $g : B \to A$ are such functions that the compositions $g \circ f : A \to A$ and $f \circ g : B \to B$ are identity maps. Show that f is a bijection.

1.13. Let $f : A \to B$, $g : B \to C$ be functions, and let $D \subseteq C$.
(a) Show that $(g \circ f)^{-1}(D) = f^{-1}(g^{-1}(D))$.
(b) If f, g are injective, then show that $g \circ f$ is injective.
(c) If f, g are surjective, then show that $g \circ f$ is surjective.
(d) If $g \circ f$ is injective, what can you say about injectivity of f and g?
(e) If $g \circ f$ is surjective, what can you say about surjectivity of f and g?

1.14. Let A, B be any sets. Prove that there exists a on-one function from A to B iff there exists a function from B onto A. [Hint: For the "if" part, you may need the axiom of choice.]

1.15. Let $A = \{1, 2, 3, \ldots, n\}$. With $O = \{B \subseteq A : |B| \text{ is odd}\}$ and $E = \{B \subseteq A : |B| \text{ is even}\}$, define a map $f : O \to E$ by $f(B) = B - \{1\}$, if $1 \in B$; else $f(B) = B \cup \{1\}$. Show that f is a bijection.

1.16. For sets A, B, define B^A as the set of all functions from A to B. Prove that for any set C, there is a one–one correspondence between $\{0, 1\}^C$ and the power set 2^C. This is the reason we write the power set of C as 2^C. [Hint: For $D \subseteq C$, define its *characteristic function,* also called the *indical function,* $\chi_D : C \to \{0, 1\}$ by "if $x \in D$, then $\chi_D(x) = 1$; else, $\chi_D(x) = 0$."]

1.17. Show that no partial order on $\mathbb{C}$ can be an extension of the $\leq$ on $\mathbb{R}$.

1.18. Recall that a nonempty set A is finite iff there is a bijection between A and $\{0, 1, \ldots, n\}$ for some $n \in \mathbb{N}$. Here, we show how cardinality of a finite set is well defined. Let A be a set; $a \in A$; $B \subsetneqq A$; and let $n \in \mathbb{N}$. Prove the following without using cardinality:
(a) There is a bijection between A and $\{0, 1, \ldots, n+1\}$ iff there is a bijection between $A - \{a\}$ and $\{0, 1, \ldots, n\}$.
(b) Suppose there is a bijection between A and $\{0, 1, \ldots, n+1\}$. Then, there exists no bijection between B and $\{0, 1, \ldots, n+1\}$. If $B \neq \emptyset$, then there exists a bijection between B and $\{0, 1, \ldots, m\}$, for some $m < n$.

1.19. Show that $\mathbb{N}$ is not a finite set.

1.20. Prove: If A is a nonempty set and $n \in \mathbb{N}$, then the following are equivalent:
(a) There is a one–one function from A to $\{0, 1, \ldots, n\}$.
(b) There is a function from $\{0, 1, \ldots, n\}$ onto A.
(c) A is finite and has at most $n + 1$ elements.

1.21. Let A, B, C be finite sets. Prove that

$$|A \cup B \cup C| = |A| + |B| + |C| - |A \cap B| - |B \cap C| - |C \cap A| + |A \cap B \cap C|.$$

Generalize this formula for n number of sets $A_1, \ldots, A_n$.

1.22. Let $B \subseteq A$. Show that if there is an injection $f : A \to B$, then $|A| = |B|$.

1.23. Prove that a denumerable union of finite sets is denumerable.

1.24. Using Cantor's theorem, show that the collection of all sets is, infact, not a set.

1.25. Prove that each infinite set contains a denumerable subset.

1.26. Show that if $f : \mathbb{N} \to A$ is a surjection and A is infinite, then A is denumerable.

1.27. Show that for each $n \in \mathbb{Z}_+$, $|\mathbb{Q}^n| = |\mathbb{Q}|$ and $|\mathbb{R}^n| = |\mathbb{R}|$.

1.28. Show that the set of algebraic numbers is denumerable. Then deduce that there are more transcendental numbers than the algebraic numbers.

1.29. Let $f : A \to B$ and $g : B \to A$ be two injections. Prove the following:
(a) Write $A_0 = A - g(B)$, $A_1 = g(B) - g(f(A))$, $B_0 = B - f(A)$, $B_1 = f(A) - f(g(B))$. Then $A_0 \cup A_1 = A - g(f(A))$ and $B_0 \cup B_1 = B - f(g(B))$. Further, $A_0 \cap A_1 = \emptyset = B_0 \cap B_1$.
(b) The relation g^{-1} from $g(A)$ to A is a function.

(c) Define $h : A_0 \cup A_1 \to B_0 \cup B_1$ by $h(x) = f(x)$ for $x \in A_0$, and $h(x) = g^{-1}(x)$ for $x \in A_1$. Then h is a bijection. Further, $h(A_0) = B_1$ and $h(A_1) = B_0$.
(d) Write $A_2 = g(f(A)) - g(f(g(B)))$, $A_3 = g(f(g(B))) - g(f(g(f(A))))$, $B_2 = f(g(B)) - f(g(f(A)))$, $B_3 = f(g(f(A))) - f(g(f(g(B))))$. Now define, by induction, the sets A_i, B_i, for each $i \in \mathbb{N}$. Then $A_{2m} \cap A_{2m+1} = \emptyset = B_{2m} \cap B_{2m+1}$.
(e) Define $\phi : \cup_{i \in \mathbb{N}} A_i \to \cup_{i \in \mathbb{N}} B_i$ by $\phi(x) = f(x)$ if $x \in A_i$, i even, and $\phi(x) = g^{-1}(x)$ for $x \in A_i$, i odd. Then ϕ is a bijection.
(f) Restrict the domain of f to the set $A - \cup_{i \in \mathbb{N}} A_i$. Then $f : A - \cup_{i \in \mathbb{N}} A_i \to B - \cup_{i \in \mathbb{N}} B_i$ is a bijection.
(g) Define $\psi : A \to B$ by $\psi(x) = \phi(x)$ if $x \in \cup_{i \in \mathbb{N}} A_i$, and $\psi(x) = f(x)$ if $x \in A - \cup_{i \in \mathbb{N}} A_i$. Then ψ is a bijection.
(h) Cantor–Schröder–Bernstein Theorem: If $|A| \le |B|$ and $|B| \le |A|$, then $|A| = |B|$.

1.30. Let A, B be sets. Is it true that there is an injection from A to B iff there is a surjection from B to A? [Hint: Does axiom of choice help? See Problem 1.10]

1.31. Define $|A| + |B| = |A \cup B|$ provided $A \cap B = \emptyset$; $|A| - |B| = |A - B|$ provided $B \subseteq A$; $|A| \times |B| = |A \times B|$; and $|A|^{|B|} = |A^B|$. Let α, β be cardinalities of infinite sets with $\alpha < \beta$. Show that $\alpha + \beta = \beta$, $\beta - \alpha = \beta$, $\alpha \times \beta = \beta$, $2^\alpha > \alpha$. Further, show that $\alpha - \alpha$ is not well defined.

1.32. Show that a denumerable product of denumerable sets is uncountable. [Hint: You may need the axiom of choice.]

1.33. Let $a \in \mathbb{R}$. Simplify the sets $\cup_{r<1}\{x \in \mathbb{R} : a - r \le x \le a + r\}$ and $\cap_{r>0}\{x \in \mathbb{R} : a - r < x < a + r\}$.

1.34. Let S denote the sentence: This sentence has no proof. Show that S is true. Conclude that there is a true sentence having no proof. [Gödel's proof of his incompleteness theorem expresses S in the system of natural numbers.]

1.35. Let S be the sentence: This sentence has no short proof. Show that if S is true, then there exists a sentence whose proof is not short, but the fact that it is provable has a short proof. [A formal version of this S expressed in the system of natural numbers is called Parikh's sentence.]

1.36. Let $P(m, n)$ denote a property involving two natural numbers. Suppose we prove that $P(0, 0)$ is true. We also prove that if $P(i, j)$ is true, then both $P(i, j + 1)$ and $P(i + 1, j)$ are true. Does it follow that $P(m, n)$ is true for any $m, n \in \mathbb{N}$?

1.37. Show that for each integer $n > 1$, $1/\sqrt{1} + 1/\sqrt{2} + 1/\sqrt{3} + \cdots + 1/\sqrt{n} > \sqrt{n}$.

1.38. Deduce the pigeon hole principle from the principle of induction.

1.39. Show that among any $n + 2$ positive integers, either there are two whose sum is divisible by $2n$ or there are two whose difference is divisible by $2n$.

1.40. Use the pigeon hole principle to show that each rational number has a recurring decimal representation.

1.41. Show that among any $n + 1$ numbers randomly chosen from $\{1, 2, \ldots, 2n\}$, there are at least two such that one divides the other.

1.42. Let $A = \{m + n\sqrt{2} : m, n \in \mathbb{Z}\}$. Show that for each $k \in \mathbb{Z}_{+}$, there is $x_k \in A$ such that $0 < x_k < 1/k$.

2 Regular Languages

2.1 Introduction

It is said that human intelligence is mainly the capability to represent a problem, its solution, or related facts in many seemingly different ways. You must have encountered it in several problem-solving situations. You first represent the problem in a known language, where you might like to eliminate or omit the irrelevant aspects and consider only the appropriate ones. The methodology is followed throughout mathematics starting from solving first arithmetic problems such as "if you already had five candies and your friend offers you one more, then how many candies do you have now?".

To give another example, the memory in a computer is only an assembly of switches which, at any moment, may be *off* or *on*. If there is a trillion of them, then it may be represented as a trillion digited binary number, when, say, *off* corresponds to 0 and *on* to 1. In some situations, we may not be interested in all possible binary numbers, but only those having a few number of digits out of the trillion, or only those having a particular pattern, such as "there is at least one 0 following every occurrence of a 1." There might arise a situation where we would like to have a representational scheme having more than two symbols. We will, however, consider only a finite number of symbols at a time, and in parallel with the existing natural languages, we will develop *formal languages* out of these symbols.

In this book, we will introduce a hierarchy of formal languages. To represent formal languages, we will study grammars. Each such type in the hierarchy will have its own type of mechanical device, which may recognize the language but not any other language from another type. In the sequel, we will have to introduce many technicalities. The technical words or phrases, as usually are, will either be defined clearly or will be left completely undefined. In the latter case, I will attempt at a description of such undefined or primitive notions so that you will be able to think about them in a certain well-intended way.

A. Singh, *Elements of Computation Theory*, Texts in Computer Science,

2.2 Language Basics

We start with the primitive notion of a **symbol**. A symbol is any written sign. We adhere to the written scripts as communication between you and me. Of course, you can even consider "spoken signs" or even "body language" or any piece in any other sign language. The implicit assumption here is that we will be able to represent other types of symbols or signs in terms of the written ones, in terms of our *symbols*.

An **alphabet** is then a nonempty finite set of symbols, where no symbol is a part of another. Notice that the phrase *is a part of* is again a primitive notion here. We will not allow the blank symbol to be in our alphabets for some technical reasons. If the blank symbol has to be used in some situation, then we will rather have some rewriting of it, say, ƀ . For example, $\{0, 1\}$, $\{a, b, c, \ldots, z\}$, $\{@, \$, !, 1, a, z, 5, >\}$ are alphabets, but $\{0, 10\}$, $\{ab, c, a\}$ are not alphabets as 0 is a part of 10 and a is a part of ab.

A **word** or a **string** over an alphabet is a finite sequence of symbols from the alphabet. For example, each word in an English dictionary is a string over the Roman alphabet. Each natural number is a string over the alphabet $\{0, 1, 2, \ldots, 9\}$. Thus a string is written as a sequence of symbols followed one after another without any punctuation marks. This way of writing a string is referred to as the operation of concatenation of symbols.

The operation can be defined for strings also. For example, concatenation of strings *alpha* and *bet* is *alphabet*, and concatenation of *bet* and *alpha* is *betalpha*. If s and t are strings (over an alphabet), then **concatenation** of s and t is the string st (over the same alphabet). There is a special string, the string containing no occurrence of any symbol whatsoever, called the **empty string**. The empty string is indeed unique, and it is a string over every alphabet. We will denote the empty string by the symbol ε. It serves as an identity of concatenation as for any string u, $u\varepsilon = \varepsilon u = u$.

The number of occurrences of symbols in a string is called its **length**. The length of *alpha* is 5 and the length of *bet* is 3. The empty string ε has length 0. Note that the length of a string depends upon the underlying alphabet as the string itself needs an alphabet, first of all. For example, the length of the string 1001 over the alphabet $\{0, 1\}$ is 4, while its length over the alphabet $\{10, 01\}$ is 2.

The vagueness in the definition of a symbol and an alphabet is removed by following a strict mathematical formalism. In this formalism, an alphabet is taken as any finite set and a string over the alphabet is taken as a map from $\{1, 2, \ldots, n\}$ to the alphabet, where n is some natural number. The natural number n is again the length of the string. For example, the string 101 over the alphabet $\{0, 1\}$ is simply the map $f : \{1, 2, 3\} \to \{0, 1\}$, where $f(1) = 1$, $f(2) = 0$, $f(3) = 1$. This string has length 3 as usual. Note that the map f here can be completely determined by its values at the integers $1, 2, 3$. That is, the map can be rewritten by noting down its values at $1, 2, 3$ one after another, and, in that order. That is how the formal definition would be connected to the informal.

When the natural number n is taken as 0, we have an empty domain for the map, and then, by convention, we will have the empty map, the empty string ε, having length 0. Moreover, the *length* function ℓ can be defined inductively over an alphabet Σ as in the following:

1. $\ell(\varepsilon) = 0$.
2. If $\sigma \in \Sigma$ and u is a string over Σ, then $\ell(u\sigma) = \ell(u) + 1$.

The **reversal** of a string s over an alphabet Σ is denoted by s^R and is defined inductively by the following:

1. $\varepsilon^R = \varepsilon$.
2. If $\sigma \in \Sigma$ and u is a string over Σ, then $(\sigma u)^R = u^R\sigma$.

See that the above definition of the reversal does really capture the notion of the reversal; for example, $(reverse)^R = esrever$. It follows that the length of the reversal of a string is same as the length of the string. It also follows that $(uv)^R = v^R u^R$ for strings u and v. Show these by induction (see Sect. 1.5.) on the length of the string!

Exercise 2.1. Define the operation of concatenation inductively and then show that this operation is associative but not commutative.

Exercise 2.2. Write $(ab)^0 = \varepsilon, (ab)^1 = ab, (ab)^2 = abab, \ldots$ for the string ab. Define $(ab)^n$ inductively for every $n \in \mathbb{N}$. Show that $\ell((ab)^n) = n\,\ell(ab)$.

A string u is a **prefix** of a string w iff there is a string v such that $w = uv$. Similarly, if for some string v, we have $w = vu$, then u is called a **suffix** of w. In general, u is a **substring** of w iff there are strings x, y such that $w = xuy$.

For example, *pre* is a prefix of *prefix* and *fix* is a suffix of *suffix* (and also of *prefix*). The string *ref* is a substring of *prefix*. Vacuously, both *pre* and *fix* are substrings of *prefix*. As the strings x, y in $w = xuy$ can be taken as the empty string ε, every string is a substring of itself. Also, every string is both a prefix and a suffix of itself.

Observe that all of ε, *p, pr, pre, pref, prefi, prefix* are prefixes of *prefix*. Out of these if you take any two, can you find any relation between them? Easy, one of them has to be a prefix of the other! If both u and v are prefixes of the same string w, then u is matching with a part of w from the left and so is v. So, the one with smaller length must be a prefix of the other. But this is not a proof!

Lemma 2.1 (Prefix Lemma). *If u and v are prefixes of a string w over an alphabet Σ, then u is a prefix of v or v is a prefix of u.*

Proof. We prove it by induction on the length of w. If $\ell(w) = 0$, then $w = \varepsilon$, and then $u = v = \varepsilon$. This shows that u is a prefix of v. Assume the induction hypothesis that for all strings w of length n, the statement holds. Let $\ell(w) = n + 1$. Write $w = z\sigma$, where $\sigma \in \Sigma$ is the last symbol of w, and then $\ell(z) = n$. Let u and v be prefixes of w. If one of u, v equals w, then the other is a prefix of it. So, suppose that neither u nor v is equal to w. Then both u and v are prefixes of z. (Why?) As $\ell(z) = n$, by the induction hypothesis, u is a prefix of v or v is a prefix of u. □

Exercise 2.3. Formulate and prove Lemma 2.1 with suffixes instead of prefixes. What about mixing prefixes and suffixes?

A **language** over an alphabet Σ is any set of strings over Σ. In particular, $\emptyset$, the empty set (of strings) is a language over every alphabet. So are the sets $\{\varepsilon\}$, the set Σ itself, and the set of all strings over Σ, which we denote by Σ^*.

Thus, any book written in English may be thought of as a language over the Roman alphabet. The set of all binary numbers starting with 1 and of length 2, that is, the set $\{10, 11\}$ is a language over $\{0, 1\}$. The binary *palindromes* (the strings that are same when read from right to left) form a language as this can be written as $\{w \in \{0, 1\}^* : w^R = w\}$, a subset of $\{0, 1\}^*$.

Theorem 2.1. *Let Σ be any alphabet. Then Σ^* is denumerable. Therefore, there are uncountable number of languages over Σ.*

Proof. Write $\Sigma_0 = \{\varepsilon\}$ and Σ_n = the set of all strings of length n over Σ, for any $n \in \mathbb{Z}_+$. If $|\Sigma| = m$, then there are m^n strings in Σ_n. As $\Sigma^* = \cup_{n \in \mathbb{N}} \Sigma_n$, a denumerable union of finite sets, it is denumerable. Each language over Σ is a subset of Σ^*. Thus, the number of such languages is the cardinality of the power set 2^{Σ^*}, which is uncountable by Theorem 1.5. □

The question is how to name all these languages? Obviously, whatever way we try to name them, the names themselves will be strings over some alphabet, and there can only be a countable number of them at the most, unless we choose to supply names from another uncountable set such as $\mathbb{R}$. So, before even attempting to name the languages, we see that any such attempt is bound to fail. But then, can we name certain interesting languages? It, of course, depends upon what our interest is and what is our naming scheme. Note that naming schemes are only finitary ways of representing the languages.

We start with some natural naming schemes. As languages are sets, we can use set operations such as union, intersection, complementation (in Σ^*) etc. We also have the operation of concatenation for strings, which can be adopted or extended to languages. Let L, L_1, L_2 be languages over an alphabet Σ. Then

$L_1L_2 = \{uv : u \in L_1 \text{ and } v \in L_2\}$ is the **concatenation** of the languages L_1 and L_2. Note the asymmetry in this *and*.

$L_1 \cup L_2 = \{w : w \in L_1 \text{ or } w \in L_2\}$ is the **union** of L_1 and L_2.

$L_1 \cap L_2 = \{w : w \in L_1 \text{ and } w \in L_2\}$ is the **intersection** of L_1 and L_2.

$L_1 - L_2 = \{w : w \in L_1 \text{ but } w \notin L_2\}$ is the **difference** of L_2 from L_1.

$\overline{L} = \{w \in \Sigma^* : w \notin L\}$ is the **complement** of the language L.

The **powers** of L, denoted by L^m, for $m \in \mathbb{N}$, are defined inductively by $L^0 = \{\varepsilon\}$ and $L^{n+1} = LL^n$.

The **Kleene star** (or the **closure** or the **asterate**) of the language L is defined as $L^* = \cup_{m \in \mathbb{N}} L^m$. Read it as L star. Notice that it goes along well with our earlier notation Σ^*, the set of all strings over the alphabet Σ.

The **Kleene plus** of L is $L^+ = LL^* = \{u_1u_2 \cdots u_k : k > 0 \text{ and } u_i \in L\}$. Read it as L plus. L^+ is also referred to as the **positive closure** of L.

Similarly other set operations will give rise to respective definitions of new languages using the old ones. Our aim is to use these symbolism for writing many interesting languages in a compact way. See the following examples.

Example 2.1. Can we represent the language $L = \{w \in \{a, b, c\}^* : w$ does not end with $c\}$ using the symbolism we have developed so far?

Solution. Will it be easy if we first try representing all strings over Σ that end with one c? Any such string will be a string from Σ^* followed by the symbol c. That is, $\overline{L} = \{uc : u \in \Sigma^*\} = \Sigma^*\{c\} = \{a, b, c\}^*\{c\}$. Thus, $L = \overline{\{a, b, c\}^*\{c\}}$. □

To develop a shorthand, we might write $\{a, b, c\} = \{a\} \cup \{b\} \cup \{c\}$ just as $a \cup b \cup c$. Here, we are dispensing with braces } and {, though we know that $a \cup b \cup c$ does not make sense at all. However, we can still use it as a shorthand, a name, with an obvious interpretation. For such an expression $a \cup b \cup c$, we will associate the language $\{a\} \cup \{b\} \cup \{c\}$.

To put it formally, we would like to use the phrase: **the language represented by** the expression so and so. Thus $a \cup b \cup c$ will be regarded as an **expression** and $\{a\} \cup \{b\} \cup \{c\}$ will be the language represented by this expression. In symbols, we will write it as $\mathcal{L}(a \cup b \cup c) = \{a\} \cup \{b\} \cup \{c\}$.

There will not be any confusion between the "string ab" and the "expression ab", as the former is only a string and the $\mathcal{L}$ of the latter is the language $\{ab\}$.

We will use, mostly in the exercises, the symbol # as a shorthand for "number of." If w is a string and σ is a symbol, we write $\#_\sigma(w)$ to denote the number of occurrences of σ in w. For example, $\#_a(babaabc) = 3$.

Problems for Section 2.2

2.1. What languages do the expressions $(\emptyset^*)^*$ and $0\emptyset$ denote?

2.2. Find all strings in $(a \cup b)^*b(a \cup ab)^*$ of length less than four.

2.3. When does $LL^* = L - \{\varepsilon\}$ happen?

2.4. Let $L = \{ab, aa, baa\}$. Which of the strings $bbbbaa$, $aaabbaa$, $abaabaaabaa$, $baaaaabaaaab$, and $baaaaabaa$ are in L^*?

2.5. Let Σ be any alphabet. Prove that $(uv)^R = v^R u^R$, for all strings $u, v \in \Sigma^+$.

2.6. Let $\Sigma = \{a, b\}$. Find strings in, and not in, $L \subseteq \Sigma^*$, where L is
(a) $\{ww^R w : w \in \Sigma\Sigma\}$.
(b) $\{w \in \Sigma^* : w^2 = w^3\}$.
(c) $\{w \in \Sigma^* : w^3 = v^2$ for some $v \in \Sigma^*\}$.
(d) $\{w \in \Sigma^* : uvw = wuv$ for some $u, v \in \Sigma^*\}$.

2.7. Let a, b, c be different symbols. Are the following true? Justify.
(a) $(a \cup b)^* = a^* \cup b^*$.
(b) $a^*b^* \cap b^*a^* = a^* \cup b^*$.
(c) $\emptyset^* = \varepsilon^*$.
(d) $\{a, b\}^* = a^*(ba^*)^*$.
(e) $a^*b^* \cap b^*c^* = \emptyset$.
(f) $(a^*b^*)^* = (a^*b)^*$.

(g) $(a \cup ab)^*a = a(a \cup ba)^*$.
(h) $a(bca)^*bc = ab(cab)^*c$.
(i) $(b \cup a^+b)(b \cup a^+b)(a \cup ba^*b)^+ = a^*b(a \cup ba^*b)^*$.
(j) $aa(a \cup b)^* \cup (bb)^*a^* = (a \cup ab \cup ba)^*$.

2.8. Let $w \in \{a, b\}^*$ be such that $abw = wab$. Show that $\ell(w)$ is even.

2.9. Does $L = L^*$ hold for the following languages:
(a) $L = \{a^n b^{n+1} : n \in \mathbb{N}\}$?
(b) $L = \{w \in \{a, b\}^* : \#_a(w) = \#_b(w)\}$?

2.10. Are there languages for which $\overline{(L^*)} = (\overline{L})^*$?

2.11. Use induction to show that $\ell(u^n) = n\ell(u)$ for all strings u and all $n \in \mathbb{N}$.

2.12. Prove that for all languages L_1, L_2, we have $(L_1 L_2)^R = L_2^R L_1^R$.

2.13. Prove or disprove the following claims:
(a) $(L_1 \cup L_2)^R = L_1^R \cup L_2^R$ for all languages L_1 and L_2.
(b) $(L^R)^* = (L^*)^R$ for all languages L.
(c) If $L_1^* = L_2^*$ then $L_1 = L_2$ for all languages L_1 and L_2.
(d) There exists a finite language L such that $L^* = L$.
(e) If $\varepsilon \in L \subseteq \Sigma^*$ and $\varepsilon \in L' \subseteq \Sigma^*$, then $(L\Sigma^* L')^* = \Sigma^*$.

2.14. Let A be a language over an alphabet Σ. Call A reflexive if $\varepsilon \in A$, and call A transitive if $A^2 \subseteq A$. Let $B \supseteq A$. Show that if B is both reflexive and transitive, then $B \supseteq A^*$.

2.15. Give a rigorous proof that $\mathcal{L}((a \cup ba)^*(b \cup \varepsilon))$ is the set of all strings over $\{a, b\}$ having no pair of consecutive b's.

2.16. Let A, B, C be languages over an alphabet Σ. Show the following properties of union, concatenation, and Kleene star:
(a) $A\{\varepsilon\} = \{\varepsilon\}A = A$.
(b) $A\emptyset = \emptyset A = \emptyset$.
(c) $A(B \cup C) = AB \cup AC$. (What about arbitrary union?)
(d) $A^*A^* = (A^*)^* = A^* = \{\varepsilon\} \cup AA^* = \{\varepsilon\} \cup A^*A$.
(e) $\emptyset^* = \{\varepsilon\}$.
(f) $(A \cup B)^* = (A^*B^*)^*$.
(g) $A(B \cap C) = AB \cap AC$ does not hold, in general.
(h) $\overline{AB} = \overline{A}\,\overline{B}$ does not hold, in general.

2.3 Regular Expressions

For the time being, we only consider the operations of concatenation, union, and the Kleene star starting from the symbols of an alphabet and the empty language $\emptyset$. We begin with such a definition of a class of expressions and the corresponding languages they represent.

A **regular expression** over an alphabet Σ and the language it represents are defined inductively by the following rules:

1. $\emptyset$ is a regular expression. $\mathcal{L}(\emptyset) = \emptyset$.
2. Each $\sigma \in \Sigma$ is a regular expression. $\mathcal{L}(\sigma) = \{\sigma\}$.
3. If α is a regular expression, then α^* is a regular expression. $\mathcal{L}(\alpha^*) = (\mathcal{L}(\alpha))^*$.
4. If α, β are regular expressions, then both $(\alpha\beta)$ and $(\alpha \cup \beta)$ are regular expressions. $\mathcal{L}((\alpha\beta)) = \mathcal{L}(\alpha)\mathcal{L}(\beta)$ and $\mathcal{L}((\alpha \cup \beta)) = \mathcal{L}(\alpha) \cup \mathcal{L}(\beta)$.
5. All regular expressions and the languages they represent are obtained by applying the rules 1–4 above.

The parentheses above are used to remove possible ambiguities such as those that occur in the expression $\alpha\beta\cup\gamma$, which may be read as $\alpha(\beta\cup\gamma)$ or as $(\alpha\beta)\cup\gamma$. However, we will dispense with many parentheses by using the following **precedence** rules:

The operation $*$ will have the highest precedence.
The operation of concatenation will have the next precedence.
The operation of union will have the least precedence.

We will also dispense with the outermost parentheses from a regular expression. This means that the regular expression $((\alpha\beta) \cup \gamma^*)$ will be rewritten (or abbreviated) as $\alpha\beta \cup \gamma^*$. Instead of writing $\mathcal{L}((\alpha\beta))$, $\mathcal{L}((\alpha \cup \beta))$, $(\mathcal{L}(\alpha))^*$, we will simply write $\mathcal{L}(\alpha\beta)$, $\mathcal{L}(\alpha \cup \beta)$, $\mathcal{L}(\alpha)^*$, respectively.

A language is called a **regular language** iff it can be represented by a regular expression, that is, when $L = \mathcal{L}(\alpha)$ for some regular expression α.

Example 2.2. Let Σ be any alphabet. Then $\emptyset$, $\{\varepsilon\}$, Σ, Σ^*, Σ^+, Σ^n for any $n \in \mathbb{N}$ are regular languages. Note that $\emptyset^0 = \{\varepsilon\}$ and $\emptyset^n = \emptyset$ for $n > 0$. Thus, $\emptyset^* = \{\varepsilon\}$, $\Sigma^+ = \Sigma\Sigma^*$, and $\Sigma^n = \Sigma \cdots \Sigma$ concatenated n times. Similarly, $L = \{a^m b^n : m, n \in \mathbb{N}\}$ is regular as $L = \mathcal{L}(a^* b^*)$. □

Sometimes, we give a description of a language in terms of its elements.

Example 2.3. Let L be the set of all strings over $\{a, b\}$ having exactly two occurrences of b, which are consecutive. Is L regular?

Solution. If two b's are consecutive in a string u, then bb is a substring of u. That is, $u = xbby$ for some strings x and y. If x has a b in it, then u will have more than two b's, similarly for y. Thus both x and y have only a's in them or they may equal the empty string ε. Hence $x, y \in \{a\}^*$. Therefore, $L = \mathcal{L}(a^*bba^*)$, and it is a regular language. □

Example 2.4. Let L be the language over $\{a, b\}$ having exactly two nonconsecutive b's. Is L a regular language?

Solution. You can write L as $\mathcal{L}(a^*ba^*ba^*) - \mathcal{L}(a^*bba^*)$. But this does not help as difference or complementation is not allowed in a regular expression. Any typical string in L has one b in its middle, followed by some a's and then another b. It may look something like $aa \cdots aba \cdots aba \cdots a$. Note that, before the first b (first from the left) there may not be an a and similarly there may not be an a after the second b. But there must be an a in between the two b's. Thus, $L = \mathcal{L}(a^*ba^+ba^*) = \mathcal{L}(a^*baa^*ba^*)$, a regular language. □

Exercise 2.4. See that $\mathcal{L}(a^*ba^*baa^*ba^*(ba^* \cup \emptyset^*)) = \{w \in \{a, b\}^* : w$ has three or four occurrences of b's in which the second and third occurrences are not consecutive$\}$.

Example 2.5. Is the complement of the language in Example 2.3 regular?

Solution. With $\Sigma = \{a, b\}$, this language L contains all the strings of the language in Example 2.4 and the strings that do not contain exactly two consecutive b's. Thus $L = \mathcal{L}(a^*baa^*ba^*) \cup \mathcal{L}(a^*) \cup \mathcal{L}(a^*ba^*) \cup \Sigma^*\{b\}\Sigma^*\{b\}\Sigma^*\{b\}\Sigma^*$. The regular expression for L is $a^* \cup a^*ba^* \cup a^*baa^*ba^* \cup (a \cup b)^*b(a \cup b)^*b(a \cup b)^*b(a \cup b)^*$. □

Example 2.6. What is $\mathcal{L}(a^*((b \cup bb)aa^*(b \cup bb))^*a^*)$?

Solution. Clearly, any string of a's is in the language. What else? There are also strings with 0 or more a's followed by one or two b's, then at least one a, and then 0 or more of b or bb, and then a string of a's. What is the middle aa^* doing? It prevents occurrences of three or more consecutive b's. The language is $\overline{\mathcal{L}((a \cup b)^*bbb(a \cup b)^*)}$. □

Hence forward, we will make the overline short; for example, we will write $\overline{\mathcal{L}}((a \cup b)^*bbb(a \cup b)^*)$ instead of $\overline{\mathcal{L}((a \cup b)^*bbb(a \cup b)^*)}$.

Exercise 2.5. Does the regular expression $a^* \cup ((a^*(b \cup bb))(aa^*(b \cup bb))^*)a^*$ represent the language $\overline{\mathcal{L}}(bbb)$?

We will also write the regular expression itself for the language it represents. This will simplify our notation a bit. For example, instead of writing $\mathcal{L}(a^*b^*)$, we will simply write the language a^*b^*. Use of the phrase "the language" will clarify the meaning.

Exercise 2.6. Does the equality $\overline{\mathcal{L}}(bbb) = (\emptyset^* \cup b \cup bb)(a \cup bb \cup abb)^*$ hold?

We will also say that two regular expressions are *equivalent* when they represent the same language. Moreover, in accordance with the last section, two equivalent regular expressions can also be written as equal. This means, for regular expressions R, E, we will use any one of the notations $R = E$ (sloppy), $\mathcal{L}(R) = \mathcal{L}(E)$ (precise), or $R \equiv E$ (R is equivalent to E, technical) to express the one and the same thing.

Problems for Section 2.3

2.17. Give regular expressions for the following languages over $\{a\}$:
(a) $\{a^{2n+1} : n \in \mathbb{N}\}$.
(b) $\{a^n : n$ is divisible by 2 or 3, or $n = 5\}$.
(c) $\{a^2, a^5, a^8, \ldots\}$.

2.18. Give a simpler regular expression for each of the following:
(a) $\emptyset^* \cup b^* \cup a^* \cup (a \cup b)^*$.
(b) $(a^*b)^* \cup (b^*a)^*$.
(c) $((a^*b^*)^*(b^*a^*)^*)^*$.
(d) $(a \cup b)^*b(a \cup b)^*$.

2.19. Find regular expressions for the following languages over $\{0, 1\}$:
(a) $\{w : w$ does not contain $0\}$.
(b) $\{w : w$ does not contain the substring $01\}$.
(c) Set of all strings having at least one pair of consecutive zeros.
(d) Set of all strings having no pair of consecutive zeros.
(e) Set of all strings having at least two occurrences of 1's between any two occurrences of 0's.
(f) $\{0^m1^n : m + n$ is even, $m, n \in \mathbb{N}\}$.
(g) $\{0^m1^n : m \geq 4, n \leq 3, m, n \in \mathbb{N}\}$.
(h) $\{0^m1^n : m \leq 4, n \leq 3, m, n \in \mathbb{N}\}$.
(i) Complement of $\{0^m1^n : m \geq 4, n \leq 3, m, n \in \mathbb{N}\}$.
(j) Complement of $\{0^m1^n : m \leq 4, n \leq 3, m, n \in \mathbb{N}\}$.
(k) Complement of $\{w \in (0 \cup 1)^*1(0 \cup 01)^* : \ell(w) \leq 3\}$.
(l) $\{0^m1^n : m \geq 1, n \geq 1, mn \geq 3\}$.
(m) $\{01^nw : w \in \{0, 1\}^+, n \in \mathbb{N}\}$.
(n) Complement of $\{0^{2m}1^{2n+1} : m, n \in \mathbb{N}\}$.
(o) $\{uwu : \ell(u) = 2\}$.
(p) Set of all strings having exactly one pair of consecutive zeros.
(q) Set of all strings ending with 01.
(r) Set of all strings not ending in 01.
(s) Set of all strings containing an even number of zeros.
(t) Set of all strings with at most two occurrences of the substring 00.
(u) Set of all strings having at least two occurrences of the substring 00. [Note: 000 contains two such occurrences.]

2.20. Write the languages of the following regular expressions in set notation, and also give verbal descriptions:
(a) $a^*(a \cup b)$.
(b) $(a \cup b)^*(a \cup bb)$.
(c) $((0 \cup 1)^*(0 \cup 1)^*)^*00(0 \cup 1)^*$.
(d) $(aa)^*(bb)^*b$.
(e) $(1 \cup 01)^*$.
(f) $(aa)^*b(aa)^* \cup a(aa)^*ba(aa)^*$.

2.21. Give regular expressions for the following languages on $\{a, b, c\}$:
(a) Set of all strings containing exactly one a.
(b) Set of all strings containing no more than three a's.
(c) Set of all strings that contain aaa as a substring.
(d) Set of all strings that do not have aaa as a substring.
(e) Set of all strings with number of a's divisible by three.
(f) Set of all strings that contain at least one occurrence of each alphabet symbol.
(g) $\{w : \ell(w) \bmod 3 = 0)\}$.
(h) $\{w : \#_a(w) \bmod 3 = 0\}$.
(i) $\{w : \#_a(w) \bmod 5 > 0\}$.

2.22. Let E_1, E_2 be arbitrary regular expressions. Are the following true? Justify.
(a) $(E_1^*)^* \equiv E_1^*$.
(b) $E_1^*(E_1 \cup E_2)^* \equiv (E_1 \cup E_2)^*$.
(c) $(E_1 \cup E_2)^* \equiv (E_1^* E_2^*)^*$.
(d) $(E_1 E_2)^* \equiv E_1^* E_2^*$.
(e) $(E_1 \cup E_2)^* \equiv E_1^* \cup E_2^*$.
(f) $(E_1 \cup E_2)^* E_2 \equiv (E_1^* E_2)^*$.
(g) $(E_1 \cup E_2)E_3 \equiv E_1 E_3 \cup E_2 E_3$.
(h) $(E_1 E_2 \cup E_1)^* E_1 \equiv E_1(E_2 E_1 \cup E_1)^*$.
(i) $(E_1 E_2 \cup E_1)^* E_1 E_2 \equiv (E_1 E_1^* E_2)^*$.
(j) $E_1(E_2 E_1 \cup E_1)^* E_2 \equiv E_2 E_2^* E_1(E_2 E_2^* E_1)^*$.

2.23. Give a general method to construct a regular expression E' from a regular expression E so that $(\mathcal{L}(E))^R = \mathcal{L}(E')$.

2.24. Let E be a regular expression that does not involve $\emptyset$ or ε. Give a necessary and sufficient condition that $\mathcal{L}(E)$ is infinite.

2.4 Regular Grammars

Consider the language $L = \{a^m b^n : m, n \in \mathbb{N}\}$. Suppose abstractly, you write any string of L as S. Then S can be ε as it happens with the choice $m = n = 0$. Or, S may be a concatenation of two strings A and B, where A is any string of a's and B is any string of b's. If both A and B can also be chosen as ε, then you may think of rewriting S as AB. Now, how to generate a string of a's from A?

Well, A can be thought of as an a followed by another string of a's. That is, A may be an a concatenated with another string of the same type as A. Let us write the last sentence by the shorthand $A \mapsto aA$; read it as A can be aA. This is analogous to an inductive step in a definition by induction. The limiting case (the basis step) "A can be ε" may be written as $A \mapsto \varepsilon$. Together they will generate the strings a^m for any $m \in \mathbb{N}$, that is, a^*, so to say. Then $L = \mathcal{L}(a^*b^*)$ may be generated by the rules:

$$S \mapsto AB, \ A \mapsto aA, \ A \mapsto \varepsilon, \ B \mapsto bB, \ B \mapsto \varepsilon.$$

To see, for example, that $a^3b^2 \in L$, we can apply the rules and derive a^3b^2 as

$$S \Rightarrow AB \Rightarrow aAB \Rightarrow aaAB \Rightarrow aaaAB \Rightarrow aaaB \Rightarrow aaabB \Rightarrow aaabbB \Rightarrow a^3b^2.$$

The symbol $\Rightarrow$ is read as "yields in one step." It means that a rule has been applied on the left side expression to obtain the one on the right. Thus the first yield $S \Rightarrow AB$ is an application of the rule $S \mapsto AB$. The second yield $AB \Rightarrow aAB$ is an application of the rule $A \mapsto aA$. Similarly, the last yield is an application of the rule $B \mapsto \varepsilon$. These successive applications of the rules above show that the string a^3b^2 can be derived from S by a sequence of applications of the above rules. We write this as

$$S \overset{*}{\Rightarrow} a^3b^2.$$

Read this as S yields (in zero or more steps) a^3b^2. We will, in fact, omit the superscript $*$ in the yield relation, and write this simply as $S \Rightarrow a^3b^2$. Such a device of rules using extra symbols (as S, A, B here) is called a grammar (a phase structure grammar, to be specific) in common parlance with the natural languages. We can, of course, find another grammar for generating exactly the same strings. For example, a grammar with the rules $S \mapsto aS, S \mapsto B, B \mapsto bB, B \mapsto \varepsilon$ will also do. Try a derivation of a^3b^2 by using these new rules.

A **regular grammar** G is a quadruple (N, Σ, R, S), where

N is a finite set, called the set of **nonterminals** or the nonterminal symbols,
Σ is an alphabet, also called the set of **terminals** or terminal symbols, $N \cap \Sigma = \emptyset$,
$R \subseteq N \times (\Sigma^* \cup \Sigma^* N)$ is a relation from N to $\Sigma^* \cup \Sigma^* N$, called the set of **productions** or production rules, and
$S \in N$ is a designated nonterminal, called the **start symbol**.

Any typical element (x, y) of R is written as $x \mapsto y$ for convenience.

For example, the grammar with the rules $S \mapsto aS, S \mapsto B, B \mapsto bB, B \mapsto \varepsilon$ is $G = (\{S, B\}, \{a, b\}, R, S)$, where $R = \{(S, aS), (S, B), (B, bB), (B, \varepsilon)\}$. You can see how the ordered pairs in the relation are obtained from the corresponding rules. Instead of writing R as a set of ordered pairs, we will write it as a set of rules using the symbol $\mapsto$. This grammar G is conveniently rewritten as $(\{S, B\}, \{a, b\}, \{S \mapsto aS, S \mapsto B, B \mapsto bB, B \mapsto \varepsilon\}, S)$. We will refer both the ordered pairs and the rules by the name "productions." The relation R says that the left hand side of a production sign $\mapsto$ is any nonterminal symbol and the right hand side string can be a string of terminals, the empty string, or a string of terminal symbols followed by only one nonterminal symbol. That is, any production in a regular grammar will have one of the following forms:

$$A \mapsto \varepsilon \text{ or } A \mapsto u \text{ or } A \mapsto uB,$$

where A, B are nonterminal symbols and u is any string of terminal symbols. Thus, the rule $S \mapsto AB$ cannot be a production in any regular grammar if both A, B are nonterminal symbols. Of course, it is enough to write only the productions and mention the start symbol S; the other components such as N and Σ can be found out from them. Sometimes, we write the four components of a grammar by using the name of the grammar as a subscript like $G = (N_G, \Sigma_G, R_G, S_G)$. This notation becomes convenient when many grammars are considered in a context.

With such a formal definition of a regular grammar, we must also say how the productions are applied to generate the strings.

Let $G = (N, \Sigma, R, S)$ be a regular grammar. If there is a string $x \in \Sigma^*$, a nonterminal $A \in N$, and a string $B \in (N \cup \Sigma)^*$ such that $A \mapsto B$ is a production in R, then we write $xA \overset{1}{\Rightarrow} xB$, and read it as xA **yields** in one step xB. Such a way of writing $xA \overset{1}{\Rightarrow} xB$ corresponding to a production $A \mapsto B$ will be referred to as an *application of the production* $A \mapsto B$.

Moreover, successive applications of productions is written as $\overset{*}{\Rightarrow}$. (In fact, $\overset{*}{\Rightarrow}$ is the transitive and reflexive closure of the relation $\overset{1}{\Rightarrow}$.) Note that for any string s, we have $s \overset{*}{\Rightarrow} s$, that is, s yields s, though possibly there would not be any production in a grammar to show that s yields s in one step. We will also abbreviate two (and then many) yields $A \overset{1}{\Rightarrow} B$ and $B \overset{1}{\Rightarrow} C$ to $A \overset{1}{\Rightarrow} B \overset{1}{\Rightarrow} C$. Such an abbreviated sequence of yields will be called a **derivation**, a derivation of the last string from the first string. For example, $A \overset{1}{\Rightarrow} B \overset{1}{\Rightarrow} C \overset{1}{\Rightarrow} D$ is a derivation of D from A.

The number of steps in the derivation, counting each "one step yield" as one, is called the **length** of the derivation. Thus the derivation $a \overset{*}{\Rightarrow} a$ has length 0, as no rule has been applied to derive the string a from itself. Note that all the productions in a regular grammar are such that a nonterminal can only occur as the rightmost symbol in any derived string. Thus an application of a rule $A \mapsto B$ to derive xB from xA simply overlooks whatever is to the left of A; it simply replaces A by B.

The yield relations (in 0 step, in m steps, or in 0 or more steps) can formally be defined by

$x \overset{0}{\Rightarrow} x$ for any $x \in (N \cup \Sigma)^*$,

$y \overset{m+1}{\Rightarrow} zB$ iff $y \overset{m}{\Rightarrow} zA$ and $A \mapsto B \in R$ for any $y, z \in (N \cup \Sigma)^*$, $A \in N$, $B \in \Sigma^* N \cup \Sigma^*$,

$y \overset{*}{\Rightarrow} z$ iff for some $m \in \mathbb{N}$, $y \overset{m}{\Rightarrow} z$.

We will write $\overset{1}{\Rightarrow}$, $\overset{m}{\Rightarrow}$, and $\overset{*}{\Rightarrow}$ as the same symbol $\Rightarrow$, if no confusion arises.

If $G = (N, \Sigma, R, S)$ is a regular grammar, then the **language generated by** G, written $\mathcal{L}(G)$, is the set of all strings of Σ^* derived from the start symbol S. That is,

$$\mathcal{L}(G) = \{w \in \Sigma^* : S \overset{*}{\Rightarrow} w\}.$$

As a shorthand, the productions $A \mapsto u$, $A \mapsto v$, $A \mapsto uB$ will be written as $A \mapsto u|v|uB$ by using the vertical bar; read it as "or."

Example 2.7. Consider the regular grammar $G = (\{S, B\}, \{a, b\}, R, S)$, where $R = \{S \mapsto aS|B,\ B \mapsto bB|\varepsilon\}$. Let us attempt a derivation.

The second production $S \mapsto B$ says that we can derive B from S. That is, $S \Rightarrow B$. The fourth production says that ε can be derived from B, that is, $B \Rightarrow \varepsilon$. Thus we have the derivation $S \Rightarrow B \Rightarrow \varepsilon$, that is, $S \Rightarrow \varepsilon$. So, $\varepsilon \in \mathcal{L}(G)$.

We also see that $a \in \mathcal{L}(G)$ and $ab^2 \in \mathcal{L}(G)$ as

$$S \Rightarrow aS \Rightarrow aB \Rightarrow a\varepsilon = a,$$

$$S \Rightarrow aS \Rightarrow aB \Rightarrow abB \Rightarrow abbB \Rightarrow abb\varepsilon = ab^2.$$

It is now clear how to construct a derivation of the string $a^m b^n$ in G. Just apply the first production m-times, then switch S to B by applying the second production, then apply the third production n-times, and finally, remove the nonterminal B by using the fourth production. However, the derivation $B \Rightarrow bB \Rightarrow b\varepsilon = b$ does not show that $b \in \mathcal{L}(G)$, as this derivation does not start with the start symbol S. Prefixing $S \Rightarrow B$ completes the derivation. □

You can see that no derivation in Example 2.7 would give a string of terminals once a b precedes an a. For example, you cannot have $S \Rightarrow ba$. Can you prove it? You may need to think a bit carefully before starting a proof.

If a string over $\{a, b\}$ is generated by G, then the last step must use the production $B \mapsto \varepsilon$, for this is the only production whose right side string does not have a nonterminal. (The derivation must terminate with terminals only.) Then, before it, B must have been generated. For this to happen, either $B \mapsto bB$ or $S \mapsto B$ has been applied. That is, once B has come into a derivation, S has gone and then a cannot enter the derivation.

As B can occur only as the rightmost symbol of any derived string, and B would eventually give a string of b's, no a can be on the right of any b. This shows that G can only generate strings of the form $a^m b^n$. You will have to show the converse that all strings of the form $a^m b^n$ are indeed generated by G. This will show that $\mathcal{L}(G) = \mathcal{L}(a^*b^*)$.

However, sometimes we will not be able to accept such reasoning as a proof, because there might be a case which we might have left unnoticed. So, how will a formal proof look like?

Let us use induction on the length of derivations. We notice that the minimum length of a derivation is 2; the particular derivation is $S \Rightarrow B \Rightarrow \varepsilon$ deriving the empty string. This forms our basis case, and there is, of course, nothing to prove here. Assume that a derivation of length n generates the string $a^r b^s$. Now, before the last step, there should be a B and the last step is an application of the rule $B \mapsto \varepsilon$. Moreover, if in any derivation B appears, then it appears as the rightmost symbol. Thus, before the last but one step ($(n-1)$th step) of the derivation of $a^r b^s$, the string appears as $a^r b^s B$. On this string, we use the production $B \mapsto bB$ and proceed to derive

$$a^r b^s B \Rightarrow a^r b^s bB \Rightarrow a^r b^s b = a^r b^{s+1}.$$

Moreover, this is the only string that can be derived from $a^r b^s B$ in the next two steps. Further, once a B is introduced to the derivation, no a can be generated. Thus, by induction, we have proved that $\mathcal{L}(G) = \mathcal{L}(a^*b^*)$.

Exercise 2.7. There are gaps in the above inductive proof of $\mathcal{L}(G) = \mathcal{L}(a^*b^*)$. Point out them and then complete the proof.

So, what is the connection between regular expressions and regular grammars? I would suggest you to wait a bit before attempting to answer this question. I ask one more related question: given a language, how do we determine whether it is regular or not? You can try to represent a given language by a regular expression. When the answer to the first question turns out to be affirmative, you know that the existence of a regular grammar for a given language would also prove that the language is regular. But suppose that neither could you construct a regular expression nor a regular grammar. Then? What do you conclude? Is the language regular or not?

Problems for Section 2.4

2.25. Find regular grammars for $\Sigma = \{a, b\}$ that generate the set of all strings with
(a) exactly one a.
(b) at least one a.
(c) no more than three a's.

2.26. Let Σ be any alphabet. Construct regular grammars to generate the languages $\emptyset$, $\{\varepsilon\}$, Σ, Σ^*, Σ^+, and Σ^n, for any $n \in \mathbb{N}$.

2.27. Let $\Sigma = \{a, b\}$. Construct regular grammars for (generating) the languages in Examples 2.3–2.5.

2.28. Find a regular grammar that generates the language
(a) $aa^*(ab \cup a)^*$.
(b) $(aab^*ab^*)^*$.

2.29. Find a regular grammar that generates the language
(a) $\{(ab)^m : m \in \mathbb{N}\}$.
(b) $\{a^{2n} : n \geq 1\}$.
(c) $\{wa : w \in \{a, b\}^*\}$.
(d) $\{bw : w \in \{a, b\}^*\}$.
(e) $\{a^m b^n c^k : m, n, k \geq 1\}$.
(f) $\{a^m b^n : m + n \text{ is even}, m, n \in \mathbb{N}\}$.
(g) $\{ab^m c^n d : m, n \geq 1\}$.
(h) $\{(ab)^m : m \geq 1\} \cup \{(ba)^n : n \geq 1\}$.

2.30. Give a regular grammar with terminals in $\{a, b, c\}$ for generating the language
(a) $\{w : w \text{ contains } aaa \text{ as a substring}\}$.
(b) $\{w : \text{number of } a\text{'s in } w \text{ is divisible by two}\}$.
(c) $\{w : w \text{ contains no run of } a\text{'s of length greater than two}\}$.
(d) $\{w : w \text{ contains at least one occurrence of each alphabet symbol}\}$.
(e) $\{w : \ell(w) \bmod 3 = 0\}$.
(f) $\{w : \#_a(w) \bmod 3 = 0\}$.
(g) $\{w : \#_a(w) \bmod 5 > 0\}$.

2.31. Show that if G is a regular grammar with $\mathcal{L}(G) \neq \emptyset$, then G must have a production of the form $A \mapsto x$ for some terminal string x.

2.32. Show that for every regular language L not containing ε, there is a regular grammar having productions in the forms $A \mapsto Ba$ or $A \mapsto a$ that generates L. [Such a grammar is a left-linear grammar. First define how a production is applied.]

2.5 Deterministic Finite Automata

What we require is a sort of an algorithm that answers "yes" or "no" according as the language is regular or not. We may think of a machine to which we give our language as an input and the machine would then tell us as an output whether the language is regular or not. But how do we give any language, possibly a set of infinite number of strings, as an input to a machine? We will rather think of a class of machines where each machine will be made for a language. Imagine a machine for the language a^*b^*. It will read any given string as an input and will let us know whether the given string is or is not in the language a^*b^*.

Once we have a machine for each regular language, we can think, in a certain sense, that the class of such machines determine the class of regular languages. Then, by arguing about machines we may be able to say whether a given language is or is not regular. Individual machines then will be tailor-made for particular languages and will be used as language recognizers.

A typical machine will have a certain finite number of **states** out of which one will be marked as the **initial state**. By default, the machine will start its operation being in this initial state. It will have a **reading head**, which will read from a one-dimensional tape, called the **input tape**.

The input tape will have a left end and will be extended to the right up to a finite extent. Imagine a paper tape, say, of 1 cm wide, one end of which you are holding in your left hand and starting from there you are using your right hand to search for the other end, which you would eventually reach. The tape is also divided into small squares (say, of dimension $1 \times 1\,\text{cm}^2$ for your 1 cm wide tape). Each square will contain an input symbol and a string will be written on the tape starting with the first square, from left to right. See Fig. 2.1.

If you think of a numbering of the squares on the tape, with the square on the left end marked as 1 and etc., then the mth symbol of a string is written on the mth square. The reading head of the machine will be kept on the leftmost square of the tape initially.

When the machine starts operating, it will be in the initial state, reading the first symbol of the input string from the leftmost square of the input tape. Depending upon the initial state and the input symbol, it will change its state to possibly another state from the initial (exactly which state, only the machine knows). It will then go to the next square. The second symbol is read similarly; but this time the machine reads it being in the new state, not (necessarily) the initial state.

Again, depending upon this new state and the input symbol on the second square, the machine will go to possibly another state and get ready for reading the symbol from the next square. This sort of reading a symbol and changing its own state continues till it finishes reading the input string. Note that upon consuming the whole string, the machine must have been in some state.

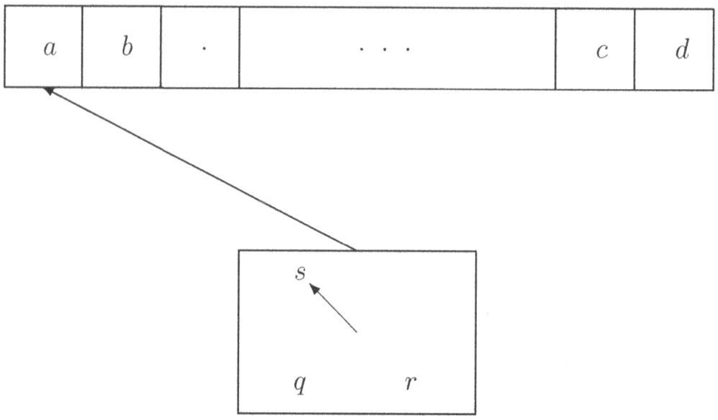

Fig. 2.1. A finite automaton

To know whether the machine has accepted or rejected the string, we will designate some of its states as the **final states**. If the last state of the machine is a final state, we will say that the machine has **accepted** the input string, otherwise, the machine has **rejected** the string. Final states are the accepting states, so to speak.

A **deterministic finite automaton** is a quintuple $D = (Q, \Sigma, \delta, s, F)$, where

Q is a finite set, called the set of **states**
Σ is an alphabet ($Q \cap \Sigma = \emptyset$), called the **input alphabet**
$\delta : Q \times \Sigma \to Q$ is a partial function, called the **transition function**
$s \in Q$ is a designated state, called the **initial state**, and
$F \subseteq Q$ is the set of **final states**.

We use the acronym **DFA** for a deterministic finite automaton. Suppose a DFA is reading the input symbol σ being in the state q. If $\delta(q, \sigma) = r$, then the DFA will go to the state r and its reading head will go to the next square. Since δ is a partial function (not necessarily a function), there might be a pair of state and input symbol for which δ is not defined at all, that is, δ does not prescribe any next state. In such a case, of course, the DFA will stop abruptly. We will see how to handle such a situation.

Suppose the DFA $M = (Q, \Sigma, \delta, s, F)$ is in operation. At any instant of time, a description of M can be completely specified by telling in which state the automaton is, and what is the remaining input string. Formally, a **configuration** (or an *instantaneous description*) of a DFA is an ordered pair (q, w), where $q \in Q$ is a state, the current state, and $w \in \Sigma^*$ is a string, that is yet to be read.

For a DFA, **one step of a computation** can be described by specifying how a description of the machine has changed. That is, a step of a computation can be specified by telling which configuration has given rise to (yields) which other configuration. For example, suppose that the DFA is currently in state q scanning the first symbol of the input string $u = \sigma v$. If the transition function of the DFA is such that in the next step of computation it goes to the state r, the remaining string to be read is v, and then the next configuration of the machine would be (r, v). In such a case, we will say that the configuration (q, u) yields the configuration (r, v) in one step.

We will use the symbol $\overset{1}{\rightsquigarrow}$ for "yields in one step." Formally, for the DFA M, the configuration (q, u) **yields in one step** the configuration (r, v), that is,

$$(q, u) \overset{1}{\rightsquigarrow} (r, v) \text{ iff } \delta(q, \sigma) = r \text{ and } u = \sigma v \text{ for some } \sigma \in \Sigma.$$

The machine **halts** (stops operating) when it encounters a state-symbol pair for which δ is not defined. Suppose that $\delta(q, \sigma)$ is left undefined in the partial function. Now the machine is in state q and the input string is σv. It halts then and there. We will also make it a convention that when the input string is totally consumed by the machine, that is, when the machine has read all the symbols on the input tape, then also it **halts**. (Cryptically, δ is not defined for (q, ε)!)

The relation "yields" in the set of all configurations is defined as the reflexive and transitive closure of the relation of "yields in one step." Explicitly, a configuration (q, u) **yields** another configuration (r, v) iff there is a sequence (of length $n \geq 0$) of configurations (q_i, w_i) such that

$$q_0 = q,\ w_0 = u,\ q_n = r,\ w_n = v \text{ and } (q_i, w_i) \overset{1}{\rightsquigarrow} (q_{i+1}, w_{i+1}).$$

Thus (q, u) yields (r, v) means that (q, u) yields, in zero or more steps, the configuration (r, v). If we have a sequence of configurations

$$(q_1, w_1) \overset{1}{\rightsquigarrow} (q_2, w_2),\ (q_2, w_2) \overset{1}{\rightsquigarrow} (q_3, w_3),\ \ldots,\ (q_{n-1}, w_{n-1}) \overset{1}{\rightsquigarrow} (q_n, w_n),$$

then (q_1, w_1) yields all the configurations $(q_1, w_1), \ldots, (q_n, w_n)$. The symbol $\overset{*}{\rightsquigarrow}$ is used to denote the yield relation just as in derivations of strings in a grammar.

A formal definition of the yield relation can be given as follows:

$(p, u) \overset{0}{\rightsquigarrow} (p, u)$.

$(p, \sigma u) \overset{1}{\rightsquigarrow} (q, u)$ iff $\delta(p, \sigma) = q$.

$(p, \sigma u) \overset{n+1}{\rightsquigarrow} (q, v)$ iff $(p, \sigma) \overset{1}{\rightsquigarrow} (r, u)$ and $(r, u) \overset{n}{\rightsquigarrow} (q, v)$ for some state r.

$(p, u) \overset{*}{\rightsquigarrow} (q, v)$ iff $(p, u) \overset{m}{\rightsquigarrow} (q, v)$ for some $m \in \mathbb{N}$.

We make it a convention to omit the superscripts $1, m, *$, and simply use the same symbol $\rightsquigarrow$ to denote all of "yields in one step," "yields in m steps," and "yields (in 0 or more steps)" if no confusion arises.

A **computation** by the DFA then will be such a sequence of yields (in one step), possibly terminating at a configuration (q, ε) for some state q. The configuration (q, ε) says that the machine is currently in state q and the whole input has already been read. We then say that the DFA M **accepts** a string $u \in \Sigma^*$ iff $(s, u) \rightsquigarrow (q, \varepsilon)$, where $q \in F$, a final state. This says that the machine starts from its initial state s; reads the input string, symbol by symbol; changes its state according to its transition function; comes to a state q after it finishes reading the whole string and then halts. If this state q is a prescribed final state, then we would interpret that the machine has accepted the string. Otherwise, the machine **rejects** the input string.

Exactly what happens when a machine rejects an input string? If the state q that the machine enters after consuming the whole string is not a final state, then the machine rejects the input string. Moreover, if the machine halts before it finishes reading the whole string, then also the input string is rejected.

The **language accepted by** the DFA is the set of all input strings accepted by the DFA. That is,

$$\mathcal{L}(M) = \{u \in \Sigma^* : (s, u) \rightsquigarrow (q, \varepsilon) \text{ for some } q \in F\}.$$

Sometimes, we also say that a DFA **recognizes** the language that it accepts. If more machines than one are discussed in a context, we will give a subscript to the yield relation; we use $(q, u) \rightsquigarrow_M (r, v)$ to say that when the machine M operates, the configuration (q, u) yields the configuration (r, v).

Example 2.8. Let $Q = \{s, q, r\}$, $\Sigma = \{a, b\}$, $F = \{s, q\}$. Define the transition δ by $\delta(s, a) = s$, $\delta(s, b) = q$, $\delta(q, a) = r$, $\delta(q, b) = r$, $\delta(r, a) = r$, $\delta(r, b) = r$. What is the language accepted by the DFA $D = (Q, \Sigma, \delta, s, F)$?

Solution. Give $u = a^2b^2$ as an input to the machines D. The DFA D starts from the initial state s while u is on its input tape. That is, the initial configuration of D is (s, a^2b^2). The reading head of D is placed on the leftmost (first) input symbol a.

When D reads the first symbol a, the **transition** (an element of the transition function) $\delta(s, a) = s$ applies and the machine enters the state s (it does not change state here), and the reading head goes to the next symbol, which is again a. By this time, the first symbol has already been read (or consumed) by the machine. The remaining of the input string is ab^2. Thus the configuration the machine is in is (s, ab^2). It reads an a and works similarly to yield in the next step the configuration (s, b^2). As $\delta(s, b) = q$, the machine now changes its state to q and the remaining of the input string is only b. As a configuration, this information is coded as (q, b). As $\delta(q, b) = q$, the machine reenters the state q and by this time, all of the input string has been read. Thus the end-configuration (q, ε) is achieved. The computation of D on the input string $u = a^2b^2$ can be written down with the help of the yield relation as follows:

$$(s, a^2b^2) \rightsquigarrow (s, ab^2) \rightsquigarrow (s, b^2) \rightsquigarrow (q, b) \rightsquigarrow (q, \varepsilon).$$

The string a^2b^2 is accepted by D as $q \in F$.

For $u = a^2ba$, computation of D is as follows:

$$(s, a^2ba) \rightsquigarrow (s, ba) \rightsquigarrow (q, a) \rightsquigarrow (r, \varepsilon).$$

The string a^2ba is rejected by D as $r \notin F$.

Now, what language is accepted by D? A look at the transition function δ says that upon reading any number of a's, D remains in the state s. Once it reads a b, it goes to the state q, and there after for more and more b's, it stays in the same state q. As both s and q are the final states, it accepts the strings of the form $a^m b^n$. Being in state q, if it encounters an a, then it goes to the state r and then there after it will remain in that state eventually rejecting the string. Therefore, the language of the machine is $\mathcal{L}(D) = a^*b^*$. □

The action of the automaton D in Example 2.8 can be represented diagrammatically as in Fig. 2.2. Such a schematic diagram is called a **transition diagram** or a **state diagram**.

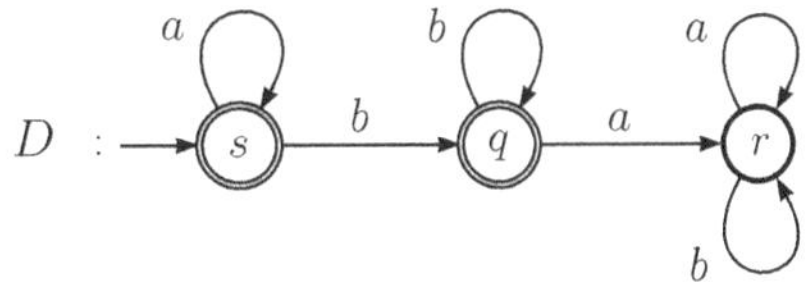

Fig. 2.2. Transition diagram of the DFA of Example 2.8.

What job does the state r do in the DFA D? As the transition function δ shows, once the machine enters the state r, it never leaves it. Moreover, it is not a final state. That means once the machine enters the state r, the input string will eventually be rejected. Is it not equivalent to omitting all the transitions with r as a state component? That is, suppose we leave δ undefined for the pairs (q, a), (r, a), and (r, b). Then there will be no change in the accepting computations. This is the reason we allowed δ to be a partial function. The new automaton $D' = (\{s, q\}, \{a, b\}, \delta', s)$ with $\delta'(s, a) = s$, $\delta'(s, b) = q$, $\delta'(q, b) = q$ would be depicted as in Fig. 2.3.

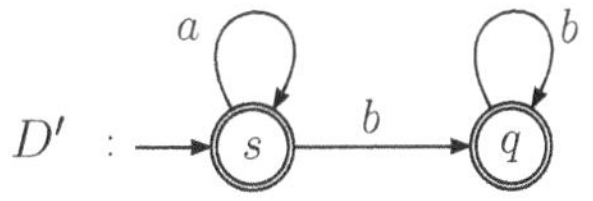

Fig. 2.3. Transition diagram of a DFA.

In D', computation with the input aba proceeds as follows:

$$(s, aba) \rightsquigarrow (s, ba) \rightsquigarrow (q, a).$$

D' halts abruptly without consuming the whole input. Thus, the string aba is rejected by D'.

By this time you must have guessed how a transition diagram is drawn corresponding to a DFA. We put the names of the states inside circles at different places. The final states are encircled once more. The initial state is shown by putting an arrow going towards it and coming from nowhere. An arrow labeled with a symbol σ is drawn from a state p to another state t iff we have $\delta(p, \sigma) = t$ as a transition for a DFA. Similarly, loops are drawn from and to the same state, provided there is a corresponding transition. Note that the labels on any edge of a DFA diagram are members of the alphabet Σ.

You can recognize such diagrams; these are simply directed labeled graphs with additional information of initial and final states. In fact, we could have defined automata as directed labeled graphs with these additional information on the initial and final states. Often we will do this by specifying automata by their diagrams instead of quintuples.

Example 2.9. Construct a DFA for accepting the language $\{w \in \{a, b\}^* : w$ has a substring bab or a substring $aba\}$.

Solution. The language can be represented by the regular expression $(a \cup b)^*(bab \cup aba)(a \cup b)^*$. A DFA is drawn in Fig. 2.4 for accepting this language.

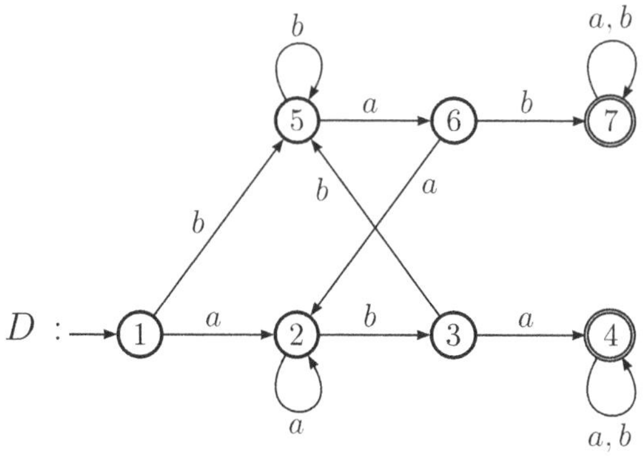

Fig. 2.4. DFA for Example 2.9.

Let us see how D proceeds with the input string $aabab$. The initial configuration of D with the input is $(s, aabab)$. The computation can be written down with the help of the yield relation. Here is the computation:

$$(1, aabab) \rightsquigarrow (2, abab) \rightsquigarrow (2, bab) \rightsquigarrow (3, ab) \rightsquigarrow (4, b) \rightsquigarrow (4, \varepsilon).$$

As the state 4 is a final state, D accepts $aabab$. Find such computations for the input strings $bbabab$, $abbaba$, $baaba$, and $abbaab$. Understand why the cross edges (transitions) from the states 3 to 5, and 6 to 2 are made. □

Exercise 2.8. Define a DFA as a directed labeled graph. Then define carefully the yield relation, the notion of a computation, and the languages accepted by them.

Sometimes a DFA is written in a tabular form by specifying its transition function. The initial state is then pointed out by an incoming arrow and the final states are put inside rectangles or circles. For example, the DFA in Fig. 2.4 can be written as in Table 2.5. I have written down the states on the top row and the symbols on the left column. You can also write the states in the left column and the symbols on the top row.

Table 2.1. Transition table for DFA of Fig. 2.4

δ	$\rightarrow 1$	2	3	$\boxed{4}$	5	6	$\boxed{7}$
a	2	2	4	4	6	2	7
b	5	3	5	4	5	7	7

Have you verified that the D in Example 2.9 is indeed a DFA? If not, do it now. Then modify it by introducing another state so that the new transition function will become a total function, but the new DFA must do the same job as D. This can be done by introducing another nonfinal state and adding all missing transitions from the relevant states to this new one, and then adding loops from this new state to itself. Such a state in a DFA is called a **state of no return**.

For defining the language an automaton accepts, we can avoid the configurations and also the yield relation. It is done by following the transitions in a sequence. What we have to do is simply extend the total function $\delta : Q \times \Sigma \to Q$ to another total function $\delta^* : Q \times \Sigma^* \to Q$ inductively. This is done by taking $\delta^*(q, \varepsilon) = q$ and $\delta^*(q, u\sigma) = \delta(\delta^*(q, u), \sigma)$ for $u \in \Sigma^*$ and $\sigma \in \Sigma$. Then the language accepted by D can be defined as $\mathcal{L}(D) = \{w \in \Sigma^* : \delta^*(s, w) \in F\}$. Computations of the DFA can be shown by the computations of δ^* on any given input string. However, we will follow the usual way of configurations and yields.

Example 2.10. What are the languages accepted by the DFAs in Fig. 2.5?

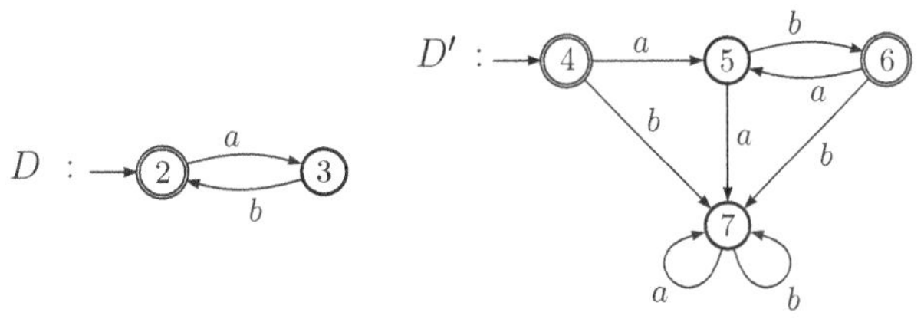

Fig. 2.5. DFAs for Example 2.10.

Solution. The initial state 2 of D is also a final state. Thus D accepts the empty string ε. Then, upon reading an a it changes its state to 3. In this state it can only read b and then it changes to the state 2, where it also accepts the already consumed input, that is, the string ab. Now, it is obvious that it will go on accepting any number of concatenations of this string ab.

Now suppose that D gets the input string as b. What does it do? It does not have a transition defined for this input, being in the initial state 2. Then it simply halts (abruptly). Though it is in a final state, namely 2, it does not accept the string because it has not consumed the input yet (and it does not know how to consume it). A similar thing happens when D is in state 3 and gets the symbol a to read. Thus it accepts the language $(ab)^*$.

As the initial state of D' is a final state, it also accepts the empty string. Moreover, being in its initial state, D' can read both a and b. If it reads b, then it enters the state 7, which is a state of no return and a nonfinal state. This leads to a halt in a nonfinal state. The string is not accepted. Thus, any string that begins with b is not accepted by D'.

However, if initially being in state 4 it reads a, then it changes to state 5, where again, an a will lead to nonacceptance. But a b at this stage leads the automaton to state 6, which is a final state. That is, the string ab is accepted. Once it enters the state 6, it is clear that the only way it would accept the string is by changing its state to 5 and then to 6, in succession. That is, it will only accept concatenations of the string ab thereafter. Thus, D' accepts the same language $(ab)^*$. □

You can make another DFA from D in Example 2.10 by introducing a state of no return as the state 7 serves D'. Then the partial function will become a total function without changing the accepted language.

Can you make a DFA to accept the language $(ab)^+$? From among the DFAs, D and D' of Example 2.10, it will be easy to modify the last one. Just do not keep the state 4 as a final state. That does the job as ε is no more accepted, and all the others will be accepted as they were. You can, of course, delete state 7.

Example 2.11. Construct a regular expression, a regular grammar, and a DFA for the language $L = \{w \in \{a, b\}^* : w$ contains an odd number of a's$\}$.

Solution. What are the strings in $\{a, b\}^*$ that are in L? It may be a string having any number of b's, followed by an a, followed by any number of b's. Or, it may be a string of the previous type and then followed by an a, then any number of b's, and then an a followed by any number of b's. Here "any number of b's" also includes zero number of b's, etc. That is, the strings can be provisionally written down as

$$b^*ab^* \cup (b^*ab^*)(ab^*ab^*) \cup (b^*ab^*)(ab^*ab^*)(ab^*ab^*) \cup \cdots .$$

A regular expression representing L is $b^*ab^*(ab^*ab^*)^*$. You can have a regular grammar to generate the language now. Starting from the nonterminal S, you derive bS; this would generate the initial b^*. Then, change S to aA and then to get the next block of b's, you take A to bA with A to ε. This will generate b^*ab^*. Instead of $A \mapsto \varepsilon$ if you take $A \mapsto aS$, this will insert an a and then will terminate in $(b^*ab^*)(ab^*ab^*)$. The latter bracketed expression should be $*$-ed. And this is achieved due to the last rule $A \mapsto aS$. Hence the grammar is $(\{S, A\}, \{a, b\}, R, S)$, where $R = \{S \mapsto bS|aA, A \mapsto \varepsilon|aS|bA\}$.

This gives us a hint for constructing the DFA. The DFA remains in its initial state after consuming any number of b's. That accounts for the first b^*. But then the initial state is not a final state as b^* does not belong to L. Upon reading an a, it changes state to another. This can be a final state as $b^m a \in L$. Here it can consume again any number of b's without changing its state. Once it gets the next a (this is the second a in the input), it changes its state, say to the initial state again. We take here the initial state instead of introducing another state. This is because, from the initial state it can consume strings in b^*ab^* and the next concatenated strings are of the form $(a)(b^*ab^*)$. It seems that the construction is complete. Look at the DFA in Fig. 2.6.

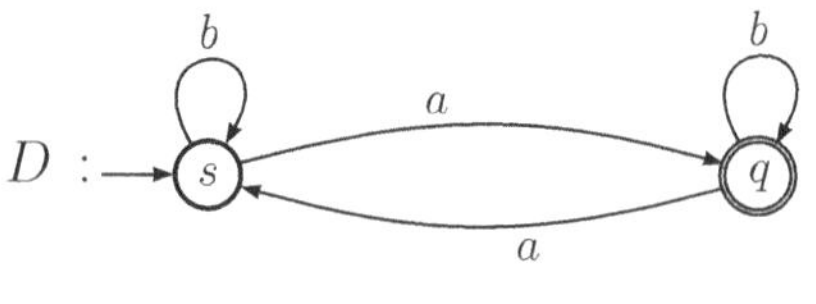

Fig. 2.6. DFA for Example 2.11.

Writing formally, the DFA is $D = (\{s, q\}.\{a, b\}, \delta, s, \{q\})$, where $\delta(s, a) = q$, $\delta(s, b) = b$, $\delta(q, a) = s$, $\delta(q, b) = b$. Computations of D with the strings $aabbabb$ and $bbbabba$ look like:

$$(s, aabbabb) \rightsquigarrow (q, abbabb) \rightsquigarrow (s, bbabb) \rightsquigarrow (s, abb) \rightsquigarrow (q, bb) \rightsquigarrow (q, \varepsilon).$$

$$(s, bbbabba) \rightsquigarrow (s, abba) \rightsquigarrow (q, bba) \rightsquigarrow (q, a) \rightsquigarrow (s, \varepsilon).$$

As q is a final state, the first string $aabbabb$ is accepted, while the second string $bbbabba$ is not accepted as s is not a final state. Convince yourself that $L(D) = L$ as claimed. Can you prove it? Also, compare the DFA with the grammar we have constructed. □

Can you see the connection between a regular expression, a regular grammar, and an automaton? Let us have some more examples.

Example 2.12. Construct an automaton to accept the language a^+b^+.

Solution. This language differs from our opening example a^*b^* only by the empty string ε. A regular grammar for the language can be constructed by having the production rules $S \mapsto aA$, $A \mapsto aA|bB$, $B \mapsto \varepsilon|bB$. Verify it. This also gives you hints for constructing a DFA. See the diagram in Fig. 2.7.

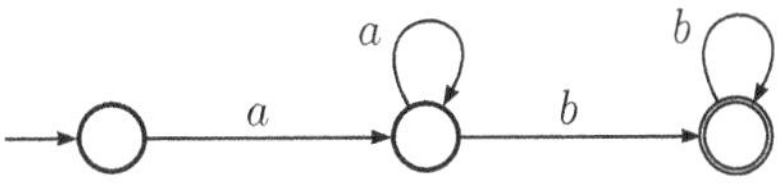

Fig. 2.7. DFA for Example 2.12.

Check whether it works correctly for some strings in the language and for some which are not in the language. □

Example 2.13. Construct a DFA that accepts the language $\{w \in \{0, 1\}^* : w$ has even number of 1's$\}$.

Solution. Convince yourself that the DFA in Fig. 2.8 does the job. Write also the regular expression and the regular grammar for the language. □

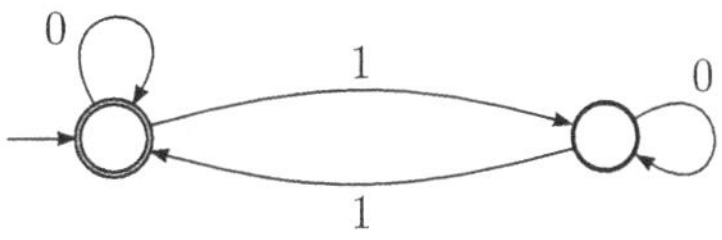

Fig. 2.8. DFA for Example 2.13.

Example 2.14. Construct a DFA for accepting the language $\{w \in \{0, 1\}^* : w$ does not contain three consecutive 1's$\}$.

Solution. You have met this language earlier where your alphabet was $\{a, b\}$ instead of $\{0, 1\}$. If you have forgotten, then construct a regular expression and a regular grammar for this. A DFA for accepting the language is given in Fig. 2.9. Write the DFA as a quintuple and then verify that it really accepts the language. □

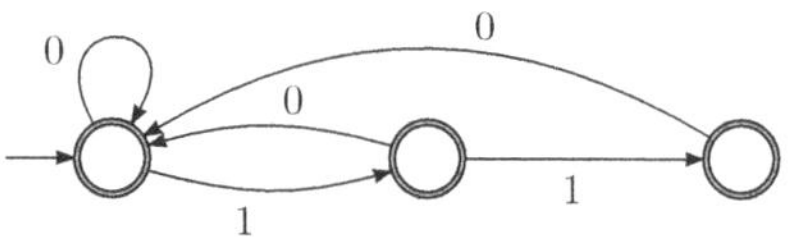

Fig. 2.9. DFA for Example 2.14.

In the problems, I will ask you to construct an automaton that accepts all strings satisfying certain constraints. What I mean there is that the language of your automaton should be the set of all strings satisfying the given constraint. Thus, it is not just enough to construct an automaton that accepts all those strings, I require that the automaton must accept those strings *and nothing else.* Similar caution goes with the grammars and regular expressions, and also with some other machines we will devise later.

Problems for Section 2.5

2.33. Exactly when does a DFA accept $\{\varepsilon\}$?

2.34. Construct a DFA that recognises any string over $\{a, b, \ldots, z\}$ having the substring abc.

2.35. For the following DFAs $(\{p, q, r, s\}, \{a, b\}, \delta, s, F)$, draw the transition diagrams and then find the languages they accept, in both the set notation and in plain English:

(a) $F = \{p\}$, $\delta(p, a) = q$, $\delta(p, b) = s$, $\delta(q, a) = r$, $\delta(q, b) = r$, $\delta(r, a) = r$, $\delta(r, b) = r$, $\delta(s, a) = p$, $\delta(s, b) = r$.

(b) $F = \{s\}$, $\delta(p, a) = q$, $\delta(p, b) = s$, $\delta(q, a) = r$, $\delta(q, b) = p$, $\delta(r, a) = r$, $\delta(r, b) = r$, $\delta(s, a) = p$, $\delta(s, b) = r$.

(c) $F = \{s\}$, $\delta(p, a) = q$, $\delta(p, b) = s$, $\delta(q, a) = q$, $\delta(q, b) = q$, $\delta(r, a) = s$, $\delta(r, b) = q$, $\delta(s, a) = p$, $\delta(s, b) = r$.

(d) $F = \{s\}$, $\delta(p, a) = q$, $\delta(p, b) = q$, $\delta(q, a) = q$, $\delta(q, b) = q$, $\delta(r, a) = s$, $\delta(r, b) = q$, $\delta(s, a) = p$, $\delta(s, b) = r$.

(e) $F = \{q\}$, $\delta(p, a) = q$, $\delta(p, b) = q$, $\delta(q, a) = q$, $\delta(q, b) = p$, $\delta(r, a) = r$, $\delta(r, b) = r$, $\delta(s, a) = p$, $\delta(s, b) = s$.

2.36. Design a DFA that accepts $(a\{a, b\}^*a)^2$.

2.37. Construct DFAs for accepting the following languages on $\Sigma = \{a, b\}$:

(a) $(a^+b)^2$.

(b) $(a^+b)^2 - (a^+b)$.

(c) $\{ab^5wb^5 : w \in \{a, b\}^*\}$.
(d) $\{uabw : u, w \in \{a, b\}^*\}$.
(e) $\{w : \ell(w) \bmod 3 = 0\}$.
(f) $\{w : \ell(w) \bmod 5 \neq 0\}$.
(g) $\{w : \#_a(w) \bmod 3 > 1\}$.
(h) $\{w : \#_a(w) \bmod 3 > \#_b(w) \bmod 3\}$.
(i) $\{w : |\#_a(w) - \#_b(w)| \bmod 3 < 2\}$.
(j) $\{w : |\#_a(w) - \#_b(w)| \bmod 3 > 0\}$.

2.38. Consider the set of all strings w on $\{0, 1\}$ defined by the requirements below. For each, construct an accepting DFA.
(a) In w, every 00 is followed immediately by a 1. For instance, 101, 0011 are in the language but 0001 is not.
(b) w contains 00 but not 000.
(c) The leftmost symbol in w differs from the rightmost symbol.
(d) Each substring of w of length at least four has at most two 0's. For instance, 001110, 011001 are in the language but 10010 is not.
(e) $\ell(w) \geq 5$ and the fourth symbol from the right end of w is different from the leftmost symbol.
(f) In w, the leftmost symbol and the rightmost two symbols are identical.

2.39. Let M be a DFA with a as an input symbol. Suppose that $\delta(q, a) = q$ for each state of M. Then show that either $a^* \subseteq \mathcal{L}(M)$ or $a^* \cap \mathcal{L}(M) = \emptyset$.

2.40. Let M be a DFA with the initial state s and a single final state r. Suppose $\delta(s, \sigma) = \delta(r, \sigma)$ for each input symbol σ. Show that if a nonempty string w is in $\mathcal{L}(M)$, then for all $n \geq 1$, the string w^n is also in $\mathcal{L}(M)$.

2.41. Let $M = (\{s, p, q\}, \{0, 1\}, \delta, s, \{s, p\})$ be a DFA with $\delta(s, 0) = p$, $\delta(s, 1) = s$, $\delta(p, 0) = q$, $\delta(p, 1) = s$, $\delta(q, 0) = q$, $\delta(q, 1) = q$. Guess the language that M accepts. Prove that your guess is correct.

2.6 Nondeterministic Finite Automata

You might have encountered plenty of nondeterministic algorithms. Whenever an algorithm says that choose an element from such and such a set, you have a choice, a nondeterministic choice. Of course, the algorithm that uses such a choice does work with whatever element you might start with. In a machine, the dependence of its next state on its current state and the current input symbol may be deterministic or nondeterministic. When the next state is completely determined, it is a deterministic machine and when there are more than one options for the next state, the machine is called nondeterministic.

To equip a finite automaton with such capability, we will modify the transition function a bit. Instead of telling a finite automaton what exactly to do from its present state when it encounters a symbol, we will give it a choice. This means, both the transitions such as $\delta(p, a) = q$ and $\delta(p, a) = r$ will be admissible. But then the transition function δ will no more be a function! Well, we will use a relation instead. To make the nondeterministic machines more flexible, we will allow them to read the empty string ε along with symbols from Σ. The following is a formal definition of a nondeterministic automaton.

A **nondeterministic finite automaton**, abbreviated to **NFA**, is again a quintuple $N = (Q, \Sigma, \Delta, s, F)$, where Q, Σ, s, F are, respectively, the set of states, the input alphabet, the initial state, and the set of final states, as in a DFA, and Δ is a finite relation from $Q \times (\Sigma \cup \{\varepsilon\})$ to Q.

As mentioned earlier, Δ is a finite subset of $(Q \times (\Sigma \cup \{\varepsilon\})) \times Q$, called the **transition relation** of N. Any element of the transition relation Δ will be called a **transition**. Instead of writing a transition as the formally correct $((q, \sigma), r)$, we will write it as the triple (q, σ, r). Moreover, we will use the term **finite automaton** to commonly refer to both a DFA and an NFA.

Suppose the NFA $N = (Q, \Sigma, \Delta, s, F)$ is currently in state q, its reading head is scanning the symbol σ and to the right of it is the remaining string v. Suppose further that we have a transition $(q, \sigma, r) \in \Delta$. Then the NFA will go to the state r with the input string left to be read as v, as usual. However, there might be several triples in Δ whose first two components are, respectively, q and σ. That is, as Δ is a relation (and not a partial function) we might have n number of triples $(q, \sigma, r_1), (q, \sigma, r_2), \cdots, (q, \sigma, r_n)$ in Δ. (Say, r is one of these r_i.) Then, we do not know exactly to which of the n states $r_1, \cdots, r_n$ the NFA goes next.

Further, the input string w can be parsed in a trivial way, that is, $w = \varepsilon w$. Now, if $(q, \varepsilon, r) \in \Delta$ (Look at the relation Δ above once again.), then the NFA may enter the state r without consuming any input. The transition (q, ε, r) says that if N is in state q, then it can change its state to r without consuming any input. Such a transition is called an **ε-transition**. We should keep all these in mind, while formalizing the notion of a computation in an NFA.

Let $N = (Q, \Sigma, \Delta, s, F)$ be an NFA. A **configuration** of N is an ordered pair (q, u), where $q \in Q$ is a state, the current state, and $u \in \Sigma^*$ is a string that is yet to be read. Let (q, u) and (r, v) be two configurations of N. We define the relation of **yields in one step**, using the same symbol $\overset{1}{\rightsquigarrow}$ by

$$(q, u) \overset{1}{\rightsquigarrow} (r, v) \text{ iff there is } \sigma \in \Sigma \cup \{\varepsilon\} \text{ such that } u = \sigma v \text{ and } (q, \sigma, r) \in \Delta.$$

The "yield" relation is defined as the reflexive and transitive closure of "yields in one step." As earlier, we use the same symbol $\rightsquigarrow$ instead of the more formal $\overset{*}{\rightsquigarrow}$ to denote the yield relation. Thus, (q, u) **yields** (r, v) iff there is a sequence of configurations each yielding in one step the next one starting from (q, u) and ending at (r, v). Also, (q, u) yields itself, in 0 steps. That is,

$$(q, u) \rightsquigarrow (q, u).$$

$$(q_i, u_i) \overset{1}{\rightsquigarrow} (q_{i+1}, u_{i+1}) \text{ for } i = 1, \ldots, n-1.$$

$$(q, u) \rightsquigarrow (r, v) \text{ iff for some } n \geq 1, q_0 = q, q_n = r, u_0 = u, u_n = v.$$

In the case of an NFA, computation, that is, the sequence of yields, is nondeterministic in the sense that starting from a configuration (q, u) it may end up at various configurations (r, v). For instance, suppose both (q, σ, r) and (q, σ, p) are members of Δ, transitions in our NFA N. In such a case, both $(q, \sigma u) \rightsquigarrow (r, u)$ and $(q, \sigma u) \rightsquigarrow (p, u)$ hold. Thus a computation starting with a configuration might end up at several configurations, and thus, in different states. However, at any one instance of computation, N would follow up one such path, but we do not know which. We will say that the NFA accepts the input string if one such computation ends up at a final state.

Formally, the NFA N **accepts** an input string $u \in \Sigma^*$ iff there is a (final) state $q \in F$ such that $(s, u) \rightsquigarrow (q, \varepsilon)$ in some computation. Then N **rejects** the input string u iff there is no possible computation yielding an end-configuration (p, ε) from the starting configuration (s, u) for any of the final states p.

Note that in both the cases of a DFA and an NFA, the automaton halts if there is no *transition* defined for any configuration met through the steps of a computation. In a DFA with a total function as its transition, the machine must have consumed the whole input and the string is thus rejected. While an NFA might follow up an alternate computation path leading to the acceptance of the string, the **language accepted by** an NFA N is

$$\mathcal{L}(N) = \{u \in \Sigma^* : (s, u) \rightsquigarrow (q, \varepsilon) \text{ in some computation, where } q \in F\}.$$

Notice that we are a bit partial towards preferring acceptance over rejection. If N has an accepting computation and also a rejecting computation on the same input, our definition says that N accepts the input.

Example 2.15. Let $N = (Q, \Sigma, \Delta, s, F)$ be the NFA with $Q = \{s, q, r\}$, $\Sigma = \{a, b\}$, $F = \{s, q\}$, and $\Delta = \{(s, a, s), (s, \varepsilon, q), (s, b, r), (q, b, q), (q, b, r)\}$. What is $\mathcal{L}(N)$?

Solution. A computation of N on the input a^2b^2 is as follows:

$$(s, a^2b^2) \rightsquigarrow (s, ab^2) \rightsquigarrow (s, b^2) \rightsquigarrow (r, b).$$

Here, the machine N halts without reading further inputs, as Δ gives no information as to what to do once it reads the input b being in the state r. However, we cannot conclude that N rejects the input string u, for there might be another accepting computation! Let us see one more possible computation of N with the same input string a^2b^2. Here is one more computation in N:

$$(s, a^2b^2) \rightsquigarrow (s, ab^2) \rightsquigarrow (s, b^2) \rightsquigarrow (q, b^2) \rightsquigarrow (q, b) \rightsquigarrow (q, \varepsilon).$$

The string a^2b^2 is thus accepted by N as $q \in F$.

Notice the yield $(s, b^2) \rightsquigarrow (q, b^2)$ above. It is justified in N as Δ contains the ε-transition (s, ε, q). This transition says that if N is in state s, then it can change its state to q without consuming any input. The above computation shows that N accepts the string a^2b^2 even though there is a rejecting computation with the same string. This is the way we are using nondeterminism in machines. I repeat: we have more liking towards acceptance than rejection!

For $u = a^2ba$, we have the following computations in N:

One: $(s, a^2ba) \rightsquigarrow (s, ba) \rightsquigarrow (r, a)$.

Two: $(s, a^2ba) \rightsquigarrow (s, aba) \rightsquigarrow (q, aba)$.

Three: $(s, a^2ba) \rightsquigarrow (q, a^2ba)$.

Any more? You see that none of the computations is an accepting computation. Hence, N rejects the string a^2ba. What does the automaton N do? The transition diagram of N is in Fig. 2.10.

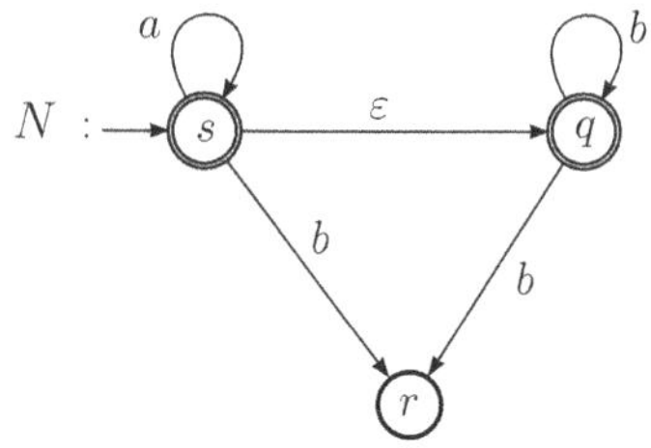

Fig. 2.10. Transition diagram of the NFA of Example 2.15.

Once N enters the state r, it never leaves the state, and as $r \notin F$ it leads to a rejecting computation. The only accepting computations are those that never enter r. It is then obvious that $\mathcal{L}(N) = a^*b^*$. Thus we can safely delete the state r from N. □

Moreover, you can also see that due to the ε-transition (s, ε, q), there is no need to keep s as a final state in Fig. 2.10. The modified NFA is drawn in Fig. 2.11.

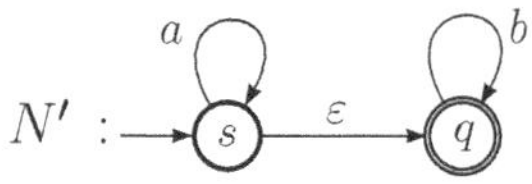

Fig. 2.11. Modified NFA.

Does N' accept the same language as N (of Fig. 2.10) or it does not? Check the computation with some strings from $\{a, b\}^*$ to answer the question and then argue abstractly.

As in a DFA, we put the names of the states inside circles at different places; the final states are encircled once more. The initial state is shown by putting an arrow going towards it and coming from nowhere. An arrow labeled with a symbol σ is drawn from a state p to another state t iff we have $(p, \sigma, t) \in \Delta$. Similarly, loops are drawn from and to the same state, provided there is a corresponding transition. The labels on any edge of an NFA-diagram can be members of the alphabet Σ, and some of them can also be ε, as well.

In case of an NFA, there can be many edges between the same pair of vertices with different labels. For example, if in an NFA, we have the transitions (q, a, r), (q, b, r), then its diagram will have an edge from q to r with label a and also another edge from q to r with label b. In such cases, instead of having multiple edges, we will draw only one edge and label it with all those strings separated by commas. For example, the transitions (q, a, r) and (q, b, r) will be depicted by a single edge labeled

"a, b." Sometimes, we will not write the names of the states inside the circles; see Example 2.16 below.

Example 2.16. Construct an NFA for accepting the language $\{w \in \{a, b\}^* : w$ has a substring bab or a substring $aba\}$.

Solution. The language can be represented by the regular expression $(a \cup b)^*(bab \cup aba)(a \cup b)^*$. See the NFA N given in Fig. 2.12 and try to connect the diagram directly with this regular expression. □

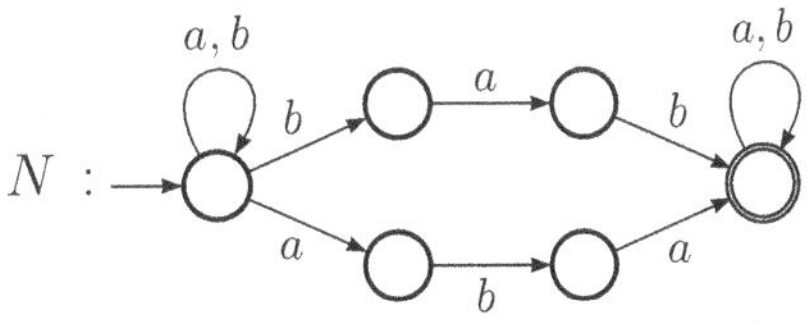

Fig. 2.12. NFA for Example 2.16.

Exercise 2.9. Define an NFA as a directed labeled graph. Then define carefully the yield relation, the notion of a computation, and the languages accepted by them.

Recollect that acceptance of languages by a DFA could also be defined by extending the transition function $\delta : Q \times \Sigma \to Q$ to another function $\delta^* : Q \times \Sigma^* \to Q$. This scheme can be adopted for NFAs. It will be easier if you look at the transition diagram of the given NFA. Suppose $N = (Q, \Sigma, s, \Delta, F)$ is an NFA. A *walk* joining the states p and q is a sequence of states and symbols $p_1, t_1, p_2, t_2, \ldots, p_m, t_m, p_{m+1}$ such that $p_1 = p$, $p_{m+1} = q$, and (p_i, t_i, p_{i+1}) is a transition. Informally, we just find out how to go to q from p by following the arrows. For such a walk, we say that its label is the concatenation of all the labels in that order, that is, the string $t_1 t_2 \cdots t_m$. The extension of Δ to Δ^* is obtained by requiring that for a string $w \in \Sigma^*$, $(p, w, q) \in \Delta^*$ iff in the transition diagram of N, there is a walk from p to q labeled w. Then the language accepted by N can be defined as $\mathcal{L}(N) = \{w \in \Sigma^* : (s, w, r) \in \Delta^*$ for some $r \in F\}$. See that such a definition of acceptance amounts to what we have defined earlier.

Example 2.17. Construct an NFA that accepts all strings of length 3 over the alphabet $\{a, b\}$.

Solution. Trivially, you can have all the strings of length 3, take only one state and have loops on this state with all these strings as labels, insert extra states to break the strings into symbols; finally declare the initial state as a final state. This will do. This also tells you how to construct a regular expression and a regular grammar for the language.

However, a better way of writing the regular expression is to use union and concatenation. It is simply $(a \cup b)(a \cup b)(a \cup b)$. Similarly, a better grammar can be constructed using the only productions as

$$S \mapsto aA|bA, \; A \mapsto aB|bB, \; B \mapsto a|b.$$

Compare the NFA in Fig. 2.13 with this grammar. Just a look at the NFA tells that it does the intended job. □

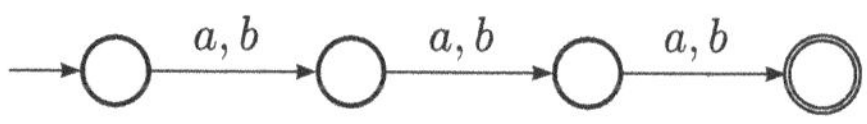

Fig. 2.13. NFA for Example 2.17.

Example 2.18. Draw the transition diagram of an NFA that accepts the language $a^+b^+ \cup a^*b$.

Solution. As $a^+ = a^*a$ you may think of a loop labeled a on the initial state and then an edge from this state to another labeled a. Similarly, from the next state, you simply have a loop and an edge to the third state both labeled b. This would account for a^+b^+, provided the third state is a final state. Notice that this itself takes care of the strings in a^+b. The only string left out is a single b. So, have an edge from the initial state to the third one. This probably does the job. The transition diagram is drawn in Fig. 2.14. Verify with some input strings and then decide whether this NFA really does the job. □

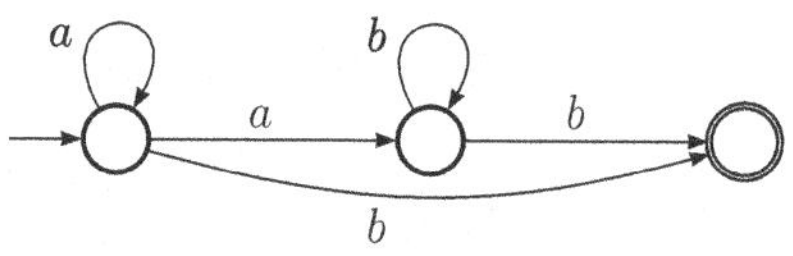

Fig. 2.14. NFA for Example 2.18.

Example 2.19. Construct an automaton that accepts the language $L = \{w \in a^* : \ell(w) = 4k + 1,\ k \in \mathbb{N}\}$.

Solution. A regular expression for L is $a(aaaa)^*$. A regular grammar for L has the productions $S \mapsto aA,\ A \mapsto \varepsilon|aaaS$. An NFA can be easily constructed by mimicking the grammar.

Take one state as s and the other as q. Being in state s, let the machine read a and go to the state q. Next, being in q, it reads aaa to come back to s. Finally, declare q as the final state. We need to modify this a bit by breaking the string into symbols. Just have two more states, say, p and r to break aaa symbol by symbol. That is, the NFA starts from s, reads a, and changes state to q. Then it reads a and changes state to p; again it reads a and changes its state to r. One more a is to be read. So, being in the state r it reads an a and changes its state back to s. Final state is q as earlier. See the diagram in Fig. 2.15. Convince yourself taking some input strings whether the automaton really does the intended job. This NFA is also a DFA. □

Exercise 2.10. Does the NFA in Fig. 2.16 accept $\{w \in \{a, b\}^* : w$ has aa or bb as substring(s)$\}$?

Fig. 2.15. DFA for Example 2.19.

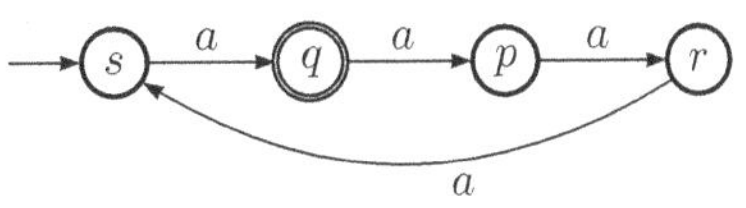

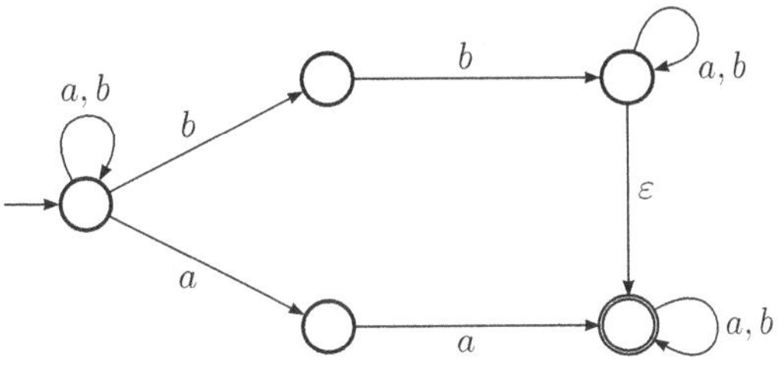

Fig. 2.16. DFA for Exercise 2.10.

In fact, nondeterminism can be used for doing very complicated jobs easily. It is because, these machines have the power to guess and then go towards choosing the correct string. The ε-transitions especially can be used for guessing a correct string. Look at Examples 2.20 and 2.21.

Example 2.20. Construct automata for accepting the language $(10 \cup 101)^*$.

Solution. A simple nondeterministic finite automaton for accepting the language is having only one state, say s, which is both initial and final. The two transitions are $(s, 10, s)$ and $(s, 101)$. However, this would not be an NFA; we need to break the strings into symbols by introducing new states in between the symbols. That is, have the set of states as $Q = \{s, p, q, r\}$. Then take the transition relation as $\Delta = \{(s, 1, p), (p, 0, s), (s, 1, q), (q, 0, r), (r, 1, s)\}$. Note that a string is accepted when there is some computation accepting the string. We can, of course, afford to have only three states. Look at Fig. 2.17.

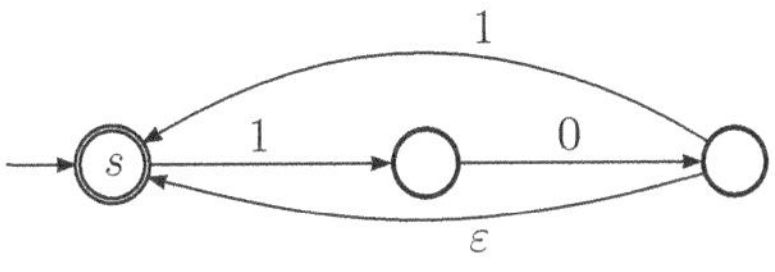

Fig. 2.17. NFA for Example 2.20.

An NFA without ε-transitions accepting the same language can also be constructed, see for example, the transition diagram in Fig. 2.18.

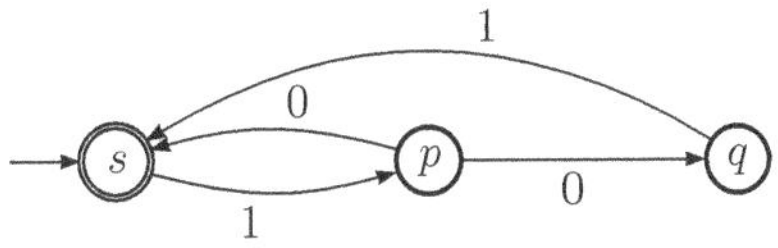

Fig. 2.18. NFA without ε-transition for Example 2.20.

You can then try constructing a DFA for accepting the same language. But first see why the NFA in Fig. 2.18 is not deterministic. It is because, we have two transitions for the state–symbol pair $(p, 0)$. The NFA could go to the state q or to the state s when it is in state p reading the symbol 0.

At this point, close the book and try to construct a DFA. Can you have a simpler DFA to accept the above language than the one drawn in Fig. 2.19? □

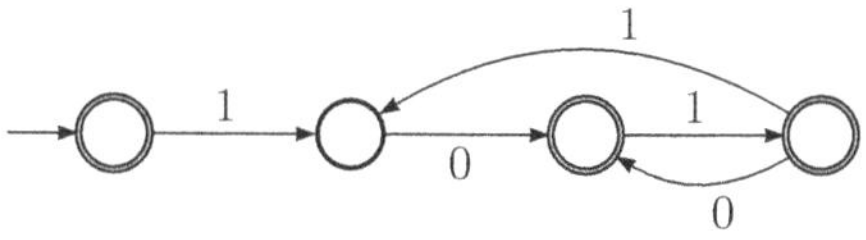

Fig. 2.19. A DFA for Example 2.20.

Example 2.21. Construct an NFA for accepting the language $L = \{w \in \{a, b, c, d\}^* :$ at least one of the symbols a, b, c, d does not appear in $w\}$.

Solution. Suppose we use the state A of the machine, from where it can accept all strings without having an occurrence of a. We can use an ε-transition to drive the machine from the initial state to A. Similarly, ε-transitions can be used to go to any of the states B, C, D, which correspondingly would accept the strings where the corresponding symbols b, c, or d does not appear. This language can be represented by the expression $(\overline{\{a\}})^* \cup (\overline{\{b\}})^* \cup (\overline{\{c\}})^* \cup (\overline{\{d\}})^*$. Look at the diagram of such a machine in Fig. 2.20. □

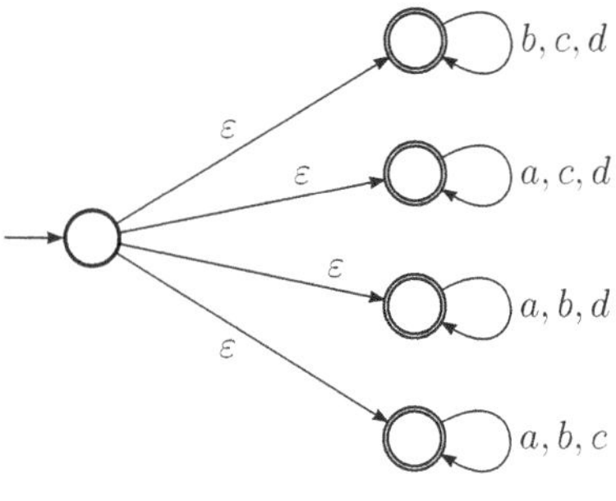

Fig. 2.20. An NFA for Example 2.21.

You can define an NFA by using a function instead of a relation. In such a case, an NFA is a quintuple $(Q, \Sigma, \Delta, s, F)$, where $\Delta : Q \times (\Sigma \cup \{\varepsilon\}) \rightarrow 2^Q$. In this definition, the ordered triples $(p, a, q_1), (p, a, q_2), \ldots, (p, a, q_n)$ are written as $\Delta(p, a) = \{q_1, q_2, \ldots, q_n\}$.

A simple answer to "why nondeterminism" is: Look at the regular grammars. They allow nondeterminism. From the same nonterminal symbol, you can derive different strings by choosing different production rules.

Problems for Section 2.6

2.42. For the following NFAs $(\{p, q, r, s\}, \{a, b\}, \Delta, p, F)$, draw the transition diagrams and then find the languages they accept, in both the set notation and in plain English:

(a) $\Delta = \{(p, a, p), (p, a, q), (p, b, p), (r, a, r), (r, b, r), (r, a, s), (s, a, s), (s, b, s)\}$, $F = \{s\}$.

(b) $\Delta = \{(p, a, q), (p, a, s), (p, b, q), (q, a, r), (q, b, q), (q, b, r), (r, a, s), (r, b, p), (s, b, p)\}$, $F = \{q, s\}$.

(c) $\Delta = \{(p, a, p), (p, a, q), (p, b, p), (q, a, r), (q, a, s), (q, b, s), (r, a, p), (r, a, r), (r, b, s)\}$, $F = \{s\}$.

(d) $\Delta = \{(p, a, p), (p, a, r), (p, b, q), (q, \varepsilon, r), (q, a, q), (q, b, r), (r, \varepsilon, q), (r, a, r), (r, a, p)\}$, $F = \{r\}$.

(e) $\Delta = \{(p, a, q), (q, b, r), (r, a, q), (q, b, s), (s, a, r)\}$, $F = \{p, r\}$.

(f) $\Delta = \{(p, \varepsilon, q), (p, a, q), (q, a, p), (q, b, q), (q, a, r), (q, b, r), (r, a, r), (r, b, q), (r, q, s), (s, a, s), (s, b, s)\}$, $F = \{q\}$.

2.43. Draw transition diagrams of NFAs that accept the following languages:

(a) $((a^*b^*a^*)^*b^*)^*$.

(b) $(ba)^*(ab)^* \cup b^+$.

(c) $((ba \cup aab) * a^*)^*$.

(d) $(a \cup ab)^* \cup (b \cup aa)^*$.

(e) $(ba \cup bba \cup aba)^*$.

(f) $(a \cup b)^*b(a \cup b)^4$.

2.44. Draw the transition diagram of the NFA with initial state p, final state q, and another state r, having transitions $(p, 0, q)$, $(p, 1, q)$, $(q, 0, p)$, $(q, 1, q)$, $(q, 0, r)$, (q, ε, r), $(r, 1, q)$. Which language does it accept?

2.45. Check whether the NFA in Fig. 2.21 accepts $\{w \in \{0, 1\}^* : w$ has a substring 11 or has a substring $101\}$. Construct a DFA to accept the same language.

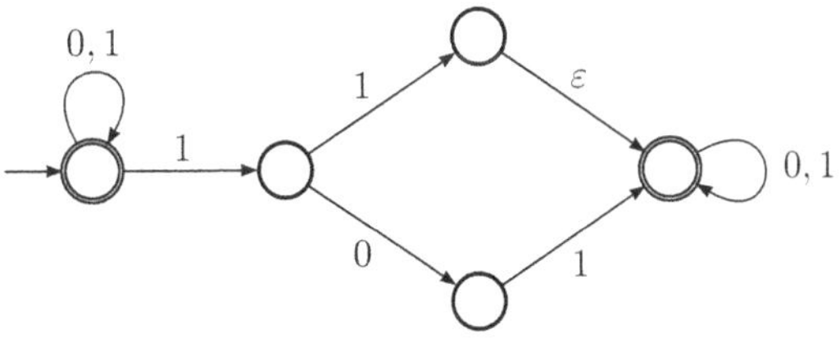

Fig. 2.21. NFA for Problem 2.45.

2.46. Construct a DFA/NFA that accepts, in each case below, the set of all strings $w \in \{a, b\}^*$ satisfying

(a) w has exactly one a.

(b) w has at least one a.

(c) w has no more than three a's.

(d) w has prefix ab.
(e) w does not contain aab as a substring.
(f) w has at least one a and exactly two b's.
(g) w has exactly two a's and no more than two b's.
(h) $\#_b(w) = 0$ and $\ell(w)$ is divisible either by 2 or by 5.
(i) w has $baba$ as a substring.
(j) w has both ba and ab as substrings.
(k) w has neither aa nor bb as a substring.
(l) Each a in w is immediately preceded by at least one b.
(m) w has an odd number of a's and an even number of b's.
(n) w has an even number of a's and an even number of b's.
(o) In w, each block of five consecutive symbols contains at least two a's.
(p) w either begins or ends with the string ab.
(q) The seventh symbol from the right of w is an a.

2.47. Prove that for each NFA with many final states, there is an NFA with a single final state accepting the same language. Is it true for DFAs also?

2.7 Summary and Additional Problems

As a uniform data structure, we have started with strings of symbols; the symbols are from a finite set, called an alphabet. Then we defined a language over an alphabet as a set of strings. For finite representation of languages, we have defined the regular expressions, which are again strings of symbols with five more new symbols such as $\emptyset$, $\cup$, $*$,), and (, obeying certain specified formation rules. We have also introduced three other mechanisms for representing languages, such as regular grammars, DFAs, and NFAs. The regular expressions are called the representation devices, the regular grammars are the generative devices, and the DFAs and NFAs are the recognizing devices. Nonetheless, all of them represent languages. Almost always it so happened that if a language is represented by any one of the four mechanisms, it is also represented by another. It raises a natural question, whether it is always so. We will show that it is indeed so in the next chapter.

Introduction of finite state transition systems go back to McCulloh and Pitts [84] in 1943. Kleene [67] used the present form of DFAs. The NFAs were introduced by Rabin and Scott [107]. The notation for automata has been used by various authors after Hopcroft and Ullman [58].

You will find in the exercises below the so-called finite state transducers first introduced in [37] and also discussed in [86] and [91]. For advanced texts on language theory, you may consult [51, 113].

The additional problems below are a bit harder. Some of them appeared in original research papers. If you cannot solve every one of them, do not get disheartened. Perhaps, after some time you will be able to solve them with the help of your teacher.

Additional Problems for Chapter 2

2.48. Let E be a regular expression with variables $x_1, \ldots, x_n$. Form a concrete regular expression by replacing each occurrence of x_i by a symbol σ_i for $i = 1, 2, \ldots, n$. Let E_1, E_2 be two regular expressions. Using the replacements, form two concrete regular expressions C_1, C_2, respectively. Show that if $\mathcal{L}(C_1) = \mathcal{L}(C_2)$, then $\mathcal{L}(E_1) = \mathcal{L}(E_2)$. [For example, if L, M are any languages, then $(L \cup M)^* = (L^*M^*)^*$ holds as $(a \cup b)^* = (a^*b^*)^*$ holds.]

2.49. The *star height* $h(E)$ of a regular expression E over an alphabet Σ is defined inductively by $h(\emptyset) = 0$, $h(\sigma) = 0$ for each $\sigma \in \Sigma$, $h(\alpha \cup \beta) = h(\alpha\beta) = \max(h(\alpha), h(\beta))$, $h(\alpha^*) = 1 + h(\alpha)$. In each case below, find an equivalent regular expression having the least star height.
(a) $(011^*0)^*$.
(b) $(a(b^*c)^*)^*$.
(c) $(0^* \cup 1^* \cup 01)^*$.
(d) $(c(ab^*c)^*)^*$.
(e) $((01)^* \cup 1^*)^*0^*$.
(f) $((abc^*)ab)^*$.

2.50. Show that each regular expression is equivalent to one in the form $(\alpha_1 \cup \alpha_2 \cdots \cup \alpha_n)$ for some $n \geq 1$, where $\cup$ does not occur in any α_i. This is called the *disjunctive normal form* of a regular expression.

2.51. Is it true that every regular expression is equivalent to one in the form $\alpha_1\alpha_2 \cdots \alpha_n$ for some $n \geq 1$, where no α_i is a concatenation of other expressions?

2.52. Find a regular expression that denotes all bit strings whose value, when interpreted as a binary integer, is greater than or equal to 40.

2.53. Find a regular expression for all bit strings with leading bit 1, which when interpreted as a binary integer, has value not between 10 and 30.

2.54. In the roman number system, numbers are represented by strings on the alphabet $\Sigma = \{M, D, C, L, X, V, I\}$. For simplicity, replace the subtraction convention in which 9 is represented as *IX*; instead write it as *VIIII*; similarly for *IV*, write *IIII* and etc. Design a DFA that accepts all strings over Σ that represents a number in roman number system.

2.55. Find an NFA that accepts the set of all strings over $\{a, b\}$ not containing the substring aba. Then construct a regular expression for the same language. Try to simplify the expression as far as possible.

2.56. Construct a DFA in each case below that accepts all binary strings w such that
(a) w is divisible by 5 when interpreted as a binary integer.
(b) Reversal of w is divisible by 5 when interpreted as a binary integer.

2.57. Show that if an NFA with m states accepts any string at all, then it accepts a string of length less than m.

2.58. Let M be a DFA with $n \geq 1$ number of states. Show that M accepts an infinite language iff M accepts a string w for which $n \leq \ell(w) < 2n$.

2.59. Construct an NFA with input alphabet $\{a\}$ that rejects some string, but the length of the shortest rejected string is more than the number of states.

2.60. Let $N = (Q, \Sigma, \Delta, s, \{q\})$ be an NFA where there is no transition to s and no transition from q. Describe the language of the modified NFA, in terms of $\mathcal{L}(N)$, using each of the following modifications to N:

(a) Add an ε-transition from q to s.
(b) Add an ε-transition from s to each state reachable from s. We say that p is reachable from s when you can follow the arrows to go from s to p.
(c) Add an ε-transition to q from each state p with the property that q is reachable from p.
(d) Modify using both (b) and (c).

2.61. The finite automata we have defined do not produce any output. Imagine a finite automaton with another tape, which is initially blank having a writing head associated with this second tape. These automata which can give output on the second tape are called *transducers*. Transducers are of two types: *Moore machines* and *Mealy machines*. In a Moore machine, symbols are written on the output tape depending on what state the machine is currently in. That is, we have an extra output alphabet, say Γ, and an extra map, say, $\lambda : Q \to \Gamma$. When the machine is in state q, the symbol $\lambda(q)$ is written on the second tape. In the Mealy machines, output are produced by the transitions, that is, $\lambda : Q \times \Sigma \to \Gamma$. These transducers do not require any final states.

(a) For a Moore machine $(Q, \Sigma, \Gamma, \delta, \lambda, q_0)$, show that the output in response to the input $\sigma_1\sigma_2 \cdots \sigma_n$ is $\lambda(q_0)\lambda(q_1) \cdots \lambda(q_n)$, where $\delta(q_{i-1}, \sigma_i) = q_i$, for $1 \leq i \leq n$. Moore machines always give output $\lambda(q_0)$ for input ε.
(b) Show that the Moore machine $M = (\{q_0, q_1, q_2\}, \{0, 1\}, \{0, 1\}, \delta, \lambda, q_0)$ with $\delta(q_0, 0) = \delta(q_1, 1), = q_0$, $\delta(q_0, 1) = \delta(q_2, 0) = q_1$, $\delta(q_1, 0) = \delta(q_2, 1) = q_2$ and $\lambda(q_0) = 0, \lambda(q_1) = 1, \lambda(q_2) = 2$ computes the residue modulo 3 for each binary string treated as a binary integer.
(c) For a Mealy machine $(Q, \Sigma, \Gamma, \delta, \lambda, q_0)$, show that the output in response to the input $\sigma_1\sigma_2 \cdots \sigma_n$ is $\lambda(q_0, \sigma_1)\lambda(q_1, \sigma_2) \cdots \lambda(q_n, \sigma_n)$, where $\delta(q_{i-1}, \sigma_i) = q_i$ for $1 \leq i \leq n$. [Mealy machines always give output ε for input ε.]
(d) Construct a Mealy machine for computing the residue modulo 3 for each binary string treated as a binary integer. [This can be trivial if you have a Moore machine for the same job.]
(e) Show that corresponding to each Moore machine, there exists a Mealy machine, which gives the same input–output mapping.
(f) Show that corresponding to each Mealy machine, there exists a Moore machine, which gives the same input–output mapping.

2.62. A serial binary adder processes two binary sequences $x = a_1a_2 \cdots a_n$ and $y = b_1b_2 \cdots b_n$, where each a_i, b_j is a bit (0 or 1). It adds x and y bit by bit. Each bit addition creates a digit for the sum as well as a carry digit for the next higher position. The rules are: $0 + 0 = 0$, $0 + 1 = 1$, $1 + 0 = 1$ without any carry digit, and $1 + 1 = 0$ with carry digit 1.

(a) A Mealy machine for the binary serial adder has three tapes, first one with x, second one with y, and the third for the output, can thus have two states, say p for *carry* and q for *no-carry*. For example, typical transitions of such a machine are $\delta(q, 1, 1) = p$ and $\lambda(q, 1, 1) = 0$. This means, if the machine is in state q and scanning 1 on its first tape (second components of δ and λ), 1 on the second tape (third components), then in its next move, it goes to state p and outputs 0 on the third tape. We depict such transitions δ and λ in diagram by an arrow from q to p labeled $(1, 1)/0$. Design such a Mealy machine for the binary serial adder.

(b) Design a Moore machine with three tapes in a similar way for the binary serial adder, where contents of the third tape gives the output.

2.63. Construct a transducer to decode a message in the form of a binary integer, which has been encoded by representing a as 00, b as 01, c as 10, and d as 11. For example, the input 101011010100 should produce the output $ccdbba$.

2.64. The two's complement of a binary string, representing a negative integer, is formed by first complementing each bit, then adding one to the lowest-order bit. For example, we write -6 as the binary 1010 instead of the conventional -110. Design a transducer for translating bit strings into their two's complement, with lower order bit as the right of the string.

2.65. Design a transducer to convert a binary integer into octal. For example, the binary string 011101100 should produce 354.

2.66. Design a transducer that computes parity of every substring of three bits of a given binary string. That is, if the input is $a_1a_2 \cdots a_n$, then the output will be $b_1b_2 \cdots b_n$, where $b_1 = b_2 = 0$ and $b_j = (a_{j-2} + a_{j-1} + a_j) \bmod 2$ for $j \geq 3$. For example, the input 1010110 gives the output 0001000.

2.67. Design a transducer that accepts bit strings $a_1a_2 \cdots a_n$ and computes the binary value of each set of three consecutive bits modulo 5. That is, it outputs $b_1b_2 \cdots b_n$, where $b_1 = 0 = b_2$ and $b_j = (4a_{j_2} + 2a_{j-1} + a_j) \bmod 5$ for $j \geq 3$.

2.68. Design a transducer to compute the maximum of two given binary integers.

2.69. A *deterministic two-tape automaton* is a device like a DFA that accepts a pair of strings. Each state is in one of the two sets, first or second; depending on which set the current state is in, the transition function refers to the first or second tape. Formally define a deterministic two-tape automaton.

2.70. Show that if a function $f : \Sigma^* \to \Sigma^*$ is computed by a transducer, then the set $\{(w, f(w)) : w \in \Sigma^*\}$ is accepted by some deterministic two-tape automaton.

2.71. Find a regular expression that generates the set of all strings of triplets defining correct binary addition.

3 Equivalences

3.1 Introduction

By now, You are familiar with regular expressions, regular grammars, deterministic finite automata, and nondeterministic finite automata. Are they talking about the same class of languages? The examples in the last chapter, at least, suggest that they might. In this chapter, we will see that it is indeed so. We will say, informally, that two mechanisms are *equivalent* if they accept the same class of languages. In fact, we solve many subproblems to arrive at these equivalences. Our route is from NFA to DFA, from DFA to regular grammar, from regular grammar to NFA, from regular expression to NFA, and finally, from NFA to regular expression.

3.2 NFA to DFA

We start with our first question. We pose a subproblem: can we construct a DFA to imitate an NFA?

Let $N = (Q, \Sigma, \Delta, s, F)$ be an NFA. If the transition relation Δ is already a partial function from $Q \times \Sigma$ to Q, then N is itself a DFA. Otherwise, suppose that Δ, a relation from $Q \times (\Sigma \cup \{\varepsilon\})$ to Q is not a partial function from $Q \times \Sigma$ to Q. Then, in Δ, we have triples of the form (q, σ, r), where

(a) $\sigma = \varepsilon$, or
(b) there exists at least one $r' \in Q, r' \neq r$ such that $(q, \sigma, r') \in \Delta$.

We will construct new NFAs to accept the same language as N but will not satisfy any of the above properties. Then, the new NFA is bound to be a DFA accepting the same language as N.

We first try to eliminate the possibility (a). This says that there may be ε-transitions, that is, edges labeled with ε. Suppose, for example, there is a triple $(q, \varepsilon, r) \in \Delta$, an ε-transition in N. Suppose further that the automaton N is currently in the state q reading some symbol. What will be the next action of N? It may continue computation from the state q or it may first go to the state r and then continue computation. While changing from the state q to r, it does not consume any input symbol due to the ε-transition. Looking from above, it amounts to telling that N is in one of the states q or r, or that it is in the state $\{q, r\}$.

Now, look at what we are trying to discuss here. The set $\{q, r\}$ can be thought of as a new naming of the states. Instead of taking the states individually by q and r and then telling that these are the possible states the machine can be in, we are simply using the set $\{q, r\}$ as a single state. In such a naming scheme of the states, the state q itself can be named as the singleton $\{q\}$.

Well, suppose that we take $\{q, r\}$ as a new state. We are going to shrink the edge labeled ε, that is, merge the states q and r and rename the merged state as $\{q, r\}$. This removes the states q and r from the picture and replaces both with $\{q, r\}$ deleting the edge from q to r labeled ε. But then, will the new machine compute the same way as the old?

There are two objections. One, the *merging* of two states q and r does not take care of the asymmetry in the directed edges. That is, when we merge q and r, does it make any difference if the edge is from r to q? No. But this should. Two, there might be another ε-transition from r to a state, say t. In that case, the automaton can go to the state t from q without consuming any input. The merging does not take care of this situation. The term *merging* is in fact, sloppy.

To overrule the objections, we have to do something more. We will have to get all those states in N that are reachable from q in place of the only other state r. That is, we must collect together all the states that are connected from q by an ε-transition, all the states connected from those latter states by ε-transitions, and so on. This renaming scheme can be formalized by defining a set of states, call it $R(q)$, the states reachable from q without consuming any input. This is defined as follows:

Let $M = (Q, \Sigma, \Delta, s, F)$ be an NFA, where $\Delta \subseteq (Q \times (\Sigma \cup \{\varepsilon\})) \times Q$. Corresponding to each state $q \in Q$, the set $R(q)$ is defined inductively by the following:

1. $q \in R(q)$.
2. If $p \in R(q)$ and $(p, \varepsilon, t) \in \Delta$, then $t \in R(q)$.
3. $R(q)$ contains all and only those states satisfying (1 and 2).

Thus, $R(q)$ is the set of all states $p \in Q$ such that $(q, \varepsilon) \rightsquigarrow (p, \varepsilon)$ in zero or more steps. The set $R(q)$ can also be defined as

$$\begin{aligned} R(q) &= \{p \in Q : (q, \varepsilon) \rightsquigarrow (p, \varepsilon)\} \\ &= \{p \in Q : (q, w) \rightsquigarrow (p, w) \text{ for every } w \in \Sigma^*\}. \end{aligned}$$

Each set $R(q)$ is nonempty as $q \in R(q)$. Moreover, $R(q)$ is the transitive closure of the set $\{q\}$ under the relation $\{(p, t) :$ there is a transition $(p, \varepsilon, t) \in \Delta\}$. You can use the while loop to construct the set $R(q)$ as in the following:

Initialize $R(q) := \{q\}$;
`while` there exists a transition $(p, \varepsilon, t) \in \Delta$ with $p \in R(q)$ and $t \notin R(q)$ `do`
    $R(q) := R(q) \cup \{t\}$ `od`

Example 3.1. Find $R(q)$ for the NFA $N = (\{p, q, r, s, t\}, \{a, b\}, \Delta, s, \{q, r\})$, where $\Delta = \{(s, a, q), (s, a, p), (q, \varepsilon, p), (q, \varepsilon, t), (q, a, s), (p, a, r), (p, \varepsilon, s), (r, \varepsilon, s), (r, b, q), (t, a, p)\}$.

Solution. To understand the example, do not just read it. First, draw a transition diagram and then proceed. Initially, we take $R(q) = \{q\}$. Now, the ε-transitions from q are (q, ε, p) and (q, ε, t). Thus, $R(q)$ is updated to $R(q) = \{q, p, t\}$. Next, the ε-transitions from p is (p, ε, s) and there is no ε-transition from t. Thus, s is again added to $R(q)$, updating it to $R(q) = \{q, p, t, s\}$. From s there is no ε-transition. Now, we also see that starting from any state in $R(q)$, if we follow up ε-transitions, we end up in a state that is also in $R(q)$. Moreover, each state in $R(q)$ is reached from q by following possibly successive ε-transitions. Thus, this is the required set. Verify that $R(p) = \{p, s\}$, $R(r) = \{r, s\}$, $R(s) = \{s\}$, and $R(t) = \{t\}$. □

Observe that whenever q is a state from which there exists no ε-transition to any other state, $R(q) = \{q\}$. Thus, we have got new states that are sets of some of the states of N. In Example 3.1, we have the new states as $R(p)$, $R(q)$, $R(r)$, $R(s)$, and $R(t)$. Now, how do we construct transitions from and to these new states? They should mimic the computations of N anyway. Well, what is the meaning attached to this new set $R(q)$?

It says that $R(q)$ contains all the states of N that are reachable from q in N by consuming no input symbol. Moreover, these are all such states. So, when N being in state q reads an a, it goes to the state s (since $(q, a, s) \in \Delta$); our new transition should be able to do it. Next, as N, from the state q, can follow an ε-transition to go to the state p and then read an a to go to the state r (since $(q, \varepsilon, p), (p, a, r) \in \Delta$), our new transitions should also be able to capture this computation.

Now, starting from q we end up in s as well as in r. Moreover, had there been an ε-transition from s, N could have gone to that state after reading a. The automaton N could take any one of these paths in any single computation. But for the acceptance of a string, it would only look for one successful computation. Thus, we must take all these states into consideration and define a transition accordingly.

Suppose that in an NFA M, we have a state q and we have already computed the sets $R(x)$ for every state x. Suppose (q, a, p) is a transition in M. This means that M can go to (p and then to) $R(p)$ by reading a. This also means that M can start from any one of $R(q)$, read an a from that state, and then as Δ permits, it goes to another state. We take care of these eventualities by defining the new transition as follows:

$$\delta(R(q), a) = \cup\{R(p) : (x, a, p) \in \Delta \text{ for some } x \in R(q)\}.$$

However, the new set $\delta(R(q), a)$ need not be any of the new states we have constructed, because the class of these $R(x)$'s may not be closed under union. Well, what if we take every subset of Q instead of only these $R(x)$'s? There, of course, the definition of δ will be all right. Moreover, δ will also be a partial function from $2^Q \times \Sigma$ to 2^Q. Notice that the nondeterminism in the transitions have already been taken care by our new naming scheme of the states. We have collected all possible next-states as a set of states, and this set of states is regarded as a new state. This will, probably, solve case (b) also. (Recollect that case (b) refers to the case of many possible transitions with the same state–symbol pair.)

What about the final states? A computation with a string becomes successful in accepting the string if it ends up in a final state. The NFA then accepts the string if there is one such successful computation. This says that if any computation with our new scheme leads to a state (a set of states now) that contains a final state, then the machine should accept the string. Similarly, the new initial state would be the set of all those states that are reachable from the old initial state by ε-transitions. That is, a new final state is a set that contains at least one final state of N. Similarly, the new initial state is simply $R(s)$.

We collect the threads together and write formally our **construction of a DFA from an NFA**, also called the *subset construction algorithm* as in the following:

ALGORITHM *Subset Construction*

Let $M = (Q, \Sigma, \Delta, s, F)$ be an NFA. The **DFA corresponding to the NFA** M is $D = (Q', \Sigma, \delta, s', F')$, where

$Q' = 2^Q$,

$s' = R(s) = \{q \in Q : (s, \varepsilon, q) \in \Delta\} \cup \{s\}$,

$F' = \{A \subseteq Q : A \cap F \neq \emptyset\}$, and

$\delta : 2^Q \times \Sigma \to 2^Q$ is defined by

$\delta(\emptyset, \sigma) = \emptyset$, and

$\delta(A, \sigma) = \cup_{p \in A}\{R(q) : (r, \sigma, q) \in \Delta \text{ for some } q \in Q, r \in R(p)\}$.

In other words, $\delta(A, \sigma) = \{q \in Q : (p, \sigma u) \rightsquigarrow (q, u) \text{ for some } p \in A\}$, and this holds for every string $u \in \Sigma^*$. The set $\delta(A, \sigma)$ is the set of all states in Q, which are reachable from some state in A by reading the symbol σ. The initial state s' of D is that subset of Q which contains all and only states in M which are reachable from s without consuming any input. The final states in F' of D are the sets of all states that contain at least one final state of M. It is easy to see that the automaton D' is indeed a DFA, as δ is a partial function. In fact, there is no need to start construction with the whole of 2^Q. We can introduce the subsets of Q as and when necessary. Look at the following example to set the idea.

Example 3.2. Construct a DFA corresponding to the NFA given in Fig. 3.1.

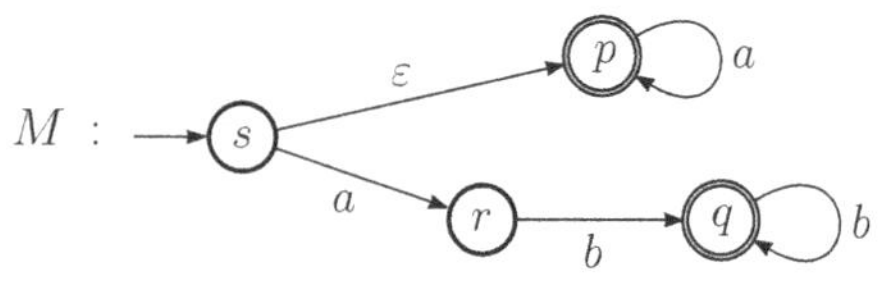

Fig. 3.1. NFA for Example 3.2.

Solution. Here, $M = (Q, \{a, b\}, \Delta, s, F)$, where $Q = \{s, p, q, r\}$, $F = \{p, q\}$, and $\Delta = \{(s, \varepsilon, p), (s, a, r), (p, a, p), (r, b, q), (q, b, q)\}$. First, we compute the sets of states reachable from the states s, p, q, r without consumption of any input. They are the following:

$$R(s) = \{s, p\}, \;\; R(p) = \{p\}, \;\; R(q) = \{q\}, \;\; R(r) = \{r\}.$$

This says that the initial state of the DFA will be $s' = \{s, p\}$. Next, we go for the computation of the transition function δ. It is done as follows:

$$\begin{aligned}
&\delta(\{s, p\}, a)\\
&= \{R(y) : (x, a, y) \in \Delta \text{ for some } y \in Q, x \in R(s)\}\\
&\qquad \cup\{R(y) : (x, a, y) \in \Delta \text{ for some } y \in Q, x \in R(p)\}\\
&= \{R(y) : (x, a, y) \in \Delta \text{ for some } y \in Q, x \in R(s)\} \ [\text{since } R(p) \subseteq R(s)]\\
&= \cup_{x\in\{s,p\}}\{R(y) : (x, a, y) \in \Delta \text{ for some } y \in Q\}.
\end{aligned}$$

We must find out the suitable triples $(x, a, y) \in \Delta$, where $x = s$ or p. They are (s, a, r) and (p, a, p). Thus,

$$\delta(\{s, p\}, a) = R(r) \cup R(p) = \{p, r\} \text{ and } \delta(\{s, p\}, b) = \emptyset.$$

The latter happens as there is no state $y \in Q$ such that $(s, b, y) \in \Delta$ or $(p, b, y) \in \Delta$. Again, as the only suitable transition to be considered for the pair $(\{p\}, a)$ is (p, a, p), we have

$$\delta(\{p\}, a) = R(p) = \{p\}.$$

Similarly, we obtain

$$\delta(\{p\}, b) = \emptyset,\ \delta(\{q\}, a) = \emptyset,\ \delta(\{q\}, b) = R(q) = \{q\},$$

$$\delta(\{r\}, a) = \emptyset,\ \delta(\{r\}, b) = R(q) = \{q\}.$$

We see that in constructing δ for the known sets, we have reached at new states such as $\{p, r\}$ and $\emptyset$. We must at least find δ for these.

$$\delta(\emptyset, a) = \emptyset,\ \delta(\emptyset, b) = \emptyset,\ \delta(\{p, r\}, a) = R(p) = \{p\},\ \delta(\{p, r\}, b) = R(q) = \{q\}.$$

You may also construct δ for other subsets of Q; however, any computation of the DFA will only use, at the most, these states. Now, what are the final states of the DFA among the subsets $\emptyset$, $\{p\}$, $\{q\}$, $\{r\}$, $\{p, r\}$.$\{s, p\}$? We had originally $F = \{p, q\}$. The subsets that intersect with F are the elements of F'. That is,

$$F' = \{\{p\}, \{q\}, \{p, r\}, \{s, p\}\}.$$

The DFA is $D = (2^Q, \Sigma, \delta, s', F')$ as given in Fig. 3.2. We do not show the ten irrelevant states in the diagram. □

In the above example, Does D accept the same language as M? The empty string ε is clearly accepted by D. If a string begins with b, then D enters the state $\emptyset$ and then halts without accepting the string. If the string starts with a, then D enters the state $\{p, r\}$. If the second symbol is b, it enters $\{q\}$ and for more b's the string is accepted, else, it is not accepted. If the third symbol happens to be a (or more a's thereafter), then also the string is accepted. Thus, $\mathcal{L}(D) = a^* \cup abb^*$, which equals $\mathcal{L}(M)$.

Fig. 3.2. DFA for Example 3.2.

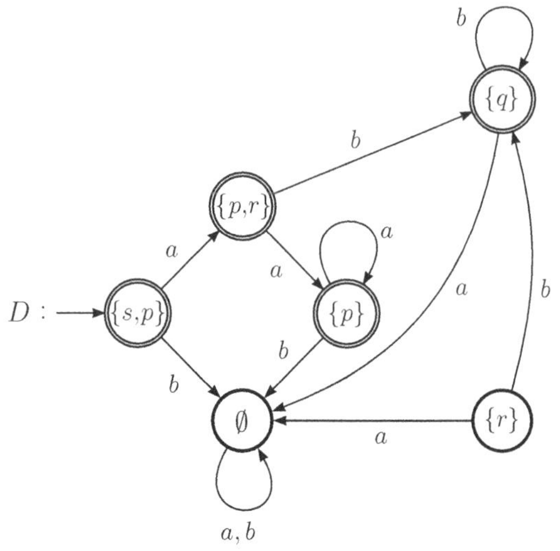

Notice that $\{r\}$ is a redundant state in D. Even though we had considered the only subsets of Q that come up during the computation of δ, it does not guarantee that all those states are really relevant. However, this does not matter as long as D accepts the same language as M. But can you prove this, in general? We prove a result looking slightly more general than this.

Lemma 3.1. *Let D be the DFA corresponding to the NFA M. Let p, q be any states of M and let $u \in \Sigma^*$. Then $(q, u) \rightsquigarrow_M (p, \varepsilon)$ iff $(R(q), u) \rightsquigarrow_D (P, \varepsilon)$ for some $P \subseteq Q$ with $p \in P$.*

Proof. We use induction on the length of u. For $u = \varepsilon$, $P = R(q)$ does the job. Lay out the induction hypothesis that the statement holds for all strings $u \in \Sigma^*$ of length m. Let $w = v\sigma$ for a string v of length m with $\sigma \in \Sigma$.

Suppose that $(q, w) \rightsquigarrow_M (p, \varepsilon)$. While computing with w, M reads v first and then the symbol σ. That is, there are states $q_1, q_2 \in Q$ such that

$$(q, v\sigma) \rightsquigarrow_M (q_1, \sigma) \rightsquigarrow_M (q_2, \varepsilon) \rightsquigarrow_M (p, \varepsilon),$$

where q_2 may also be equal to p. But $(q, v\sigma) \rightsquigarrow_M (q_1, \sigma)$ gives $(q, v) \rightsquigarrow_M (q_1, \varepsilon)$. By the induction hypothesis, $(R(q), v) \rightsquigarrow_D (Q_1, \varepsilon)$ for some $Q_1 \subseteq Q$ with $q_1 \in Q_1$. Thus, $(R(q), v\sigma) \rightsquigarrow_D (Q_1, \sigma)$. As $(q_1, \sigma) \rightsquigarrow_M (q_2, \varepsilon)$, we have $(q_1, \sigma, q_2) \in \Delta$. By the construction of D, we have $P_1 = \delta(Q_1, \sigma) \supseteq R(q_2)$. Combining this with $(R(q), v\sigma) \rightsquigarrow_D (Q_1, \sigma)$, we see that

$$(R(q), v\sigma) \rightsquigarrow_D (Q_1, \sigma) \rightsquigarrow_D (P_1, \varepsilon).$$

However, $(q_2, \varepsilon) \rightsquigarrow_M (p, \varepsilon)$ shows that $p \in R(q_2)$. Thus, $p \in \delta(Q_1, \sigma) = P_1$. Taking $P = P_1$, we obtain the required computation as $(R(q), v\sigma) \rightsquigarrow_D (P, \varepsilon)$.

Conversely, suppose that $(R(q), v\sigma) \rightsquigarrow_D (P, \varepsilon)$ for some $P \subseteq Q$ containing p. Suppose that D enters the state R_3 after consuming v. Then $(R(q), v\sigma) \rightsquigarrow_D (R_1, \sigma) \rightsquigarrow_D (P, \varepsilon)$. Clearly,

$$p \in P = \delta(R_1, \sigma) = \cup_{x \in R_1} \{R(y) : (z, \sigma, y) \in \Delta, z \in R(x)\}.$$

Then there are states $r_2 \in Q$, $p \in R(r_2)$, $r_3 \in R_1$ such that $(r_3, \sigma, r_2) \in \Delta$. The condition $p \in R(r_2)$ says that $(r_2, \varepsilon) \rightsquigarrow_M (p, \varepsilon)$. By the induction hypothesis, $(q, v) \rightsquigarrow_M (r_3, \varepsilon)$. Combine all these to obtain $(q, v) \rightsquigarrow_M (r_3, \varepsilon) \rightsquigarrow_M (r_2, \varepsilon) \rightsquigarrow_M (p, \varepsilon)$. □

We thus have the following result:

Lemma 3.2. *Let D be the DFA corresponding to the NFA M obtained by the subset construction algorithm. Then $\mathcal{L}(M) = \mathcal{L}(D)$.*

Proof. Let $M = (Q, \Sigma, \Delta, s, F)$ be an NFA. Construct $D = (Q', \Sigma, \delta, s', F')$ corresponding to M using the subset construction algorithm. Take a string $w \in \Sigma^*$. By Lemma 3.1, we have

$$\begin{aligned}
&(s, w) \rightsquigarrow_M (p, \varepsilon) \text{ for some } p \in F \\
&\text{iff } (R(s), w) \rightsquigarrow_D (P, \varepsilon) \text{ for some } P \subseteq Q \text{ containing } p \\
&\text{iff } (R(s), w) \rightsquigarrow_D (P, \varepsilon) \text{ for some } P \in F',
\end{aligned}$$

as F' contains all subsets of Q each containing at least one final state of M. This shows that $\mathcal{L}(M) = \mathcal{L}(D)$ completing the proof. □

When two automata accept the same language, we say that they are *equivalent*. The above result shows that corresponding to each NFA there is an equivalent DFA.

Problems for Section 3.2

3.1. Consider the NFA $M = (\{s, p, q\}, \{0, 1\}, \Delta, s, \{q\})$ with $\Delta = \{(s, 0, s), (s, 1, s), (s, 1, p), (p, 0, q), (p, 1, q), \}$. Find a DFA equivalent to M using the subset construction algorithm. How many states does the DFA have?

3.2. Fix an $n > 1$. Define the NFA $M = (\{s, q_1, \ldots, q_n\}, \{0, 1\}, \Delta, s, \{q_n\})$ with $\Delta = \{(s, 0, s), (s, 1, s), (s, 1, q_0)\} \cup \{(q_i, \sigma, q_{i+1}) : \sigma \in \{0, 1\}, i = 1, 2, \ldots, n-1\}$. Use the subset construction algorithm to obtain a DFA equivalent to M. What is $\mathcal{L}(M)$?

3.3. Let M be the NFA with input alphabet $\{0, 1\}$, states p, q, r, initial state p, final state p, and transitions $(p, 1, q)$, (p, ε, r), $(q, 0, p)$, $(q, 0, r)$, and $(q, 1, r)$. What is $\mathcal{L}(M)$? Construct a DFA accepting $\mathcal{L}(M)$.

3.4. Let $M = (\{p, q, r, s\}, \{a, b\}, \Delta, s, \{s\})$ be the NFA with $\Delta = \{(s, a, s), (s, a, p), (s, a, q), (s, a, r), (r, b, q), (q, b, p), (p, b, s)\}$. Give a string that starts with an a and that is not accepted by M. Construct a DFA D equivalent to M.

3.5. Construct a DFA equivalent to the NFA with initial state p, final state q, and transitions $(p, a, p), (p, a, q), (q, b, q), (q, b, p)$. Omit inaccessible states. Also give an equivalent regular expression.

3.6. Construct DFAs equivalent to the NFAs $(\{s, p, q, r\}, \{a, b\}\Delta, s, \{r\})$, with Δ as given below. Show clearly which subsets correspond to the states s, p, q, r. Omit inaccessible states.
(a) $\Delta = \{(s, a, s), (s, b, s), (s, a, p), (p, a, q), (q, b, r)\}$.
(b) $\Delta = \{(s, a, s), (s, b, s), (s, a, p), (p, a, q), (q, b, r), (r, a, r), (r, b, r)\}$.

3.7. Design an NFA M that accepts a^* with the property that there is a unique transition which when removed from M, the resulting NFA accepts $\{a\}$. Can you design a DFA with the same property instead of an NFA? Why?

3.8. Find a DFA equivalent to the NFA with initial state p, final state q, and another state r having transitions
(a) $(p, a, q), (q, a, q), (q, \varepsilon, r), (r, b, p)$.
(b) $(p, 0, p), (p, 0, q), (p, 1, q), (q, 0, r), (q, 1, r), (r, 1, r)$.
(c) $(p, 0, q), (p, 1, q), (q, 0, p), (q, 1, q), (q, 0, r), (q, \varepsilon, r), (r, 1, q)$.
(d) $(p, 0, q), (p, \varepsilon, q), (q, 0, p), (q, 1, q), (q, 0, r), (q, 1, r), (r, o, r), (r, 1, q)$.

3.9. Each DFA is an NFA. Apply subset construction algorithm to construct a DFA from a given DFA M, treating M as an NFA. What do you obtain?

3.3 Finite Automata and Regular Grammars

Does there exist a similar connection between regular grammars and finite automata? Can we mimic the operation of a DFA by a regular grammar?

The first point we observe is that the states in a DFA and the nonterminals in a grammar are doing almost the same job. Suppose that $D = (Q, \Sigma, \delta, s, F)$ is a DFA. Let us use the state symbols of D as the nonterminals of a grammar, s being the start symbol. What about the final states? After reading a string when D enters a final state, computation stops and the string is accepted. In terms of a derivation in a grammar, if a final state, a nonterminal symbol now, is met, then we must accept the string without generating any further string. That is, corresponding to each final state q, we may have a production of the form $q \mapsto \varepsilon$. This gives a hint as to how to define other productions from δ.

Formally, corresponding to the DFA D, we construct the regular grammar $G_D = (N, \Sigma, R, S)$, where $N = Q$, $S = s$, and

$$R = \{X \mapsto \sigma Y : \delta(X, \sigma) = Y \text{ for } X, Y \in Q\} \cup \{X \mapsto \varepsilon : X \in F\}.$$

Lemma 3.3. *For the DFA D, let G_D be the grammar as constructed earlier. Then* $\mathcal{L}(G_D) = \mathcal{L}(D)$.

Proof. Let $u, v \in \Sigma^*, \sigma \in \Sigma$, and $p, q \in Q$. Now, writing $\rightsquigarrow$ for $\rightsquigarrow_D$ and $\Rightarrow$ for $\Rightarrow_{G_D}$, we see that

$(p, \sigma u) \rightsquigarrow (q, u)$ in one step

iff there is a production $p \mapsto \sigma q$ in R.

iff $vp \Rightarrow v\sigma q$ in one step, for all $v \in \Sigma^*$.

Now, if $u = \sigma_1\sigma_2 \cdots \sigma_m \in \mathcal{L}(D)$, then there are states $q_1, q_2, \ldots, q_m$, with $q_m \in F$, such that

$$(s, \sigma_1\sigma_2 \cdots \sigma_m) \rightsquigarrow (q_1, \sigma_2 \cdots \sigma_m) \rightsquigarrow \cdots (q_{m-1}, \sigma_m) \rightsquigarrow (q_m, \varepsilon)$$

$$\text{iff } s \Rightarrow \sigma_1 q_1 \Rightarrow \sigma_1\sigma_2 q_2 \Rightarrow \cdots \sigma_1 \cdots \sigma_m q_m \Rightarrow \sigma_1 \cdots \sigma_m.$$

This shows that $u \in \mathcal{L}(G_D)$, that is, $\mathcal{L}(D) \subseteq \mathcal{L}(G_D)$. As the above steps are reversible, $\mathcal{L}(G_D) \subseteq \mathcal{L}(D)$. This completes the proof. □

Instead of a DFA, had you started with an NFA, a similar construction of production rules from the transitions could have served the purpose.

Does it give any hint for constructing a DFA from a regular grammar? There is a minor problem. We would like to have a transition such as $\delta(q, \sigma) = p$ from the production $q \mapsto \sigma p$. However, all the productions in the grammar need not be in this form. For example, a regular grammar allows a production of the form $q \mapsto \sigma\tau p$ for nonterminals p, q, whereas the analog transition $\delta(q, \sigma\tau) = p$ is not allowed in a DFA.

At this stage you can think of constructing another grammar from the original where each production will be in the form $q \mapsto \sigma p$ (or of the form $q \mapsto \varepsilon$) and which accepts the same language as the original. However, a similar construction can give rise to an NFA.

Observe that we also must take care of the productions of the form $q \mapsto u$ for any string u. As this kind of productions have no nonterminal symbol on the right hand side, we introduce a new final state; in fact, this will be the only final state of our NFA.

Formally, let $G = (N, \Sigma, R, S)$ be a regular grammar. Define $Q = N \cup \{f\}$ where $f \notin N$, and

$$\Delta = \{(X, u, Y) : X \mapsto uY \in R \text{ for } X, Y \in N \text{ and } u \in \Sigma^*\}$$
$$\cup\{(X, u, f) : X \mapsto u \in R \text{ for } X \in N \text{ and } u \in \Sigma^*\}.$$

Here also, (X, u, Y) is not allowed as a transition in an NFA; we must break the string u into symbols. This is done by introducing new states as earlier. Corresponding to every element $(X, \sigma_1\sigma_2 \cdots \sigma_m, Y) \in \Delta$, we introduce the triples

$$(X, \sigma_1, X_1), (X_1, \sigma_2, X_2), \ldots, (X_{m-1}, \sigma_m, Y),$$

and remove the original triple (X, u, Y), where $X_1, \ldots, X_{m-1}$ are new states added to Q. Notice that for each distinct u appearing in a transition in Δ, these X_i's are added fresh. That means, for precision, we should have written X_{u_i} instead of X_i. After this replacement is over, we have possibly a bigger set of states and a bigger transition relation. Call the updated set of states as Q' and the updated transition relation Δ'. The NFA corresponding to the grammar G is defined as $N_G = (Q', \Sigma, \Delta', S, \{f\})$.

Lemma 3.4. *For a regular grammar G, let N_G be the NFA as constructed earlier. Then $\mathcal{L}(N_G) = \mathcal{L}(G)$.*

Proof. (Outline) It is clear from the construction of N_G that if $Y_1, \ldots, Y_n \in N$ and $u_1, \ldots, u_n \in \Sigma^*$, then

$$Y_1 \Rightarrow u_1 Y_2 \Rightarrow \cdots \Rightarrow u_1 u_2 \ldots u_{n-1} Y_n \Rightarrow u_1 u_2 \ldots u_n$$
$$\text{iff } (Y_1, u_1 \ldots u_n) \rightsquigarrow (Y_2, u_2, \ldots u_n) \rightsquigarrow \cdots \rightsquigarrow (Y_n, u_n) \rightsquigarrow (f, \varepsilon).$$

Now, you can complete the proof. □

Lemma 3.4 is often expressed as: corresponding to each regular grammar there exists an *equivalent* NFA.

Problems for Section 3.3

3.10. Let N be an NFA with input alphabet $\{0, 1\}$, states p, q, r, initial state p, final state p, and transitions $(p, 1, q)$, (p, ε, r), $(q, 0, p)$, $(q, 0, r)$ and $(q, 1, r)$. Construct a regular grammar generating $\mathcal{L}(M)$.

3.11. Construct a DFA that accepts the same language generated by the grammar with productions $S \mapsto aA,\ A \mapsto abS|b$.

3.12. Construct a DFA having three states that accepts the language accepted by the DFA $(\{p, q, r, s, t\}, \{0, 1\}, \delta, p, \{t\})$, where $\delta(p, 0) = q,\ \delta(p, 1) = s,\ \delta(q, 0) = r,\ \delta(q, 1) = t,\ \delta(r, 0) = q,\ \delta(r, 1) = t,\ \delta(s, 0) = r,\ \delta(s, 1) = t,\ \delta(t, 0) = \delta(t, 1) = t$. Can you find a regular grammar for the same language?

3.13. Find a regular grammar that generates all those strings w over $\{a, b\}$ for which $2 \cdot \#_a(w) + 3 \cdot \#_b(w)$ is even.

3.14. Construct regular grammars for the following languages on $\{a, b\}$:
(a) $\{w : \#_a(w) \text{ and } \#_b(w) \text{ are both even}\}$.
(b) $\{w : (\#_a(w) - \#_b(w)) \mod 3 = 1\}$.
(c) $\{w : (\#_a(w) - \#_b(w)) \mod 3 \neq 0\}$.
(d) $\{w : \#_a(w) - \#_b(w) \text{ is an odd integer}\}$.

3.15. Construct, in each case, a DFA/NFA that recognises $\mathcal{L}(G)$, where G is a regular grammar with productions
(a) $S \mapsto b|aS|aA,\ A \mapsto a|aA|bS$.
(b) $S \mapsto aS|bS|aA,\ A \mapsto bB,\ B \mapsto aC,\ C \mapsto \varepsilon$.

3.16. Construct a DFA with six states that accepts all and only strings over $\{a, b\}$ that start with the prefix ab.

3.4 Regular Expression to NFA

What about finite automata and regular expressions? Regular expressions (over Σ) are generated from the symbols $\emptyset$ and those in the alphabet Σ inductively by using the operations of $\cup$, $*$ and concatenation. To begin with, we must say how to construct NFAs that accept the languages (represented by) $\emptyset$ and $\{\sigma\}$ for each $\sigma \in \Sigma$. Trivially, an NFA with no final states accepts the language $\emptyset$. The NFA in Fig. 3.3 accepts the language $\{\sigma\}$.

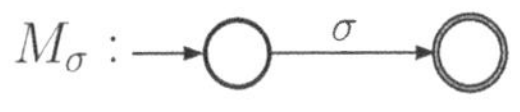

Fig. 3.3. NFA accepting $\{\sigma\}$.

We next try to simulate the operations of union, Kleene star, and concatenation. What does this simulation mean? The question is:

Suppose, for two regular expressions α, β, we have the corresponding NFAs, say M_1 and M_2, that is, we have $\mathcal{L}(M_1) = \mathcal{L}(\alpha)$, $\mathcal{L}(M_2) = \mathcal{L}(\beta)$. How do we construct NFAs for the languages $\mathcal{L}(\alpha \cup \beta)$, $\mathcal{L}(\alpha\beta)$, and $\mathcal{L}(\alpha^*)$?

For simulating union, assume that we have NFAs M_1, M_2 accepting the languages $\mathcal{L}(\alpha)$, $\mathcal{L}(\beta)$, respectively. We want to construct an NFA M to accept $\mathcal{L}(\alpha) \cup \mathcal{L}(\beta)$, which is the same language as $\mathcal{L}(\alpha \cup \beta)$. In M, we take a new initial state, say, s and try simulating the computations of M_1 and M_2. Think of M_1 and M_2 as diagrams. M starting from its initial state can go to the initial state of either of the automata M_1 or M_2. Then M will compute as that automaton does. It is now obvious that we should have ε-transitions from s to the initial states of the other machines. Final states of both of them can serve as the final states of M.

Formally, let $M_1 = (Q_1, \Sigma, \Delta_1, s_1, F_1)$, $M_2 = (Q_2, \Sigma, \Delta_2, s_2, F_2)$, with $Q_1 \cap Q_2 = \emptyset$. For the union of $\mathcal{L}(M_1)$ and $\mathcal{L}(M_2)$, define $M = (Q, \Sigma, \Delta, s, F)$, where

$Q = Q_1 \cup Q_2 \cup \{s\}$ for a new state $s \notin Q_1 \cup Q_2$,
$F = F_1 \cup F_2$, and
$\Delta = \Delta_1 \cup \Delta_2 \cup \{(s, \varepsilon, s_1), (s, \varepsilon, s_2)\}$.

Speaking in terms of diagrams, we just take a new initial state s and add an edge from s to s_1, and another edge from s to s_2, label the edges with ε, remove the initial arrows from s_1 and s_2, keeping the earlier diagrams of M_1 and M_2 as they were. The new automaton is M. It is clear that $\mathcal{L}(M) = \mathcal{L}(M_1) \cup \mathcal{L}(M_2)$.

Example 3.3. Construct an NFA that accepts the language $ab \cup baa$.

Solution. By the above construction, we see that M is the NFA in Fig. 3.4. It is also obvious that M accepts the language $ab \cup baa$. □

For concatenation, suppose that a string u is accepted by an NFA M_1 and a string v is accepted by an NFA M_2. We want to construct an NFA M that would accept the string uv. The obvious choice is that M works as M_1 does from the start till u is over,

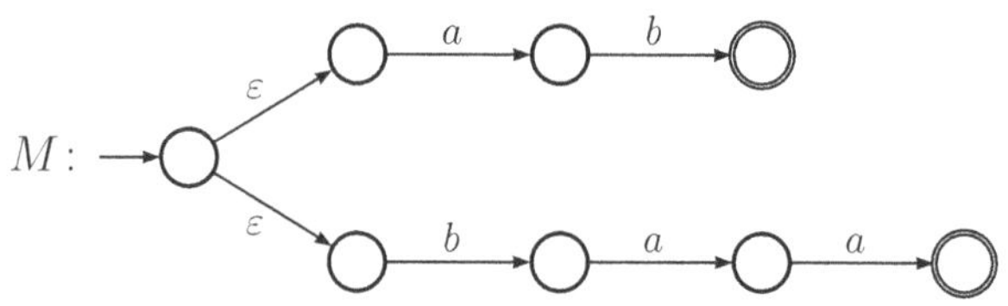

Fig. 3.4. NFA for Example 3.3.

and then M mimics M_2. So, the initial state of M should be that of M_1 and the final states of M should be the final states of M_2. Moreover, M should not stop operating just after computing with u. Observe that after computing with u, the NFA M is in a final state of M_1. To continue with the computation, M must go to the initial state of M_2 from any final state of M_1 without consuming any input. That is, we must have ε-transitions from the final states of M_1 to the initial state of M_2.

Formally, let $M_1 = (Q_1, \Sigma, \Delta_1, s_1, F_1)$, $M_2 = (Q_2, \Sigma, \Delta_2, s_2, F_2)$, with $Q_1 \cap Q_2 = \emptyset$. For the concatenation of $\mathcal{L}(M_1)$ and $\mathcal{L}(M_2)$, define $M = (Q_1 \cup Q_2, \Sigma, \Delta, s_1, F_2)$, where $\Delta = \Delta_1 \cup \Delta_2 \cup \{(q, \varepsilon, s_2) : q \in F_1\}$.

In terms of the transition diagrams, we simply put M_1, M_2 side by side, remove the arrow sign from the initial state of M_2, remove the extra circle from the final states of M_1, and join ε-transitions from the final states of M_1 to the initial state of M_2. It is clear that $\mathcal{L}(M) = \mathcal{L}(M_1)\mathcal{L}(M_2)$; but show it formally.

Example 3.4. Construct an NFA to accept the language $(ab \cup baa)ba$.

Solution. You can construct such an NFA easily. However, I want you to see the above construction at work. You have the NFA M_1 as the M of Example 3.3 accepting the language $ab \cup baa$. First, construct an NFA, say, M_2 for accepting ba. From these two, construct an NFA M by using the above idea; see Fig. 3.5. □

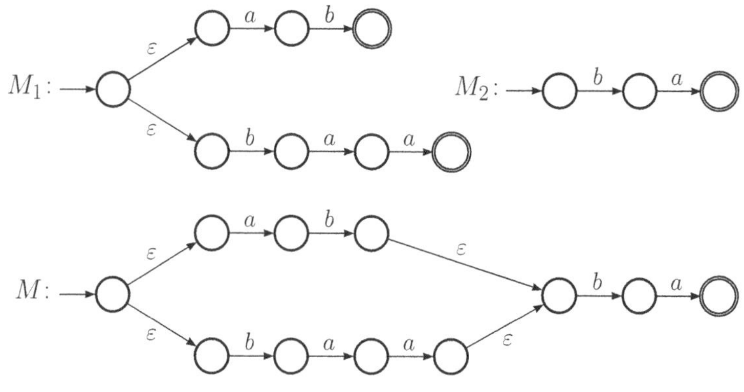

Fig. 3.5. Construction of an NFA for Example 3.4.

For Kleene star, it looks straightforward that we add an edge from every final state to the initial state with label as ε. That is, without consuming any input let the automaton change its state from any final state to its initial state. One minor problem: this new automaton will not accept the empty string ε if the initial state is originally not a final state. There are many remedies.

We add a new state; make this new state our initial state and also an extra final state. This will now force the automaton to accept the empty string. We add ε-transitions from old final states to this new state; and then add one ε-transition from this new state (a final state) to the old initial state, which is no more an initial state. Further, we can afford to have only one final state, namely, the new initial state, as there are already ε-transitions from the old final states to this new initial state. (What are the other remedies?) See the following example.

Example 3.5. Construct an NFA for accepting the language $(aba \cup abb)^*$.

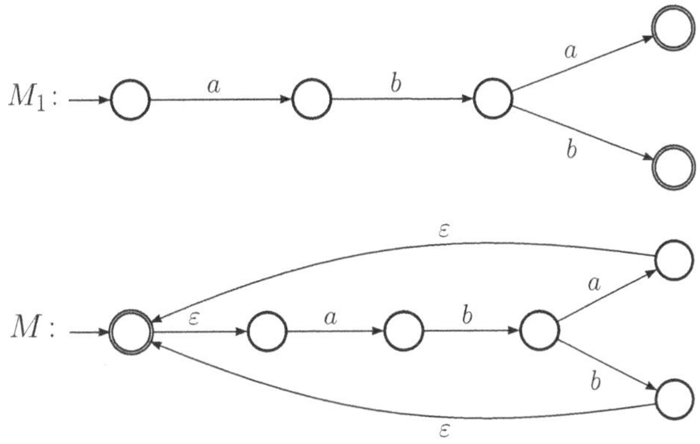

Fig. 3.6. Construction of an NFA for Example 3.5.

Solution. Observe the construction of M from M_1 in Fig. 3.6. □

Formally, let $M_1 = (Q_1, \Sigma, \Delta_1, s_1, F_1)$ be an NFA. For the Kleene star of $\mathcal{L}(M_1)$, define the NFA $M = (Q_1 \cup \{s\}, \Sigma, \Delta, s.\{s\})$, where s is a new state not in Q_1, and $\Delta = \Delta_1 \cup \{(s, \varepsilon, s_1)\} \cup \{(q, \varepsilon, s) : q \in F_1\}$.

With that you have proved the following:

Lemma 3.5. *If L is a regular language, then $L = \mathcal{L}(M)$ for some NFA M.*

Problems for Section 3.4

3.17. Design an NFA that accepts the language $(abab^* \cup aba^*)$. How many states you require at the minimum?

3.18. Construct an NFA with three states that accepts $\{ab, abc\}^*$. Can you have an NFA with fewer than three states that accepts the same language?

3.19. Use the construction of an NFA for regular expressions on the expressions $a\emptyset$ and $\emptyset^*$. What do you observe?

3.20. Design NFAs and DFAs that accept the following languages:
(a) $a^+ \cup b^*a^*$.
(b) $aa^*(a \cup b)$.
(c) $(a \cup bb)^*(ba^* \cup \varepsilon)$.
(d) $(ab^*aa \cup bba^*ab)$.
(e) Complement of $(ab * aa \cup bba^*ab)$.
(f) $(aa^* \cup aba^*b^*)$.
(g) $(a \cup b)^*b(a \cup bb)^*$.
(h) $(abab)^* \cup (aaa^* \cup b^*)$.
(i) $((aa^*)^*b)^*$.
(j) $(ab^*a^*) \cup ((ab)^*ba)$.
(k) $(ab^*a^*) \cap ((ab)^*ba)$.
(l) $(a \cup b)a^* \cap baa^*$.
(m) $(aa \cup bb)^*(ab \cup ba)(aa \cup bb)^*$.
(n) $(aaa)^*b \cup (aa)^*b$.
(o) $(a(ab)^*(aa \cup b) \cup b(ba)^*(a \cup bb))^*$.

3.21. Design two NFAs each with four states that accept the languages $a^* \cup b^+a$ and $(ab \cup abb \cup abbb)^*$.

3.5 NFA to Regular Expression

What about a language that is accepted by an NFA? Is it always a regular language? The answer is in the affirmative as you have guessed. To show this, we will generalize the idea of an NFA a bit. In any NFA, with input alphabet Σ, the edges are labeled with a symbol from $\Sigma \cup \{\varepsilon\}$. In our generalization, we will allow the edges to be labeled with regular expressions over Σ. If we now have an edge from a state q to another state r with label as $R_{q,r}$, where $R_{q,r}$ is a regular expression, then this would mean that being in the state q, the NFA can read any string from the language represented by $R_{q,r}$ and then changes its state to r. For example, assume that you have an NFA with only two states, say, s and p, where s is the initial state, p is the final state, and only one edge from s to p labeled a^*b^*. Then this NFA accepts the language a^*b^*. In fact, starting from any given NFA, we will slowly go towards an NFA of this type, so that it will tell us immediately the required language. Here is the construction, usually called as the **state elimination algorithm** or the **node elimination algorithm**.

Phase 1: In the first phase of this algorithm, we want an NFA whose initial state is a *source*. That is, we want an NFA where there would not be any transition from any state to the initial state. We must also see to it that the language accepted by such an NFA is the same as our given one. Let M be an NFA. Call its initial state by the name q. Add a new state to M, call it s. Make s as the initial state and add an ε-transition from s to q.

Clearly, the language accepted by the new NFA is the same as that accepted by M. Note that if there is no transition from any state to q, then we can afford to keep q as the initial state. But then, it may ensue some modification to the other phases. Try doing that after you understand the algorithm.

Phase 2: In the second phase, we want an NFA where the final state is a *sink*, that is, there will be a single final state from which no transition would go to any state. This is achieved in a way very much similar to the first phase. Introduce a new final state f, add ε-transitions from every old final state to f, and then do not regard the old final states as final states any more.

By the end of Phase 2, we have an NFA; call it M' with only one initial state s to which no transition comes in, and only one final state f from which no transition goes out. Moreover, there is no transition from s to itself, and none from f to itself. It is easy to see that $\mathcal{L}(M) = \mathcal{L}(M')$. As in Phase 1, if already there is only one final state in M with no transition going from it, then we can afford to refrain from introducing another new state. However, try this modification only after completely understanding the full algorithm.

Phase 3: In this phase, we want an NFA, or rather a generalized NFA, where we should have exactly one transition between any (ordered) pair of states. That is, we should have only one transition from p to q and exactly one more from q to p, for any states p, q of our required generalized NFA. Let p, q be any two (different) states in M'. If there is no transition from p to q, then add a transition and label it with $\emptyset$. If there are $n \geq 2$ transitions from p to q, say with labels $\sigma_1, \ldots, \sigma_n$ for symbols $\sigma_1, \ldots, \sigma_n$ from $\Sigma \cup \{\varepsilon\}$, then remove all those transitions and add a new transition with label as $\sigma_1 \cup \ldots \cup \sigma_n$. Call the new generalized NFA as M''. Notice again that the languages accepted by this new generalized NFA M'' and by the old M' are the same.

The next phase is the proper node-elimination process for which the earlier phases form a background.

Phase 4: We have a generalized NFA M'' with the initial state s and only one final state f where there is no transition from any state to s and no transition from f to any state. There are not even any loops around s and f. Further, if x, y are any states of M'' not equal to any of s or f, then there is exactly one transition from x to y with label as a regular expression, say, $R_{x,y}$.

Pick any state r other than s, f. Any such state is an original state of N. Corresponding to this state r, find all pairs of states (p, q) such that a transition comes from p to r, a transition goes to q from r, and there is a transition from p to q. There might possibly be a loop around r. But loops around p, q are overlooked.

The relevant portion of the NFA M'' now looks like the left side diagram in Fig. 3.7 with the respective labels on the transitions. For each such pair (p, q) of states in M'' corresponding to the state r, replace the label of the transition $R_{p,q}$ by $R_{p,q} \cup R_{p,r} R_{r,r}^* R_{r,q}$. Look at Fig. 3.7; the diagram on the left is replaced by the diagram on the right (inside M'').

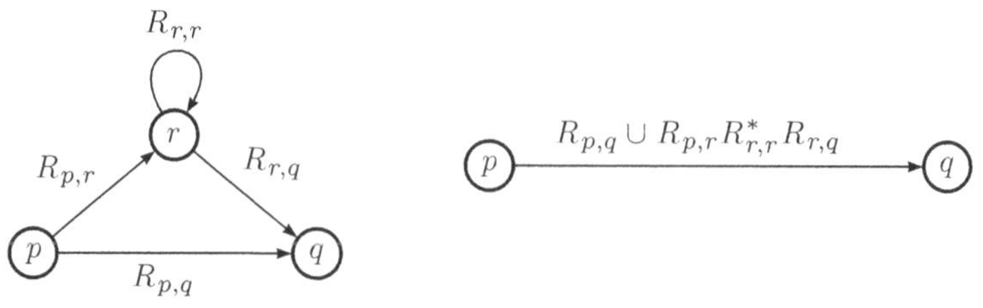

Fig. 3.7. Phase 4 of state elimination algorithm.

This replacement must be done simultaneously for each such pair (p, q). This means that you do not immediately remove r going for the replacement that calls for the pair (p, q), but explore first whether there are other such pairs for the same r. Compute replacements for all such pairs, and then remove r by effecting all these replacements. The pairs (p, q) also include the possibility (p, p). This subcase and the corresponding replacement is drawn schematically in Fig. 3.8.

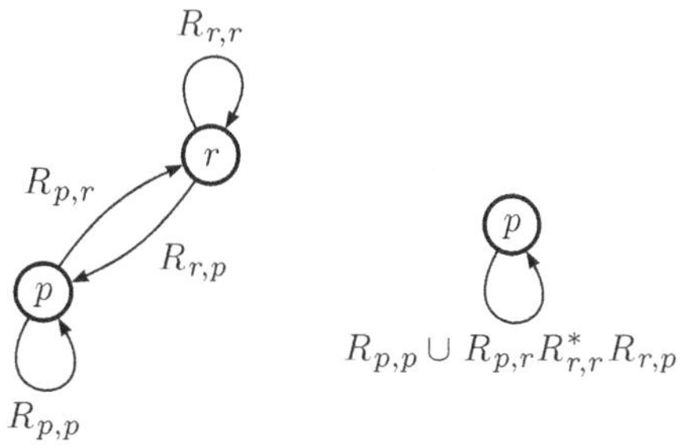

Fig. 3.8. Phase 4 of state elimination algorithm.

Notice that the process of replacement starts with choosing a state r from M'' and then replacing simultaneously many transitions. After all these replacements are over for this state r, delete the state r and the transitions from and to r. The new generalized NFA now has one state less than that of M''.

Repeat this phase to eliminate all the states except the initial state s and the final state f.

Caution: While considering pairs (p, q) corresponding to a state r, do not exclude the states s and f. The state s can serve as a p and the state f can also be a q. You must consider both the possibilities (p, q) and (q, p) separately. They are not the same pair. The order is important because for a given r, the state p is the one from which there is a transition to r and q is a state to which there is a transition from r. Observe that such replacements have an effect only when there are transitions from p to r and from r to q in the original NFA M. You can also follow up many simplifications such as when $R_{r,r} = \emptyset$, the expression $R_{p,q} \cup R_{p,r} R^*_{r,r} R_{r,q}$ is simply $R_{p,q} \cup R_{p,r} R_{r,q}$, and when $R_{p,q} = \emptyset$, we can have $R_{p,r} R^*_{r,r} R_{r,q}$ instead of $R_{p,q} \cup R_{p,r} R^*_{r,r} R_{r,q}$. Further, the order in which we pick the pairs (p, q) is irrelevant, although it might affect the size of the final expression. Similarly, the order in which we choose different r's from the automaton M'' is also irrelevant. This phase is over when the only remaining states are s and f. One more point to remember: we write ε as a regular expression instead of the official expression $\emptyset^*$.

You must first understand why at each phase, the language accepted by the new generalized NFA is the same as that accepted by the old one. It will be helpful to first duplicate the states in another place and then slowly add or delete the required transitions as asked in various phases. Look at the following two examples. I would suggest you to try to do the examples yourself before looking at the solutions. Then, of course, verify that your solution is, indeed, the same as the one given here. You may not get the same regular expression because of the order of the states r and the corresponding state-pairs (p, q).

Do not be discouraged; you can show that it is the same language that you get from your regular expression as the one given in the solution. But try to solve the examples again by choosing the same order of the states as done in the solutions. This reworking is sometimes tedious. But this is the easiest way to learn!

Example 3.6. Apply the state elimination algorithm on the NFA M to obtain the regular expression representing the language accepted by it, where $M = (\{A, B, C\}, \{a, b\}, \Delta, A, \{C\})$ with $\Delta = \{(A, a, A), (A, b, A), (A, b, B), (B, a, C), (B, b, C)\}$.

Solution. The transition diagram of M is in Fig. 3.9 (find A, B, C).

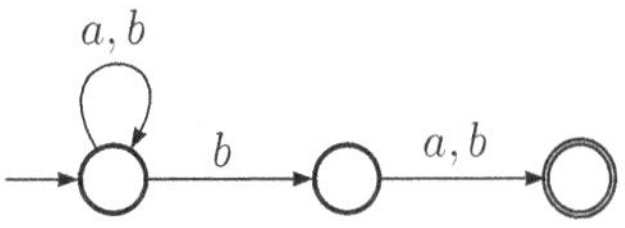

Fig. 3.9. The NFA M.

In Phase 1, we add the initial states s, final state f, and introduce the ε-transitions as in Fig. 3.10.

In Phase 2, we replace the multiple transitions by regular expressions and then add transitions with label as $\emptyset$ whenever there is no transition between states. In fact, we do not show all these $\emptyset$-transitions; rather, we take only some relevant ones and suppress others thinking that they are there. We then obtain the generalized NFA M' as in Fig. 3.11.

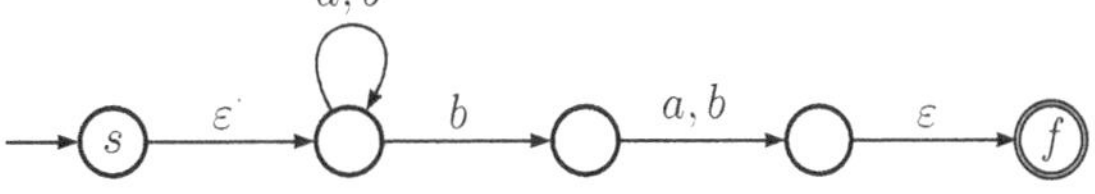

Fig. 3.10. M after Phase 1.

In Phase 3, we choose a state as r, which is already shown as "r" in Fig. 3.11. Now, taking s as p and the third (from the left) as q, we see that the transition labeled $\emptyset$ must be replaced by $\varepsilon(a \cup b)^*b$. Next, with the same r we may choose the third state as p and the fourth state as q, but then due to $\emptyset$-transitions (not shown in figures) the regular expressions will be same as $\emptyset$, and thus this is not relevant. Similarly, all other possibilities of choosing different pairs (p, q) for this r give only $\emptyset$-transitions.

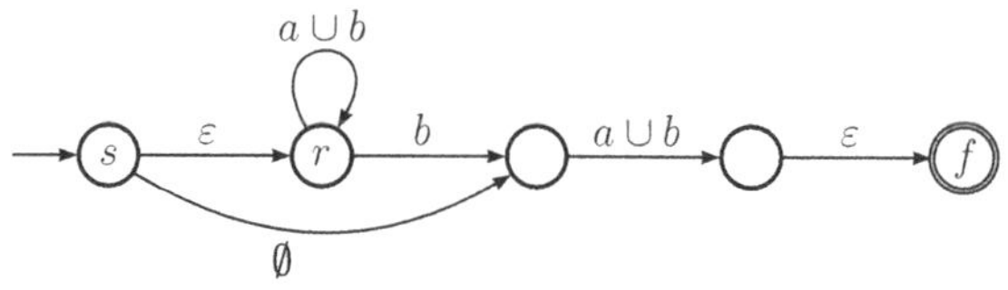

Fig. 3.11. M after Phase 2.

At the end of this phase, we must delete r and all the transitions from and to it. The r-state in Fig. 3.11 is deleted to obtain the diagram in Fig. 3.12.

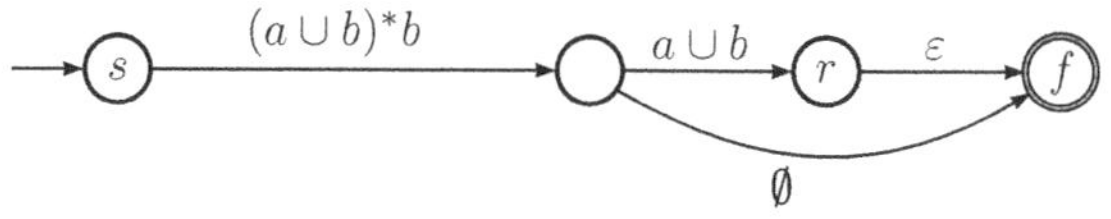

Fig. 3.12. M in Phase 3.

In Fig. 3.12, we also show another relevant $\emptyset$-transition with our next choice of r. Note that the r-state in the succeeding diagram is freshly chosen; it is not the same as in the previous diagram. As shown in Fig. 3.12, we choose our state r afresh. Then among various other pairs of (p, q), first we take the state to the left of r as p and f as q. This is the reason we have shown the $\emptyset$-transition in Fig. 3.12. Now, with this choice of (p, q), we see that the $\emptyset$-transition is to be replaced by $a \cup b$ and then the r-state to be deleted. Thus the generalized NFA is updated to the NFA in Fig. 3.13.

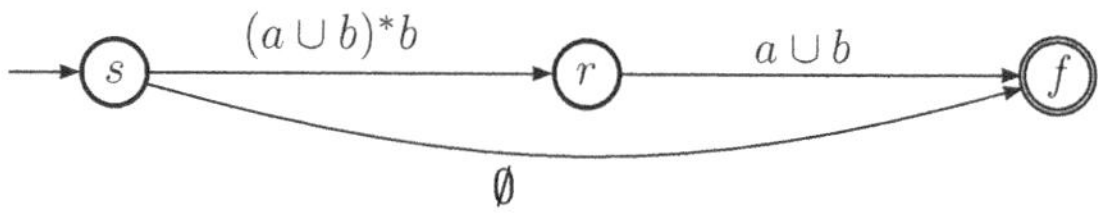

Fig. 3.13. M after Phase 3.

We have again chosen our new r as the middle state (the only choice here), and have shown the relevant $\emptyset$-transition from s to f. Eliminating the state r gives us the new transition from s to f labeled $(a \cup b)^*b(a \cup b)$ as in Fig. 3.14.

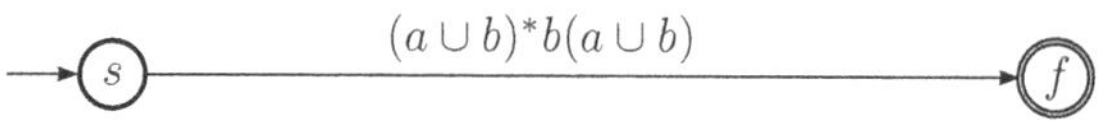

Fig. 3.14. M after Phase 4.

The state elimination algorithm suggests that the language accepted by the NFA M is $(a \cup b)^*b(a \cup b)$. □

Example 3.7. Apply the state elimination algorithm on the NFA in Fig. 3.15 to obtain the regular expression that represents the language accepted by it.

Fig. 3.15. An NFA.

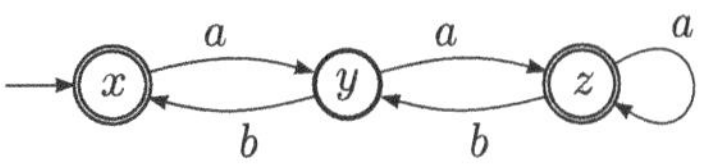

Solution. We redraw the diagram and add the ε-transitions changing the initial and final states. The result of these operations in Phase 1 gives us the left hand side transition diagram in Fig. 3.16.

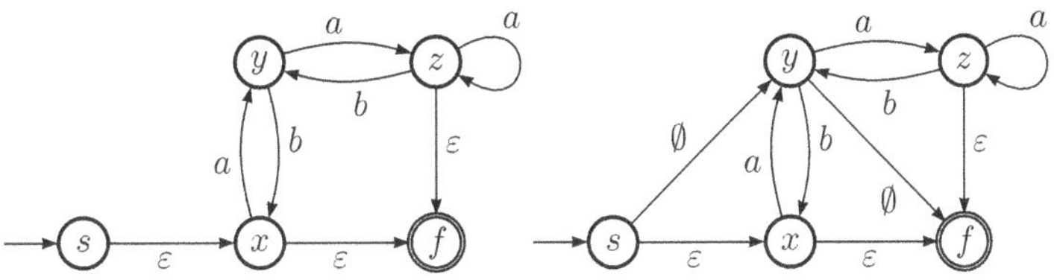

Fig. 3.16. Example 3.7.

We want to eliminate the state x from the NFA. This state as the r-state, we have the pairs (p, q) as (s, y), (y, y), and (y, f). Then the two $\emptyset$-transitions as drawn on the right hand side transition in Fig. 3.16 become relevant.

For the pair (s, y), we replace the transition from s to y by following the sequence of states s to x to y; the other sequence is from s to y, which is $\emptyset$. So union of these is simply a. Thus the $\emptyset$-transition from s to y is relabeled as a.

For the pair (y, y), we follow the sequence of states y, x, y to get the string (expression) ba. Thus we introduce a loop at y with label ba.

For the pair (y, f), we follow the states as y, x, f or as y, f directly. These alternatives give us the expressions $b\varepsilon$ or $\emptyset$, respectively. Their union is simply b. This is the label for the edge from y to f.

We delete the state x along with the transitions from and to it to obtain the generalized automaton as in the left hand side diagram of Fig. 3.17. Also look at the relevant $\emptyset$-transitions added to it as on the right hand side diagram of Fig. 3.17.

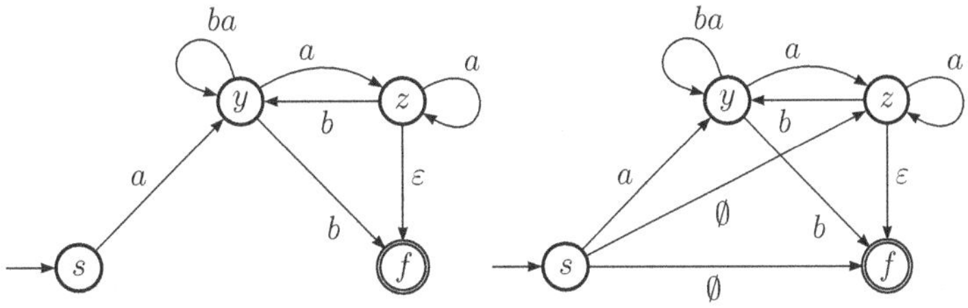

Fig. 3.17. Example 3.7 continued.

For eliminating y, the relevant pairs (p, q) are (s, f), (s, z), and (z, z). Considering the pair (s, f), we get the sequence of states from s to f through y. This way, the NFA accepts the expression $a(ba)^*b$. The other alternative is an $\emptyset$-transition contributing nothing to the union. Thus the new label on the transition from s to f is $a(ba)^*b$. The pair (s, z) gives the new label on the transition from s to z as $a(ba)^*a$. The pair (z, z) gives similarly a loop around z with label $a \cup (b(ba)^*a)$. Therefore, the elimination of the state y results in the generalized NFA as drawn on the left hand side of Fig. 3.18.

Next, we eliminate the state z, for which the only p, q pair is (s, f). We just have to get the regular expression and delete this state. It is shown on the right hand side of Fig. 3.18.

The required regular expression is $R = a(ba)^*b \cup a(ba)^*a(a \cup (b(ba)^*a))^*$. □

Once you think you have understood the state elimination algorithm, the following result should be obvious. It says that corresponding to each NFA, there exists an equivalent regular expression. However, its proof has to be done by induction.

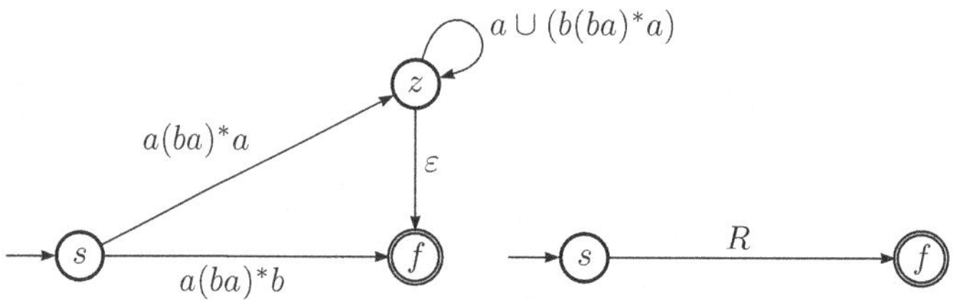

Fig. 3.18. Example 3.7 completed.

Lemma 3.6. *Let α be the regular expression constructed by the state elimination algorithm from the NFA M. Then $\mathcal{L}(\alpha) = \mathcal{L}(M)$.*

Problems for Section 3.5

3.22. In Example 3.7, eliminate the states in the order z, y, x instead of x, y, z. As a check, see that you get the corresponding p, q pairs as $(y, y), (y, f)$ for z, $(x, x), (x, f)$ for y, and (s, f) for x. Also, check whether you get the regular expression $(a(aa^*b)^*b)^*(\varepsilon \cup a(aa^*b)aa^*)$.

3.23. Recall that instead of a transition relation for an NFA, we can have a transition function. In this definition, an NFA is taken as a five tuple: $M = (Q, \Sigma, \delta, q_0, F)$, where Q is a finite set of states, Σ is an alphabet, $q_0 \in Q$ is the initial state, $F \subseteq Q$ is the set of final states, and $\delta : Q \times (\Sigma \cup \{\varepsilon\}) \to 2^Q$. Here, for instance, the transitions (p, σ, q) and (p, σ, r) in the transition relation Δ are written as $\delta(p, \sigma) = \{q, r\}$. The transition function δ is extended to $\delta^* : Q \times \Sigma^* \to 2^Q$ in such a way that $\delta^*(q, w)$ contains $p \in Q$ iff there is a walk in the transition graph from q to p labeled w. The language accepted by M is defined to be $\mathcal{L}(M) = \{w \in \Sigma^* : \delta^*(q_0, w) \cap F \neq \emptyset\}$.

(a) Define δ^* set theoretically and then show that our definition of $\mathcal{L}(M)$ and this definition are equivalent.

(b) Define the operation of *yield* between configurations using this definition of an NFA.
(c) Define acceptance in this case by extending δ^* for strings in place of symbols.
(d) Connect this definition to the State Elimination Algorithm.
(e) Is it true that $\overline{\mathcal{L}}(M) = \{w \in \Sigma^* : \delta^*(q_0, w) \cap F = \emptyset\}$? Prove or give a counter example.
(f) Is it true that $\overline{\mathcal{L}}(M) = \{w \in \Sigma^* : \delta^*(q_0, w) \cap (Q - F) \neq \emptyset\}$? Prove or give a counter example.

3.24. Design an NFA with a single final state and without ε-transitions that accepts $\{a\} \cup \{b^n : n \geq 1\}$.

3.25. If the NFA happens to be a DFA, the state elimination algorithm becomes simplified. Write the state elimination algorithm for DFAs in a step-by-step manner.

3.26. Match each NFA $(\{p, q, r\}, \Delta, p, F)$ with an equivalent regular expression given below.

NFAs:

(a) $F = \{p\}, \Delta = \{(p, a, q), (q, a, p), (q, b, r), (r, b, q), (r, a, r)\}$.
(b) $F = \{p\}, \Delta = \{(p, a, q), (q, b, p), (q, b, r), (r, b, q), (r, a, r)\}$.
(c) $F = \{p\}, \Delta = \{(p, a, q), (q, a, p), (q, a, r), (r, b, q), (r, b, r)\}$.
(d) $F = \{q\}, \Delta = \{(p, a, q), (p, b, r), (q, b, q), (q, b, p), (r, a, r), (r, a, p)\}$.
(e) $F = \{q\}, \Delta = \{(p, a, q), (p, b, r), (q, a, q), (q, b, p), (r, a, r), (r, a, p)\}$.
(f) $F = \{q\}, \Delta = \{(p, b, q), (p, a, r), (q, a, q), (q, b, p), (r, a, r), (r, b, p)\}$.
(g) $F = \{q, r\}, \Delta = \{(p, a, q), (p, a, r), (q, a, q), (q, a, p), (r, a, r), (r, b, p)\}$.
(h) $F = \{q, r\}, \Delta = \{(p, b, q), (p, b, r), (q, a, q), (q, a, p), (r, a, r), (r, b, p)\}$.
(i) $F = \{p, r\}, \Delta = \{(p, a, q), (q, a, p), (q, a, r), (r, b, q), (r, b, r)\}$.
(j) $F = \{p, r\}, \Delta = \{(p, a, q), (q, b, p), (q, b, r), (r, b, q), (r, a, r)\}$.

Regular Expressions:

(i) $\emptyset^* \cup a(ab^*b \cup aa)^*ab^*$.
(ii) $\emptyset^* \cup a(ba^*b \cup ba)^*ba^*$.
(iii) $\emptyset^* \cup a(ba^*b \cup aa)^*a$.
(iv) $\emptyset^* \cup a(ab^*b \cup aa)^*a$.
(v) $\emptyset^* \cup a(ba^*b \cup ba)^*b$.
(vi) $(aa^*b \cup ba^*b)^*ba^*$.
(vii) $(ba^*a \cup ba^*b)^*ba^*$.
(viii) $(ba^*a \cup aa^*b)^*aa^*$.
(ix) $(ba^*a \cup ab^*b)^*ab^*$.
(x) $(aa^*a \cup aa^*b)^*aa^*$.

3.27. Let M be an NFA with initial state p, final state q, and another state r having transitions $(p, a, q), (q, \varepsilon, r), (r, \varepsilon, p)$. What is the complement of the language accepted by M?

3.28. Consider a generalized NFA having transitions $(p, a, p), (p, a \cup b, q), (q, ab, r)$, $(r, a, p), (r, \varepsilon, q), (r, bb, r)$, where p is the initial state and r is the final state.

(a) Find an equivalent generalized NFA with only two states.
(b) What is the language accepted by the generalized NFA?

3.29. What is $\mathcal{L}(N)$, where N is a generalized NFA with initial state p, final state r, and transitions
(a) $(p, b, p), (p, a, q), (q, a, q), (q, a, s), (r, b, s), (s, a, r)$?
(b) $(p, a, q), (p, b, r), (q, b, p), (q, a, r), (r, b, q), (p, \varepsilon, r)$?
(c) $(p, a, p), (p, b, q), (p, b, r), (q, a, r), (q, \varepsilon, r), (r, a, q)$?
(d) $(p, a, p), (p, a, q), (p, a^*b \cup c, r), (q, a \cup b, q), (q, a \cup b, r), (r, a \cup b^*, r)$?

3.30. Find regular expressions equivalent to the DFAs $(\{p, q, r\}, \{a, b\}, \delta, p, F)$, with
(a) $\delta(p, a) = \delta(p, b) = q, \delta(q, a) = \delta(r, a) = \delta(r, b) = p, \delta(q, b) = r;\ F = \{r\}$.
(b) $\delta(p, a) = \delta(q, b) = q, \delta(p, b) = \delta(r, a) = r, \delta(q, a) = \delta(r, b) = p;\ F = \{q, r\}$.

3.31. Give regular expressions for the languages accepted by the DFA $(Q, \Sigma, \delta, a, F)$, where $Q = \{a, b, c, d, e, f, g\}$, $\Sigma = \{0, 1\}$ and
(a) $\delta(a, 0) = b, \delta(a, 1) = e, \delta(b, 0) = a, \delta(b, 1) = d, \delta(c, 0) = g, \delta(c, 1) = b,$
$\delta(d, 0) = e, \delta(d, 1) = c, \delta(e, 0) = d, \delta(e, 1) = c, \delta(f, 0) = c, \delta(f, 1) = f,$
$\delta(g, 0) = c, \delta(g, 1) = a;\ F = \{a, b\}$.
(b) $\delta(a, 0) = b, \delta(a, 1) = e, \delta(b, 0) = a, \delta(b, 1) = f, \delta(c, 0) = d, \delta(c, 1) = c,$
$\delta(d, 0) = g, \delta(d, 1) = a, \delta(e, 0) = f, \delta(e, 1) = g, \delta(f, 0) = e, \delta(f, 1) = d,$
$\delta(g, 0) = d, \delta(g, 1) = b;\ F = \{a, b\}$.
(c) $\delta(a, 0) = b, \delta(a, 1) = f, \delta(b, 0) = a, \delta(b, 1) = g, \delta(c, 0) = e, \delta(c, 1) = b,$
$\delta(d, 0) = b, \delta(d, 1) = c, \delta(e, 0) = c, \delta(e, 1) = a, \delta(f, 0) = g, \delta(f, 1) = c,$
$\delta(g, 0) = f, \delta(g, 1) = e;\ F = \{a, b\}$.
(d) $\delta(a, 0) = a, \delta(a, 1) = c, \delta(b, 0) = f, \delta(b, 1) = c, \delta(c, 0) = e, \delta(c, 1) = g,$
$\delta(d, 0) = f, \delta(d, 1) = a, \delta(e, 0) = a, \delta(e, 1) = g, \delta(f, 0) = b, \delta(f, 1) = g,$
$\delta(g, 0) = e, \delta(g, 1) = c;\ F = \{b, d, e\}$.

3.6 Summary and Additional Problems

In this chapter, we have fulfilled the promise of the previous chapter. Our sole concern has been to show that the devices of regular expressions, regular grammars, and the finite automata are equivalent. The sense of equivalence here is in terms of the class of languages the devices represent. Our path has been the following: DFA to NFA (trivial), NFA to DFA, DFA to regular grammar, regular grammar to NFA, regular expression to NFA, and finally, NFA to regular expression.

Equivalence of DFAs and NFAs were shown by Rabin and Scott [107]. Kleene [67] proved the equivalence of DFAs and regular expressions; and a shorter proof was given by McNaughton and Yamada [85]. Chomsky and Miller [17] observed the equivalence of regular grammars and regular expressions. The state elimination algorithm as presented here closely follows that in [34]. The equivalence problem for two-tape automata has been examined in [10]. The equivalence of two-way automata with the standard ones has been shown in [117].

Additional Problems for Chapter 3

3.32. Let $L = \{w \in \{a, b\}^* : \#_a(w)$ is even and $\#_b(w)$ is odd$\}$. Find a DFA D and then a regular expression E such that $\mathcal{L}(D) = L = \mathcal{L}(E)$.

3.33. Fix an $n > 1$. Let $N = (\{q_0, q_1, \ldots, q_n\}, q_0, \Delta, \{q_n\})$, where $\Delta = \{(q_0, 0, q_0)\} \cup \{(q_{i-1}, 0, q_i), (q_{i-1}, 1, q_i) : 1 \le i \le n\}$. Show that
(a) $\mathcal{L}(N) = \{w \in \{0, 1\}^*$: the nth symbol from the right end of w is $1\}$.
(b) If D is a DFA such that $\mathcal{L}(D) = \mathcal{L}(N)$, then D has at least 2^n states.

This shows that the subset construction cannot be improved to have less than 2^n states, in general, where n is the number of states in an NFA.

3.34. Consider a modification of the definition of an NFA, where there are possibly many initial states, that is, $N = (Q, \Sigma, \delta, Q_0, F)$, where $Q_0 \subseteq Q$ is the set of initial states. Define $\mathcal{L}(M)$ as the set of strings from Σ^*, which drive at least one initial state to at least one final state. Show that for every NFA with multiple initial states, there is one NFA with a single initial state so that they accept the same language. What if NFAs are replaced by DFAs here?

3.35. Let L be a regular language that does not contain ε. Show that there exists an NFA without ε-transitions and with a single final state that accepts L.

3.36. Give formal proofs of Lemmas 3.4–3.6.

3.37. If a DFA has n states $q_1, q_2, \ldots, q_n$, then its transition function δ can be specified by an $n \times n$ matrix T, called its *transition matrix*, whose *ij*th entry is given by the set of input symbols taking state q_i to q_j, that is, $(T)_{ij} = \{\sigma \in \Sigma : \delta(q_i, \sigma) = q_j\}$. Given a set of states Q having n elements and an input alphabet Σ, consider the collection $\mathcal{C}$ of all $n \times n$ matrices, which are the transition matrices of some DFAs with state-set as Q. Define the identity matrix $I \in \mathcal{C}$ as one whose *ij*th entry I_{ij} = "$\{\varepsilon\}$ for $i = j$ and $\emptyset$ for $i \ne j$". Define addition and multiplication of matrices in $\mathcal{C}$ by $(A + B)_{ij} = A_{ij} \cup B_{ij}$ and $(AB)_{ij} = \cup_{1 \le k \le n} A_{ik} B_{kj}$. Then, the powers of matrices are defined by taking $A^0 = I,\ A^{n+1} = A^n A$. Similarly, the asterate of a matrix is defined by $(A^*)_{ij} = \cup_{n \ge 0} (A^n)_{ij}$. Let $M = (Q, \Sigma, \delta, s, F)$ whose transition matrix is A, as constructed from δ. Also, let δ^* denote the extended transition of δ that is defined for the strings in Σ^*. Then, prove that
(a) $(A^n)_{ij} = \{u \in \Sigma^* : \ell(u) = n$ and $\delta^*(q_i, u) = q_j\}$.
(b) $\mathcal{L}(M) = \cup_{q_k \in F} (A^*)_{ik}$.

3.38. It is claimed that one can construct a regular expression from a finite automaton by solving a system of linear equations of the form $Ax + b = x$, where A is an $n \times n$ matrix with entries a_{ij} as sets of strings, b is an $n \times 1$ vector with entries b_j as sets of strings, and the unknown $n \times 1$ vector x has entries, also, sets of strings. Here, $+$ denotes union and multiplication denotes concatenation. Prove the claim and give an algorithm to solve such a linear system.

3.39. The *Hamming distance* between two strings $u, v \in \{0, 1\}^*$ of the same length is defined as $H(u, v)$ = number of places where they differ. Moreover, if $\ell(u) \neq \ell(v)$, then $H(u, v) = \infty$. If A is a language over $\{0, 1\}$ and $x \in \{0, 1\}^*$, then $H(x, A) = \min_{y \in A} H(x, y)$. For any $L \subseteq \{0, 1\}^*$ and $n \in \mathbb{N}$, define $M_n(L) = \{w : H(w, L) \leq n\}$, the set of all strings whose Hamming distance from L is at most n. Prove that if L is regular, then so is $M_2(L)$.

3.40. Design an algorithm which takes a DFA and a natural number n as its inputs and computes the number of strings of length n that are accepted by the DFA.

3.41. For a regular language $L \subseteq \Sigma^*$ and $a \in \Sigma$, define $L_a = \{uav : uv \in L\}$.
(a) Given an NFA for L, show how to construct an NFA for L_a.
(b) How do you use an NFA for L_a to write a pattern recognition program using a regular expression for L as an input?

3.42. For $n \geq 1$, define the NFAs $M_n = (Q_n, \Sigma_n, \Delta_n, s_n, F_n)$ with $Q_n = \{q_1, \ldots, q_n\}$, $\Sigma_n = \{a_{ij} : 1 \leq i, j \leq n\}$, $s_n = q_1$, $F_n = \{q_1\}$, and $\Delta_n = \{(i, a_{ij}, j) : 1 \leq i, j \leq n\}$.
(a) What is $\mathcal{L}(M_n)$? Describe it in English.
(b) What are the regular expressions for $\mathcal{L}(M_2)$, $\mathcal{L}(M_3)$, and $\mathcal{L}(M_5)$?
(c) Can you see the plausibility in the conjecture that for each polynomial p there is an n such that no regular expression for $\mathcal{L}(M_n)$ has length smaller than $p(n)$? I am not asking for a proof.

3.43. A run in a string is a substring of length at least two, as long as possible and consisting entirely of the same symbol. For instance, the string $abaaaba$ has a run aaa. Find DFAs for the following languages over $\{a, b\}$:
(a) $L = \{w : w$ contains no runs of length less than four$\}$.
(b) $L = \{w :$ every run of a's has length either two or three$\}$.
(c) $L = \{w :$ there are at most two runs of a's of length three$\}$.
(d) $L = \{w :$ there are exactly two runs of a's of length three$\}$.

3.44. Give a regular expression for the set of all strings over $\{a, b, c\}$ in which all runs of a's have lengths that are multiples of three.

4 Structure of Regular Languages

4.1 Introduction

To take stock, look up at the statements in Lemmas 3.2–3.6. They assert that regular languages can be defined in many ways: by regular expressions, as languages generated by regular grammars, as languages accepted by DFAs, and as languages accepted by NFAs. We summarize these results in the following statement:

Theorem 4.1. *Let L be any language. Then the following are equivalent:*

1. *There is a regular expression* α *such that* $\mathcal{L}(\alpha) = L$.
2. *There is an NFA M such that* $\mathcal{L}(M) = L$.
3. *There is a DFA D such that* $\mathcal{L}(D) = L$.
4. *There is a regular grammar G such that* $\mathcal{L}(G) = L$.

This is a nice statement to have, because to know whether a language is regular or not, we may use any one of the four machinery. Sometimes one is easier than the other and that helps.

Example 4.1. Is $L = \{w \in \{0, 1\}^* : \#_0(w) \bmod 3 = 1\}$ a regular language?

Solution. The DFA in Fig. 4.1 accepts the language and hence it is regular. □

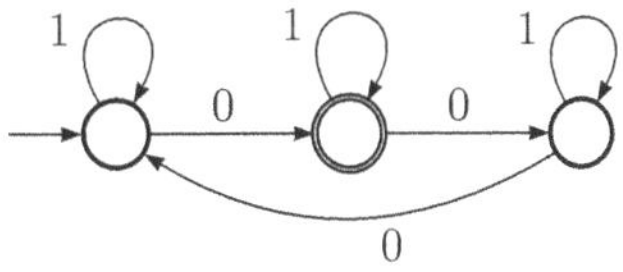

Fig. 4.1. DFA for Example 4.1.

Exercise 4.1. Let L be the language of Example 4.1.
(a) Show that $L = 1^*01^* \cup (1^*01^*0)(\emptyset^* \cup 1 \cup 01^*01^*0)^*(01^*01^*)$.
(b) Construct a regular expression for L from the given DFA by using the state elimination algorithm. Also find a simpler regular expression for L.

A. Singh, *Elements of Computation Theory*, Texts in Computer Science,

(c) Show that L is generated by the regular grammar with the following productions: $S \mapsto 0A$, $S \mapsto 1S$, $A \mapsto 0B$, $A \mapsto 1A$, $B \mapsto 0S$, $B \mapsto 1B$.
(d) Can you find a regular grammar with less than six productions for generating L?
(e) Can you find a regular grammar for L with only two nonterminals?

4.2 Closure Properties

To see whether a given language is not regular is altogether a different question. The methods we develop, in this section and the next, will enable us to answer this question in most of the cases. In this section, we will look into some closure properties of regular languages; they will help us in showing some languages to be nonregular once we know that certain other languages are nonregular. We begin with an example.

Example 4.2. Is $L = \{w \in \{0, 1\}^* : \#_0(w) \bmod 3 \neq 1\}$ a regular language?

Solution. Yes; the DFA in Fig. 4.2 accepts the language. □

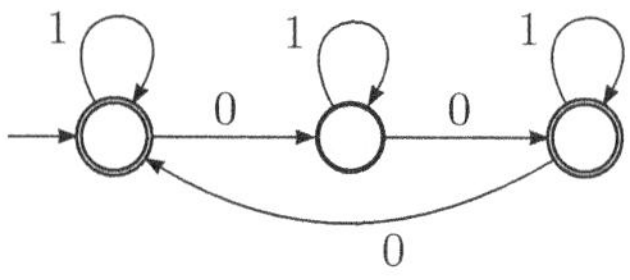

Fig. 4.2. DFA for Example 4.2.

Can you show the claim in the solution of Example 4.2? Compare this DFA with that in Example 4.1. The final states are simply interchanged. You can, of course, do that for any DFA, and then complement of every regular language will be regular. However, you must take care to make the transition function of the original DFA a total function. See the following statement:

Theorem 4.2 (Closure Properties of Regular Languages). *Let L, L' be regular languages over an alphabet Σ. Then the languages $L \cup L'$, LL', L^*, $\overline{L}$, $L \cap L'$, $L - L'$ are all regular languages.*

Proof. It follows from the definition of regular languages that $L \cup L'$, LL', L^* are regular. For complementation, just interchange the final states in a DFA. That is, if $D = (Q, \Sigma, \delta, s, F)$ is a DFA that accepts L, then define the DFA $\overline{D} = (Q, \Sigma, \delta, s, Q - F)$. Here we assume, without loss of generality, (Show it.) that the transition function δ of D is a total function. Let $w \in \Sigma^*$ and $q \in Q$. As the transition function δ is same for both the DFAs, we see that

$$(s, w) \rightsquigarrow_D (q, \varepsilon) \text{ iff } (s, w) \rightsquigarrow_{\overline{D}} (q, \varepsilon).$$

Now, $w \in \overline{L}$ iff $w \notin L$ iff $q \notin F$ iff $q \in Q - F$ iff $w \in \mathcal{L}(\overline{D})$. Therefore, $\overline{L} = \mathcal{L}(\overline{D})$. Thus, $\overline{L}$ is a regular language.

As $L \cap L' = \overline{(\overline{L} \cup \overline{L'})}$, $L \cap L'$ is also regular. Similarly $L - L' = L \cap \overline{L'}$ shows that $L - L'$ is regular. □

Exercise 4.2. What happens if we do not assume the transition function δ in the DFA D to be a total function in the proof of Theorem 4.2? [Hint: Can you have a DFA where each state is a final state but the DFA does not accept every string?]

Example 4.3. Is $L = \{a^m b^n : m \neq 3k + 1 \text{ for any } k \in \mathbb{N}\}$ a regular language?

Solution. As $L_1 = \{a^m b^n : m, n \in \mathbb{N}\} = a^* b^*$, it is regular. We have seen that the language $L_2 = \{w \in \{a, b\}^* : \text{number of } a\text{'s is } 3k + 1 \text{ for some } k \in \mathbb{N}\}$ is regular in Example 4.1. Hence the given language, which equals $L_1 - L_2$, is a regular language. Alternatively, use the language of Example 4.2. □

Exercise 4.3. Show that if L is regular, then so is $L^R = \{u^R : u \in L\}$.

We have shown that intersection of two regular languages is regular by using the closure properties of regular languages with respect to union and complement, which are themselves proved via regular expressions and finite automata. However, there is a more direct approach to showing this using DFAs. It is the approach of construction of a product automaton.

Suppose you have two DFAs M_1 and M_2 that accept the languages L_1 and L_2, respectively. Now, given an input x, imagine feeding it to both the automata simultaneously. M_1, upon reading σ, the first symbol of x starting from its initial state s_1 goes to the state p, say. Similarly, M_2 goes to the state q reading σ starting from its own initial state s_2. For monitoring just this one step of computations of both the automata, you must take into account both the new states p and q. In other words, imagine a new machine M_3, which starts from the state (s_1, s_2) and goes to the state (p, q) upon reading σ. This new machine is able to keep track of the computations of both M_1 and M_2 at least for one step. Now, computation in M_3 may proceed analogously by simulating both the automata simultaneously this way.

Formally, let $M_1 = (Q_1, \Sigma, \delta_1, s_1, F_1)$ and $M_2 = (Q_2, \Sigma, \delta_2, s_2, F_2)$ be two DFAs, where both δ_1 and δ_2 are total functions. The **product DFA** of M_1 and M_2 is the DFA

$$M_1 \times M_2 = (Q_1 \times Q_2, \Sigma, \delta, (s_1, s_2), F_1 \times F_2),$$

where $\delta : (Q_1 \times Q_2) \times \Sigma \to Q_1 \times Q_2$ with $\delta((p, q), \sigma) = (\delta_1(p, \sigma), \delta_2(q, \sigma))$.

Exercise 4.4. Show that $\mathcal{L}(M_1 \times M_2) = \mathcal{L}(M_1) \cap \mathcal{L}(M_2)$. [Hint: You may either follow the configurations-yield approach or δ^*-approach. In the latter case, first show that $\delta^*((p, q), x) = (\delta_1^*(p, x), \delta_2^*(q, x))$ for $x \in \Sigma^*$.]

Besides the set theoretical closures, regular languages admit of rewriting of symbols as other symbols. The notion of rewriting the symbols as some other symbols is formally achieved by defining a homomorphism. In this context, a homomorphism is a concatenation preserving map. Let Σ, Γ be two alphabets. A **homomorphism** is a function $h : \Sigma^* \to \Gamma^*$ that satisfies $h(uv) = h(u)h(v)$ for all strings $u, v \in \Sigma^*$. For example, rewriting every symbol in Σ as a amounts to the homomorphism $h : \Sigma^* \to \{a\}^*$ defined by $h(u) = a^{\ell(u)}$. It is easy to see that h preserves concatenation, as $\ell(uv) = \ell(u) + \ell(v)$.

Let $h : \Sigma^* \to \Gamma^*$ be a homomorphism, $A \subseteq \Sigma^*$, and $B \subseteq \Gamma^*$. The **image** of A under h is the set $h(A) = \{h(u) \in \Gamma^* : u \in A\}$; it is called the **homomorphic image** of A. Similarly, the **preimage of** $B \subseteq \Gamma^*$ under h is the set $h^{-1}(B) = \{u \in \Sigma^* : h(u) \in B\}$; it is called the **homomorphic preimage** of B.

Theorem 4.3. *Homomorphic images and homomorphic preimages of regular languages are regular.*

Proof. (Outline) Suppose E is a regular expression. As $h(E)$ is a regular expression whenever E is, it is enough to show that $\mathcal{L}(h(E)) = h(L(E))$. Use structural induction. That is, in the basis case, if E is either $\emptyset$ or a symbol σ, then see that $\mathcal{L}(h(E)) = h(L(E))$ holds. Next, assuming it to hold for regular expressions E and E', show that it also holds for $E \cup E'$, EE', and E^*. This will prove one part of the theorem that homomorphic images of regular languages are regular.

Next, take a DFA $D = (Q, \Sigma, \delta, s, F)$ accepting L. To complete the proof, we construct a DFA $M = (Q, \Gamma, \delta', s, F)$ for accepting the preimage $h^{-1}(L)$. The idea is to compute h of a given input and then use D on the result. This is realized by taking $\delta'(q, \sigma) = q'$, where $(q, h(\sigma)) \rightsquigarrow_D (q', \varepsilon)$. If you use the extension $\delta^*$of δ for computations in D instead of configurations, then you might write $\delta'(q, \sigma) = \delta^*(q, h(\sigma))$. In either way, you must then show that D accepts w iff M accepts $h(w)$ for any string $w \in \Sigma^*$. This can be done by induction on $\ell(w)$. □

Theorem 4.3 is also stated as: Regular languages are closed under *homomorphism* and *inverse homomorphism.*

If $h : \Sigma_1^* \to \Sigma_2^*$ is a homomorphism and $L_1 \subseteq \Sigma_1^*$, $L_2 \subseteq \Sigma_2^*$ are regular languages, then Theorem 4.3 says that both $h(L_1)$ and $h^{-1}(L_2)$ are regular. This does not guarantee that if $h(L) \subseteq \Sigma_2^*$ is regular for some language $L \subseteq \Sigma_1^*$, then L is regular. For example, let h be the map that rewrites both a, b as a. That is, $h : \{a, b\}^* \to \{a\}^*$ is defined as $h(u) = a^{\ell(u)}$. It is easy to verify that h is a homomorphism.

In other words, Theorem 4.3 does not say that A is regular whenever $h(A)$ is regular; it is false. Let $L = \{a^n b^n : n \in \mathbb{N}\}$. Then $L' = h(L) = \{a^{2n} : n \in \mathbb{N}\}$ is a regular language, but L is not. We will show, in Sect. 4.3, that this L is not a regular language. However, what Theorem 4.3 guarantees is that the preimage of L', that is, $h^{-1}(L') = \{u \in \{a, b\}^* : \ell(u) \text{ is even }\}$ is a regular language.

Problems for Section 4.2

4.1. Find a regular expression for each of the following languages:
(a) $\{w : \#_a(w) \text{ and } \#_b(w) \text{ are both even}\}$.
(b) $\{w : (\#_a(w) - \#_b(w)) \mod 3 = 1\}$.
(c) $\{w : (\#_a(w) - \#_b(w)) \mod 3 \neq 0\}$.
(d) $\{w : (2 \cdot \#_a(w) + 3 \cdot \#_b(w)) \text{ is even}\}$.

4.2. Without using Theorem 4.2, show that if L is regular, so is $L - \{\varepsilon\}$.

4.3. Give a construction of a product automaton for proving that union of two regular languages is regular.

4.4. For each pair of DFAs $M = (Q, \{a, b\}, \delta, p, F)$, $M' = (Q, \{a, b\}, \delta', p, F')$ given below, construct product automata accepting (i) $\mathcal{L}(M) \cap \mathcal{L}(M')$ and (ii) $\mathcal{L}(M) \cup \mathcal{L}(M')$.

(a) $Q = \{p, q\}$, $F = \{q\} = F'$, $\delta(p, a) = \delta(p, b) = q$, $\delta(q, a) = \delta(q, b) = p$, and $\delta'(p, a) = \delta'(q, b) = p$, $\delta'(q, a) = \delta'(p, b) = q$.

(b) $Q = \{p, q, r\}$, $F = \{q, r\}$, $F' = \{p, r\}$, $\delta(p, a) = \delta(r, b) = p$, $\delta(q, a) = \delta(p, b) = r$, $\delta(r, a) = \delta(q, b) = p$, and $\delta'(p, a) = \delta'(q, b) = r$, $\delta'(p, b) = \delta'(r, a) = q$, $\delta'(q, a) = \delta'(r, b) = p$.

(c) $Q = \{p, q\}$, $F = \{q\} = F'$, $\delta(p, a) = \delta(p, b) = q$, $\delta(q, a) = \delta(q, b) = p$, and $\delta'(p, a) = \delta(q, b) = q$, $\delta'(q, a) = \delta'(p, b) = p$.

4.5. Show that the following languages are regular:

(a) $L = \{vwv : v, w \in \{a, b\}^*, \ell(v) = 2\}$.

(b) $L = \{a^n : n \geq 4\}$.

(c) $L = \{a^n : n \in \mathbb{N}, n \neq 4\}$.

(d) $L = \{a^n : n = i + jk,\ i, k \text{ fixed},\ j \in \mathbb{N}\}$.

(e) The set of all real numbers in the programing language C.

4.6. Without using Theorem 4.2, show that if L is regular, so is $L - \{\varepsilon\}$.

4.7. What happens to the acceptance of languages when we interchange the final and nonfinal states of an NFA?

4.8. Instead of DFAs, if we take NFAs and construct their product, and do similar modifications to the final states, do they accept the union or intersection of the languages?

4.9. Define an operation *tr* (a shorthand for *truncate*) that removes the rightmost symbol of a string. For instance, $tr(abbab) = abba$. Extend the operation to languages by $tr(L) = \{tr(w) : w \in L\}$. Show how to construct a DFA that accepts $tr(L)$ from a DFA that accepts L. Does it prove that if L is regular, then $tr(L)$ is regular?

4.10. Construct a DFA that accepts the language generated by the grammar with productions $S \mapsto abA$, $A \mapsto baB$, $B \mapsto aA|bb$.

4.11. Give a formal proof of Theorem 4.3.

4.12. Show that if L is a regular language, then so are the following:

(a) $\{uv : u \in L, \ell(v) = 2\}$,

(b) $\{uv : u \in L, v \in L^R\}$.

4.13. Let h be a homomorphism. Which of the following are true for all regular languages?

(a) $h(L \cup L') = h(L) \cup h(L')$.

(b) $h(L \cap L') = h(L) \cap h(L')$.

(c) $h(LL') = h(L)h(L')$.

(d) $h(L^*) = (h(L))^*$.

4.14. Show that if the sentence "for all languages L, L', if L and $L \cup L'$ are regular, then L' is regular" were true, then all languages would be regular.

4.3 Nonregular Languages

Is it not obvious that there exist nonregular languages? Well, not to intrigue you, suppose Σ is an alphabet. Any regular expression is a string over $\Gamma = \Sigma \cup \{\cup, \emptyset, *,), (\}$. By Theorem 2.1, there are a denumerable (countably infinite) number of strings over Γ. Consequently, there are only a countable number of regular expressions over Σ. On the other hand, by Theorem 2.1, there are uncountable number of languages over Σ. Hence, there are uncountable number of nonregular languages. However, do we have one such to demonstrate?

Well, is $L = \{a^m b^m : m \in \mathbb{N}\}$ a regular language? How do you construct a regular expression to represent L? Had it been $a^m b^n$, you would have taken $a^* b^*$. It is not at all clear how to get L by a regular expression. It looks we have to go on adding a's on the left and simultaneously adding that many b's on the right. But with the help of a nonterminal, in a regular grammar, we can only add on the right. So, it looks we will fail in every attempt. But it is not a proof that we cannot have a regular grammar for generating L.

What about a DFA or an NFA? How many states should we start with? One state will not accept L any way. (Why?) Suppose we have two states, say, s and q. To accept ε, we must make s a final state. To accept ab, we must have a transition $\delta(s, a) = q$. This is so, because, if we have a transition $\delta(s, a) = s$, then any number of a's can be taken so that an accepted string may not have equal number of a's and b's. Then, we may have $\delta(q, b) = s$ so that the string ab is accepted. To accept $aabb$, these transitions will not be enough. It may need introducing another state. It looks this process will continue for ever. To accept a bigger string, we may have to add another state. But is that a DFA, with infinite number of states?

Suppose we have a DFA with n states. Consider the string $a^{n+1}b^{n+1}$, for our construction might fail there. Now, the DFA changes its state after reading the first a (including the possibility to remain in the same state). By reading a^{n+1}, it must have changed its state $n + 1$ times. But there are only n states in the machine. So? By the pigeon hole principle (See Sect. 1.5.), it must have entered a state at least twice. Suppose the state q_k has been entered (at least) twice during the reading of the input a^{n+1}.

To fix notation, assume that the DFA entered first time the state q_k after reading a^i. It entered second time after consuming the next a^j, and then it consumes $a^{n+1-i-j}b^{n+1}$, finally anyway, it enters a final state, say, f to accept the string $a^{n+1}b^{n+1}$. Then, it is clear that the DFA also accepts $a^i a^j a^j a^{n+1-i-j} b^{n+1}$. How? The extra j-number of a's in the middle will again drive the machine back to q_k and then it follows the same line of computation as earlier for the succeeding string $a^{n+1-i-j}b^{n+1}$. However, this new string $a^i a^j a^j a^{n+1-i-j} b^{n+1} \notin L$. Therefore, the DFA does not accept L.

Do you see what we have done? We have started from the assumption that a certain DFA accepts the language, and then we find that the same DFA does not accept it. That means there cannot be any DFA that accepts the language. The language is not regular.

We have an interesting observation. If a string is accepted by a DFA, then certain other string of possibly bigger length should also be accepted by the DFA. This new bigger (or smaller) string is obtained from the old by *pumping* another smaller string into the old one. Moreover, this pumping can be done as often as we please. However, we must start with a suitable string and the language must be an infinite language.

Theorem 4.4 (Pumping Lemma for Regular Languages). *Let L be an infinite regular language over an alphabet Σ. Then there is an $m \geq 2$ such that any string in L of length more than m can be written as xyz for $x, y, z \in \Sigma^*$, $y \neq \varepsilon$, $\ell(xy) \leq m$ and that $xy^n z \in L$ for every $n \in \mathbb{N}$.*

Proof. The idea of the proof is contained in the discussion above about determining whether $\{a^m b^m : m \in \mathbb{N}\}$ is regular or not. It is a matter of formalization only. Why don't you try it yourself? Here it is.

Let L be a regular language. Let $D = (Q, \Sigma, \delta, s, F)$ be a DFA that accepts L. (See Theorem 4.1.) Let $m = |Q|$, the number of states in D, if D has already more than one state, else, take $m = 2$. Let $u \in L$ be of length more than m.

Writing explicitly, let $u = \sigma_1\sigma_2 \ldots \sigma_{m+k}$, $k \geq 1$. The automaton D undergoes a sequence of change of states while reading u. Let the sequence of states be $q_0, q_1, \ldots, q_{m+k}$. That is, D starts from q_0, reads σ_1, then changes to q_1, and so on. Then, $q_0 = s$, $q_{m+k} \in F$ and $\delta(q_i, \sigma_i) = q_{i+1}$ for $0 \leq i \leq m + k - 1$. As there are only m states in Q, by pigeon hole principle, some state appears at least twice in this sequence of states. Moreover, a repetition of states occurs within reading of the first m symbols of u. To fix notation, let $q_j = q_l$ for $0 \leq j < l < m$. Then take

$$x = \sigma_1\sigma_2 \ldots \sigma_j,\ y = \sigma_{j+1} \ldots \sigma_l,\ z = \sigma_{l+1} \ldots \sigma_{m+k},\ j + l \leq m.$$

Observe that this notation allows x to be the empty string. This happens when $j = 0$, that is, when the state $q_0 = s$ is repeated. Writing $u = xyz$ says that

$$(q_0, x) \rightsquigarrow (q_j, \varepsilon),\ (q_j, y) \rightsquigarrow (q_l, \varepsilon) \text{ and } (q_l, z) \rightsquigarrow (q_{m+k}, \varepsilon).$$

We must see that the string $xy^n z$ is also accepted by D for any $n \in \mathbb{N}$. As $q_0 = s$ is the start state of D, we show that $(q_0, xy^n z) \rightsquigarrow (q_{m+k}, \varepsilon)$. Here is how a computation of D goes with the string $xy^n z$. For $n = 0$ or 1 this is obvious. For $n \geq 2$, we have the following computation:

$$\begin{aligned}
&(q_0, xy^n z)\\
&= (q_0, \sigma_1 \ldots \sigma_j(\sigma_{j+1} \ldots \sigma_l)^n \sigma_{l+1} \ldots \sigma_{m+k})\\
&\rightsquigarrow (q_j, (\sigma_{j+1} \ldots \sigma_l)^n \sigma_{l+1} \ldots \sigma_{m+k})\\
&\rightsquigarrow (q_l, (\sigma_{j+1} \ldots \sigma_l)^{n-1} \sigma_{l+1} \ldots \sigma_{m+k})\\
&= (q_j, (\sigma_{j+1} \ldots \sigma_l)^{n-1} \sigma_{l+1} \ldots \sigma_{m+k})\\
&\rightsquigarrow (q_j, \sigma_{l+1} \ldots \sigma_{m+k}) \text{ (here } \rightsquigarrow \text{ in } n - 1 \text{ steps)}\\
&\rightsquigarrow (q_{m+k}, \varepsilon).
\end{aligned}$$

As $q_{m+k} \in F$, the string $xy^n z$ is accepted by D, and thus this string is in L. It remains to show that we can afford to have the length of the pumped string y bounded by m. This requires a special choice in the repeated state.

Choose the repeated state, say, q in such a way that no other state is repeated within these two occurrences of q; also q is not repeated any more within these two occurrences.

Obviously such a choice can always be made. With this choice of q, we see that the length of the string y has to be less than or equal to the total number of states. This completes the proof. □

Exercise 4.5. Write the proof of Theorem 4.4 as a proof by induction.

The proof of the pumping lemma can be summarized easily. It says that once the prefix x of u has been read by D, the DFA enters a certain state, say, q. Then it reads the nonempty substring y and finishing this portion, it reenters the state q. Thus any number of copies of y can be pumped into the string, and then D would accept all such new strings.

Observe that the number m depends upon the regular language L as the pumping lemma shows. The proof uses the number of states in a DFA accepting L as this m. This m then is not unique, for you can always construct a DFA with more number of states to accept the same language. However, there might be a minimum of such m, which gives a hint that perhaps a DFA of minimum number of states can be constructed to accept a given regular language. This problem is addressed under the banner of *minimization of states*.

Example 4.4. Use pumping lemma to show that $L = \{a^k b^k : k \in \mathbb{N}\}$ is not a regular language.

Solution. Let L be a regular language and $m \geq 2$ be as in the pumping lemma. Let $u = a^{m+1}b^{m+1} \in L$. Then $u = xyz,\ y \neq \varepsilon$. What can be y? As $\ell(xy) \leq m$, and xy is a prefix of u, y consists of only a's. That is, $y = a^j,\ x = a^i$, and $z = a^{m+1-j-i}b^{m+1}$ for some $j,\ i$. Now,

$$xy^2z = a^i a^{2j} a^{m+1-j-i} b^{m+1} = a^{m+j+1} b^{m+1} \notin L.$$

As pumping lemma is violated, L is not a regular language. □

Sometimes it is easier to appeal to the proof of the pumping lemma rather than its statement. There are thus many strengthening of the pumping lemma; some of them are in the exercises at the end of the chapter. However, I would suggest you to apply the proof method itself rather than any strengthening of pumping lemma to individual problems. With a good choice of the string xyz, pumping lemma makes the things easy. See the following example.

Example 4.5. Show that $L = \{w \in \{a, b\}^* : w^R = w\}$, the set of all palindromes over $\{a, b\}$, is not a regular language.

Solution. Assume that L is regular. Let m be the number provided by the pumping lemma. Choose $w = a^m b a^m$. If $w = xyz,\ y \neq \varepsilon,\ \ell(xy) \leq m$, then $y = a^k$ for some $k \geq 1$. Now, the string xy^2z is not a palindrome. This violates the pumping lemma, and therefore, L is not regular. □

Example 4.6. Show that $L = \{ww^R : w \in \{a, b\}^*\}$ is not regular.

Solution. If L is regular, let D be a DFA having $m > 1$ (Why?) states that accepts L. The string $a^{m+1}bba^{m+1} \in L$. The proof of the pumping lemma says that some nonempty substring of a^{m+1} can be pumped into the string and the new string will also be accepted by D. In that case, if the substring is a^k for some $k \geq 1$, then the string $a^{m+1+k}bba^{m+1}$ must also be accepted by D. But this string is not in L. This contradiction shows that L is not regular. □

Sometimes, the closure properties (Theorem 4.2) also help in proving that certain languages are not regular. This is only a method of reduction, where you must know beforehand that certain other languages are not regular.

Example 4.7. Is $L = \{u \in \{a, b\}^* : \#_a(u) = \#_b(u)\}$ regular?

Solution. Suppose that L is regular. As a^*b^* is regular, $L_1 = L \cap a^*b^*$ must also be regular by Theorem 4.2. But $L_1 = \{a^m b^m : m \in \mathbb{N}\}$ is not regular. Hence L is not a regular language. □

Example 4.8. Show that $\{a^m b^n : m \leq n\}$ is not a regular language.

Solution. You can apply the pumping lemma directly. Here is an alternative. If L is regular, by Exercise 4.3, its reversal L^R is also regular. Then there is a regular grammar G that generates L^R. Construct a grammar G' from G by interchanging the terminal symbols a and b. The new grammar G' is also a regular grammar. Then, $\mathcal{L}(G') = L' = \{a^n b^m : m \leq n\}$ is a regular language. As $L \cap L' = \{a^k b^k : k \in \mathbb{N}\}$, it must also be regular. But it is not; therefore, L is not regular. □

Exercise 4.6. Show that $\{a^m b^n : m \geq n\}$ is not a regular language.

Example 4.9. Show that $L = \{a^p : p \text{ is a prime number}\}$ is not regular.

Solution. Assume that L is regular. Let m be the number for L as given by the pumping lemma. Let $u \in L$ be of length more than m. Then $u = a^i a^j a^k$, where $0 \leq i \leq m$, $1 \leq j \leq m$, $i + j \leq m$ and that $a^i a^{jn} a^k \in L$ for every $n \in \mathbb{N}$. This means that $i + j + k$ is a prime number and $i + jn + k$ is also a prime number for every $n \in \mathbb{N}$. However, for $n = i + 2j + k + 2$, we have

$$\begin{aligned} i + jn + k &= i + j(i + 2j + k + 2) + k = i + 2j + k + j(i + 2j + k) \\ &= (i + 2j + k)(j + 1), \end{aligned}$$

which is not a prime number as both $j + 1 \geq 2$ and $i + 2j + k \geq 2$. Hence L is not a regular language. □

Exercise 4.7. Show that $\{a^{n^2} : n \in \mathbb{N}\}$ is not a regular language.

Example 4.10. Show that $L = \{a^{n!} : n \in \mathbb{N}\}$ is not a regular language.

Solution. Assume that L is a regular language. Let m be the number for L provided by the pumping lemma. Let $u = a^{(m+3)!}$ so that $\ell(u) > m$. Then $u = a^p a^q a^r$, where $0 \leq p$, $1 \leq q$, $0 \leq r$, $p + q + r = (m + 3)!$. Now,

$$(m + 3)! = p + q + r < p + 2q + r \leq 2(m + 3)! < (m + 4)!.$$

Hence $p + 2q + r \neq (m + k)!$ for any $k \in \mathbb{N}$, contradicting the pumping lemma. Therefore, L is not regular. □

Application of the pumping lemma is viewed as a *game with the demon.* Given a language $L \subseteq \Sigma^*$, you want to show that it is not regular while the demon claims that it is regular. The demon chooses m. You choose $w \in L$ such that $\ell(w) > m$. The demon picks strings $x, y, z \in \Sigma^*$ in such a way that $w = xyz$ and $y \neq \varepsilon$. You choose $n \in \mathbb{N}$. Now, you win if $xy^nz \notin L$, and the demon wins otherwise. Here, of course, we assume that each of you and the demon makes the best choice possible. That means, it is not enough for you to win, but to have a strategy to win.

For instance, in Example 4.10 above, the demon might choose any m. Then, you choose w as $a^{(m+3)!}$. It is a strategical choice and not a particular one. Next, the demon chooses $x = a^p, y = a^q, z = a^r$ with $q > 0$. You choose $n = 2$. As $xy^2z \notin L$, you have own; and the language is not regular.

For regular languages over a single-letter alphabet (say, $\{a\}$), the pumping lemma can be strengthened in the sense that the suffix z in the rewriting of u can be chosen to be ε. This is so, because a DFA that accepts such a language has a diagram with the property that from any state only one arrow can go to another state (if at all any arrow goes from it). Now just follow the arrows from the initial state successively. If there is a string of length more than the number of states, then a state has to repeat. Take the first such occurrence of a repetition. This is actually the second time occurrence of that state. Any string that is accepted must be accepted inside this loop. That is, one of the states inside this loop is a final state.

Write a^i as the string that corresponds to the sequence of states before the first occurrence of this state, the string a^j for the loop, and the string a^k from the first occurrence of the repeated state to one such final state. Then the string a^{i+k} is accepted without completing a loop. But when we go through the loop, say, n times, the string a^{i+nj+k} is also accepted. This amounts to asserting that for any such language L, if for some i, $a^m \in L$ for any $m \geq i$ (take $m = i + k$), then we have also a j such that $a^{m+nj} \in L$ for all $n \in \mathbb{N}$.

Surprisingly, its converse also holds. That is, for a language over a single-letter alphabet, if this property holds, then the language must be regular. To show this, you just start with the given numbers i, j, k with $k \leq j$, and then construct a DFA with a cycle after joining $(i + 1)$th state to itself, which goes through j-states with $(i + k)$th state as a final state.

This result is put formally by defining an ultimately periodic subset of $\mathbb{N}$. A subset $P \subseteq \mathbb{N}$ is called **ultimately periodic**, provided there exist natural numbers $n \geq 0$, $k \geq 1$ such that for all $m \geq n$, we have $m \in P$ iff $m + k \in P$. For example, the set of all even natural numbers is ultimately periodic, as, with $n = 0, k = 2$, we see that for each natural number m, m is even iff $m + 2$ is even. We have thus proved that

Theorem 4.5. *$L \subseteq a^*$ is regular iff $\{\ell(u) : u \in L\}$ is ultimately periodic.*

What happens if a language is not over a one-letter alphabet? Can Theorem 4.5 help in getting some idea about the set of lengths of its strings? Suppose L is any regular language and we have a DFA D that accepts it. If you rename every symbol in the alphabet of L as a, then D would become an NFA. The language L' accepted by this NFA is regular. Moreover, L' is simply obtained from L by rewriting every symbol as a, which preserves the lengths of strings in L. Therefore, the set of lengths of strings of L and that of L' coincide. However, L' is regular; so, this set is ultimately periodic. We have thus proved the following result:

Theorem 4.6. *If L is a regular language, then $\{\ell(u) : u \in L\}$ is ultimately periodic.*

Problems for Section 4.3

4.15. Are the following true? Prove or give a counter example.
(a) Each subset of a regular language is regular.
(b) Each regular language has a regular proper subset.
(c) If L is regular, then so is $\{xy : x \in L, y \notin L\}$.
(d) If R is any set of regular languages, then $\cup R$ is also regular.
(e) The language $\{wu^R : u, w \in \Sigma^*\}$ is regular.

4.16. Show that there is no DFA that accepts all (and only) palindromes over $\{a, b\}$.

4.17. Let D be the transition diagram of a DFA M. Prove the following:
(a) If $\mathcal{L}(M)$ is infinite, then D must have at least one cycle for which there is a path from the initial vertex to some vertex in the cycle, and a path from some vertex in the cycle to some final vertex.
(b) If $\mathcal{L}(M)$ is finite, then there exists no such cycle in D.

4.18. Show that the following languages over $\{a, b\}$ are not regular:
(a) $\{a^n b a^n : n \geq 1\}$.
(b) $\{a^n : n$ is a perfect cube $\}$.
(c) $\{w : \#_a(w) < \#_b(w)\}$.
(d) $\{wb^n : w \in \{a, b\}^*, \ell(w) = n\}$.
(e) $\{(ab)^m b^n : m > n \geq 0\}$.
(f) $\{a^m b^n : m \neq n, m.n \in \mathbb{N}\}$.
(g) $\{a^m b^n a^k : k \geq m + n\}$.
(h) $\{a^n b^{2n} : n \geq 1\}$.
(i) $\{a^m b^n : 0 < m < n\}$.
(j) $\{a^m b^n a^k : k \neq m + n\}$.
(k) $\{a^m b^n a^k : m = n$ or $n \neq k\}$.
(l) $\{a^m b^n : m \leq n\}$.
(m) $\{w \in \{a, b\}^* : \#_a(w) \neq \#_b(w)\}$.
(n) $\{ww : w \in \{a, b\}^*\}$.
(o) $\{w^R w : w \in \{a, b\}^*\}$.
(p) $\{wwww^R : w \in \{a, b\}^*\}$.
(q) $\{a^m b^n : m > n\} \cup \{a^m b^n : m + 1 \neq n\}$.
(r) $\{uww^R v : u, v, w \in \{a, b\}^+\}$.
(s) $\{ww^n v : v, w \in \{a, b\}^+, n \geq 1\}$.
(t) $\{w\overline{w} : w \in \{a, b\}^*\}$, where $\overline{w}$ is the string obtained from w by changing a to b, and b to a simultaneously.

4.19. Are the following languages over $\{a\}$ regular?
(a) $\{a^n : n = m^2 \text{ for some } m \in \mathbb{N}\}$.
(b) $\{a^n : n = 2^m \text{ for some } m \in \mathbb{N}\}$.
(c) $\{a^{p-1} : p \text{ is a prime number}\}$.
(d) $\{a^{mk} : m \text{ and } k \text{ are prime numbers}\}$.
(e) $\{a^n : n \text{ is either a prime or a product of two or more primes}\}$.

4.20. Are the following languages over $\{a, b\}$ regular?
(a) $L = \{w \in \{a, b\}^* : \#_a(w) = \#_b(w)\}$.
(b) $\{a^m b^{2n} : m, n \in \mathbb{N}\}$.
(c) $\{a^m b^n : m = 2n; m, n \in \mathbb{N}\}$.
(d) $\{a^m b^n : m \neq n\}$.
(f) $\{a^m b^{m+2009} : m \in \mathbb{N}\}$.
(g) $\{a^m b^n : m - n \leq 2009\}$.
(h) $\{a^m b^n : m \geq n \text{ and } n \leq 2009\}$.
(i) $\{a^m b^n : m \geq n \text{ and } n \geq 2009\}$.
(j) $\{(a^m b)^n a^k : m, n, k \in \mathbb{N}\}$.
(k) $\{w : \#_a(w) = 2 \cdot \#_b(w)\}$.
(l) $\{w : \#a(w) - \#b(w) < 9\}$.
(m) $\{w : \#a(w) \cdot \#b(w) \text{ is even}\}$.
(n) $\{a^m b^n : m \text{ is even or } m > n\}$.
(o) Any infinite subset of $\{a^m b^m : m \in \mathbb{N}\}$.
(p) $\{ab(bbaa)^n bba(ba)^n : n \in \mathbb{N}\}$.
(q) The set of all syntactically correct C programs.

4.21. Are the following languages over $\{a, b, c\}$ regular?
(a) $\{w : w^R = w\}$.
(b) $\{w \in \{a, b, c\}^* : w = a^m b^n c^{m+n}, m, n \in \mathbb{N}\}$.
(c) $\{ucv : u, v \in \{a, b\}^*, u \neq v\}$.
(d) $\{ww : w \in \{0, 1\}^*\}$.
(e) $\{a^{mn} : m, n \in \mathbb{N}\}$.
(f) $\{wcw : w \in \{a, b\}^*\}$.
(g) $\{vcw : v, w \in \{a, b\}^*\}$.
(h) $\{a^m b^m c^m : m \in \mathbb{N}\}$.
(i) $\{a^m b^n c^k : m, n, k \in \mathbb{N}; m + n = k\}$.

4.22. Let L, L' be any languages over an alphabet Σ. Prove or disprove the following:
(a) If L, L' are nonregular languages, then $L \cup L'$ is nonregular.
(b) If L, L' are regular, then $L'' = \{w : w \in L, w^R \in L'\}$ is regular.
(c) Infinite union of regular languages is regular.
(d) Infinite intersection of regular languages is regular.
(e) If L and $L \cup L'$ are regular, then L' is regular.
(f) If L and LL' are regular, then L' is regular.
(g) If L^* is regular, then L is regular.

4.23. Construct nonregular languages $L_1, L_2 \subseteq \{0, 1\}^*$ such that
(a) L_1^* is regular.
(b) L_2^* is not regular.

4.24. Are the following languages over $\{a, b\}$ regular?
(a) $\{a^m b^n a^k : m + n + k > 6\}$.
(b) $\{a^m b^n a^k : m > 6, n > 3, k \leq n\}$.
(c) $\{a^m b^{mn} : m > 1, n > 1\}$.
(d) $\{a^m b^n : m + n \text{ is a prime number}\}$.
(e) $\{a^m b^n : m \leq n \leq 2n\}$.
(f) $\{a^m b^n : m \geq 2009, n \leq 2009\}$.
(g) $\{a^m b^n : m - n = \pm 2\}$.

4.25. Show that the set of balanced parentheses is not a regular language.

4.26. Formal languages can be used to describe various two-dimensional figures. For example, let $\Sigma = \{u, d, l, r\}$. Interpreting the symbols as drawing a line segment of one unit from the current position in the direction of "up, down, left, right," respectively. A string such as $urdl$ draws a square.
(a) Draw figures corresponding to the expressions $(rd)^*$, $(urddru)^*$, and $(ruldr)^*$.
(b) Find a necessary and sufficient condition on an expression E over Σ so that the figures it represents are closed contours.
(c) Let L be the set of all $w \in \{u, d, l, r\}^*$ that describe rectangles. Show that L is not a regular language.

4.27. Use Theorem 4.3 to derive Theorem 4.6 from Theorem 4.5.

4.28. What about the converse of Theorem 4.6?

4.4 Myhill–Nerode Theorem

Consider the language $L = (10 \cup 101)^*$ of Example 2.20. I have intrigued you by claiming that possibly no simpler DFA can be constructed to accept the same language L than that given in Fig. 1.19. What I mean by "simpler" is that the number of states is smaller. Let us name the states of the DFA as in Fig. 4.3.

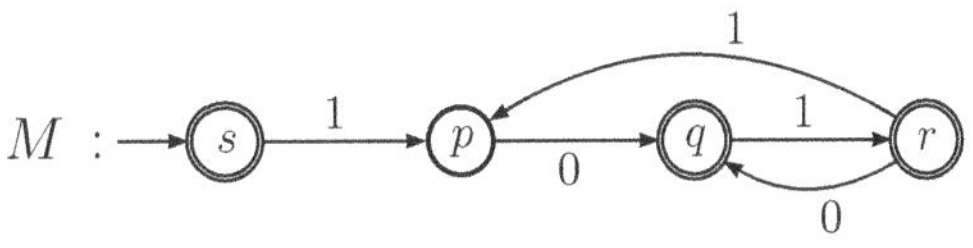

Fig. 4.3. States of Fig. 1.19 named.

The DFA in Fig. 4.3 is $M = (Q, \Sigma, \delta, s, F)$, where $Q = \{s, p, q, r\}$, $\Sigma = \{0, 1\}$, $F = \{s, q, r\}$, and the transition function $\delta : Q \times \Sigma \to Q$ is given by

$$\delta(s, 1) = p, \ \delta(p, 0) = q, \ \delta(q, 1) = r, \ \delta(r, 0) = q, \ \delta(r, 1) = p.$$

What about the DFA $M' = (Q', \Sigma, \delta', s, F')$, where $Q' = Q \cup \{t\}$, $F' = F \cup \{t\}$, and $\delta' = \delta$ on $Q \times \Sigma$ with $\delta'(t, 0) = t$, $\delta'(t, 1) = q$? Look at Fig. 4.4.

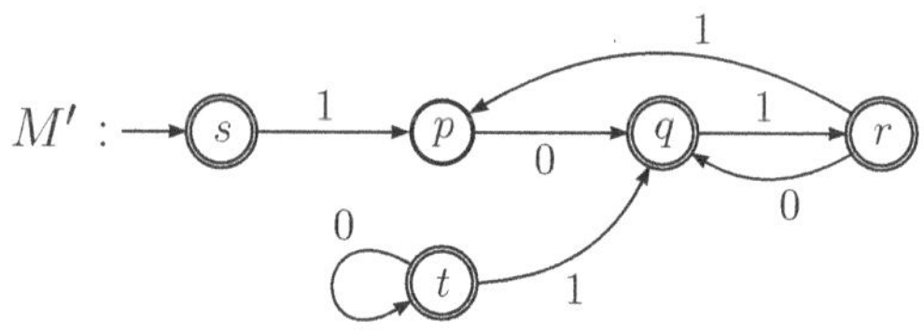

Fig. 4.4. The DFA M'.

Clearly, $\mathcal{L}(M') = \mathcal{L}(M)$. Reason? No string can drive the state s to state t in M'! In the transition diagram you see that there is no path from the initial state to t. In such a case, the state t is one that is *inaccessible* from the initial state.

Formally, we say that a state in a DFA $D = (Q, \Sigma, \delta, s, F)$ is an **accessible state** if there is some string $u \in \Sigma^*$ such that $(s, u) \rightsquigarrow (q, \varepsilon)$ (equivalently, $\delta^*(s, u) = q$). A state that is not accessible is said to be an **inaccessible state**. It is obvious that a DFA can be cleaned up by simply deleting all its inaccessible states and the corresponding transitions from it. The resulting DFA still accepts the same language as the old one. If two automata accept the same language, we say that they are equivalent, as usual. That is, the DFA (or NFA) D_1 **is equivalent to** the DFA (or NFA) D_2 iff $\mathcal{L}(D_1) = \mathcal{L}(D_2)$.

As long as we are interested in the equivalence of DFAs, we assume, without loss of generality, that no DFA has an inaccessible state. Further, we will assume that all our DFAs have transition functions as total functions. That is, $\delta : Q \times \Sigma \to Q$ is a total function. This assumption again does not affect generality, as a *state of no return* can always be added to one where δ is a partial function. For example, to the DFA of Fig. 4.3, we can add another state, say, t with all missing transitions directed to it and a loop on t with labels as all symbols from Σ. For example, see how the DFA M of Fig. 4.5 works on the inputs 1010, 1110, 10101, and 10101101. You find that

$$(s, 1010) \rightsquigarrow (q, \varepsilon),\ (s, 1110) \rightsquigarrow (t, \varepsilon),\ (s, 10101) \rightsquigarrow (r, \varepsilon), (s, 10101101) \rightsquigarrow (r, \varepsilon).$$

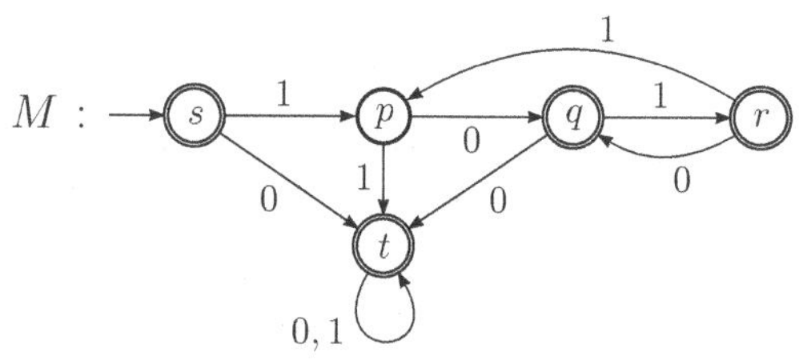

Fig. 4.5. The DFA M.

Notice that the computation of M that ends in the state t would correspond to an abrupt halt in the earlier DFA of Fig. 4.3. Take, for example, the computation with input 1110. In the DFA of Fig. 4.3, we would have an abrupt halt with the configuration $(p, 110)$. The converse also holds, that is, when the earlier DFA halts abruptly, the new DFA ends up with t, the added state of no return.

We are interested in minimizing the states of a DFA having no inaccessible states and where the transition function is a total function. Let us call such DFAs as **simple** DFAs.

This assumption of simplicity helps in presenting the minimization in a rather easier way. The ease comes from the fact that computation in such a DFA with any string in Σ^* will terminate normally leaving the DFA in some state. It suggests that we can possibly categorize the strings in Σ^* by looking at the terminal state of the DFA. For example, the above computation suggests that the strings 10101 and 10101101 are in the same category as they drive the DFA M (of Fig. 4.5) to the state r, starting from the initial state. Notice that the categories or classes depend on the DFA in consideration. We may start with a relation, say, $\simeq_D$, induced by the DFA D on Σ^*.

Let $D = (Q, \Sigma, \delta, s, F)$ be a simple DFA. Let $u, v \in \Sigma^*$ be any strings. We write $u \simeq_D v$ iff $(s, u) \rightsquigarrow_D (t, \varepsilon)$ and $(s, v) \rightsquigarrow_D (t, \varepsilon)$ for the same state $t \in Q$. Equivalently, $u \simeq_D v$ iff $\delta^*(s, u) = \delta^*(s, v)$. The binary relation $\simeq_D$ is said to be the relation induced by D on Σ^*. You may read $u \simeq_D v$ as u is related to v by D.

Whenever, we have only one DFA in question, we will omit the subscript D from the symbol $\simeq_D$ and write it as $\simeq$; similarly, let us abbreviate $\rightsquigarrow_D$ to $\rightsquigarrow$. The relation $\simeq$ is reflexive as δ is a total function. Trivially, $\simeq$ is symmetric. Now, if $(s, u) \rightsquigarrow (t, \varepsilon)$ and $v \simeq w$, then $(s, w) \rightsquigarrow (t, \varepsilon)$ for the same t, and thus $\simeq$ is a transitive relation. We conclude that $\simeq$ is an equivalence relation on Σ^*. We thus say that $\simeq$ that is, $\simeq_D$, is the equivalence relation on Σ^* induced by the simple DFA D.

The relation $\simeq$ then decomposes Σ^* into equivalence classes. A typical equivalence class will have all strings that drive the DFA D from its initial state to some particular state in Q. Thus, there are exactly $|Q|$ number of such equivalence classes, because inaccessible states have already been cleaned up from D. Thus, corresponding to each state, there is at least one string in Σ^* that drives the DFA to this state.

An equivalence relation on Σ^* that has a finite number of equivalence classes is said to be a **relation of finite index**, the index being the number of such equivalence classes. In this terminology, we see that the relation $\simeq$ is of finite index, the index being $|Q|$.

If both $u, v \in \Sigma^*$ are in the same equivalence class, then there is some state, say, $t \in Q$ such that $(s, u) \rightsquigarrow (t, \varepsilon)$ and $(s, v) \rightsquigarrow (t, \varepsilon)$. We will write this equivalence class as $[t]$. We see that

$$\Sigma^* = \cup_{p \in Q}[p], \text{ and } [p] \neq \emptyset \text{ for any } p \in Q.$$

This notation of an equivalence class $[x]$ is unlike the usual notation where x is an element of it. But it is convenient for us.

The relation $\simeq$, moreover, satisfies another elegant property. Let $u, v, w \in \Sigma^*$ with $u \simeq v$. Let $(s, u) \rightsquigarrow (t, \varepsilon)$ and $(s, v) \rightsquigarrow (t, \varepsilon)$ for a state $t \in Q$. Now, $(s, uw) \rightsquigarrow (t, w) \rightsquigarrow (q, \varepsilon)$ for some $q \in Q$. Also, $(s, vw) \rightsquigarrow (t, w) \rightsquigarrow (q, \varepsilon)$. Therefore, $uw \simeq vw$.

A relation R on Σ^* having the property

$$\text{for all } u, v, w \in \Sigma^*, \text{ if } uRv, \text{ then } uwRvw$$

is called a **right invariant** relation. We have shown that $\simeq$ is a right invariant relation.

Using the fact that $\mathcal{L}(D)$ is the set of all strings that drive the DFA D from its initial state to one of the final states, we obtain the following:

Lemma 4.1. *Let $D = (Q, \Sigma, \delta, s, F)$ be a simple DFA. Then the relation $\simeq$ on Σ^* induced by D is a right invariant equivalence relation of finite index. It decomposes Σ^* into $|Q|$ number of equivalence classes; that is, index of $\simeq$ is $|Q|$ and $\Sigma^* = \cup_{q \in Q}[q]$. Moreover, $\mathcal{L}(D) = \cup_{t \in F}[t]$.*

Each equivalence class $[t]$ is regular as the DFA $D_t = (Q, \Sigma, \delta, s, \{t\})$ accepts it. You can now view Lemma 4.1 in a different way. If L is a regular language, then there is a DFA D such that $\mathcal{L}(D) = L$. This D gives rise to a right invariant equivalence relation of finite index, such as $\simeq_D$, where each equivalence class is a regular language. Remarkably, the converse of this statement also holds.

Lemma 4.2. *If $\sim$ is a right invariant equivalence relation of finite index on Σ^*, then each equivalence class of $\sim$ is a regular language.*

Proof. Let $\sim$ be a right invariant equivalence relation of index n on Σ^*. It decomposes Σ^* into n equivalence classes, say, $C_1, C_2, \ldots, C_n$. For notational convenience, assume that C_1 is the equivalence class containing the empty string ε. Write $C_i w = \{uw : u \in C_i\}$ for any $w \in \Sigma^*$ and $1 \leq i \leq n$. We first show the following statements:

1. For every $w \in \Sigma^*$ and $1 \leq i \leq n$, $w \in C_i$ iff $C_1 w \subseteq C_i$.
2. For every $w \in \Sigma^*$ and $1 \leq i \leq n$, there is a unique j with $1 \leq j \leq n$ such that $C_i w \subseteq C_j$. Moreover, $C_i w \cap C_k = \emptyset$ for $j \neq k$.

For (1), if $C_1 w \subseteq C_i$, then $w = \varepsilon w \in C_i$ as $\varepsilon \in C_1$. Conversely, suppose $w \in C_i$. Let $u \in C_1$. As $\varepsilon \in C_1$ and C_1 is an equivalence class of $\sim$, $u \sim \varepsilon$. Right invariance of $\sim$ says that $uw \sim \varepsilon w = w$. As $w \in C_i$, we see that $uw \in C_i$. As u is an arbitrary element of C_1, we conclude that $C_1 w \subseteq C_i$.

For (2), let $w \in \Sigma^*$ and $1 \leq i \leq n$. Let $u, v \in C_i$. As $u \sim v$, by right invariance of $\sim$, $uw \sim vw$. That is, all elements of the form xw belong to the same equivalence class whenever $x \in C_i$. This shows that $C_i w \subseteq C_j$ for some j. Moreover, if some $uw \in C_k$ and $vw \in C_j$, then $uw \sim vw$ would imply that $C_j = C_k$. This proves (2).

We want to show that each equivalence class C_i is a regular language. To this end, define DFAs $D_i = (\{q_1, \ldots, q_n\}, \Sigma, \delta_i, q_1, \{q_i\})$, where the function $\delta_i : Q \times \Sigma \to Q$ is given by

$$\delta_i(q_i, \sigma) = q_j \text{ iff } C_i \sigma \subseteq C_j.$$

The function δ_i is well defined due to (2). If $\delta_i(q_1, \sigma) = q_j$ and $\delta_i(q_j, \tau) = q_k$, then $(q_1, \sigma\tau) \rightsquigarrow (q_j, \tau) \rightsquigarrow (q_k, \varepsilon)$. This shows, by induction, that

3. $(q_1, u) \rightsquigarrow (q_j, \varepsilon)$ iff $C_1 u \subseteq C_j$.

Notice that statement (3) can also be rephrased as $\delta_i^*(q_i, u) = q_j$ iff $C_i u \subseteq C_j$.

We show that $\mathcal{L}(D_i) = C_i$ for $1 \le i \le n$. So, let $u \in C_i$. As $\varepsilon \in C_1$, $u \in C_1 u$. By (2), we have $C_1 u \subseteq C_i$. By (3), we see that $(q_1, u) \rightsquigarrow (q_i, \varepsilon)$. (i.e., $\delta_i^*(q_1, u) = q_i$). As q_i is a final state, $u \in \mathcal{L}(D_i)$. Conversely, let $u \in \mathcal{L}(D_i)$. Then $(q_1, u) \rightsquigarrow (q_i, \varepsilon)$ (i.e., $\delta_i^*(q_1, u) = q_i$). By (3), $C_1 u \subseteq C_i$. By (1), $u \in C_i$. □

Lemma 4.2 was essentially due to J. Myhill and A. Nerode. We combine Lemmas 4.1 and 4.2 and summarize the above discussion as follows:

Theorem 4.7 (Myhill–Nerode Theorem). *A language $L \subseteq \Sigma^*$ is regular iff L is a union of some of the equivalence classes of a right invariant equivalence relation of finite index on Σ^*.*

Myhill–Nerode theorem can be used to show that certain languages are not regular; see the following example:

Example 4.11. Use Myhill–Nerode theorem to show that $L = \{a^n b^n : n \in \mathbb{N}\}$ is not a regular language over $\{a, b\}$.

Solution. Let L be regular. By Myhill–Nerode theorem, we have a right invariant equivalence relation $\sim$ on $\{a, b\}^*$ of finite index such that L is the union of some of its equivalence classes. The set $A = \{a^n : n \in \mathbb{Z}_+\} \subseteq \{a, b\}^*$ is an infinite set. Thus, there are at least two strings in A, say, a^i, a^j, which are in the same equivalence class. That is, $a^i \sim a^i$ but $i \ne j$. As $\sim$ is right invariant, $a^i b^i \sim a^j b^i$. L is the union of some of the equivalence classes of $\sim$; thus, once a string is in L, all strings equivalent to it (i.e., related by $\sim$) must also be in L. As $a^i b^i \in L$ we see that $a^j b^i \in L$. But this is a contradiction as $i \ne j$. Therefore, L is not a regular language. □

Problems for Section 4.4

4.29. Design a DFA with fewer states equivalent to one with initial state p, final states q, r, s, t, and having transitions:
$\delta(p, 0) = q$, $\delta(p, 1) = t$, $\delta(q, 0) = \delta(r, 0) = r$, $\delta(q, 1) = \delta(r, 1) = s$, $\delta(s, 0) = u$, $\delta(s, 1) = \delta(t, 0) = \delta(t, 1) = \delta(u, 1) = u$, $\delta(u, 0) = \delta(v, 0) = \delta(v, 1) = v$.

4.30. Construct a DFA with fewer states that accepts the same language as the DFA $D = (\{p, q, r, s, t\}, \{0, 1\}, \delta, p, \{s, t\})$ with $\delta(p, 0) = q$, $\delta(p, 1) = \delta(q, 0) = \delta(r, 0) = r$, $\delta(q, 1) = \delta(s, 0) = \delta(s, 1) = s$, $\delta(r, 1) = \delta(t, 0) = \delta(t, 1) = t$.

4.31. Let $M = (Q, \Sigma, \delta, s, F)$ be a DFA. Define a relation R on Q by pRq iff for some $\sigma \in \Sigma$, $\delta(p, \sigma) = \delta(q, \sigma)$. Is R an equivalence relation?

4.32. Show that if a language has a string of minimum length n, then any DFA that accepts this language must have at least $n + 1$ states.

4.5 State Minimization

Myhill–Nerode theorem can be used to prove the existence of a minimal DFA, one with minimum number of states. Let L be a regular language over an alphabet Σ. We then have a right invariant equivalence relation $\sim$ on Σ^* of finite index such that L is the union of some of the equivalence classes of $\sim$. Let $C_1, C_2, \ldots, C_n$ be the equivalence classes of $\sim$. Notice that our construction of such a relation was done through a DFA accepting L. In that case, the relation $\sim$ is simply the relation induced by the DFA, and n is the number of states in the DFA. However, there is no guarantee that any such relation having the properties in Myhill–Nerode Theorem must be one associated with a DFA. Is there?

We will try to find another relation of finite index looking at the perspective of choosing one where the index is minimum. Later, we will connect this minimum index to a DFA. The idea is to look at the definition of right invariance. It demands only a conditional statement such as

$$\text{if } uRv, \text{ then } uwRvw, \text{ for all strings } w \in \Sigma^*$$

to be satisfied so that R is right invariant. The trick is to demand a biconditional statement. We thus define the **Myhill–Nerode relation** ρ_L on Σ^* by

$$\text{for any } u, v \in \Sigma^*, \; u\,\rho_L\,v \text{ iff (for all } w \in \Sigma^*, \; uw \in L \text{ iff } vw \in L).$$

We will write ρ_L as ρ omitting the subscript L if the language L is clear from the context. Our aim is to show that such a relation would generate minimum number of equivalence classes among all those given in Theorem 4.7.

First of all, is ρ an equivalence relation? Trivially, for all $w \in \Sigma^*$, $uw \in L$ iff $uw \in L$, showing that ρ is reflexive. Symmetry of ρ is also equally trivial. If for all $w \in \Sigma^*$, $uw \in L$ iff $vw \in L$, and for all $w \in \Sigma^*$, $vw \in L$ iff $zw \in L$, then clearly, for all $w \in \Sigma^*$, $uw \in L$ iff $zw \in L$. Thus, ρ is transitive. Therefore, ρ is an equivalence relation. Further, ρ is right invariant from its very definition. In general, ρ need not be of finite index. But when L is a regular language, ρ (i.e., ρ_L) is a relation of finite index; this is shown in the following statement:

Lemma 4.3. *Let $D = (Q, \Sigma, \delta, s, F)$ be a simple DFA and $L = \mathcal{L}(D)$. Then, $\simeq_D \subseteq \rho_L$, that is, if $u \simeq_D v$, then $u\,\rho_L\,v$. Thus, ρ_L is of index less than or equal to $|Q|$.*

Proof. Assume that $u \simeq_D v$ for strings $u, v \in \Sigma^*$. We show that $u\,\rho_L\,v$. Let $w \in \Sigma^*$ be any string. We want to show that $uw \in L$ iff $vw \in L$.

Suppose $uw \in L$. Then $(s, uw) \rightsquigarrow_D (t, \varepsilon)$ for some $t \in F$. As $u \simeq_D v$ and $\simeq_D$ is right invariant, we have $(s, vw) \rightsquigarrow_D (t, \varepsilon)$. This says that $vw \in L$. Conversely, if $vw \in L$, then $(s, vw) \rightsquigarrow_D (t, \varepsilon)$ for some $t \in F$. Again, owing to the right invariance of $\simeq_D$ and $u \simeq_D v$ (thus $v \simeq_D u$), we have $(s, uw) \rightsquigarrow_D (t, \varepsilon)$. This proves that $uw \in L$. This completes the proof of the fact that $\simeq_D \subseteq \rho_L$.

But both $\simeq_D$ and ρ_L are equivalence relations. Thus any equivalence class of ρ_L is a union of some of the equivalence classes of $\simeq_D$. As L is regular, by Myhill–Nerode theorem, index of $\simeq_D$ is finite. Therefore, index of ρ_L is finite and is less than or equal to the index of $\simeq_D$, which is equal to $|Q|$. □

The question is whether there exists a DFA M such that $\rho_L = \simeq_M$? If such a thing happens, then M would be one DFA having minimum number of states. The answer is in the affirmative as the following theorem shows.

Theorem 4.8. *Let L be a regular language over an alphabet Σ. Then there exists a simple DFA M such that M has the minimum number of states among all simple DFAs accepting L.*

Proof. From Lemma 4.3, we only show that there is a DFA M such that $\simeq_M = \rho_L$. Let $C_1, C_2, \ldots, C_m$ be the equivalence classes of ρ_L. Using Myhill–Nerode theorem, we write

$$L = C_{i1} \cup \cdots \cup C_{ik} \text{ for some } k \leq m,\ 1 \leq i1 \leq \cdots \leq ik \leq m.$$

Define $M = (Q, \Sigma, \delta, q_1, F)$, where $Q = \{q_1, \ldots, q_m\}$, $F = \{q_{i1}, \ldots, q_{ik}\}$, and $\delta : Q \times \Sigma \to Q$ is given by $\delta(q_j, \sigma) = q_l$ iff $C_j\sigma \subseteq C_l$. Here, $C_j\sigma = \{x\sigma : x \in C_j\}$.

It is now obvious that the relation $\simeq_M$ induced by M on Σ^* is the same as ρ_L. Therefore, M is a minimal DFA accepting L. □

Though Theorem 4.8 guarantees the existence of a minimal DFA, the construction of the minimal DFA seems useless. A better way is to start from a DFA and then choose or delete some of the states and modify the transitions so that a minimal DFA may be obtained. To choose the states that may be kept (and others deleted), we will define another equivalence relation on the set of states. Our plan is to keep only one state out of an equivalence class of states so that the resulting DFA might accept the same language. Our motivation is to identify two states when the DFA starting from either of them ends in a final state, whatever string we may feed to it. This approach leads to quotient automata, which we introduce next via another equivalence relation. See the following definition.

Let $D = (Q, \Sigma, \delta, s, F)$ be a simple DFA. For any states $p, q \in Q$ write $p \approx_D q$ iff (for every $u \in \Sigma^*$, $(p, u) \rightsquigarrow_D (r, \varepsilon)$ for some $r \in F$ iff $(q, u) \rightsquigarrow_D (t, \varepsilon)$ for some $t \in F$). The relation $\approx_D$ on Q is called the relation of **state equivalence**. If $p \approx_D q$, we say that p is state equivalent to q. We will write $\approx_D$ simply as $\approx$ if no confusion arises.

Exercise 4.8. Show that $\approx_D$ is an equivalence relation on Q and that if $p \approx_D q$, then $\delta(p, \sigma) = \delta(q, \sigma)$ for any $\sigma \in \Sigma$. Further, show that $p \approx_D q$ iff for all $u \in \Sigma^*$, $\delta^*(p, u) \in F$ exactly when $\delta^*(q, u) \in F$.

We will employ state equivalence to define the so-called **quotient automaton**, which would serve our purpose. The trick is to identify all equivalent states as one state and then redefine the transitions using those of the old DFA. To this end, let Q' be the set of equivalence classes of $\approx$ on Q. Denote the states in Q' by primed ones, meaning that q' is the equivalence class that contains the state q. Let $F' = \{q' : q \in F\}$. Similarly, s' is the set of all states in Q that are equivalent to the state s, the initial state of the old DFA D. Notice that, in general, $|Q'| \leq |Q|$ and $|F'| \leq |F|$.

Define the function δ' as δ modulo $\approx$, that is, $\delta' : Q' \times \Sigma \to Q'$ is defined by $\delta'(q', \sigma) = (\delta(q, \sigma))'$. The function δ' is, indeed, well defined. For, if $q_1, q_2 \in q'$, then

$q_1 \approx q_2$. By Exercise 4.8, $\delta(q_1, \sigma) \approx \delta(q_2, \sigma)$, and then $(\delta(q_1, \sigma))' = (\delta(q_2, \sigma))'$. Our aim is to show that the DFA $D' = (Q', \Sigma, \delta', s', F')$ is a minimal DFA accepting the same language as D.

First, we must check that $\mathcal{L}(D') = \mathcal{L}(D)$. To see this, suppose $u \in \mathcal{L}(D)$. Then $(s, u) \rightsquigarrow_D (t, \varepsilon)$ for some $t \in F$. From the definition of δ' (using induction on length of u), it is obvious that $(s', u) \rightsquigarrow_{D'} (t', \varepsilon)$. Further, $t' \in F'$ as $t \in F$. That is, $u \in \mathcal{L}(D')$. Conversely, suppose $u \in \mathcal{L}(D')$. Then $(s', u) \rightsquigarrow_{D'} (t', \varepsilon)$ for some $t' \in F'$. Again, from the definition of δ', we have $(s_1, u) \rightsquigarrow_D (t_1, \varepsilon)$, where $t_1 \in F$, $s \approx s_1$, and $t \approx t_1$. It then follows that $(s, u) \rightsquigarrow_D (t_1, \varepsilon)$ for $t_1 \in F$, that is, $u \in \mathcal{L}(D)$.

We next compare the number of equivalence classes of ρ_L and the number of states in Q'. Recollect that ρ_L is supposed to give us a minimal DFA by its equivalence classes.

Now, suppose $|Q'| \geq n$, where n is the index of ρ_L. Then there are two distinct states (equivalence classes, looking from Q), say, p' and q' in Q' and strings $x, y \in \Sigma^*$ such that when D' gets inputs x and y, it reaches p' and q', respectively, starting from its initial state s'. Moreover, the strings x and y are in the same equivalence class of ρ_L. This means that $(s', x) \rightsquigarrow_{D'} (p', \varepsilon)$ and $(s', y) \rightsquigarrow_{D'} (q', \varepsilon)$. As we have defined the transition function δ', the yields can be seen in D itself. That is, we have

$$(s, x) \rightsquigarrow_D (p, \varepsilon),\ (s, y) \rightsquigarrow_D (q, \varepsilon),\ p \not\approx q,\ x\, \rho_L y.$$

As $p \not\approx q$, there is $z \in \Sigma^*$ such that $(p, z) \rightsquigarrow_D (r, \varepsilon)$ and $(q, z) \rightsquigarrow_D (t, \varepsilon)$, where one of r or t is in F and the other is not. Now,

$$(s, xz) \rightsquigarrow_D (p, z) \rightsquigarrow_D (r, \varepsilon) \text{ and } (s, yz) \rightsquigarrow_D (q, z) \rightsquigarrow_D (t, \varepsilon).$$

As only one of r, t is in F, this shows that xz and yz are not related by ρ_L. This contradicts the right invariance of ρ_L. Hence $|Q'| \leq$ the index of ρ_L. But the index of ρ_L is the minimum number of states required for any DFA for accepting L (Theorem 4.8). Therefore, D' is a minimal DFA.

We summarize the above discussion as in the following:

Theorem 4.9. *Let $D = (Q, \Sigma, \delta, s, F)$ be a DFA and $D' = (Q', \Sigma', \delta', s', F')$ be its quotient DFA as constructed earlier, modulo state equivalence. Then $\mathcal{L}(D') = \mathcal{L}(D)$ and D' has the minimum number of states for accepting $L(D)$.*

The construction of the quotient DFA that uses state equivalence can be written as an algorithm. The idea is to tick out the nonequivalent states and keep the equivalent ones by starting from all pairs of states. As we are computing with an equivalence relation, there is no need to start with all ordered pairs. Only half of them would be sufficient, taking symmetry into consideration. The algorithm goes as follows:

ALGORITHM *State Minimization*

Suppose, we have ordered the set of all states Q of the simple DFA D, say, $Q = \{q_1, \ldots, q_n\}$. We consider the set $E = \{(q_i, q_j) : 1 \leq i < j \leq n\}$. We want to mark off all those ordered pairs in E, the components of which are not state equivalent. Notice that when a state p is a final state and a state q is not a final state, then they cannot be state equivalent. So, initially, we mark all the pairs (p, q) where one of them is in F and the other is not. These are the states that are shown easily to be nonequivalent by using the empty string as the input to D. Next, we consider the inputs as symbols from Σ, and take an unmarked pair (p, q). Consider the pair $(\delta(p, \sigma), \delta(q, \sigma))$, taking in turn, each $\sigma \in \Sigma$. If any such pair $(\delta(p, \sigma), \delta(q, \sigma))$ is already marked, then p, q are not equivalent, and we mark them now. This step is repeated till no new markings can be done. The unmarked pairs correspond to equivalent states, that is, if a pair (p, q) remains unmarked, it means that the states p, q are state equivalent to each other.

State minimization algorithm takes a constant times kn^2 units of time, where $k = |\Sigma|$ and $n = |Q|$, in the worst case.

Example 4.12. Construct a minimal DFA equivalent to the simple DFA whose transition diagram is given in Fig. 4.6.

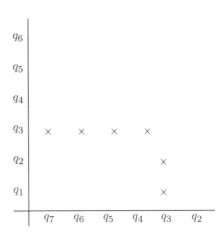

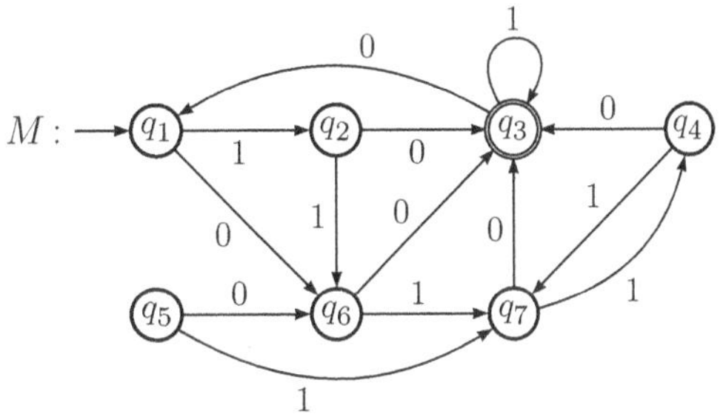

Fig. 4.6. DFA of Example 4.12.

Solution. We take the same ordering of the states as their subscripts and form the unordered pairs (q_i, q_j) for $1 \le i < j \le 7$. Initially, we mark the pairs (p, q), where $p \in F$ and $q \in Q - F$. Owing to symmetry, we also mark the pairs (p, q), where $p \in Q - F$ and $q \in F$.

While doing it manually, as we are doing now, it is convenient to construct a triangular array keeping blank slots (to be marked) corresponding to rows and columns. See Fig. 4.7.

q_6						
q_5						
q_4						
q_3	×	×	×	×		
q_2					×	
q_1					×	
	q_7	q_6	q_5	q_4	q_3	q_2

Fig. 4.7. Working array for Example 4.12.

The rows correspond to the states q_1 to q_6 and the columns correspond to states q_2 to q_7. We omit the first state in columns and the last state in the rows; they are not required as any state is always equivalent to itself. Moreover, we need not consider any entry beyond the triangular region bounded by the horizontal line, the vertical line, and the line joining the top most point of the vertical line and the right most point of the horizontal line. We want to mark those pairs (p, q) where p is not equivalent to q (and q is not equivalent to p). After the initial marking (with a $\times$), our array looks like that in Fig. 4.7.

As q_3 is the only final state, we have marked the pairs

$$(q_1, q_3), (q_2, q_3), (q_3, q_4), (q_3, q_5), (q_3, q_6), (q_3, q_7).$$

Next, we check the pairs (q_1, q_i) for $i = 2, 4, 5, 6, 7$. To this end, we consider

$$(\delta(q_1, 0), \delta(q_2, 0)) = (q_6, q_3) = (q_3, q_6) \quad \text{[due to symmetry]}.$$

The pair (q_3, q_6) has already been marked. Therefore, the pair (q_1, q_2) is also marked. Similarly, the pairs (q_1, q_4), (q_1, q_6), and (q_1, q_7) are also marked as

$$(\delta(q_1, 0), \delta(q_4, 0)) = (q_6, q_3), \; (\delta(q_1, 0), \delta(q_6, 0)) = (q_2, q_3),$$

$$(\delta(q_1, 0), \delta(q_7, 0)) = (q_2, q_3).$$

For the states q_1 and q_5, we see that

$$(\delta(q_1, 0), \delta(q_5, 0)) = (q_6, q_6), \; (\delta(q_1, 1), \delta(q_5, 1)) = (q_2, q_7).$$

Next, we consider the input symbols. The input symbol 0 forces us to end up with (q_6, q_6), where nothing can be concluded. However, the input symbol 1 gives some information. It says that

if the states q_2, q_7 are not equivalent, then q_1, q_5 are also not equivalent.

So, we must maintain this relation among the pairs, which can be used in a future time. For example, if we find later, during the execution of the algorithm, that the pair (q_2, q_7) are marked, then this will lead us to mark the pair (q_1, q_5) also. Let us write this association by $(q_2, q_7) \twoheadrightarrow (q_1, q_5)$.

Next, we consider the pairs (q_2, q_i) for $i = 4, 5, 6, 7$. We see that

$$(\delta(q_2, 0), \delta(q_4, 0)) = (q_3, q_3), \; (\delta(q_2, 1), \delta(q_4, 1)) = (q_6, q_7),$$

$$(\delta(q_2, 0), \delta(q_5, 0)) = (q_3, q_6), \; (\delta(q_2, 0), \delta(q_6, 0)) = (q_3, q_3),$$

$$(\delta(q_2, 1), \delta(q_6, 1)) = (q_6, q_7), \; (\delta(q_2, 0), \delta(q_7, 0)) = (q_3, q_3),$$

$$(\delta(q_2, 1), \delta(q_7, 1)) = (q_6, q_4).$$

Thus, the pair (q_2, q_5) is marked and the association of pairs is updated to

$$(q_2, q_7) \twoheadrightarrow (q_1, q_5), \; (q_6, q_7) \twoheadrightarrow (q_2, q_4), \; (q_6, q_7) \twoheadrightarrow (q_2, q_6), \; (q_6, q_4) \twoheadrightarrow (q_2, q_7).$$

For pairs (q_4, q_i) with $i = 5, 6, 7$, we find that

$$(\delta(q_4, 0), \delta(q_5, 0)) = (q_3, q_6), \ (\delta(q_4, 0), \delta(q_6, 0)) = (q_3, q_3),$$

$$(\delta(q_4, 1), \delta(q_6, 1)) = (q_7, q_7), \ (\delta(q_4, 0), \delta(q_7, 0)) = (q_3, q_3),$$

$$(\delta(q_4, 1), \delta(q_7, 1)) = (q_7, q_4).$$

This tells that the pair (q_4, q_5) is marked and no new update of the association of the pairs is possible.

For pairs $(q_5, q_6), (q_5, q_7)$, we see that

$$(\delta(q_5, 0), \delta(q_6, 0)) = (q_6, q_3), \ (\delta(q_5, 0), \delta(q_7, 0)) = (q_6, q_3).$$

Therefore, the pairs (q_5, q_6) and (q_5, q_7) are marked. For the pair (q_6, q_7), we find that

$$(\delta(q_6, 0), \delta(q_7, 0)) = (q_3, q_3), \ (\delta(q_6, 1), \delta(q_7, 1)) = (q_7, q_4).$$

From this, nothing can be concluded. However, the last equality says that the association of pairs must be updated to

$$(q_2, q_7) \twoheadrightarrow (q_1, q_5), \ (q_6, q_7) \twoheadrightarrow (q_2, q_4), \ (q_6, q_7) \twoheadrightarrow (q_2, q_6),$$

$$(q_6, q_4) \twoheadrightarrow (q_2, q_7), \ (q_7, q_4) \twoheadrightarrow (q_6, q_7).$$

At the end of this round, the triangular array looks like that in Fig. 4.8.

q_6						
q_5	×	×				
q_4			×			
q_3	×	×	×	×		
q_2			×		×	
q_1	×	×		×	×	×
	q_7	q_6	q_5	q_4	q_3	q_2

Fig. 4.8. Working array for Example 4.12 continued.

In the final round, we must exploit the associations, using them recursively. As (q_2, q_7) is unmarked, the first association cannot be used rightnow. The pair (q_6, q_7) is unmarked; this also cannot be used rightnow. Other pairs such as (q_6, q_4) and (q_7, q_4) are again unmarked. That is, none of the pairs to the left of $\twoheadrightarrow$ is marked.

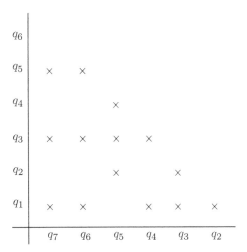

This round does not mark any of the earlier unmarked pairs, and then we conclude that the vacant entries in the triangular array correspond to the equivalent states. But what are they?

Look at the vacant entries in the column labeled q_7. It says that the state q_7 is equivalent to all of (from the rows now) q_2, q_4, q_6. Similarly, q_6 is equivalent to q_2, and q_4; q_5 is equivalent to q_1, and q_4 is equivalent to q_2. That is, the equivalence classes are $p = \{q_1, q_5\}$, $q = \{q_3\}$, and $r = \{q_2, q_4, q_6, q_7\}$.

To construct the DFA with these equivalence classes, we have to look at the transition diagram of Fig. 4.7. We just add transitions from an equivalence class to another if there is a transition from at least one element of the first class to one of the second, keeping the label as the same. For example, there is a transition from state q_2 to q_3 labeled 0. Thus, we add a transition from the equivalence class $\{q_2, q_4, q_6, q_7\}$ to $\{q_3\}$ labeled 0. Again, the final states will be those of which an element is a final state in the original DFA. We then have the minimized DFA as in Fig. 4.9. □

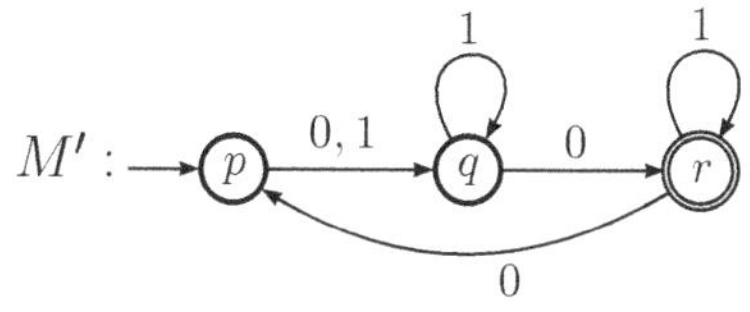

Fig. 4.9. Minimized DFA for Example 4.12.

Example 4.13. Construct a DFA with fewest states that accepts $(10 \cup 101)^*$.

Solution. A simple DFA accepting the language is given in Fig. 4.10.

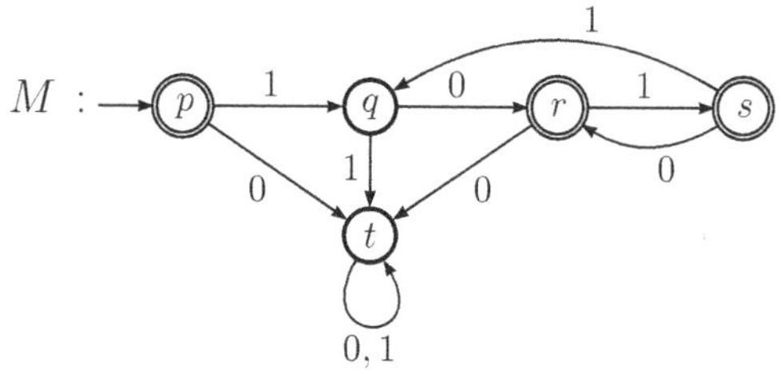

Fig. 4.10. A simple DFA accepting $(10 \cup 101)^*$.

To minimize the number of states in the DFA M of Fig. 3.10, we start with a triangular array and mark the pairs $(p, q), (p, t), (r, q), (r, t), (s, q), (s, t)$ as $F = \{p, r, s\}$ and $Q - F = \{q, t\}$. At this point, make a triangular array and start the process yourself. Then come back to this page. Look at Fig. 4.11.

Next, compute $(\delta(q_i, \sigma), \delta(q_j, \sigma))$ for marking the nonequivalent states. You see that

$$(\delta(p, 0), \delta(s, 0)) = (t, r),\ (\delta(p, 0), \delta(r, 0)) = (t, t),\ (\delta(p, 1), \delta(r, 1)) = (q, s),$$

$$(\delta(q, 0), \delta(t, 0)) = (r, t),\ (\delta(r, 0), \delta(s, 0)) = (t, r).$$

Fig. 4.11. Working array for the simple DFA of Fig. 4.10.

s	×			
r	×			
q		×	×	
p	×			×
	t	s	r	q

For the pair (p, r), we see that $(\delta(p, 0), \delta(r, 0)) = (t, t)$ gives no information. So we had to compute $(\delta(p, 1), \delta(r, 1)) = (q, s)$. We see that (t, r) and (q, s) are already marked. Hence, all the pairs $(p, s), (p, r), (q, t), (r, s)$ are marked nonequivalent. (Here are no associations of pairs for future marking.) That is, no equivalent pair exists in the DFA M. This means that the simple DFA M is already minimized.

However, this may not be a DFA with the fewest states; for, there might be one with fewer states that is not a simple DFA. This may happen due to a state of no return in the simple DFA. For example, the DFA M has the state t, a state of no return. Remove t from M and call this new DFA M'.

Now, is M' a DFA with fewest states? If not, then there is another DFA, say, M' with less than four states that accepts the language $(10 \cup 101)^*$. If there are redundant states (unreachable from the initial state) in M', delete them. Then, add to the resulting DFA a state of no return to make a simple DFA. Call the new DFA M''. The number of states in M'' is no more than 4. But M'' accepts the same language as M. This is a contradiction to the fact that M is a minimal DFA accepting the language $(10 \cup 101)^*$. Therefore, M' is a DFA with fewest states. Check that this is the DFA of Example 1.19. □

Problems for Section 4.5

4.33. Suppose in a DFA the states p, q, r are such that p, q are state equivalent but p, r are not. Show that q, r are not state equivalent.

4.34. Design a minimal DFA that accepts $L = \mathcal{L}(a^*b^* \cup b^*a^*)$. Also, give a regular expression describing each of the regular classes of the Myhill–Nerode relation ρ_L.

4.35. Let L be the set of strings of balanced parentheses. What are the equivalence classes of the relation ρ_L?

4.36. Write a precise algorithm of state minimization employing the idea of making a triangular array and the association of pairs via $\twoheadrightarrow$.

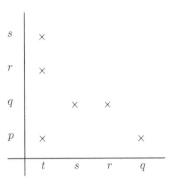

4.37. Construct DFAs with minimum number of states that accept the following languages and argue why the DFAs are minimal:
(a) $\{a^m b^n : m \geq 2, n \geq 1\}$.
(b) $\{a^n b : n \geq 0\} \cup \{b^n a; n \geq 1\}$.
(c) $\{a^n : n \neq 2 \text{ and } n \neq 4\}$.
(d) $\{a^n : n \geq 0, n \neq 3\}$.

4.38. Minimize the states of the DFA $D = (\{p, q, r, s, t, u\}, \{0, 1\}, \delta, p, \{t, u\})$, where $\delta(p, 0) = q, \delta(p, 1) = t, \delta(q, 0) = p, \delta(q, 1) = t, \delta(r, 0) = q, \delta(r, 1) = s, \delta(s, 0) = \delta(s, 1) = t, \delta(t, 0) = \delta(t, 1) = u, \delta(u, 0) = \delta(u, 1) = u$.

4.39. Prove or disprove: Let $D = (Q, \Sigma, \delta, p, F)$ be a minimal DFA that accepts a language L. If δ is a total function, then the DFA $D' = (Q, \Sigma, \delta, p, Q - F)$ is a minimal DFA that accepts $\overline{L}$.

4.40. Show that state equivalence is an equivalence relation, but state inequivalence is not an equivalence relation.

4.41. Minimize the DFAs $(\{a, b, c, d, e, f, g, h\}, \{0, 1\}, \delta, a, F)$, where F and δ are as follows. Indicate which equivalence class corresponds to which state of the new DFA.

(a) $\delta(a, 0) = f, \delta(a, 1) = c, \delta(b, 0) = e, \delta(b, 1) = f, \delta(c, 0) = d, \delta(c, 1) = e, \delta(d, 0) = c, \delta(d, 1) = b, \delta(e, 0) = b, \delta(e, 1) = a, \delta(f, 0) = a, \delta(f, 1) = d, F = \{c, d\}$.
(b) $\delta(a, 0) = b, \delta(a, 1) = c, \delta(b, 0) = e, \delta(b, 1) = f, \delta(c, 0) = a, \delta(c, 1) = d, \delta(d, 0) = f, \delta(d, 1) = c, \delta(e, 0) = b, \delta(e, 1) = a, \delta(f, 0) = e, \delta(f, 1) = d, F = \{c, d\}$.
(c) $\delta(a, 0) = d, \delta(a, 1) = f, \delta(b, 0) = c, \delta(b, 1) = e, \delta(c, 0) = g, \delta(c, 1) = d, \delta(d, 0) = g, \delta(d, 1) = g, \delta(e, 0) = a, \delta(e, 1) = c, \delta(f, 0) = b, \delta(f, 1) = g, \delta(g, 0) = c, \delta(g, 1) = g, F = \{e, f\}$.
(d) $\delta(a, 0) = d, \delta(a, 1) = c, \delta(b, 0) = d, \delta(b, 1) = f, \delta(c, 0) = c, \delta(c, 1) = g, \delta(d, 0) = c, \delta(d, 1) = b, \delta(e, 0) = f, \delta(e, 1) = e, \delta(f, 0) = f, \delta(f, 1) = d, \delta(g, 0) = f, \delta(g, 1) = a, F = \{a, b\}$.
(e) $\delta(a, 0) = c, \delta(a, 1) = f, \delta(b, 0) = g, \delta(b, 1) = c, \delta(c, 0) = g, \delta(c, 1) = g, \delta(d, 0) = g, \delta(d, 1) = e, \delta(e, 0) = f, \delta(e, 1) = a, \delta(f, 0) = e, \delta(f, 1) = d, \delta(g, 0) = b, \delta(g, 1) = g, F = \{e, f\}$.
(f) $\delta(a, 0) = d, \delta(a, 1) = f, \delta(b, 0) = g, \delta(b, 1) = d, \delta(c, 0) = g, \delta(c, 1) = e, \delta(d, 0) = g, \delta(d, 1) = g, \delta(e, 0) = a, \delta(e, 1) = f, \delta(f, 0) = c, \delta(f, 1) = e, \delta(g, 0) = b, \delta(g, 1) = g, F = \{e, f\}$.
(g) $\delta(a, 0) = f, \delta(a, 1) = b, \delta(b, 0) = c, \delta(b, 1) = f, \delta(c, 0) = b, \delta(c, 1) = d, \delta(d, 0) = e, \delta(d, 1) = c, \delta(e, 0) = d, \delta(e, 1) = a, \delta(f, 0) = a, \delta(f, 1) = e, \delta(g, 0) = a, \delta(g, 1) = h, \delta(h, 0) = h, \delta(h, 1) = g, F = \{c, d\}$.
(h) $\delta(a, 0) = f, \delta(a, 1) = d, \delta(b, 0) = g, \delta(b, 1) = e, \delta(c, 0) = b, \delta(c, 1) = h, \delta(d, 0) = a, \delta(d, 1) = h, \delta(e, 0) = b, \delta(e, 1) = f, \delta(f, 0) = c, \delta(f, 1) = a, \delta(g, 0) = e, \delta(g, 1) = b, \delta(h, 0) = d, \delta(h, 1) = b, F = \{a, b\}$.

4.6 Summary and Additional Problems

In this chapter, we have discussed the closure properties of regular languages with respect to the usual set theoretical operations, concatenation, asterate (Kleene star), homomorphism, and inverse homomorphism. We have also tried to discover the periodicity structures of regular languages, including the pumping lemma. These results help us in showing the existence of nonregular languages constructively, as well as in proving that some languages are regular provided some others are. We have also discussed an important result, the Myhill–Nerode theorem. An application of this result has led us to a solution of the problem of minimizing the number of states in a DFA.

Closure properties of regular languages were studied by Ginsburg and some of his collaborators [38, 41, 42], McNaughton and Yamada [85], Rabin and Scott [107], and Stearns and Hartmanis [127], among others. For a general treatment of ultimate periodicity and some other functions that preserve regularity, see [116]. Myhill–Nerode theorem was proved independently, in different forms, by Myhill [93] and Nerode [95]. For minimization of DFAs, you may like to see [57, 60, 91] and the references therein.

The pumping lemma for regular languages as treated here was found by Bar-Hillel et al. [8]. This only gives a necessary condition for a language to be regular. For stronger forms of pumping lemma giving both necessary and sufficient conditions, see [29, 61, 126].

We have not discussed here the algebraic theory of automata and regular languages. In one of the exercises below, you will find Kleene algebra and its complete axiomatization. This was introduced by Kozen [70]. The first complete axiomatization of the algebra of regular languages was given by Salomaa [112]. For an extensive treatment of the topic, see Conway [22]. See also [134] for algebraic recognizability of languages. You may like to see [6] and the references there in.

As minimal NFAs are not necessarily unique up to isomorphisms, we have not discussed it here. In one of the exercises below we ask some pertinent questions. A part of Myhill–Nerode theorem can be generalised to NFAs [71] using the notion of *bisimulation*, which is an important notion for concurrency; see for example, [88]. You may also like the presentations about term automata and two way finite automata in Kozen [71].

An important application of DFAs is pattern matching with regular expressions; see [4, 130]. For a nice example and an algorithm for search using regular expressions, see [69].

Additional Problems for Chapter 4

4.42. Suppose L is a finite language and $L \cup L'$ is regular. Does it follow that L' is regular?

4.43. Assume that all our NFAs have only a single final state. (Can you show that there is no loss in generality in assuming this?) There are some simple modifications for taking care of the operations of union, concatenation, and Kleene star as given in the following:

(a) For union, merge the two initial states into one with all the transitions of both the NFAs. Likewise, merge the two final states having all transitions to both of them to go to the merged state.

(b) For concatenation, merge the final state of the first NFA with the initial state of the second.

(c) For the Kleene star, add ε-transitions from the final state to the initial state.

Show that each of these simplifications do the intended job.

4.44. Show that the set of all triples of bits that represent correct multiplication is not a regular language.

4.45. Let L be a regular language over Σ. Show that $\{w \in \Sigma^* : vw \in L \text{ and } \ell(v) = (\ell(w))^2 \text{ for some } v \in \Sigma^*\}$ is a regular language.

4.46. Let L, L' be regular languages over an alphabet Σ. Show that the following languages are regular, or give a counter example:

(a) $L \ominus L' = \{w \in \Sigma^* :$ either $w \in L$ or $w \in L'\}$. (Symmetric Difference)
(b) $nor(L, L') = \{w \in \Sigma^* : w \notin L$ and $w \notin L'\}$. (Neither–Nor)
(c) $cor(L, L') = \{w \in \Sigma^* : w \notin L$ or $w \notin L'\}$. (Complementary Or)
(d) $L/L' = \{w \in \Sigma^* : wv \in L$ for some $v \in L'\}$. (Right Quotient)
(e) $L' \setminus L = \{w : u \in L', uw \in L\}$. (Left Quotient)
(f) $head(L) = \{w : wu \in L$ for some $u \in \Sigma^*\}$.
(g) $tail(L) = \{w : uw \in L$ for some $u \in \Sigma^*\}$.
(h) $prune2(L) = \{w : \sigma\tau w \in L$ for some $\sigma, \tau \in \Sigma\}$.
(i) $shift(L) = \{\sigma_2\sigma_3 \cdots \sigma_n\sigma_1 : \sigma_1\sigma_2 \cdots \sigma_n \in L\}$.
(j) $swap(L) = \{\tau u\sigma : \sigma u\tau \in L$ for some $\sigma, \tau \in \Sigma, u \in \Sigma^*\}$.
(k) $shuffle(L_1, L_2) = \{w_1v_1w_2v_2 \cdots w_nv_n : w_1w_2 \cdots w_n \in L_1, v_1v_2 \cdots v_n \in L_2\}$.
(l) $minus5(L) = \{abcdw : abcdew \in L\}$. (Removing the fifth symbol)
(m) $rem2(L) = \{w \in \Sigma^* : abw \in L$ for some $a, b \in \Sigma\}$.
(n) $min(L) = \{w \in L :$ there is no $u \in L, v \in \Sigma^*$ with $w = uv\}$.
(o) $max(L) = \{w \in L :$ if $v \neq \varepsilon$, then $wv \notin L\}$.
(p) $left(L) = \{w : ww^R \in L\}$.
(q) $half(L) = \{u : uv \in L$ and $\ell(u) = \ell(v)\}$.
(r) $1third(L) = \{u : uv \in L$ and $2\ell(u) = \ell(v)\}$.
(s) $mthird(L) = \{v : uvw \in L$ and $\ell(u) = \ell(v) = \ell(w)\}$. (Middle–Third)
(t) $del1(L) = \{uv : u1v \in L\}$ when $\Sigma = \{0, 1\}$.
(u) $odd(L) = \{a_1a_3 \cdots a_{2n-1} : a_1a_2a_3 \cdots a_k \in L$ where $k = 2n - 1$ or $2n\}$.
(v) $even(L) = \{a_1a_3 \cdots a_{2n} : a_1a_2a_3 \cdots a_k \in L$ where $k = 2n$ or $2n + 1\}$.
(w) $cycle(L) = \{uv : vu \in L$ for some strings u and $v\}$.

4.47. Find $(aba^*)/(a^*baa^*)$. Construct two languages L, L' to show that $L'L/L$ is not necessarily equal to L. [See part (d) of the last problem.]

4.48. The set of regular languages $REG(\Sigma)$ over an alphabet Σ is closed under union, concatenation, intersection, and complementation. Moreover, each finite

subset of Σ^* is regular. Is it true that $REG(\Sigma)$ is the least class of subsets of Σ^* that contains all finite subsets of Σ^* and is closed under
(a) union, intersection, and complementation?
(b) union, concatenation, and complementation?
(c) intersection, complementation, and concatenation?

4.49. A *substitution* is a generalization of a homomorphism, in the sense that instead of associating a string, it associates a language to a symbol. Formally, a substitution from an alphabet Σ to an alphabet Γ is a map $h : \Sigma \to 2^{\Gamma^*}$, that is, $h(\sigma)$ is a language over Γ. We extend such an h to Σ^* by taking $h(\varepsilon) = \varepsilon$ and $h(w\sigma) = h(w)h(\sigma)$ for strings $w \in \Sigma^*$. Show that $REG(\Sigma)$ is closed under substitution.

4.50. A language is called *definite* if there is $k \in \mathbb{N}$ such that whether any string w is in L or not depends only on the last k symbols of w. Suppose L, L' are any definite languages over an alphabet Σ. Show that
(a) L is regular.
(b) $L \cup L'$ is definite.
(c) $\Sigma^* - L$ is definite.
(d) LL' is not necessarily definite.
(e) L^* is not necessarily definite.

4.51. For $i = 1, 2$, let $M_i = (\{0, 1, 2, 3, 4\}, \{0, 1\}, \delta_i, 0, \{0\})$ be two DFAs, where for $q \in \{0, 1, 2, 3, 4\}$ and $\sigma \in \{0, 1\}$, the transitions δ_1, δ_2 are given by

$$\delta_1(q, \sigma) = (q^2 - \sigma) \bmod 5, \text{ and } \delta_2(q, \sigma) = (q^2 + \sigma) \bmod 5.$$

Prove that $\mathcal{L}(M_1)$ = The set of binary strings containing an even number of 1's. What is $\mathcal{L}(M_2)$? Give minimal DFAs equivalent to both M_1 and M_2. [Note: Here, we write the states as $0, 1, 2, 3, 4$ instead of q_0, q_1, q_2, q_3, q_4, for convenience.]

4.52. Prove that any equivalence class of a right congruence of finite index on Σ^* is a regular subset of Σ^*.

4.53. The usual way of writing numbers is called the decimal notation; this uses the digits 0 to 9. In unary notation a number n is written as 0^n. Are the following languages regular? Justify.
(a) $\{w : w$ is the unary representation of a number divisible by $11\}$.
(b) $\{w : w$ is the decimal representation of a number divisible by $11\}$.
(c) $\{w : w$ is the unary representation of 10^n for some $n \in \mathbb{N}\}$.
(d) $\{w : w$ is the decimal representation of 10^n for some $n \in \mathbb{N}\}$.
(e) $\{w : w$ is a sequence of digits in the decimal expansion of $1/7\}$.
(f) $\{w : w$ is the unary representation of a number n such that there is a pair of twin primes bigger than $n\}$ [Numbers $p, p+2$ are called twin primes if both of them are primes.]

4.54. Construct a nonregular language that satisfies the conclusions of the pumping lemma (for regular languages).

4.55. Let $A \subseteq \mathbb{N}$. Define $binary(A) = \{$binary representations of numbers in $A\}$, and $unary(A) = \{0^n : n \in A\}$. Which one of the following sentences is true and which one is false? Justify.

(a) For each $A \subseteq \mathbb{N}$, if $binary(A)$ is regular then so is $unary(A)$.
(b) For each $A \subseteq \mathbb{N}$, if $unary(A)$ is regular then so is $binary(A)$.

4.56. Recall that an *arithmetic progression* is the set $\{m + nk : k \in \mathbb{N}\}$ for some $m, n \in \mathbb{N}$. An arithmetic progression is also called a *linear set.* A finite union of linear sets is called a *semilinear set.* Let L be any language over an alphabet Σ.
(a) Show that if L is regular, then the set $\{\ell(w) : w \in L\}$ is a semilinear set.
(b) Is it true that if $\{\ell(w) : w \in L\}$ is a semilinear set, then L is regular?

4.57. Let M be a DFA with n states. Let $w \in \mathcal{L}(M)$ with $\ell(w) \geq n$. Show that there are strings x, y, z such that $w = xyz$, $\ell(y) \geq (\ell(w) - n + 1)/n$, and $xy^kz \in \mathcal{L}(M)$ for each $k \in \mathbb{N}$. Why is this a stronger version of the pumping lemma?

4.58. Prove the following stronger versions of the pumping lemma:
(a) If L is a regular language over Σ, then there exists an $m > 1$ such that the following holds for any sufficiently long string w and for each one of its rewritings $w = uvu'$ with $u, u' \in \Sigma^*$, $\ell(v) \geq m$; the middle string v can be written as $v = xyz$ with $\ell(xy) \leq m$, $\ell(y) \geq 1$, such that $uxy^kzu' \in L$ for all $k \in \mathbb{N}$.
(b) (Jaffe [61]) A language $L \subseteq \Sigma^*$ is regular iff there exists $k \in \mathbb{N}$ such that for each $y \in \Sigma^*$ with $\ell(y) = k$, there exist strings $u, v, w \in \Sigma^*$ such that $y = uvw$, $v \neq \varepsilon$, and for all $z \in \Sigma^*$, for all $j \in \mathbb{N}$, we have $yz \in L$ iff $uv^jwz \in L$.
(c) (Stanat and Weiss[126]) A language $L \subseteq \Sigma^*$ is regular iff there exists $k \in \mathbb{N}$ such that for each $y \in \Sigma^*$ with $\ell(y) \geq k$, there exist $u, v, w \in \Sigma^*$ such that $y = uvw$, $v \neq \varepsilon$, and for all $x, z \in \Sigma^*$ and for all $j \in \mathbb{N}$, we have $xuz \in L$ iff $xuv^jz \in L$.

4.59. Are the following languages regular?
(a) $\{uww^Rv : u, v, w \in \{a, b\}^+, \ell(u) \geq \ell(v))\}$.
(b) $\{a^mba^nba^{m+n} : m, n \geq 1\}$.

4.60. Let L be a regular language. Consider the two languages:

$$L_1 = \{w : w^n \in L \text{ for some } n \in \mathbb{N}\} \text{ and } L_2 = \{w^n : w \in L \text{ for some } n \in \mathbb{N}\}.$$

Which one of L_1, L_2 is regular and which is not? Justify.

4.61. Let $\Sigma = \{a, b, c\}$. Consider the two languages:

$$L = \{uv : \#_a(u) = \#_b(v)\} \text{ and } L' = \{ucv : \#_a(u) = \#_b(v)\}.$$

Which one of L, L' is regular and which is not? Justify.

4.62. Let L be a regular language. Define three languages L_1, L_2, L_3 as follows:

$$L_1 = \{w : w^{\ell(w)} \in L\}.$$
$$L_2 = \{u : uv \in L \text{ and } \ell(v) = 2^{\ell(u)} \text{ for some } v\}.$$
$$L_3 = \{u : uv \in L \text{ and } \ell(v) = \log \ell(u) \text{ for some } v\}.$$

Which two of L_1, L_2, L_3 are regular and which is not? Justify.

4.63. Does there exist a regular language L over $\{0, 1\}$ such that both L and $\overline{L}$ have infinite subsets that are not regular languages?

4.64. Let $p(\cdot)$ be a polynomial of degree n whose coefficients are natural numbers. Show that if L is a regular language, then so is $L' = \{u : \ell(v) = p(\ell(u))$ for some $v \in L\}$.

4.65. For a language L, define a map $d_L : \mathbb{N} \to \mathbb{N}$, called the *density of* L by: $d_L(n)$ is the number of strings in L whose length is less than or equal to n, i.e., $d_L(n) =$ cardinality of $\{w \in L : \ell(w) \le n\}$.
(a) What is the density of $((ab \cup abb)^*)^*$?
(b) Show that the density of any regular language is either bounded above by a polynomial or bounded below by an exponential.
(c) Does the converse of the statement in (b) hold?

4.66. Let $\mathcal{L} \subseteq a^*$. Prove that $L^* = \{a^m n : m \in \mathbb{N}\} - A$, where A is some finite set and n is the greatest common divisor of all natural numbers k with $a^k \in L$. Then, show that L^* is regular. This would prove that if L is any language over a singleton alphabet, then L^* is regular; how?

4.67. For any $w \in \{a, b\}^*$, let $tail(w) = \{uw : u \in \{a, b\}^*\}$; the set of strings which end with the string w. For example, the NFA $(\{p, q, r\}, \{a, b\}, \Delta, p, \{r\})$ with $\Delta = \{(p, a, p), (p, b, p), (p, a, q), (q, b, r)\}$ accepts $tail(ab)$. $tail(w)$ is also written as $tail(\{w\})$. Show that the minimal DFA for $tail(w)$ has $n + 1$ states if $\ell(w) = n$.

4.68. (Greibach) Let N be an NFA with a single final state. First, reverse the transitions and interchange the initial and final states to get an NFA for $\mathcal{L}(N)^R$. Next, use the subset construction to get an equivalent DFA for $\mathcal{L}(N)^R$. Finally, repeat the previous two steps once again. Prove that the DFA so obtained is a state-minimized DFA for $\mathcal{L}(N)$.

4.69. A string $w \in \Sigma^*$ is called *square-free* if w cannot be written as $uvvx$ for some strings $u, v, x \in \Sigma^*$. Show that if Σ contains at least two symbols, then the set of all square-free strings in Σ^* is not a regular language.

4.70. For strings u, v over an alphabet Σ define $u \sqsubset v$ if u can be obtained from v by deleting 0 or more occurrences of symbols. A language $L \subseteq \Sigma^*$ is called *closed downward under* $\sqsubset$ if $u \in L$ whenever $u \sqsubset v$ and $v \in L$. Prove the following:
(a) *Higman's Lemma*: The set of $\sqsubset$-minimal elements of any $L \subseteq \Sigma^*$ is finite. That is, if $L \subseteq \Sigma^*$, then there is a finite subset $L' \subseteq L$ such that for each $v \in L$, there exists $u \in L'$ where $u \sqsubset v$.
(b) Any language $L \subseteq \Sigma^*$ closed downward under $\sqsubset$ is regular.

4.71. Show that minimal-state NFAs are not necessarily unique.

4.72. (Arden) Let P, Q be two regular expressions over an alphabet Σ, where $\varepsilon \notin \mathcal{L}(P)$. Then the equation $R = Q \cup RP$ has the only solution $R = QP^*$ in the set of regular expressions over Σ.

4.73. Give a construction of NFAs having arbitrarily large number of states, where the length of the shortest rejected string can be exponential in the number of states.

4.74. Write a computer program that produces a minimal DFA for a given DFA.

4.75. Two interesting particular cases of quotients of languages are obtained by taking one of the languages as a singleton. Fix an alphabet Σ. Let L be a language and a be a symbol. Define $L/a = \{w \in \Sigma^* : wa \in L\}$. For example, if $\Sigma = \{a, b\}$ and $L = \{ab, baa\}$, then $L/a = \{b\}$. Prove or disprove the following:
(a) If L is regular, then so is L/a.
(b) $(L/a)a = L$.
(c) $(La)/a = L$.
Define similarly the left quotient $a/L = \{w \in \Sigma^* : aw \in L\}$ and then ask and answer questions similar to (a–c).

4.76. Let L be a language and a be a symbol. The quotient $a \setminus L$ is also called a derivative. That is, define the *derivative* $\mathrm{d}L/\mathrm{d}a = \{w : aw \in L\}$. For example, if $L = \{b, ba, ab\}$, then $\mathrm{d}L/\mathrm{d}b = \{\varepsilon, a\}$. Prove or disprove the following:
(a) If L is regular, then so is $\frac{\mathrm{d}L}{\mathrm{d}a}$.
(b) $a\frac{\mathrm{d}L}{\mathrm{d}a} = L$.
(c) $\frac{\mathrm{d}(aL)}{\mathrm{d}a} = L$.
(d) $\frac{\mathrm{d}(L \cup L')}{\mathrm{d}a} = \frac{\mathrm{d}L}{\mathrm{d}a} \cup \frac{\mathrm{d}L'}{\mathrm{d}a}$.
(e) If $\varepsilon \notin L$, then $\frac{\mathrm{d}(LL')}{\mathrm{d}a} = \frac{\mathrm{d}(L)}{\mathrm{d}a}L'$.
(f) If $\varepsilon \in L$, then $\frac{\mathrm{d}(LL')}{\mathrm{d}a} = \frac{\mathrm{d}(L)}{\mathrm{d}a}L' \cup \frac{\mathrm{d}(L')}{\mathrm{d}a}$.

4.77. With the derivative as defined in the previous problem, answer the following:
(a) Give an expression for $\frac{\mathrm{d}(L^*)}{\mathrm{d}a}$.
(b) Characterize all languages L for which $\frac{\mathrm{d}(L)}{\mathrm{d}a} = L$.
(c) What are the languages L for which $\frac{\mathrm{d}(L)}{\mathrm{d}a} = \emptyset$?

5 Context-free Languages

5.1 Introduction

You have tried to construct a regular grammar for the language $\{a^k b^k : k \in \mathbb{N}\}$, but failed. However, the strings of the form $a^k b^k$ can be generated by two productions such as $S \mapsto \varepsilon, S \mapsto aSb$. For example, with such productions, we can have a derivation for a^3b^3 as

$$S \Rightarrow aSb \Rightarrow aaSbb \Rightarrow aaaSbbb \Rightarrow aaabbb.$$

The production $S \mapsto aSb$ violates the regularity condition that a nonterminal on the right hand side of a production can appear as only the rightmost symbol. To accommodate these general type of productions, we will define context-free grammars.

5.2 Context-free Grammars

A **context-free grammar (CFG)** is a quadruple $G = (N, \Sigma, R, S)$, where

N is an alphabet, called the set of nonterminal symbols,
Σ is another alphabet, called the set of terminal symbols, $N \cap \Sigma = \emptyset$,
$S \in N$ is the start symbol, and
R is a finite subset of $N \times (N \cup \Sigma)^*$.

The finite relation R from N to $(N \cup \Sigma)^*$ is the set of productions, where each element (A, u) is rewritten in the form $A \mapsto u$ as earlier. An application of a production rule is formalized by the relation of derivation in one step, denoted by $\stackrel{1}{\Rightarrow}$. We write $xAy \stackrel{1}{\Rightarrow} xuy$ (read as xAy gives in one step xuy) iff $x, y \in (N \cup \Sigma)^*$, $A \in N$, $u \in (N \cup \Sigma^*)$, and $A \mapsto u$ is a production in R. We denote by $\stackrel{*}{\Rightarrow}$, the transitive and reflexive closure of the one-step yield $\stackrel{1}{\Rightarrow}$.

We simply write $\Rightarrow$ for both $\stackrel{1}{\Rightarrow}$ and $\stackrel{*}{\Rightarrow}$ if no confusion arises. That is, we overuse this symbol for derivation: $u \Rightarrow v$ (read as u gives v) iff there is a sequence of derivations $u \stackrel{1}{\Rightarrow} u_1 \stackrel{1}{\Rightarrow} u_2 \stackrel{1}{\Rightarrow} \cdots \stackrel{1}{\Rightarrow} u_n \stackrel{1}{\Rightarrow} v$ or when $u = v$. Such a sequence is also called a **derivation**. The **length** of this derivation is either $n + 1$ or 0, accordingly. If many context-free grammars are involved in a particular context, then we will write $\Rightarrow_G$ to say that the derivation is done in the grammar G. A string $w \in \Sigma^*$ is **generated by** the grammar G iff $S \Rightarrow w$ in G.

The **language** generated by G is $\mathcal{L}(G) = \{w \in \Sigma^* : S \Rightarrow w \text{ in } G\}$. A language is called a **context-free language (CFL)** if it is generated by a context-free grammar.

The word "context-free" comes from the mode of application of a production. The production $A \mapsto u$ is applied on the string xAy, where the context x (on the left) along with y (on the right) remains unchanged. That is, it can be applied without bothering what x and y are. When this mode of application is violated, a grammar is called context-sensitive. We will come across such grammars later in an appropriate place.

Example 5.1. Let $G = (\{S\}, \{a, b\}, \{S \mapsto \varepsilon, S \mapsto aSb\}, S)$. What is $\mathcal{L}(G)$?

Solution. This is the grammar that we discussed in the first paragraph of this chapter. It is, of course, obvious that $\mathcal{L}(G) = \{a^n b^n : n \in \mathbb{N}\}$. But how do you prove it? Induction?

For any $n \in \mathbb{N}$, if we apply the production $S \mapsto aSb$, n-times, and then $S \mapsto \varepsilon$, we will have a derivation of $a^n b^n$ as

$$S \Rightarrow aSb \Rightarrow a^2Sb^2 \Rightarrow \cdots \Rightarrow a^nSb^n \Rightarrow a^nb^n.$$

Conversely, if any string is generated by G, then the last production applied must be $S \mapsto \varepsilon$, as the only production in G having no nonterminal symbol on its right hand side is $S \mapsto \varepsilon$. Moreover, $S \mapsto \varepsilon$ could not have been applied anywhere other than the last step of the derivation for obtaining a string of terminal symbols. Hence, the production $S \mapsto aSb$ has only been applied repeatedly before the application of $S \mapsto \varepsilon$. Thus, the only strings that could be generated by G are of the form a^nb^n. □

For an arbitrary context-free grammar, the proof may not be so straight forward. It may demand the use of induction either on the length of a string or on the length of a derivation. Sometimes we may need to use, in such proofs, whatever that can be derived on from the start symbol S in a grammar rather than the derived strings of terminals. These more general strings are called *sentential forms*. In this terminology, a string generated by a grammar is a sentential form consisting of terminal symbols only.

Exercise 5.1. Show that a CFG G with productions $S \mapsto aSa \mid bSb \mid a \mid b \mid \varepsilon$ generates the set of all palindromes over $\{a, b\}$, that is, $\mathcal{L}(G) = \{w \in \{a, b\}^* : w^R = w\}$. Is $\mathcal{L}(G)$ regular?

Each regular grammar is a context-free grammar. What do you conclude from this about regular and context-free languages?

Example 5.2. Let $G = (\{S\}, \{a, b\}, \{S \mapsto SS|aSb|\varepsilon\}, S)$. What is $\mathcal{L}(G)$?

Solution. You must first construct a few derivations, and then look at the production rules to generalize from these typical strings. If required, you will prove your conjecture by induction. Here are a few derivations:

$$S \Rightarrow SS \Rightarrow aSbS \Rightarrow ab.$$
$$S \Rightarrow SS \Rightarrow aSbS \Rightarrow abS \Rightarrow abab, \text{ (using the previous derivation)}$$
$$S \Rightarrow SS \Rightarrow SSS \Rightarrow SSaSb \Rightarrow SSaaSbb \Rightarrow SaSbaaSbb \Rightarrow abaabb.$$

We observe from the derivations that the second and third production rules develop a string of interlaced a's and b's. If you think of a as "(" and b as ")," then the string is simply a correct sequence of closing parentheses. The closing of parentheses (we work with a, b instead) can be conveniently described in an inductive way by the following formation rules:

1. If $x, y \in \mathcal{L}(G)$, then $xy \in \mathcal{L}(G)$.
2. If $x \in \mathcal{L}(G)$, then $axb \in \mathcal{L}(G)$.
3. The empty string $\varepsilon \in \mathcal{L}(G)$.
4. Only those which can be generated by the above rules are in $\mathcal{L}(G)$.

Now you see that the grammar is only a translation of the formation rules. Do you see or not? We have avoided an inductive proof by the idea of these formation rules. However, this is only a shifting of responsibility; it is no proof! □

Exercise 5.2. Show that the language is as claimed in the solution of Example 5.2.

Example 5.3. Let $G = (\{S, A, B\}, \{a, b\}, R, S)$ be the CFG with $R = \{S \mapsto aA \mid bB,\ A \mapsto b \mid bS \mid aAA,\ B \mapsto a \mid aS \mid bBB\}$. What is $L(G)$?

Solution. What are the strings generated by G? Have some derivations before thinking abstractly.

$$S \Rightarrow aA \Rightarrow ab.$$

$$S \Rightarrow aA \Rightarrow abS \Rightarrow abab.$$

$$S \Rightarrow aA \Rightarrow aaAA \Rightarrow aababbab.$$

If we follow the path of B instead of A, we might have to interchange a's with b's, from that point onward, for example,

$$S \Rightarrow aA \Rightarrow abS \Rightarrow abbB \Rightarrow abbbBB \Rightarrow abbbabaaba.$$

So, what do you guess? No good patterns? Well, it seems that number of a's must be equal to the number of b's. Look at the production rules again and convince yourself why this guess is more suggestive than others. Even if it is obvious, we do require a proof, and naturally, by induction. All the most, as it is not very obvious, we must give a proof. The proof might also give more insight into the pattern. Note that the guess might turn out to be wrong, and then we may have to modify our guess and try a fresh proof. So, let us try a proof by induction.

Can ε be generated by G? Obviously not; the rules do not mention of ε. Moreover, no string of length 1 can be generated by G. So, let $w \in \{a, b\}^*$ be of length 2. We have seen that $ab \in \mathcal{L}(G)$. Similarly, $ba \in \mathcal{L}(G)$; give a derivation. Also, argue that $aa \notin \mathcal{L}(G)$ and $bb \notin \mathcal{L}(G)$. This completes the basis step; our basis step starts with $\ell(w) = 2$.

Lay out the induction hypothesis that if $w \in \{a, b\}^*$ and $\ell(w) < n$, then $w \in \mathcal{L}(G)$ iff $\#_a(w) = \#_b(w)$. Let $w \in \{a, b\}^*$ with $\ell(w) = n$. We must show that

(a) if $w \in \mathcal{L}(G)$, then $\#_a(w) = \#_b(w)$.
(b) if $\#_a(w) = \#_b(w)$, then $w \in \mathcal{L}(G)$.

How to show (a)? Look at the derivations above. You see that the last rule applied will be either $A \mapsto b$ or $B \mapsto a$, and that will replace the only nonterminal in the last but one step of any derivation. According as w has the last symbol a or b, you would have to apply the second or first rule at the last step. If your guess is correct, then the symbol A must be generating one more b than required and the symbol B must be generating one more a. That is, we may have to prove these statements first. Do not be disheartened; add two more statements to be proved by induction. The revised statements are

(i) $S \Rightarrow w$ iff $\#_a(w) = \#_b(w)$.
(ii) $A \Rightarrow w$ iff $\#_a(w) = \#_b(w) - 1$.
(iii) $B \Rightarrow w$ iff $\#_b(w) = \#_a(w) - 1$.

For this new statements (i–iii), if $\ell(w) = 1$, then $w = a$ or $w = b$. In that case, all of (i–iii) hold. That proves the basis step of induction. For the inductive step, assume that for any $v \in \{a, b\}^*$ with $\ell(v) \leq n$, all of (i–iii) hold. Let $w \in \{a, b\}^*$ have length $n + 1$. Write $w = au$ if it starts with a, else write $w = bv$. In either case, $\ell(u) = n = \ell(v)$, though only one of them is appropriate. Now, in the derivation of w, it must start with $S \Rightarrow aA$ or with $S \Rightarrow bB$. Then, $A \Rightarrow u$ or $B \Rightarrow v$. By our assumption, $\#_a(u) = \#_b(u) - 1$. In the other case, $\#_b(v) = \#_a(v) - 1$. As $w = au$ or $w = bv$, $\#_a(w) = \#_b(w)$. This proves (i). For (ii), look at the relevant productions. They are $A \mapsto b$, $A \mapsto bS$, and $A \mapsto aAA$. Repeat the above argument. The statement (iii) is similar to (ii). Hence $\mathcal{L}(G) = \{w \in \{a, b\}^* : w \neq \varepsilon \text{ and } \#_a(w) = \#_b(w)\}$. □

Exercise 5.3. Consider the context-free grammar $G = (\{a, b\}, \{S, A\}, R, S)$ with $R = \{S \mapsto AA, A \mapsto b \mid aA \mid Aa \mid AAA\}$. Show by induction on the number of derivation steps that if $S \Rightarrow w$ (for $w \in \{a, b, S, A\}^*$), in at least one step, then $\#_a(w) + \#_b(w)$ is even. What is $\mathcal{L}(G)$?

An important context-free language is the language of balanced parentheses. It is pivotal in the sense that each CFL has to do something with this language. We will see this point later at an appropriate place. But, what is the language of balanced parentheses? In a string involving the only symbols "[" and "]," is it enough to say that number of "[" matches with the number of "]?" Certainly not; for example,] [is not a string with balanced parentheses. What we need is the restriction that the number of left parentheses does not exceed the number of right parentheses in any prefix of a string of balanced parentheses. Suppose x is any string over the alphabet $\{[,\ ]\}$. Let us use the following notation:

$$left(x) = \text{ number of "[" in } x, \quad right(x) = \text{ number of "]" in } x.$$

Then a string $w \in \{[,\ ]\}^*$ is called a *string of balanced parentheses* iff

$$left(w) = right(w) \text{ and for each prefix } y \text{ of } w,\ left(y) \geq right(y).$$

The fact is that the language of balanced parentheses is generated by the CFG with productions $S \mapsto \varepsilon|SS|[\,S\,]$. A proof requires that if a string is at all generated by the CFG, then it satisfies the conditions above. This is proved by induction on the length of a derivation. The other implication that any string over the alphabet $\{[,\ ]\}$ that satisfies the above conditions is generated by the CFG is proved by induction on the length of the string. Why don't you try it?

Problems for Section 5.2

5.1. Is it rue that if a CFG generates ε, then it does so via a derivation of length 1?

5.2. Which of the strings $aaaaba$, $aabaab$, $abaaba$, $aabbaa$ are generated by the CFG, with productions $S \mapsto AB|ABS$, $A \mapsto b|bA$, $B \mapsto aA$, and which are not? Give reasons.

5.3. Show that the grammar with productions $S \mapsto SS|\varepsilon|aSb|bSa$ generates the language $L = \{w \in \{a, b\}^* : \#_a(w) = \#_b(w)\}$.

5.4. Let $L_i = \mathcal{L}(G_i)$, where the CFGs G_i, for $i = 1, 2, 3, 4, 5$, have productions as in the following:

$G_1 : S \mapsto \varepsilon|00S0.$
$G_2 : S \mapsto 0|0S00.$
$G_3 : S \mapsto \varepsilon|00|S000.$
$G_4 : S \mapsto 00A,\ A \mapsto \varepsilon|0S.$
$G_5 : S \mapsto \varepsilon|0|00A,\ A \mapsto 0S.$

Determine all pairs (i, j) such that
(a) $L_i \cap L_j = \emptyset$.
(b) $L_i \subseteq L_j$.
(c) $L_i = L_j$.

5.5. Let G be a CFG with productions $A \mapsto BAB|C$, $C \mapsto aDb|bDa$, $D \mapsto BDB|B|\varepsilon$, $B \mapsto a|b$. What could be its start symbol? What then is $\mathcal{L}(G)$?

5.6. Design a CFG for generating $\{a^m b^n : 0 \leq n \leq m \leq 4n\}$. Give a derivation of the string $aaaaaaaabbb$.

5.7. In the CFG with productions $S \mapsto AA,\ A \mapsto a|bA|Ab|AAA$, give a derivation of the string $b^m ab^n ab^k$ for any $m, n, k > 1$.

5.8. Show that the string $aabbabba$ is not generated by the CFG that has the productions $S \mapsto aaA,\ A \mapsto Ba,\ B \mapsto \varepsilon|bAb$.

5.9. Give a verbal description of the language generated by the CFG with productions $S \mapsto \varepsilon|AB,\ A \mapsto aB,\ B \mapsto Sb$.

5.10. Give a derivation of the string $abbaaa$ in the CFG that has the productions $S \mapsto \varepsilon|aAa|bAb,\ A \mapsto SS$. Try to describe, in English, the language generated by this grammar.

5.11. What is $\mathcal{L}(G)$, where $G = (\{a, b\}, \{S, A\}, R, S)$ with $R = \{S \mapsto a|aAS,\ A \mapsto ba|SS|SbA\}$? Prove your claim.

5.12. What is the shortest string that can be derived in the CFG with productions $S \mapsto aB|aaabb,\ A \mapsto a|B,\ B \mapsto bb|AaB$? Prove your claim.

5.13. Show that $\{w \in \{a, b\}^* : \#_a(w) = \#_b(w)\}$ is generated by the CFG with productions $S \mapsto aA|bB,\ A \mapsto b|bS|BAA,\ B \mapsto a|aS|ABB$.

5.14. Show that the CFG with productions $S \mapsto a|abA|BC,\ A \mapsto c|abA,\ B \mapsto b$ does not generate the language $\{w \in \{a, b, c\}^* : \#_a(w) = \#_b(w)\}$.

5.15. Show that $\mathcal{L}(G)$ is regular, where G has the productions
(a) $S \mapsto \varepsilon|aSa|aSb|bSa|bSb$.
(b) $S \mapsto AbB,\ A \mapsto \varepsilon|aA,\ B \mapsto \varepsilon|aB|bB$.

5.16. What is the language generated by the CFG with productions $S \mapsto AA|B$, $A \mapsto aA|Aa|b,\ B \mapsto aBaa|b$? Is it regular?

5.17. Let G_1, G_2 have the sets of productions R_1, R_2, respectively, where $R_1 = \{S \mapsto ab|aSb\}$ and $R_2 = \{S \mapsto AB|AC,\ A \mapsto a,\ B \mapsto b,\ C \mapsto SB\}$. Is $\mathcal{L}(G_1) = \mathcal{L}(G_2)$?

5.18. What is the language generated by the CFG given by the productions
(a) $S \mapsto ab|abS$?
(b) $S \mapsto ab|SS$?
(c) $S \mapsto aA,\ A \mapsto a|bS$?
(d) $S \mapsto \varepsilon|a|b,\ A \mapsto \varepsilon|B,\ B \mapsto b|A$?
(e) $S \mapsto aSb|aAb,\ A \mapsto b|bA$?
(f) $S \mapsto aSb|aAb,\ A \mapsto ba|bAa$?
(g) $S \mapsto a|b|aSb|aA|bB,\ A \mapsto a|aA,\ B \mapsto b|bB$?
(h) $S \mapsto aA|bS|a|b,\ A \mapsto bA|bS|b$?
(i) $S \mapsto AC|CB,\ C \mapsto \varepsilon|aCb,\ A \mapsto a|aA,\ B \mapsto b|bB$?
(j) $S \mapsto abB,\ B \mapsto bbAa,\ A \mapsto \varepsilon|aaBb$?
(k) $S \mapsto \varepsilon|aSb|SS$?

5.19. Find CFGs that generate the following languages over $\{a, b\}$:

(a) $\{a^{n+1}b^2ab^n : n \in \mathbb{N}\}$.
(b) $\{ww^R : w \in \{a, b\}^+\}$.
(c) $\{a^n b^{n+1} : n \in \mathbb{N}\}$.
(d) $\{a^m b^n : m \neq n - 1; m, n \in \mathbb{N}\}$.
(e) $\{a^m b^n : 2m \leq n \leq 3m; m, n \in \mathbb{N}\}$.
(f) $\{a^m b^n : m \geq n; m, n \in \mathbb{N}\}$.
(g) $\{a^m b^n : m \leq 2n; m, n \in \mathbb{N}\}$.
(h) $\{a^m b^n : m \neq 2n; m, n \in \mathbb{N}\}$.
(i) $\{a^m b^n : m = n \text{ or } m = 2n; m, n \in \mathbb{N}\}$.
(j) $\{a^m b^n : m > 2n; m, n \in \mathbb{N}\}$.
(k) $\{a^m b^n : m \leq n + 3; m, n \in \mathbb{N}\}$.
(l) $\{w \in \{a, b\}^* : \#_a(w) \neq \#_b(w)\}$.
(m) $\{w \in \{a, b\}^* : \#_a(w) \geq 3\}$.
(n) $\{w \in \{a, b\}^* : \#_a(x) \leq \#_b(x)$ for any prefix x of $w\}$.
(o) $\{w \in \{a, b\}^* : \#_b(w) = 2\#_a(w)\}$.
(p) $\{w \in \{a, b\}^* : \#_b(w) = 2\#_a(w) + 1\}$.
(q) $\{a^n b^{2n} : n \geq 1\}$.
(r) $\{a^m b^n : m, n \geq 1, m \neq n\}$.
(s) $\{w \in \{a, b\}^* : \#_a(w) = 2 \cdot \#_b(w))\}$.
(t) $\{uawb : u, w \in \{a, b\}^*,\ \ell(u) = \ell(w)\}$.
(u) $\{a^m b^m : m \geq 1\} \cup \{b^m a^m : m \geq 1\}$.
(v) $\{a^m b^n a^m : m, n \geq 1\} \cup \{a^m b^n c^n : m, n \geq 1\}$.
(w) $\{a^m b^n : 1 \leq m \leq n\}$.

5.20. Construct a CFG to generate the language

(a) $\{a^m b^n c^k : k = n + m; m, n, k \in \mathbb{N}\}$.
(b) $\{a^m b^n c^n : m \geq 0, n \geq 1\}$.
(c) $\{a^m b^n c^k : k = m + 2n; m, n, k \in \mathbb{N}\}$.
(d) $\{a^m b^m c^n : m \geq 1, n \geq 0\}$.
(e) $\{a^m b^n c^k : k \neq m + n; m, n, k \in \mathbb{N}\}$.
(f) $\{a^m b^n c^{m+n} : m, n \in \mathbb{N}\}$.
(g) $\{a^m b^n c^k : k = |m - n|; m, n, k \in \mathbb{N}\}$.
(h) $\{a^m b^n c^k : m = n \text{ or } n \neq k; m, n, k \in \mathbb{N}\}$.
(i) $\{w \in \{a, b, c\}^* : \#_a(w) + \#_b(w) \neq \#_c(w)\}$.
(j) $\{a^m b^n c^k : m \leq n, \text{ or } n \leq k; m, n, k \in \mathbb{N}\}$.
(k) $\{a^m b^n c^k :$ two of m, n, k are equal and the other is $1\}$.
(l) $\{a^m b^n c^k d^r : m + n = k + r,\ m, n, k, r \in \mathbb{N}\}$.

5.21. Give a simple description of the language generated by the grammar with productions $S \mapsto aA|\varepsilon$, $A \mapsto bS$.

5.22. What language is generated by the grammar with productions: $S \mapsto Aa$, $A \mapsto B$, $B \mapsto Aa$? Prove your claim.

5.23. Find a CFG for generating the language $\{a^m ww^R b^m : m \geq 1, w \in \{a, b\}^*\}$.

5.24. Let $L = \{a^n b^n : n \in \mathbb{N}\}$ and $k \geq 1$. Show that the languages L^k, L^*, $\{a, b\}^* - L$ are context-free.

5.25. Consider two grammars: G_1 with productions $S \mapsto aSb|ab|\varepsilon$ and G_2 with productions $S \mapsto aAb|ab$, $A \mapsto aAb|\varepsilon$. Is $\mathcal{L}(G_1) = \mathcal{L}(G_2)$?

5.26. Show that G_1, G_2 generate the same language, where their productions are $G_1 : S \mapsto SS|SSS|aSb|bSa|\varepsilon$ and $G_2 : S \mapsto SS|aSb|bSa|\varepsilon$.

5.27. Show that the CFGs G_1 with productions $S \mapsto a|abC|BA$, $B \mapsto b$, $C \mapsto c|abC$, and G_2 with productions $S \mapsto a|b|cD$, $D \mapsto c|cD$ do not generate the same language.

5.28. Find a CFG that generates all regular expressions over the alphabet $\{a, b\}$.

5.29. Find a CFG that generates all production rules for the CFGs with a, b as terminals and A, B, C as nonterminals.

5.30. Define precisely what one means by properly nested parentheses when two types of parentheses are involved. For example, ( ), [ ], ([( ) ([[ ]])]) are properly nested but ([ )], (( ]) [ ] are not. Using your definition, give a CFG for generating all properly nested parentheses, when two types of parentheses are involved.

5.3 Parse Trees

There is a peculiarity of derivations in context-free grammars, which we do not find in regular grammars. Certainly, the peculiarity should come from the differences between the two types of grammars. What is the difference? In a context-free grammar, the right hand side of a production can have many nonterminals, while in a regular grammar, only one nonterminal is allowed on the right hand side, and that must occur as the rightmost symbol. This has impact on how a string is derived. In a regular grammar, there is essentially a unique derivation of a string with regard to "at what step, which nonterminal grows." While in a context-free grammar, a string can be derived in several essentially different ways.

Example 5.4. Let G be the grammar of Example 5.3. It has the productions:

$$S \mapsto aA \mid bB,\ A \mapsto b \mid bS \mid aAA,\ B \mapsto a \mid aS \mid bBB.$$

The following are two derivations of the same string $aabbba$:

$$S \Rightarrow aA \Rightarrow aaAA \Rightarrow aaAbS \Rightarrow aaAbbB \Rightarrow aaAbba \Rightarrow aabbba,$$

$$S \Rightarrow aA \Rightarrow aaAA \Rightarrow aabA \Rightarrow aabbS \Rightarrow aabbbB \Rightarrow aabbba.$$

These derivations are depicted in Fig. 5.1. Both the diagrams are trees, but we had also placed numbers 1–6 below the derived symbols according to the step in which the corresponding string is derived. Without these numbers, the trees are not different.

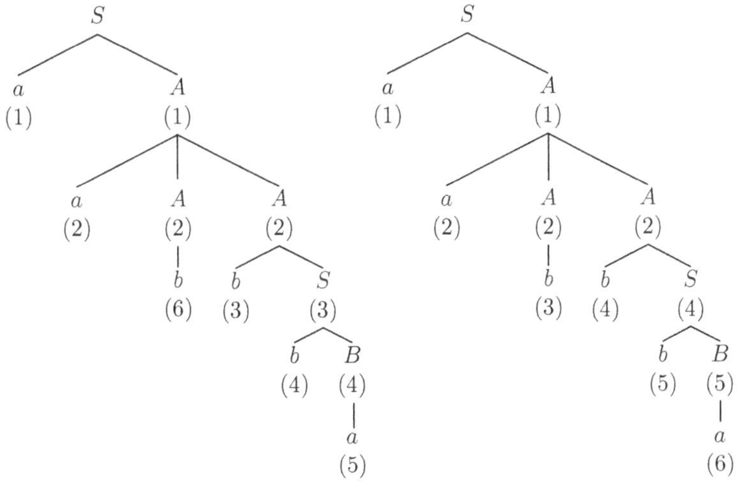

Fig. 5.1. Numbered parse trees for Example 5.4.

The numbering in the right hand side tree in Fig. 5.1 has a nice property. In this derivation, only the first-from-left nonterminal symbols are rewritten by applying a production. Look at the derivation tree again and its numbering. The first-from-left nonterminal is expanded first in the tree. In the first derivation (the left tree in Fig. 5.1), this heuristic is not followed. □

A derivation that *grows on* the leftmost nonterminal symbol is called a **leftmost derivation**.

Exercise 5.4. Give a leftmost derivation of the string $aabbabaababb$ in the context-free grammar $(\{S\}, \{a, b\}, \{S \mapsto \varepsilon \mid SS \mid aSb\}, S)$, and also label the nodes of the tree correspondingly.

Are there context-free grammars where a string can be derived but it cannot be derived by any leftmost derivation? The **derivation trees** above suggest that this should not be the case. For, if we have a derivation of a string, we draw its derivation tree and then reassign the numbers to various nodes. We can always have a renumbering that would correspond to a leftmost derivation. This is the content of the following lemma; you may try a formal proof by induction on the number of derivation steps.

Lemma 5.1. *Let* $G = (N, \Sigma, R, S)$ *be a context-free grammar and* $w \in \Sigma^*$ *be a string. Then,* $w \in \mathcal{L}(G)$ *iff there is a leftmost derivation of* w *in* G.

Proof. If w has been derived by a leftmost derivation in G, then clearly $w \in \mathcal{L}(G)$. Conversely, let $w \in \mathcal{L}(G)$. Then there is a derivation in G for w. We go on checking from the beginning of the derivation whether the leftmost nonterminal is rewritten in a step or not. Once we find the first discrepancy, say, on the nth step, we see that we have a string of the form $uAXBYv$, where $u \in \Sigma^*$, $A \in N$, $X \in (N \cup \Sigma)^*$, $B \in N$, $Y \in (N \cup \Sigma)^*$, and $v \in \Sigma^*$, such that instead of applying a production on A, a production has been applied on B. However, at the end of the derivation, A does not appear. That is, somewhere latter in the derivation, a production on A has been applied leading to a string of terminal symbols alone.

We bring this whole sequence of derivations starting with this occurrence of A to the string of terminals that has grown from this A before applying the production on B. As G is context-free, when a production is applied on a nonterminal, all other symbols in the string remain untouched. If we just switch the productions that have been applied on the symbols A and B, we will not get a derivation. The trick is to switch the sequence of steps between these two applications of productions, one on A and the other on B. This means that in the $(n+1)$th step of the derivation, we apply a production on A instead of B and then follow it up. This will construct a new derivation from the old, which is leftmost up to $n+1$ steps. Now you can complete the proof by induction. □

Exercise 5.5. Use the construction in the proof of Lemma 5.1 and induction on "the least n such that at nth step the leftmost condition is violated" to give a formal proof of the existence of a leftmost derivation.

The first derivation in Example 5.4 (left tree in Fig. 5.1) was:

$$S \Rightarrow aA \Rightarrow aaAA \Rightarrow aaAbS \Rightarrow aaAbbB \Rightarrow aaAbba \Rightarrow aabbba.$$

Scanning from the left (Here, take a piece of paper, write the above derivation and follow up.), you find that at the third step of the derivation (not before it), the leftmost condition is violated. We should have grown on the first A instead of the second. We proceed further to check where this first A has been grown in the derivation. This is done in the sixth step, giving rise to one b. Thus we switch the sequence of steps. That is, the last step is brought to the third. This means that we first construct

$$S \Rightarrow aA \Rightarrow aaAA \Rightarrow aabA \Rightarrow$$

and then continue as earlier, omitting the sixth and the last step, to obtain

$$S \Rightarrow aA \Rightarrow aaAA \Rightarrow aabA \Rightarrow aabbS \Rightarrow aabbbB \Rightarrow aabbba.$$

Again, we scan from the left of whatever derivation we have got thus far, looking for any violation of the leftmost condition. We see that there is no violation as such. We stop, and we have got a leftmost derivation.

Revisit the derivation trees of Example 5.4 in Fig. 5.1. Omitting the step numbers assigned to them, you find that they are the same tree. Abstractly speaking, the derivation trees are, in a way, fixed by the grammar. The derivation trees where we do not mention the step numbers of the derivation are also called **parse trees**.

Many derivation trees can give rise to the same parse tree; this corresponds to many different derivations having the same structure. Consider once again the grammar G of Example 5.4, which has productions

$$S \mapsto aA \mid B,\ A \mapsto b \mid bS \mid aAA,\ B \mapsto a \mid aS \mid bBB.$$

A parse tree in G is given in Fig. 5.2.

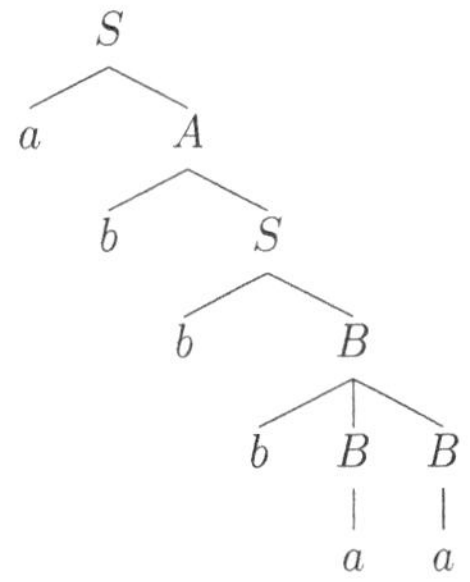

Fig. 5.2. Another parse tree for Example 5.4.

In constructing the parse tree in Fig. 5.2, we have used the productions $S \mapsto aA$, $A \mapsto bS$, $S \mapsto bB$, $B \mapsto bBB$ in that order. The parse tree shows a derivation (many derivations at a time) of the string $abbbaa$. In fact, other rules could have been used and also in a different order, giving rise to many parse trees. Look at the parse trees in Fig. 5.3.

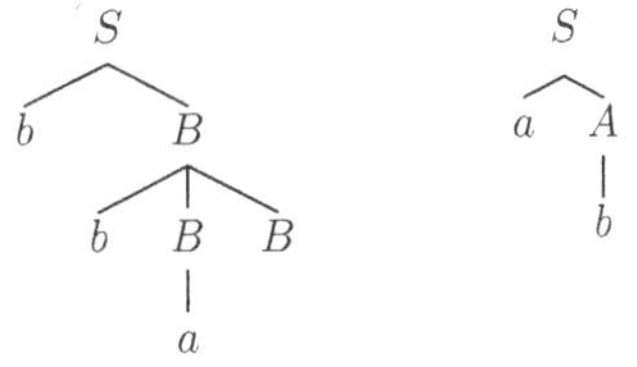

Fig. 5.3. Two parse trees.

The first parse tree is a parse tree for the string $bbaB$ and the second is for the string ab. You see that the string for which a parse tree has been constructed is not necessarily a string over the underlying alphabet of the grammar; it may contain some nonterminal symbols. We will refer to all these strings as the *yields* or *harvests* of the tree. We can have a formal definition of a parse tree.

Let $G = (N, \Sigma, R, S)$ be a context-free grammar. The **parse trees**, their **roots**, **leaves**, and **yields** can be defined inductively as in the following:

(a) For each $A \in N \cup \Sigma$, the single vertex parse tree is A, whose root is A, whose set of leaves is $\{A\}$, and its yield is A.

(b) For each production of the form $A \mapsto \varepsilon$ in R, the parse tree is

whose root is A, set of leaves is $\{\varepsilon\}$, and yield is ε.

(c) Let $A \mapsto A_1 \cdots A_n$ be a production in R. For each i with $1 \le i \le n$, let T_i be a parse tree whose root is A_i, set of leaves is L_i, and yield is y_i. Then T is a parse tree with A as its root, $L_1 \cup \cdots \cup L_n$ as the set of leaves, and $y_1 y_2 \cdots y_n$ as the yield. This says that if we have a production in the form $A \mapsto A_1 \ldots A_n$ and existing parse trees have roots A_i with yields y_i, then the parse tree obtained by using this production looks like

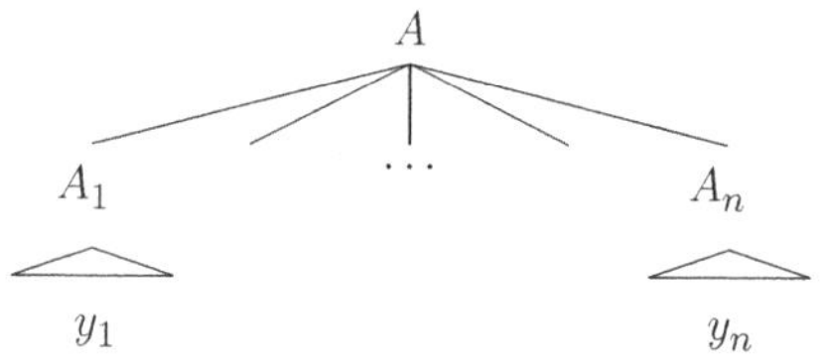

This parse tree has the yield as the concatenated string $y_1 \ldots y_n$, and root A.

(d) The parse trees are generated using the only rules as (a–c) above.

In this definition, each derivation is represented by a parse tree. A parse tree, in fact, represents many derivations. When the yield y of a parse tree is a string of terminal symbols, it is a string generated by the grammar, that is, $y \in \mathcal{L}(G)$. In general, for any $A \in N$ and for any $\alpha \in (N \cup \Sigma)^*$, if $A \Rightarrow \alpha$, then there is a parse tree whose root is A with yield as α; conversely, a parse tree with root A and yield α gives rise to a derivation of $A \Rightarrow \alpha$.

Exercise 5.6. Prove the last statement by using induction on the length of a derivation or on the height of a parse tree.

Example 5.5. Let $G = (\{S, A\}, \{a, b\}, \{S \mapsto a \mid aAS, A \mapsto ba \mid SS \mid SbA\}, S)$. Some of the parse trees in G are drawn in Fig. 5.4. The corresponding yields are a, $abaS$, and $aaASaa$.

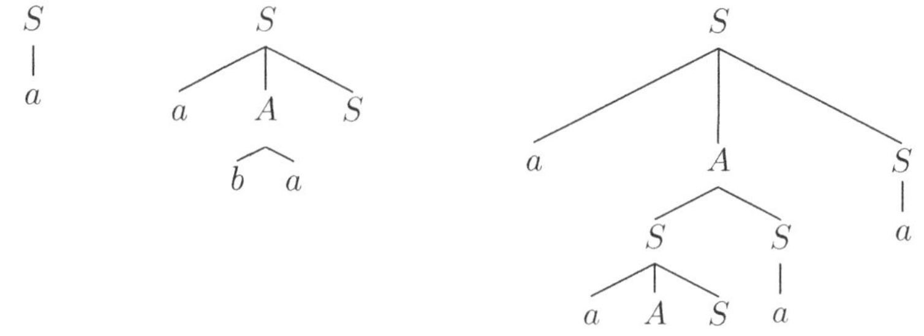

Fig. 5.4. Trees for Example 5.5.

They represent the following derivations, respectively,

$$S \Rightarrow a.$$

$$S \Rightarrow aAS \Rightarrow abaS.$$

$$S \Rightarrow aAS \Rightarrow aSSS \Rightarrow aaASSS \Rightarrow aaASaS \Rightarrow aaASaa.$$

However, the last parse tree in Fig. 5.4 also represents the derivation

$$S \Rightarrow aAS \Rightarrow aAa \Rightarrow aSSa \Rightarrow aSaa \Rightarrow aaASaa.$$

This derivation, of course, generates the same string. We see that the former derivation was a leftmost derivation while the latter is not. See Fig. 5.4. □

It is easy to see that there is a correspondence between parse trees and leftmost derivations. Though a parse tree represents possibly many derivations of a string, among them there can be only one leftmost derivation.

Problems for Section 5.3

5.31. Draw all possible parse trees for the string $aabbaa$ in the CFG with productions $S \mapsto a|aSA,\ A \mapsto ba|SS|SbA.$

5.32. In the CFG with productions $S \mapsto aB|bA,\ A \mapsto a|aS|bAA,\ B \mapsto b|bS|aBB,$ give leftmost and rightmost derivation trees with proper numbering of the nodes with the same yield as the string $aabbabab.$

5.33. In the CFG with productions $S \mapsto a|b|SS$, give two strings each of which has exactly two leftmost derivations. Give two strings each of which has a unique leftmost derivation.

5.4 Ambiguity

Just like many possible derivations of a string, is it possible to have many parse trees with the same yield? You have already seen that each leftmost derivation gives rise to a unique parse tree and each parse tree also represents a unique leftmost derivation. In this light, we ask: whether there are grammars where a string has different leftmost derivations? You can rephrase the question via parse trees. It thus reads as whether there are grammars in which there can be constructed distinct parse trees with the same yield? See the following example.

Example 5.6. Let $G = (\{a, b, +, \times\}, \{S\}, R, S)$, where $R = \{S \mapsto a \mid b \mid S{+}S \mid S{\times}S\}$. Clearly, $a \times b + b \in \mathcal{L}(G)$. Here are two different derivations:

$$S \Rightarrow S + S \Rightarrow S \times S + S \Rightarrow a \times S + S \Rightarrow a \times b + S \Rightarrow a \times b + b.$$
$$S \Rightarrow S \times S \Rightarrow a \times S \Rightarrow a \times S + S \Rightarrow a \times b + S \Rightarrow a \times b + b.$$

Moreover, both the derivations are leftmost derivations. What are the parse trees? Look at Fig. 5.5. The parse trees are clearly different. □

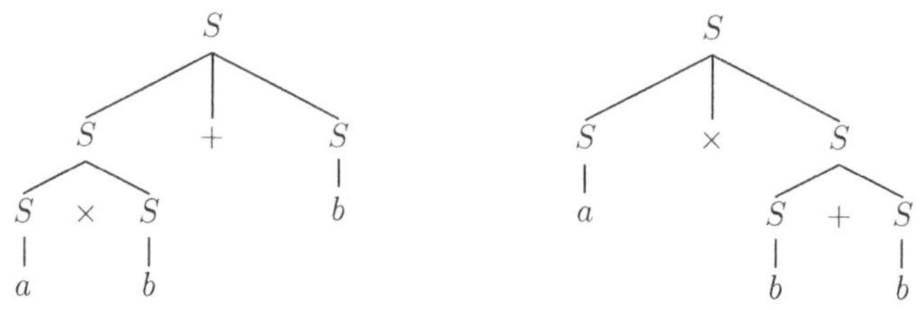

Fig. 5.5. Trees for Example 5.6.

In Example 5.6, if you read the derivations and look at the parse trees, you may see how bracketing of the symbols is represented by parse trees, though there are no brackets used in the grammar.

Suppose that a, b are some numbers and $+$, $\times$ are the usual operations of plus and product of numbers. Suppose that you evaluate a string such as the above by using its parse tree. In this method, you start looking at the leaves. The value of a leaf is passed to its parent. If there are three leaves somewhere, of which the middle one is an operation, say, $+$, then this node, the parent node of the leaves, stores the value of first leaf plus the third leaf. The evaluation proceeds toward the root. In the first tree above, the root node stores the value $(a \times b) + b$, while the root node of the second tree stores the value $a \times (b + b)$.

This is why we would be interested in understanding when a grammar permits many parse trees with the same yield. It is, of course, dangerous to have such a grammar for any fragment of a programing language.

Exercise 5.7. Show that for the CFG $G = (\{a, b, +, \times,), (\}, \{S\}, R, S)$ with $R = \{S \mapsto a \mid b \mid (S + S) \mid (S \times S)\}$, there cannot be two parse trees for the same yield.

If we interpret the strings of the language of Example 5.6 as ordinary arithmetic expressions, then the strings with proper bracketing as done in Exercise 5.7 disambiguates the possible confusion. It clearly says how to evaluate the expressions in a unique way. However, as formal languages, the language of Example 5.6 and of this exercise are different. For, trivially, there is no parentheses in the former language. Can we have a grammar for the same language where there is a unique parse tree for each string?

Let $G = (N, \Sigma, R, S)$ be a context-free grammar. We say that G is **ambiguous** iff there is a string $w \in \mathcal{L}(G)$, which is the yield of at least two different parse trees in G. The grammar G is called **unambiguous** iff it is not ambiguous, that is, iff each string in $\mathcal{L}(G)$ is the yield of a unique parse tree. Thus, in an unambiguous context-free grammar, each string is generated in a unique leftmost derivation. For example, the grammar of Example 5.6 is ambiguous, while that in Exercise 5.7 is unambiguous. But can we have an unambiguous context-free grammar for the language of the grammar in Example 5.6?

Example 5.7. Let $G = (\{S, A\}, \Sigma, R, S)$, where $R = \{S \mapsto A \mid S + A, A \mapsto a \mid b \mid A \times A\}$. What is $\mathcal{L}(G)$? Is G ambiguous?

Solution. From S, we can derive either A or $S + A$. If we go on applying this rule, we will end up at a string of the form $S + A + \cdots + A$ and then $A + A + \cdots + A$. Once S is gone, no more $+$ can be generated. Then the A's can be rewritten as a or b for generating a string, or, some of the A's can generate $\times$, eventually along with a's and b's. G generates the same language as in Example 5.6.

However, this grammar is unambiguous. This is because a leftmost derivation will have to expand S first for deriving all the summands in the required string. Then the A rules will start acting from left to right till all the products are obtained. There seems to be no possibility for generating the same string in two different leftmost derivations. A formal proof of this fact will again use induction for showing that any yield determines the parse tree. Note that the trick lies in defining the precedence of $\times$ over $+$ somewhat implicitly in the grammar. □

Example 5.8. Let $G = (\{S, A, B\}, \{a\}, R, S)$ be a CFG, where $R = \{S \mapsto a \mid SA \mid BS, A \mapsto a, B \mapsto a\}$. Clearly, G is an ambiguous grammar. For example, the distinct parse trees in Fig. 5.6 have the same yield aa:

Fig. 5.6. Trees for Example 5.8.

However, it is easy to disambiguate the grammar. For example, the productions $S \mapsto a \mid aS$ generate the same language $\{a^m : m \geq 1\}$. □

Ambiguity is not only a property of grammars, it pervades languages also. The question whether every ambiguous context-free grammar can be disambiguated leads us to discuss ambiguous languages. We say that a context-free language is **unambiguous** if there is an unambiguous context-free grammar which generates the language. If no such grammar exists, then that would mean that each and every grammar for the language is ambiguous. In such a case, we say that the language is **inherently ambiguous**.

There are inherently ambiguous languages, but to show that they are, indeed, inherently ambiguous is a tedious job. For such a language, if you construct any grammar, then there will be at least two distinct parse trees giving the same yield (for some yield, not necessarily for each and every yield). This means that given any grammar to generate such a language, there is always a string in the language admitting two distinct leftmost derivations in the grammar. See the following example.

Example 5.9. Let $L = \{a^m b^m c^k d^k : m, k \in \mathbb{Z}_+\} \cup \{a^m b^k c^k d^m : m, k \in \mathbb{Z}_+\}$. Construct a context-free grammar G for L and show that G is ambiguous.

Solution. You can generate $a^m b^m$ by a context-free grammar. Similarly, you can have one for $c^k d^k$. Then just add a new start symbol, have a production expressing "the new start symbol can be the start symbol of the first followed by the start symbol of the second." This will give you a grammar for $a^m b^m c^k d^k$. Similarly, you can try for the second one in the union and then try to combine the grammars for obtaining the desired one.

We have one such grammar $G = (\{S, A, B, C, D\}, \{a, b, c, d\}, R, S)$, where $R = \{S \mapsto AB \mid C, A \mapsto aAb \mid ab, B \mapsto cBd \mid cd, C \mapsto aCd \mid aDd, D \mapsto bDc \mid bc\}$. Is G really ambiguous? Yes, as we have two distinct parse trees for the string $a^2b^2c^2d^2$ in G. See Fig. 5.7. In fact, any context-free grammar for L is ambiguous. But the proof of this fact is far from obvious, and we will not indulge in this tedious task. □

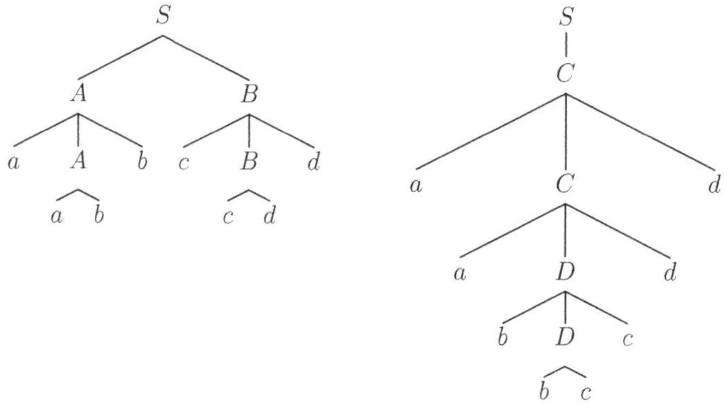

Fig. 5.7. Trees for Example 5.9.

Problems for Section 5.4

5.34. Show that the CFG with productions $S \mapsto \varepsilon | SS | [S]$ generating all balanced parentheses is ambiguous. Construct an unambiguous CFG for the same language.

5.35. The CFG G with productions $S \mapsto +SS | \times SS | - SS | a | b$ generates arithmetic expressions with the operations $+$, $\times$, and $-$ and operands a and b, in *prefix* notation. Find leftmost and rightmost derivation trees for the string $- \times +baab$. Is G ambiguous?

5.36. Let $L = \mathcal{L}(G)$, where G has productions $S \mapsto \varepsilon | aS | aSbS$. Show that

(a) G is ambiguous.

(b) L consists of strings whose every prefix has, at the least, as many a's as b's.

(c) L is not inherently ambiguous.

5.37. Show that the CFG with productions $S \mapsto a|aAb|abSb$, $A \mapsto bS|aAAb$ is ambiguous.

5.38. Is the CFG with productions $S \mapsto ab|aA$, $A \mapsto a|aAB$, $B \mapsto b|ABb$ ambiguous?

5.39. A CFG is called *simple* if each of its productions is in the form $A \mapsto ax$, where A is a nonterminal, a is a terminal, x is any string of terminals and nonterminals, and the pair (A, a) occurs at most once in the productions. For example, the CFG with productions $S \mapsto \varepsilon|aS|bSS$ is simple, while the CFG with productions $S \mapsto \varepsilon|aS|aSS|bSS$ is not.
(a) Find a simple CFG to generate the language $\{b\} \cup aaa^*b$.
(b) Find a simple CFG for $\{a^n b^n : n \in \mathbb{N}\}$.
(c) Find a simple CFG for $\{a^n b^{n+1} : n \in \mathbb{N}\}$.
(d) Show that every simple CFG is unambiguous.
(e) Show that if G is a simple CFG, then any $w \in \mathcal{L}(G)$ can be parsed with an effort proportional to $\ell(w)$.
(f) For a simple CFG G, give an upper bound for the number of productions in terms of the number of terminals and the number of nonterminals.

5.40. Show that the CFGs with productions given below are ambiguous, but the languages they generate are not inherently ambiguous:
(a) $S \mapsto AB|aaB$, $A \mapsto a|Aa$, $B \mapsto b$.
(b) $S \mapsto \varepsilon|SS|aSb$.

5.41. Can a regular grammar be ambiguous? Justify.

5.42. Show that a regular language cannot be inherently ambiguous.

5.43. Construct an unambiguous CFG for generating $\{ww^R : w \in \{a, b\}^*\}$.

5.44. Construct an unambiguous CFG for generating all regular expressions over $\{a, b\}$.

5.45. Are the CFGs with the productions as given below ambiguous? Justify.
(a) $S \mapsto \varepsilon|aSbS|bSaS$.
(b) $S \mapsto \varepsilon|SS|aSb|bSa$.
(c) $S \mapsto aAB$, $A \mapsto bBb$, $B \mapsto \varepsilon|A$.
(d) $S \mapsto \varepsilon|bCa$, $C \mapsto \varepsilon|bCa$.

5.5 Eliminating Ugly Productions

While deriving strings in a CFG, you might have noticed an irritant. It is the absence of any information as to whether you are going in the right direction. The irritation is bound to persist, so to speak. For example, in a CFG, it can very well happen that after some steps of a derivation, you get a string of much bigger length than aimed at string of terminals. That is, after some more steps, some of the nonterminals just give the empty string.

To avoid such a situation, you may try to construct a grammar without the productions of the form $A \mapsto \varepsilon$, the **empty productions** or the **ε-productions**.

In that case, the empty string will not be generated. But can you get everything else? If so, then later you might just add to it the production $S \mapsto \varepsilon$, where S is the start symbol. Thus, we require a CFG whose language does not contain ε and where no step of a derivation will decrease the length of the derived string. Moreover, given any CFG, can we construct one such grammar for generating the same language, of course, leaving the empty string as an exception?

The answer is, in fact, in the affirmative, and we have at least two such normal forms for any CFG. These normal forms are some restricted versions of context-free grammars. In this section, we will prepare for arriving at the two normal forms. In the next section, we will show how to reach at the normal forms.

If a derivation gives a string (of terminals and/or nonterminals) of length less than the current one, then empty productions are present in the CFG. We must thus try to eliminate all empty productions. To eliminate an empty production $A \mapsto \varepsilon$ means that except the string ε we must be able to generate all other strings that could have been generated by using the empty production, and nothing more. The problem is, such productions might be used in conjunction with others in deriving nonempty strings. In that case, how are such productions used?

Let $G = (N, \Sigma, R, S)$ be a CFG, where $A \mapsto \varepsilon \in R$. The only way it can be used nontrivially (trivial when $A = S$) is that A must occur on the right hand side of other productions. So, suppose $B \mapsto xAy \in R$ is another production where $x, y \in (N \cup \Sigma)^*$. In any derivation, their conjunction gives us $B \Rightarrow xy$. This amounts to considering the production $A \mapsto \varepsilon$ in R and then adding $B \mapsto xy$ to R for all such productions $B \mapsto xAy \in R$. Once R is updated this way, we may remove $A \mapsto \varepsilon$ from R. But, wait a bit. First let us check whether the new grammar generates the same language as G. Define the new set of productions R_1 as in the following:

For each $A \in N$ and $A \mapsto \varepsilon \in R$, construct

$R_A = \{B \mapsto xy : B \mapsto xAy \in R,\ x, y \in (N \cup \Sigma)^*\}$.

Next, construct $R_\varepsilon = \cup_{A \in N} R_A$.

Finally, take $R_1 = R \cup R_\varepsilon$.

Let $G_1 = (N, \Sigma, R_1, S)$. As $R \subseteq R_1$, any string that is derived in G is also derived in G_1. Thus $\mathcal{L}(G) \subseteq \mathcal{L}(G_1)$. Moreover, any string that is derived in G_1 using one production, say, $B \mapsto xy$ in R_ε can also be derived in G by using the productions $B \mapsto xAy$ and $A \mapsto \varepsilon$ for some $A \in N$. This shows that $\mathcal{L}(G_1) \subseteq \mathcal{L}(G)$, and therefore, $\mathcal{L}(G) = \mathcal{L}(G_1)$.

Now, we can proceed to removing all empty productions from R_1. Let

$$R_2 = R_1 - \{A \mapsto \varepsilon : A \in N\} \text{ and } G_2 = (N, \Sigma, R_2, S).$$

Obviously, ε is never generated in G_2 and $\mathcal{L}(G_2) \subseteq \mathcal{L}(G_1)$. We want to see that everything else is generated in G_2.

Let $u \neq \varepsilon$ be in $\mathcal{L}(G_1)$. Consider a derivation of u in G_1. If in the derivation a production of the form $A \mapsto \varepsilon$ has been used, then the derivation looks like

$$S \Rightarrow_{G_1} xAy \stackrel{1}{\Rightarrow}_{G_1} xy \Rightarrow_{G_1} u, \text{ for } x \in (N \cup \Sigma)^+ \text{ or } y \in (N \cup \Sigma)^+.$$

The condition $x \in (N \cup \Sigma)^+$ or $y \in (N \cup \Sigma)^+$ is satisfied because otherwise, $xy = \varepsilon$ will contradict $u \neq \varepsilon$. As $S \Rightarrow_{G_1} xAy$, there must have been a production of the form $B \mapsto x'Ay'$ applied somewhere earlier. Then the derivation has the form

$$S \Rightarrow_{G_1} \alpha B\beta \Rightarrow_{G_1} \alpha x'Ay'\beta = xAy \stackrel{1}{\Rightarrow}_{G_1} xy \Rightarrow_{G_1} u.$$

Notice that we have found out two productions $A \mapsto \varepsilon$ and $B \mapsto x'Ay'$ in G_1, and $A \mapsto \varepsilon$ is not in R_2. By the construction of R_2, we have the production $B \mapsto x'y'$ in G_2. Instead of the above derivation, we could have the derivation

$$S \Rightarrow_{G_2} \alpha B\beta \Rightarrow_{G_2} \alpha x'y'\beta = xy \Rightarrow_{G_2} u.$$

Thus, we have proved that if a string $u \neq \varepsilon$ is derived in G_1 using an ε-production at some step, then we can have an alternate derivation of u in G_2. This proves the following statement:

Lemma 5.2 (Elimination of Empty Productions). *Let G, G_1, G_2 be the CFGs as constructed earlier, where G_2 has no ε-productions. Then $\mathcal{L}(G_2) = \mathcal{L}(G_1) - \{\varepsilon\} = \mathcal{L}(G) - \{\varepsilon\}$.*

Another irritant is a production of the form $X \mapsto Y$, where both X and Y are nonterminals. It is because another production $Y \mapsto xXy$ may cause loops, and also structurally, it becomes difficult to see the strings of terminals that might be generated with the presence of such productions. These productions are named **unit productions**. Elimination of unit productions is relatively easier than the empty productions.

Suppose $X \mapsto Y$ and $Y \mapsto Z$ are in R. Then both the productions can be effected by the single production $X \mapsto Z$, even when Z is not just a nonterminal, but a string of terminals and nonterminals. So, let $G = (N, \Sigma, R, S)$ be a CFG. We construct the CFG $G_3 = (N, \Sigma, R_3, S)$ by:

> For each $X \in N$ and a unit production $X \mapsto Y \in R$, construct
>
> $R_X = \{X \mapsto Z : Y \mapsto Z \in R,\ Z \in (N \cup \Sigma)^*\}$.
>
> Next, construct $R_u = \cup_{X \in N} R_X$ and take $R_3 = R \cup R_u - \{X \mapsto Y \in R : X, Y \in N\}$.

Proceeding along the lines of elimination of empty productions, we get the following result for the elimination of unit productions:

Lemma 5.3 (Elimination of Unit Productions). *Let G_3 be the context-free grammar as constructed above from the context-free grammar G by eliminating the unit productions. Then $\mathcal{L}(G_3) = \mathcal{L}(G)$.*

As a matter of elegance, you may not like to have a production such as $S \mapsto Ab$ if all the productions are $S \mapsto Ab|\varepsilon$. It is because the production $S \mapsto Ab$ is never used in any derivation leading to a string of terminals. Such productions are called *useless productions*. The useless production here involves a *useless nonterminal* A. The useless nonterminals are of two types: those which can never be reached from S, and those which never generate any string of terminals.

In general, we call a symbol A in a CFG as *useless* if there is no derivation of the form $S \Rightarrow \alpha A\beta \Rightarrow w$. A useless symbol may be a terminal or a nonterminal. As the name suggests, such symbols never turn up in any derivation of a string of terminals. For example, in the CFG with productions $S \mapsto a|AB$, $A \mapsto b$ the symbols A, B, b are useless. This is not so obvious. But if you try deriving strings of terminals, that is, of a's and b's, then you find that the only string generated by this grammar is a. And other symbols are useless. Useless terminals are easier to find than useless nonterminals.

Let A be a nonterminal in a CFG with start symbol S. The nonterminal A is a **generating nonterminal** if $A \Rightarrow w$ for some string w of terminals. A is a **reachable nonterminal** if $S \Rightarrow \alpha A\beta$ for some strings α, β of terminals and/or nonterminals. Combining both, we say that A is a **useful nonterminal** if there are strings α, β of terminals and/or nonterminals and a string w of terminals such that $S \Rightarrow \alpha A\beta \Rightarrow w$. A **useless nonterminal** is a nonterminal that is not useful. A **useless production** is one that is never used in any derivation of a terminal string.

Let $G = (N, \Sigma, R, S)$ be a CFG. To eliminate the nonreachable nonterminals from G, we compute the set of all reachable nonterminals iteratively by the following:

Mark S reachable.
If $X \in N$ has been marked reachable, and $X \mapsto \alpha Y\beta \in R$, then mark Y reachable.

You can show easily, by induction, that the set of reachable symbols is correctly computed by the above marking scheme. To eliminate the nongenerating nonterminals, we assume that $\mathcal{L}(G) \neq \emptyset$. This assumption guarantees that the start symbol S is a generating nonterminal. We compute the set of generating nonterminals by the following iterative scheme:

Mark ε and each terminal in Σ as generating.
If $X \mapsto \alpha \in R$ and each symbol in α is generating, then mark X to be generating.

The useless symbols are then eliminated by first throwing away nongenerating symbols, and all productions wherever they occur, and then by eliminating all symbols that are not reachable. For example, consider the CFG with productions $S \mapsto a|AB$, $A \mapsto b$. We see that S is generating as $S \Rightarrow a$, A is generating as $A \Rightarrow b$; terminals A, b are generating, by default. The only nongenerating symbol is B. We through away the production $S \mapsto AB$ updating the productions to $S \mapsto a$, $A \mapsto b$. Now, we see that the symbol A (as well as b) are not reachable. Eliminating it, we are left with the only production $S \mapsto a$. And that is it; we have a CFG with no useless symbols. That is, we proceed as follows:

Compute the generating nonterminals and mark the nongenerating ones. Delete from R all productions involving one or more of these marked nongenerating nonterminals.
Compute the reachable nonterminals and mark the nonreachable ones. Delete from R all productions involving one or more of these marked nonreachable nonterminals.
Delete all terminals that do not occur in the updated productions.
Update the CFG G and call it G_4.

Notice that the order of elimination in the above procedure does matter. If you eliminate the nonreachable nonterminals first, and then eliminate the nongenerating ones, you may not, in general, reach at a CFG equivalent to G.

An easy induction proof yields the following result:

Lemma 5.4 (Elimination of Useless Productions). *Let G be a CFG such that $\mathcal{L}(G) \neq \emptyset$. Let G_4 be the context-free grammar as constructed above from the context-free grammar G by eliminating the useless productions. Then $\mathcal{L}(G_4) = \mathcal{L}(G)$.*

Exercise 5.8. What happens if we first eliminate the nonreachable symbols and then nongenerating symbols? Will the resulting CFG be without useless symbols? [Hint: Carry out the steps on the CFG with productions $S \mapsto a,\ A \mapsto b$.]

Elimination of useless symbols is not only a matter of elegance; it also helps in saving time while converting a CFG to its normal forms. The normal forms we will be discussing in the next section are not only helpful for allowing nice progress indicators in a derivation, but also they help in proving certain results in relatively shorter way. One more point of elegance: do not be carried away by thinking of the empty productions, unit productions, and the useless productions as ugly as the heading to this section suggests.

Problems for Section 5.5

5.46. Find a CFG without ε-productions equivalent (except generating ε) to the CFG with productions $S \mapsto ABaC,\ A \mapsto BC,\ B \mapsto \varepsilon|b,\ C \mapsto \varepsilon|D,\ D \mapsto c$.

5.47. Find a CFG without unit productions equivalent to the CFG with productions $S \mapsto B|Aa,\ B \mapsto A|bb,\ A \mapsto a|bc|B$.

5.48. Let G be a CFG with productions $S \mapsto aBC,\ B \mapsto bCb,\ C \mapsto \varepsilon|B$. Find a CFG equivalent to G, which does not have any production in the forms $A \mapsto \varepsilon$ or $A \mapsto B$ for any nonterminals A, B.

5.49. Show that the CFGs G, G' with the productions as given below are equivalent:

$$G:\ S \mapsto ab|baAB,\ A \mapsto bbb,\ B \mapsto aa|bA,$$
$$G':\ S \mapsto ab|baAbb,\ baAbA,\ A \mapsto bbb.$$

5.50. Eliminate all ε-productions from the CFG with productions
(a) $S \mapsto bbA|BaA,\ A \mapsto \varepsilon|bbB,\ B \mapsto \varepsilon$.
(b) $S \mapsto AB,\ A \mapsto \varepsilon|bAA,\ B \mapsto \varepsilon|aBB$.

5.51. Eliminate all unit productions from the CFG with productions $S \mapsto A|S + A$, $A \mapsto B|A \times B,\ B \mapsto C|(S),\ C \mapsto a|b|Ca|Cb|C0|C1$.

5.52. Consider the CFG G with productions $S \mapsto AB|CD,\ A \mapsto a|BC,\ B \mapsto C|AC,\ C \mapsto AB|CD,\ D \mapsto d|AC$.
(a) Find all generating symbols in G.
(b) Eliminate all useless productions from G.
(c) From the resulting CFG, eliminate all nonreachable symbols.

5.53. Eliminate useless symbols from the CFG with productions $S \mapsto AB|CA$, $A \mapsto a,\ B \mapsto AB|BC,\ C \mapsto b|aB$.

5.54. Eliminate all unit productions and ε-productions from the CFG with productions $S \mapsto bA|bBB,\ A \mapsto \varepsilon|bbA,\ B \mapsto aB|aaC,\ C \mapsto B$. Try simplifying the resulting grammar further by eliminating useless symbols. What is $\mathcal{L}(G)$?

5.55. Eliminate all unit productions from the CFG that has the productions $S \mapsto c|A|B,\ A \mapsto D|Ca,\ B \mapsto E|Cb,\ C \mapsto d|D|E,\ D \mapsto S|Ea,\ E \mapsto S|Db$.

5.56. In a CFG G, a symbol is called *nullable* if it derives ε in one or more steps. A *unit pair* in G is a pair (X, Y) of nonterminals, such that there is a derivation $X \Rightarrow_G Y$ using only unit productions of G. Consider the CFG G with productions $S \mapsto AB|CD,\ A \mapsto z|BG,\ B \mapsto \varepsilon|AD,\ C \mapsto b|CD,\ D \mapsto E|BB,\ E \mapsto Bb|AF,\ F \mapsto aC|EG,\ G \mapsto AG|BD$.
(a) Find all nullable symbols in G.
(b) Modify the productions in G so that the resulting equivalent (except generating of ε) CFG has no ε-productions.
(c) Find all unit pairs in G.
(d) Eliminate all unit productions from G.

5.57. Let G be a CFG with productions $S \mapsto A|bSaa,\ A \mapsto \varepsilon|S|aAbb$. Construct a CFG G' without unit productions and without ε-productions such that $\mathcal{L}(G') = \mathcal{L}(G) - \{\varepsilon\}$.

5.58. Consider the CFG G with productions $A \mapsto BC|DE,\ C \mapsto DE|F,\ E \mapsto GH|I,\ G \mapsto HI|C$, where the usual convention of taking S as the start symbol, and upper-case symbols as nonterminals, and others as terminals is not followed. We also mention that each production in G is used in some derivation of a string of terminals. Now, identify the start symbol, the nonterminals, and the terminals in G.

5.59. Give an example of a CFG where elimination of ε-productions introduces previously nonexistent unit productions.

5.6 Normal Forms

In this section, we will discuss two normal forms, the *Chomsky Normal Form* and the *Greibach Normal Form*, named after their originators, N. Chomsky and S. A. Greibach, respectively.

A context-free grammar G is said to be in **Chomsky normal form (CNF)** if all productions in G are of the form $A \mapsto BC$ or $A \mapsto \sigma$, for nonterminals A, B, C, and terminals σ.

Once you have eliminated the empty productions and the unit productions, you can convert the resulting grammar into Chomsky normal form. The productions of the form $A \mapsto \sigma$ for $\sigma \in \Sigma$ can be left as they are. The problematic productions are now in the form $A \mapsto A_1 A_2 \cdots A_m$, where $A_i \in N \cup \Sigma$. (Why?) If all of $A_i \in N$, then we can introduce new nonterminals so that there will be exactly two nonterminals on the right hand side of $\mapsto$. For example, the production $A \mapsto XYZ$ can be replaced by the pair $A \mapsto XB$ and $B \mapsto YZ$, provided B is a new nonterminal.

On the other hand, if an A_i is a terminal symbol, then we introduce another new nonterminal, say I corresponding to it and add a production accordingly. For example, the production $A \mapsto aB$ can be replaced by the pair $A \mapsto XB$ and $X \mapsto a$, where X is again a new nonterminal. We summarize the discussion in the following statement:

Theorem 5.1 (Chomsky Normal Form). *Corresponding to every CFG G, there exists a CFG G′ in Chomsky normal form such that* $\mathcal{L}(G') = \mathcal{L}(G) - \{\varepsilon\}$.

Proof. (Outline) We merely give the construction of G'. Using Lemmas 5.2 and 5.3, you can give a formal proof. Let $G = (N, \Sigma, R, S)$ be a CFG. The construction, to be carried out step-by-step, is as follows:

1. Eliminate the empty productions from G to get G'.
2. Eliminate the unit productions from G' to get G''.
3. Corresponding to each $\sigma \in \Sigma$, introduce a new nonterminal X_σ and a new production $X_\sigma \mapsto \sigma$.
4. Replace all occurrences of σ by X_σ on the right hand side of each production, leaving the productions of the form $B \mapsto \sigma$.
5. Replace each production of the form $A \mapsto A_1 A_2 \cdots A_m$, $m > 2$, A, A_i nonterminals, including X_σ's, by the m productions $A \mapsto A_1 B_1$, $B_1 \mapsto A_2 B_2, \ldots, B_{m-1} \mapsto A_m B_m$, where $B_1, \ldots, B_m$ are new nonterminals. □

As a matter of fact, empty productions are to be eliminated first. If you eliminate unit productions and then eliminate the empty productions, you may not get the desired effect. Can you see why? Moreover, you can eliminate useless nonterminals (or even useless terminals) after eliminating the empty productions and the unit productions.

Example 5.10. Convert $G = (\{S\}, \{a, b\}, \{S \mapsto aSb \mid \varepsilon\}, S)$ to CNF.

Solution. The first step is to eliminate the empty productions. The only empty production is $S \mapsto \varepsilon$. Then,

$$R_S = \{B \mapsto xy : B \mapsto xSy \in R,\ x, y \in (N \cup \Sigma)^*\} = \{S \mapsto ab\}.$$

as $S \mapsto aSb$ is the only production of the form $B \mapsto xSy$. Next,

$$R_\varepsilon = \cup R_S = R_S,\ R_1 = R \cup R_\varepsilon = \{S \mapsto aSb \mid ab \mid \varepsilon\}.$$

$$R_2 = R_1 - \{S \mapsto \varepsilon\} = \{S \mapsto aSb \mid ab\}.$$

As there is no unit production in R_2, we just introduce the new nonterminals corresponding to each terminal (Steps 4 and 5). This updates, writing A for X_a and B for X_b, the new productions along with the modified old ones to

$$A \mapsto a,\ B \mapsto b,\ S \mapsto ASB,\ S \mapsto AB.$$

The only production having more than two symbols on the right hand side is $S \mapsto ASB$. This requires introducing another nonterminal, say, C and replacing this by the pair $S \mapsto AC$ and $C \mapsto SB$. We then take together all the productions to obtain

$$R' = \{A \mapsto a,\ B \mapsto b,\ S \mapsto AC,\ C \mapsto SB,\ S \mapsto AB\}.$$

The required CFG in CNF is $G' = (\{S, A, B, C\}, \{a, b\}, R', S)$. □

Exercise 5.9. Show that a CNF for the CFG $(\{S\}, \{a, b\}, \{S \mapsto aSb \mid SS \mid \varepsilon\}, S)$ is $(\{S, A, B, C\}, \{a, b\}, \{S \mapsto AB,\ S \mapsto AC,\ S \mapsto SS,\ C \mapsto SB,\ A \mapsto a,\ B \mapsto b\}, S)$.

A derivation of the string $aabb$ in the CNF G' of Example 5.10 looks like

$$S \Rightarrow AC \Rightarrow aC \Rightarrow aSB \Rightarrow aABB \Rightarrow aaBB \Rightarrow aabB \Rightarrow aabb.$$

In every step of this derivation, some progress towards the string $aabb$ is clearly visible. The indicator of progress is the length of the current string (of terminals and nonterminals taken together) compared with the earlier. The length grows as we go for more and more steps of any derivation. A better progress indicator would be the string itself. That is, imagine a derivation where in each new step, you get one more symbol from the aimed at string. In such a case, you would first get a, next, another a, next, b, and finally, the last b. To get a CFG where this can be realized, we may have to do more work. Notice that in such a grammar, each production must start with a terminal symbol, and then each application of a production would give rise to one more relevant symbol of the string. The requirements are formalized by the following:

A context-free grammar G is said to be in **Greibach normal form (GNF)** if each production in G is of the form $A \mapsto \sigma A_1 A_2 \cdots A_m$ or $A \mapsto \sigma$, where $A_1, A_2, \ldots, A_m$ are nonterminals and σ is a terminal.

To convert a CFG to one in GNF, we start with eliminating the productions of the form $A \mapsto A\alpha$, where the symbol on the left hand side of a production is the same as the first symbol on the right hand side. For example, if $A \mapsto AC$ is a production in a given CFG, then possibly a straight away loop might hinder progress. Our aim here is to replace such productions with a set of productions which do not show this peculiarity.

Let us call any production of the form $A \mapsto A\beta$, $\beta \in (N \cup \Sigma)^*$ a **self-production**. Notice that $A \mapsto \beta A$ is not a unit production. To eliminate self-productions, let us look at the mechanism of their applications.

Suppose we have a self-production $A \mapsto AC$. The only way it eventually generates a terminal is that somewhere in the derivation we apply another production of the form $A \mapsto \alpha$, where α is any string over $N \cup \Sigma$ not having an occurrence of A. Now, can we just replace the two productions by $A \mapsto \alpha C$? No. For example, a derivation such as

$$A \Rightarrow AC \Rightarrow ACC \Rightarrow \alpha CC$$

cannot be generated from the single production $A \mapsto \alpha C$. The trick is to introduce another nonterminal, say, B, and then have the productions

$$A \mapsto \alpha, \; A \mapsto \alpha B, \; B \mapsto C, \; B \mapsto CB.$$

Notice that we run the risk of introducing unit productions; we will see how to tackle those later (You have at least Lemma 5.3.). For example, the string αCC can now be derived from A as

$$A \Rightarrow \alpha B \Rightarrow \alpha CB \Rightarrow \alpha CC.$$

Lemma 5.5 (Elimination of Self-productions). *Let A be a nonterminal in a context-free grammar G. Let $A \mapsto A\alpha_1 \mid A\alpha_2 \mid \cdots \mid A\alpha_m$ be all self-productions in G and $A \mapsto \beta_1 \mid \beta_2 \mid \cdots \mid \beta_n$ be all other productions in G that have A on the left hand side. Let B_A be a new nonterminal. Let $G_A = (N \cup \{B_A\}, \Sigma, R_A, S)$ with*

$$R_A = (R - \{A \mapsto A\alpha_1 \mid A\alpha_2 \mid \cdots \mid A\alpha_m\}) \cup$$
$$\{A \mapsto \beta_1 B_A \mid \cdots \mid \beta_n B_A, \; B_A \mapsto \alpha_1 \mid \cdots \mid \alpha_m, \; B_A \mapsto \alpha_1 B_A \mid \cdots \mid \alpha_m B_A\}.$$

Then $\mathcal{L}(G_A) = \mathcal{L}(G)$.

Proof. As $A \in N$, any leftmost derivation of any string in Σ^* must have an application of a production of the form $A \mapsto \beta_j$. Without loss of generality, we thus assume that any derivation that uses a production of the form $A \mapsto A\alpha$ would have in it a derivation of the following form:

$$A \Rightarrow A\alpha_{i1} \Rightarrow A\alpha_{i2}\alpha_{i1} \Rightarrow \cdots \Rightarrow A\alpha_{ik} \ldots \alpha_{i1} \Rightarrow \beta_j \alpha_{ik} \ldots \alpha_{i1}. \tag{5.1}$$

Such a (portion of a) derivation can be mimicked in G_A by

$$A \Rightarrow A\beta_j B_A \Rightarrow \beta_j \alpha_{ik} B_A \Rightarrow \beta_j \alpha_{ik} \alpha_{i\,k-1} \ldots \alpha_{i2} B_A \Rightarrow \cdots \Rightarrow \beta_j \alpha_{ik} \ldots \alpha_{i2}\alpha_{i1}. \tag{5.2}$$

Hence $\mathcal{L}(G) \subseteq \mathcal{L}(G_A)$.

Conversely, any derivation in G_A that involves B_A must have come from an earlier A. Any derivation that involves A might use directly a production $A \mapsto \beta_j$ or one of the other three types of productions (that we have introduced while eliminating $A \mapsto A\alpha$) involving B_A. The first type of productions $A \mapsto \beta_j$ are already in G. On the other hand, if a derivation brings in B_A from A, then it must eventually end in an application of a production of the type $B_A \mapsto \alpha_i$. Any such derivation in G_A is of the form (5.2) above. And each of these (portions of) derivations can be replaced by a derivation of the type (5.1) in G. Thus $\mathcal{L}(G_A) \subseteq \mathcal{L}(G)$. □

We want to guarantee that in each derivation some progress is made towards the string of terminals that is being derived. By eliminating self-productions, we see that from an A, we would not get $A\alpha$. But we may get $C\alpha$ from A. Again, somewhere in a derivation, we may get $A\beta\alpha$ from $C\alpha$, for example, if we have a production $C \mapsto A\beta$. This would result in, essentially, a self-production again. To prevent such a situation, let us follow an ordering of the nonterminals, say, N has the elements $A_1, A_2, \ldots, A_m$. Our aim is to have no production of the form $A_i \mapsto A_j\alpha$, where $j < i$. If $A = A_1$, $C = A_3$, then we cannot have a production $C \mapsto A\beta$ after this aim is fulfilled.

For this purpose, let us call a production of the form $A_i \mapsto A_j\alpha,\ j < i$, a **dominated production**. Now, how to eliminate the dominated productions? This is done by applying the following result inductively:

Lemma 5.6 (Elimination of Dominated Productions). *Let $G = (N, \Sigma, R, S)$ be a context-free grammar. Let $A \mapsto C\beta \in R$ and $C \mapsto \alpha_1 \mid \alpha_2 \mid \cdots \mid \alpha_n$ be all the productions in R with C on the left hand side. Let*

$$R_d = (R - \{A \mapsto C\beta\}) \cup \{C \mapsto \alpha_1\beta \mid \cdots \mid \alpha_n\beta\} \text{ and } G_d = (N, \Sigma, R_4, S).$$

Then $\mathcal{L}(G) = \mathcal{L}(G_d)$.

Proof. Consider any derivation involving A and then proceed as in the proof of Lemma 5.5. □

For eliminating the dominated productions, assume that all self-productions have been eliminated. Then order the nonterminals. Suppose the ordered nonterminals are $A_1, A_2, \ldots, A_m$. As we do not have any production of the form $A_1 \mapsto A_1\alpha$, trivially, the following is satisfied for $i = 1$:

$$\text{If } A_i \mapsto A_j\alpha \text{ is a production, then } j > i. \tag{5.3}$$

Inductively, assume that for $1 \leq i < k$, the statement (5.3) is satisfied. If $A_k \mapsto A_{k-1}\alpha$ is a production, then consider all productions of the form $A_{k-1} \mapsto \beta$. In β, the first symbol, if at all a nonterminal, must be one of $A_k, A_{k+1}, \ldots, A_m$, due to our assumption. Using the replacement in Lemma 5.6, we see that $A_k \mapsto A_{k-1}\alpha$ can be replaced by productions of the form $A_k \mapsto \beta\alpha$, where the first symbol on the right hand side (i.e., of β), if at all a nonterminal, is one among $A_k, A_{k+1}, \ldots, A_m$. If it is A_k, then use Lemma 5.5 to eliminate the self-production. Thus, the production $A_k \mapsto A_{k-1}\alpha$ is eliminated. Similarly, if $A_k \mapsto A_{k-r}\alpha,\ r > 0$ is a production, then we repeat this construction, at the most, r times to eliminate it. After this elimination, any production with A_k on the left hand side will satisfy (5.3) for $i = k$.

Notice that this might require eliminating self-productions which, in turn, will introduce new nonterminals. Such nonterminals, call them $B_1, B_2, \ldots, B_l$, are not in the ordered ones such as $A_1, \ldots, A_m$. We will see how to tackle these new nonterminals.

What we have done is that if (5.3) is satisfied for $i = k-1$, then we can modify the productions in such a way that (5.3) will be satisfied for $i = k$. From Lemmas 5.4–5.6, the language generated by a grammar with the modified productions remains the same. Thus, inductively, this shows that we can have a CFG where all the productions will satisfy (5.3) without changing the language.

Our argument shows that by the end of this step, all productions are in one of the following forms:

$$A_i \mapsto A_j\alpha,\ A_i \mapsto \sigma\beta,\ \sigma \in \Sigma,\ \text{or } B_t \mapsto \gamma, \tag{5.4}$$

where $\alpha, \beta, \gamma \in (N \cup \{B_1, B_2, \ldots, B_l\} \cup \Sigma)^*$, $\sigma \in \Sigma$, and $j > i$.

We observe that no production with A_m on the left hand side can be of the form $A_m \mapsto A_i\beta$, due to (5.4). Thus, it has to be of the form $A_m \mapsto \sigma\beta$ for some $\sigma \in \Sigma$. If we have a production of the form $A_{m-1} \mapsto A_m\alpha$, then we replace it with $A_{m-1} \mapsto \sigma\beta\alpha$ using Lemma 5.6 on the pair of productions $A_m \mapsto \sigma\beta$ and $A_{m-1} \mapsto A_m\alpha$. Now, for $i = m, m-1$, we have all our productions of the form $A_i \mapsto \sigma\beta$.

Next, we proceed for the productions of which A_{m-2} is the left hand side replacing similarly all those of the form $A_{m-2} \mapsto A_{m-1}\alpha_1$ and $A_{m-2} \mapsto A_m\alpha_2$. Proceeding further backward, we replace each production of the form $A_i \mapsto A_j\alpha$ by several productions of the form $A_i \mapsto \sigma\beta$. Notice that the language generated by the new grammar with the updated set of productions remains the same. The updated productions are in one of the following forms:

$$A_i \mapsto \sigma\beta \text{ for } \sigma \in \Sigma,\ \text{ or }\ B_t \mapsto \gamma,\ \text{ for } \beta, \gamma \in (N \cup \{B_1, B_2, \ldots, B_l\} \cup \Sigma)^*. \tag{5.5}$$

What about the B_t's? A look at Lemma 5.5 shows that no production with a B_t on the left hand side can begin with any B_p. Therefore, any production of the form $B_t \mapsto \gamma$ has either a terminal or some A_i as the first symbol on the right hand side (i.e., of γ). If it is A_i, then use Lemma 5.6 again to replace it with productions of the form $B_t \mapsto \tau\alpha$, where $\tau \in \Sigma$. Thus the updated productions are

$$A_i \mapsto \sigma\beta \text{ for } \sigma \in \Sigma,\ \text{ or } B_t \mapsto \tau\gamma \text{ for } \tau \in \Sigma,\ \beta, \gamma \in (N \cup \{B_1, B_2, \ldots, B_l\} \cup \Sigma)^*. \tag{5.6}$$

If in any of these productions, any of the β or γ has a terminal symbol in it, then we can introduce new nonterminals and replace them with pairs of productions as in Chomsky normal form. That is, suppose we have a production of the form $A_i \mapsto \sigma A_1 \cdots A_q$, where $A_1 = a \in \Sigma$. Then we introduce a new nonterminal, say, C_1 and replace the production $A_i \mapsto \sigma A_1 \cdots A_q$ with the two productions $A_i \mapsto \sigma C_1 \cdots A_q$ and $C_1 \mapsto a$.

Repeating this for each such production, and in it, for each such occurrence of a terminal on the right hand side except the first one, we see that each updated production is of the form

$$A_i \mapsto \sigma\beta \text{ for } \sigma \in \Sigma, \text{ or } B_t \mapsto \tau\gamma \text{ for } \tau \in \Sigma, \ \beta, \gamma \in (N \cup \{B_1, B_2, \ldots, B_l\})^*. \tag{5.7}$$

This step is unnecessary if you start in the beginning with a grammar in CNF. (Why?) If each production is in one of the forms of (5.7), then you have got a grammar in Greibach normal form. We summarize the above discussion as follows:

Theorem 5.2 (Greibach Normal Form). *Let $G = (N, \Sigma, R, S)$ be a context-free grammar. There is a context-free grammar $\tilde{G} = (\tilde{N}, \Sigma, \tilde{R}, S)$ in Greibach normal form such that $\mathcal{L}(\tilde{G}) = \mathcal{L}(G) - \{\varepsilon\}$.*

Example 5.11. Give a GNF for $G = (\{S\}, \{a, b\}, R, S)$, where $R = \{S \mapsto aSb \mid SS \mid \varepsilon\}$.

Solution. First, we eliminate the empty production $S \mapsto \varepsilon$. This results in the productions

$$S \mapsto aSb, \ S \mapsto SS, \ S \mapsto ab.$$

For eliminating the self-production $S \mapsto SS$, we introduce a new nonterminal, say, B (See Lemma 5.5) and update the productions to

$$S \mapsto aSb, \ S \mapsto ab, \ S \mapsto aSbB, \ S \mapsto abB, \ B \mapsto S, \ B \mapsto SB.$$

As S is the only old nonterminal, there is no dominated production. (Alternatively, you can choose the ordering S and then B.) We replace the productions involving new nonterminals on the left using Lemma 5.6. Here, the first symbol on the right of such productions is B. For the production $B \mapsto S$, we have the corresponding productions

$$B \mapsto aSb, \ B \mapsto ab, \ B \mapsto aSbB, \ B \mapsto abB.$$

For the other production $B \mapsto SB$, we have

$$B \mapsto aSbB, \ B \mapsto abB, \ B \mapsto aSbBB, \ B \mapsto abBB.$$

We thus have the updated productions

$$S \mapsto aSb \mid ab \mid aSbB \mid abB,$$

$$B \mapsto aSb \mid ab \mid aSbB \mid abB, \ B \mapsto aSbB \mid abB \mid aSbBB \mid abBB.$$

In the final step, we just introduce new nonterminals for taking away the terminals from all the right hand sides of productions, leaving the first one. Suppose we use A for b. (For a, we need not introduce any new nonterminal here.) Then the updated set of productions is

$$\tilde{R} = \{S \mapsto aSA \mid aA \mid aSAB \mid aAB, \ B \mapsto aSA \mid aA \mid aSAB \mid aAB,$$
$$B \mapsto aSAB \mid aAB \mid aSABB \mid aABB, \ A \mapsto b\}.$$

The GNF corresponding to G is the grammar $\tilde{G} = (\{S, A, B\}, \{a, b\}, \tilde{R}, S)$. □

Exercise 5.10. Write the conversion of a grammar to a GNF in a step-wise manner as in the proof of Theorem 5.1.

Example 5.12. Construct a GNF for $G = (\{S\}, \{a, b\}, R, S)$, where

$$R = \{S \mapsto AB \mid AC \mid SS,\ C \mapsto SB,\ A \mapsto a,\ B \mapsto b\}.$$

Solution. Introduce a new nonterminal E for eliminating the self-production $S \mapsto SS$. By Lemma 5.5, we have the updated productions as

$$S \mapsto AB \mid AC,\ C \mapsto SB,\ A \mapsto a,\ B \mapsto b,\ S \mapsto ABE \mid ACE,\ E \mapsto S \mid SE.$$

Suppose we choose the ordering of the nonterminals as E, C, S, A, B, from smallest to biggest, that is, $A_1 = E$, $A_2 = C$, $A_3 = S$, $A_4 = A$, $A_5 = B$, then there is no dominated production. (For each production of the form $A_i \mapsto A_j\alpha$, $j > i$ is satisfied.) We thus use Lemma 5.6 to get rid of the "first symbol on the right which may be a nonterminal".

We start with the biggest nonterminal, that is, B. There is only one production with B as the left hand side. It is $B \mapsto b$. But the first symbol on its right hand side is a terminal. The same happens for A, the next biggest in our ordering. For S, we have the productions

$$S \mapsto AB \mid AC \mid ABE \mid ACE.$$

Lemma 5.6 gives the corresponding updated productions as

$$S \mapsto aB \mid aC \mid aBE \mid aCE.$$

Similarly, for C and E, the production $C \mapsto SB$ and $E \mapsto S \mid SE$ give

$$C \mapsto aBB \mid aCB \mid aBEB \mid aCEB;$$

$$E \mapsto aB \mid aC \mid aBE \mid aCE \mid aBE \mid aCE \mid aBEE \mid aCEE.$$

You may drop one of the two instances of the production $E \mapsto aCE$, if you wish. There is no production where a terminal is in the middle of the right hand side; so we need not introduce any new nonterminal. The GNF is

$$\tilde{R} = \{S \mapsto aB \mid aC \mid aBE \mid aCE, A \mapsto a,\ B \mapsto b, C \mapsto aBB \mid aCB \mid aBEB \mid aCEB,$$
$$E \mapsto aB \mid aC \mid aBE \mid aCE \mid aBE \mid aBEE \mid aCEE\}.$$ □

Exercise 5.11. In Example 5.12, if you take the ordering of nonterminals as E, S, C, A, B instead of E, C, S, A, B, then you will eliminate $C \mapsto SB$ by replacing it with the productions $C \mapsto ABB \mid ACB \mid ABEB \mid ACEB$. Proceed and construct the GNF. Is it same as above?

Given a CFG in GNF and a string of terminals, it is relatively easy to visualize how to determine whether the string can be generated in the given grammar or not. In the first place, we start with all the productions having the start symbol on the left hand side and the first symbol of the given string as the first symbol on the right hand side. If there are m such productions, we have m choices for one such production.

Next, we look for the second symbol of the given string, and choose an appropriate production and proceed. For example, if our string is aba and our grammar has the productions $S \mapsto aAB \mid aBA \mid bAB$, $S \mapsto a$, then in the first place, we have the choices $S \mapsto aAB$ and $S \mapsto aBA$ (leaving here $S \mapsto a$ which will not lead us towards the string).

If we take up the first choice, then we have derived aAB. We look for the next symbol in the given string, which happens to be b. So, we need a production that would have the left hand side as A and the right hand side would begin with a b. There is no such production; so this way we cannot derive the string. We go back and take up the left out choice in the first step, that is, we start a fresh derivation by choosing $S \mapsto aBA$.

Clearly, this gives an algorithm for the decision problem of determining whether a given string of terminals is at all generated by a given grammar in GNF or not. However, this process is exponential in the length of the given string. There is, of course, a better algorithm that takes only $O(n^3)$ time, where n is the length of the given string. It is the so-called CYK algorithm; see Problem 10.88 in the additional problems to Chap. 10.

Problems for Section 5.6

5.60. Let G be the CFG with productions $S \mapsto \varepsilon|ASB$, $A \mapsto b|bAS$, $B \mapsto aa|A|SaS$. Eliminate all
(a) ε-productions,
(b) unit productions,
(c) useless symbols, and finally
(d) convert to CNF.

5.61. Prove: If each production of the form $A \mapsto xy$ in a CFG G is replaced by the pair of productions $A \mapsto By$, $B \mapsto x$, where B is a new nonterminal, then the language generated by the new CFG is $\mathcal{L}(G)$.

5.62. Eliminate self-productions from the CFG having productions $S \mapsto \varepsilon|Sa|aBc$, $B \mapsto bc|Bc$.

5.63. Convert the CFGs with the following productions to equivalent CFGs in CNF:

(a) $S \mapsto a|b|cSS$.
(b) $S \mapsto Ab$, $A \mapsto aB$, $B \mapsto a|Ab$.
(c) $S \mapsto \varepsilon|A|ASA$, $A \mapsto \varepsilon|aa$.
(d) $S \mapsto CDa$, $C \mapsto aab$, $D \mapsto Cc$.
(e) $S \mapsto A|bSbA$, $A \mapsto a|baA$.
(f) $S \mapsto bB|AB$, $A \mapsto \varepsilon|bba$, $B \mapsto aaA$.
(g) $S \mapsto baAA$, $A \mapsto \varepsilon|aAB$, $B \mapsto \varepsilon|A|b$.
(h) $S \mapsto aB|ASA$, $A \mapsto S|B$, $B \mapsto \varepsilon|b$.

5.64. Let G be a CFG in CNF and $w \in \mathcal{L}(G)$ such that $\ell(w) = n \geq 1$. Show that
(a) there is a derivation of length $2n - 1$ of w in G.
(b) each parse tree for w has $2n - 1$ interior nodes.

5.65. Show that each CFG can be converted to one whose productions are in the forms $A \mapsto \varepsilon$, $A \mapsto BC$, or $A \mapsto aBC$.

5.66. Convert the CFGs with the following productions to equivalent GNFs:
(a) $S \mapsto ab|aSb$.
(b) $S \mapsto a|b|cSS$.
(c) $S \mapsto bb|baSa$.
(d) $S \mapsto a|b|aSb|bSa$.
(e) $S \mapsto a|AA,\ A \mapsto b|SS$.
(f) $S \mapsto Ab,\ A \mapsto aB,\ B \mapsto a|Ab$.
(g) $S \mapsto ba|bS|bbS$.
(h) $S \mapsto b|ABa,\ A \mapsto bbA|B,\ B \mapsto aAa$.
(i) $S \mapsto AB,\ A \mapsto a|aB|bA,\ B \mapsto a$.

5.67. Construct CFGs in CNF and in GNF for generating the following languages:
(a) $\{w \in \{a,b\}^* : w$ is not a palindrome$\}$.
(b) $\{a^m b^n a^m : m, n \geq 1\}$.
(c) $\{a^m b^n c^k : m, n, k \geq 1,\ k \leq 2m\}$.
(d) $\{a^m b^{2m} c^n : m, n \geq 1\}$.
(e) The language of balanced parentheses.
(f) $\{w \in \{0,1\}^* : w^R = \overline{w},\ \ell(w) \geq 1\}$, where $\overline{w}$ is obtained from w by interchanging 0's and 1's.

5.68. Prove that the CFG in GNF with productions $S \mapsto aA|bB$, $A \mapsto b|bS|aAA$, $B \mapsto a|aS|bBB$ generates the language $\{w \in \{a,b\}^* : \#_a(w) = \#_b(w) \geq 1\}$.

5.69. Suppose that the CFG G is in CNF and the CFG G' is in GNF. If $\mathcal{L}(G) = \mathcal{L}(G')$, then what can you say about the lengths of derivations of a string w in G and in G'; which one is shorter?

5.7 Summary and Additional Problems

Removing certain restrictions on regular grammars, we arrived at context-free grammars. The derivations in CFGs naturally led us to study parsing and the ambiguities in parsing. As the derivations are nondeterministic, seeking for some sort of progress indicators landed us at the normal forms. We discussed two important normal forms, one due to N. Chomsky and the other due to S.A. Greibach. Conversion to normal forms involves eliminating unit productions, useless productions, self-productions, and dominated productions.

Context-free languages were created by Chomsky [14]. Similar ideas were introduced by Backus [7] for Fortran, and by Naur [94] for Algol. Ambiguities in CFGs were first studied in [11, 18, 33]. Elimination of ε-productions and unit productions were first considered in [8]. Chomsky normal form conversion is from [15] and Greibach normal form conversion is from [47]. We have just introduced the concept of parse trees, but have not given any details regarding parsing in actual programming languages. You will find some references on parsing in the summary to next chapter. For advanced texts on context-free languages, see [38, 51].

Additional Problems for Chapter 5

5.70. What language is generated by the CFG with productions $S \mapsto \varepsilon|aS|Sb|bSa$? Prove your claim.

5.71. Show that the context-free grammar with productions $S \mapsto \varepsilon|aAB|aBA|bAA$, $A \mapsto aS|bAAA$, $B \mapsto bS|aABB|aBBA|aBAB$ generates all and only strings having exactly twice as many a's as b's.

5.72. Write an algorithm to construct a CFG from a given regular expression. Note that the CFG need not be a regular grammar. Prove the correctness of the algorithm.

5.73. Suppose in the CFG G no production has the right side as ε. Let $w \in \mathcal{L}(G)$ have $\ell(w) = n$. Show that if w has a derivation of length m in G, then w has a parse tree with $m + n$ nodes.

5.74. Suppose $w \in \mathcal{L}(G)$ have $\ell(w) = n \geq 1$, where G is any CFG. Show that if w has a derivation of length m in G, then w has a parse tree with $2m + n - 1$ nodes.

5.75. Show that the following algorithm for testing membership of a string (input string), in the language L generated by the CFG with productions $S \mapsto \varepsilon|SS|aS|aSbS$, is correct:

1. If the current string begins with b, then the input string is not in L.
2. If the current string has no b's, then the input string is in L.
3. Otherwise delete the first b and the a that occurs to its immediate left; and repeat the three steps on the new string.

5.76. Some strings in $\mathcal{L}(G)$, where G is the CFG with productions $S \mapsto \varepsilon|aS|aSbS$, have unique left-derivations and some do not have. Give an algorithm to determine whether a given string is of one type or the other.

5.77. Show that the CFG G with productions $S \mapsto AbB$, $A \mapsto \varepsilon|aA$, $B \mapsto \varepsilon|aB|bB$ is unambiguous. Construct an ambiguous CFG for $\mathcal{L}(G)$.

5.78. Let $L = \{a^m b^n : m \leq n + 3,\ m, n \in \mathbb{N}\}$. Find a CFG that generates $head(L) = \{w : wu \in L \text{ for some } u \in \Sigma^*\}$.

5.79. Let G be a CFG and $L = \mathcal{L}(G)$. How do you construct a CFG G' from G so that $\mathcal{L}(G') = head(L)$?

5.80. Let G be a CFG where each production is of the form $A \mapsto x$ with $\ell(x) = k > 1$. Show that the height h of the parse tree for any $w \in \mathcal{L}(G)$ satisfies $(k-1)\log_k \ell(w) \le h(k-1) \le \ell(w) - 1$.

5.81. Let $G = (N, \Sigma, R, S)$ be a CFG having no productions in either of the forms $A \mapsto \varepsilon$ or $A \mapsto B$ for any $A, B \in N$. Show that parsing the breadth-first way can be made into an algorithm which either produces a parsing of a given string or reports that no parsing is possible. This method is called the *method of exhaustive search parsing*.

5.82. Use the exhaustive search parsing to parse the string $baaaaaa$ in the CFG with productions $S \mapsto bAB,\ A \mapsto aBa,\ B \mapsto \varepsilon|A$.

5.83. Show that if the method of exhaustive search parsing is used, there can be, at the most, $n + n^2 + \cdots n^{2\ell(w)}$ number of nodes in a parse tree for a string w, where n is the number of production rules in the CFG G.

5.84. Show that for every CFG G, there exists an algorithm that parses any $w \in \mathcal{L}(G)$ in a number of steps proportional to $(\ell(w))^3$.

5.85. Prove: If G is a CFG in which each nonterminal occurs on the left side of at most one production, then G is unambiguous.

5.86. Show that if L is a CFL over Σ and $a \in \Sigma$, then
(a) $L/a = \{w : wa \in L\}$ is a CFL.
(b) $cycle(L) = \{xy : yx \in L,\ \text{for } x, y \in \Sigma^*\}$ is a CFL.

5.87. Eliminate all useless productions from the CFGs whose productions are
(a) $S \mapsto AB|bS,\ A \mapsto aA,\ B \mapsto AA.$
(b) $S \mapsto b|bA|B|C,\ A \mapsto \varepsilon|bB,\ B \mapsto Ab,\ C \mapsto aCD,\ D \mapsto ccc.$

5.88. Prove that the marking method for eliminating useless productions works correctly.

5.89. In the marking algorithm for eliminating useless productions, we essentially eliminate all nonterminals that do not yield a terminal string, and then eliminate those which cannot be reached from the initial state. Suppose we reverse the order of elimination. Will the new procedure work correctly? Justify.

5.90. Show that if elimination of unit productions is done first, and then the elimination of useless symbols, then the CFG is better simplified, than otherwise.

5.91. Define the difficulty level or *complexity* of a CFG G by first defining complexity of a production as $complexity(A \mapsto x) = 1 + \ell(x)$ and then taking $complexity(G)$ as the sum of complexities of all productions of G. Show that elimination of useless productions reduces the complexity of a CFG.

5.92. A CFG G is said to be *minimal* for a given language L if $complexity(G) \le complexity(G')$ for each CFG G' with $\mathcal{L}(G) = \mathcal{L}(G') = L$. Show that elimination of useless productions does not necessarily produce a minimal CFG.

5.93. Our self-productions can be termed as left-self-productions. You can define right-self-productions similarly and try to eliminate them. Let the set of productions in a CFG G involving the nonterminal A on the left side be divided into two disjoint subsets:

$A \mapsto x_1 A | x_2 A | \cdots | x_m A$ and $A \mapsto y_1 | y_2 | \cdots y_n$,

where A is not a suffix of any y_j. Replace these productions with

$X \mapsto x_1 | \cdots x_m | X x_1 | \cdots X x_m$ and $A \mapsto y_1 | \cdots y_n | X y_1 | \cdots X y_n$,

respectively, to obtain the grammar G', where X is a new nonterminal. Show that $\mathcal{L}(G') = \mathcal{L}(G)$.

5.94. Let G be a CFG with m number of productions, n number of nonterminals, and k be the maximum number of symbols on the right of any production. If G does not have an ε-production, then show that there is a CFG in CNF equivalent to G with at most $n + m(k - 1)$ number of productions.

5.95. Let G be a CFG in CNF having k nonterminals. Show that, if G generates a string using a derivation of length 2^k or more, then $\mathcal{L}(G)$ is infinite.

5.96. Let G be a CFG without an ε-production. Assume that the right hand side of each production has length at most n. Convert G to G' in CNF. Show that

(a) G' has at the most a constant times n^2 number of productions.

(b) Considering the elimination of unit productions, show that it is possible for G' to have a number of productions actually proportional to n^2.

5.97. Prove or disprove: Each CFL not containing ε is generated by a grammar having all its productions in either of the forms $A \mapsto BCD$ or $A \mapsto a$.

5.98. Let G be a CFG with m productions. Assume that the right side of each production in G has length at most n. Show that if $A \Rightarrow_G \varepsilon$, then there is a derivation of ε from A in at most $(n^m - 1)/(n - 1)$ steps.

6 Structure of CFLs

6.1 Introduction

What can be the additions to a finite automaton so that it may accept any context-free language? Clearly, some addition is required as each regular language is context-free and there are context-free languages that are not regular. For example, the language $L = \{a^n b^n : n \in \mathbb{N}\}$ is context-free but it is not regular. Here you can see that somehow the automaton must remember how many a's it has read, and then it has to consume b's one after another matching to that number. A finite automaton does not remember any thing. So, we do not see how an automaton can accept the language $\{w\overline{w} : w \in \{a, b\}^*\}$, where $\overline{w}$ is the string obtained from w by interchanging a's with b's. We add some sort of *memory* to a finite automaton so that it remembers the first half and then match that with the second half, symbol by symbol.

Moreover, how would it know that it has come across the first half of the input string? Well, we will keep that undetermined so that the automaton would guess and then will do the right thing when it has the right guess. If it is not the right guess, then the machine will not be penalized for it. A nondeterministic machine with some kind of memory might do the job.

6.2 Pushdown Automata

We start with such a nondeterministic machine with a very primitive type of memory, called a *stack* or a *pushdown* storage. A stack resembles keeping books on your table, where you can add one more on the top or take out one from the top. We plan to add such a stack to a finite automaton. These two operations of adding one more on the top, called *pushing to* the stack, and deleting one from the top, called *popping off* from the stack, are inbuilt to any stack.

To see how the language $\{a^n b^n : n \in \mathbb{N}\}$ can be accepted by an automaton with a stack, think of an automaton that accepts $\{a^m b^n : m, n \in \mathbb{N}\}$ and then give it the capability of putting one pebble on the stack when an a is read and throwing one pebble out when a b is read. So, the stack must be empty before the automaton starts

A. Singh, *Elements of Computation Theory*, Texts in Computer Science,

and it must also be empty at the end of its work for accepting the string. This stack along with a finite automaton for a^*b^* will accept $\{a^m b^m : m \in \mathbb{N}\}$.

Such a machine has a finite input tape containing its input. The reading head reads from left to right one symbol at a time, or even it reads nothing sometimes. The control unit has a device to point out the current state of the machine. Moreover, the control unit can change the state of the machine by following certain instructions that are given by a transition relation.

The stack-head is capable of reading only one symbol (or can pretend not to read anything, i.e., reading the empty string) from the top of the stack. It can also sense when the stack (-string) is empty. It can push a string $B_1 B_2 \cdots B_k$ down the stack in one move by pushing B_k first and B_1 last, in that order. The stack can store any finite amount of information; therefore, it is potentially infinite. Look at the schematic drawing in Fig. 6.1.

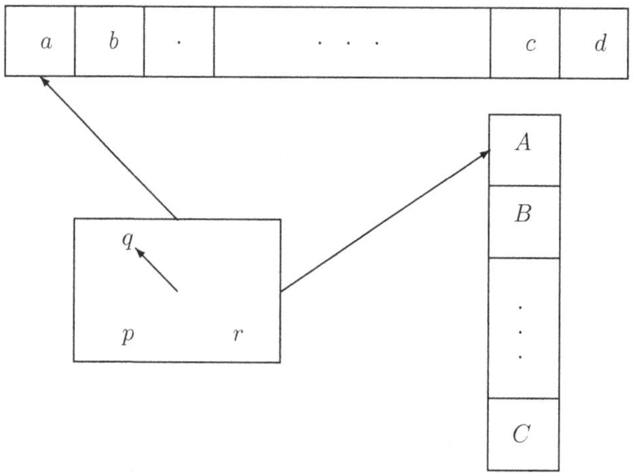

Fig. 6.1. A pushdown automaton.

A **pushdown automaton** or **PDA** is a six-tuple $P = (Q, \Sigma, \Gamma, \Delta, s, F)$, where

Q is a finite set of symbols, the *set of states*,
Σ is a finite set of symbols, the input alphabet,
Γ is a finite set, the *stack alphabet*, with which the stack operates,
$s \in Q$ is the initial state,
$F \subseteq Q$ is the set of final states, and
Δ, the transition relation is a finite relation from $Q \times (\Sigma \cup \{\varepsilon\}) \times (\Gamma \cup \{\varepsilon\})$ to $Q \times \Gamma^*$.

We usually take the sets Q, Σ, Γ to be pairwise disjoint, so that there will be no confusion in the input symbols, states, and stack symbols.

Thus, Δ is a finite subset of $(Q \times (\Sigma \cup \{\varepsilon\}) \times (\Gamma \cup \{\varepsilon\})) \times (Q \times \Gamma^*)$, each member of which looks like $((q, \sigma, A), (r, \alpha))$. We will write such an element of Δ as a quintuple $(q, \sigma, A, r, \alpha)$ and call it a *transition* of the PDA.

Intuitively, this means that when the pushdown automaton P is in the state q, reading the symbol σ on the input tape, having the symbol A on the top of the stack, it acts by changing its state to r, pops off A from the top of the stack, and pushes α to the stack (replaces A by α on the stack so that α from left to right matches with α from top to bottom on the stack). The input symbol σ has thus been read, and the reading head of P is placed on the square next to the last symbol σ.

Notice that our pushdown automaton is nondeterministic; it has the *freedom* of taking one of the many allowed actions even from the same state, input-symbol, and stack-symbol tuple. And all this is done in a single move. This happens due to the transition relation Δ, which is a relation and not (necessarily) a partial function. During its operation it may find zero or more number of transitions, which may be applicable at a particular instant. Let us see the degenerate cases when one or many of the possibilities $\sigma = \varepsilon$, $A = \varepsilon$, or $\alpha = \varepsilon$ happens.

For $\sigma = \varepsilon$, the automaton does not read anything but does all the other things it is supposed to do. This means that if $(q, \varepsilon, A, r, \alpha) \in \Delta$, the machine is currently in state q, the symbol on the top of the stack is A, and if it chooses to follow this transition (there might be other applicable transitions), then without moving its reading head it changes its state to r, pops off the symbol A from the top of the stack, and pushes the string α to the stack.

For $A = \varepsilon$, the quintuple $(q, \sigma, \varepsilon, r, \alpha)$ says that if the machine is currently in the state q reading the symbol σ on the input string, by following this transition, it changes its state to r, moves its reading head to the square next to the currently scanned square (unless $\sigma = \varepsilon$), and simultaneously, pushes the string α to the stack.

Similarly, when $\alpha = \varepsilon$, the quintuple $(q, \sigma, A, r, \varepsilon)$ says that if the machine is in state q, reading σ on the input tape, and finding A on the top of the stack, then it changes its state to r, moves the reading head one square to the right (unless $\sigma = \varepsilon$), and does not do anything with the stack. Here, after this transition is followed, the top of the stack remains the same as earlier.

Exercise 6.1. What do the transitions $(q, \varepsilon, \varepsilon, r, \alpha)$, $(q, \varepsilon, A, r, \varepsilon)$, $(q, \sigma, \varepsilon, r, \varepsilon)$, and $(q, \varepsilon, \varepsilon, r, \varepsilon)$ mean?

At any instant of time during its operation, a description of a PDA must inform us in what state it is, what is the remaining part of the input string yet to be read, including the currently scanned symbol, and what string is there on the stack, from top to bottom. Such an instantaneous description of a PDA is called a **configuration**. Formally, a configuration of P is any element of $Q \times \Sigma^* \times \Gamma^*$. For example, the initial configuration of P with an input u is (s, u, ε).

The one-step operation of P can be described by telling which configuration gives rise to which other configuration. That is, an operation in one step can be described by determining "how the PDA looks like after one step of operation." This can be formally captured by the *yield in one step* relation from configuration to configuration. Suppose that currently P is in state q, reading the input symbol σ (i.e., σv on the tape for some $v \in \Sigma^*$) while the symbol A is on the top of the stack (i.e., you have $A\beta$ on the stack from top to bottom, for some string $\beta \in \Gamma^*$). This means that P has the current configuration $(q, \sigma v, A\beta)$.

Suppose further that there is a transition $(q, \sigma, A, r, \alpha)$. Then by using this transition, P would change its state from q to r, move its reading head to the next square on the right (of that containing σ), and replace the symbol A by α on the stack by popping off A and pushing α. Which means that the next configuration of P will be $(r, v, \alpha\beta)$. In such a case, we will say that the configuration $(q, \sigma v, A\beta)$ yields in one step the configuration $(r, v, \alpha\beta)$. We will use the symbol $\overset{1}{\rightsquigarrow}$ for "yield in one step." The reflexive and transitive closure of the relation of yield in one step will be referred to as **yield** and we will denote it by the symbol $\overset{*}{\rightsquigarrow}$. Formally,

$$(q, \sigma v, A\beta) \overset{1}{\rightsquigarrow} (r, v, \alpha\beta) \text{ if } (q, \sigma, A, r, \alpha) \in \Delta.$$

For configurations C, C', $C \overset{*}{\rightsquigarrow} C'$ iff either $C = C'$ or

there is a sequence of configurations $C_1, \cdots, C_n,\ n \geq 1$, such that

$$C \overset{1}{\rightsquigarrow} C_1,\ C_1 \overset{1}{\rightsquigarrow} C_2, \ldots, C_{n-1} \overset{1}{\rightsquigarrow} C_n,\ C_n \overset{1}{\rightsquigarrow} C'.$$

Such a sequence of configurations is called a **computation** of the PDA. The computation $C \rightsquigarrow C$ is an empty computation. The degenerate cases of the configurations and the corresponding yield relation are taken care as in the following:

(a) If $(q, \varepsilon, A, r, \alpha) \in \Delta$, then $(q, v, A\beta) \overset{1}{\rightsquigarrow} (r, v, \alpha\beta)$.
(b) If $(q, \sigma, \varepsilon, r, \alpha) \in \Delta$, then $(q, \sigma v, \beta) \overset{1}{\rightsquigarrow} (r, v, \alpha\beta)$.
(c) If $(q, \sigma, A, r, \varepsilon) \in \Delta$, then $(q, \sigma v, A\beta) \overset{1}{\rightsquigarrow} (r, v, \beta)$.
(d) If $(q, \varepsilon, \varepsilon, r, \alpha) \in \Delta$, then $(q, v, \beta) \overset{1}{\rightsquigarrow} (r, v, \alpha\beta)$.
(e) If $(q, \sigma, \varepsilon, r, \varepsilon) \in \Delta$, then $(q, \sigma v, \beta) \overset{1}{\rightsquigarrow} (r, v, \beta)$.

Perhaps you do not have to remember the above degenerate cases in toto. All that you understand is that in a transition, an empty second component means that in one step of computation the automaton does not read anything, an empty third component says that it does not pop off anything from the stack, and the empty fifth component means nothing to be pushed to the stack.

As earlier, we will simply write $\rightsquigarrow$ instead of $\overset{1}{\rightsquigarrow}$ and $\overset{*}{\rightsquigarrow}$ if no confusion arises. If many PDAs occur in some context, then we will have a subscript in the symbol $\rightsquigarrow$ to mention "yield in which PDA."

Starting from its initial configuration, a PDA can halt in two ways, just like a finite automaton. The first way is when the input string is over, and the second way is when there is not an appropriate transition, though there might be a nonempty string on the input tape yet to be read. In the second case, we say that the PDA has *halted abruptly*. The string is **accepted** only when the PDA halts the first way, and on the top of that, the last state where the PDA halts must be a final state, and the stack must also be empty.

Formally, let $P = (Q, \Sigma, \Gamma, \Delta, s, F)$ be a PDA. P **accepts** w iff there exists a computation of P in which $(s, w, \varepsilon) \rightsquigarrow_P (q, \varepsilon, \varepsilon)$, for some state $q \in F$. The **language accepted by** P is

$$\mathcal{L}(P) = \{w \in \Sigma^* : P \text{ accepts } w\}.$$

Notice that we write the same ε for the empty input string and also for the empty string on the stack; it is the empty string any way. If the PDA halts abruptly, then the input string has not been read completely; consequently, it is not accepted. As you see, an NFA is a PDA that never operates on its stack.

Example 6.1. What is $\mathcal{L}(P)$ if $P = (\{s, q, r\}, \{a, b\}, \{\alpha, \beta\}, \Delta, s, \{s\})$ is a PDA with $\Delta = \{(s, a, \alpha, q, \alpha), (q, b, \alpha, s, \alpha), (q, b, \alpha, r, \alpha), (r, a, \alpha, s, \alpha)\}$?

Solution. As earlier, let us have some computations. What happens if P is given the input string ab? The initial configuration of P with the input ab is (s, ab, ε). We see that every transition requires α to be on the top of the stack. Hence no transition is applicable and P halts, that is, $(s, ab, \varepsilon) \rightsquigarrow (s, ab, \varepsilon)$. The machine does not read any nonempty string at all. However, $(s, \varepsilon, \varepsilon) \rightsquigarrow (s, \varepsilon, \varepsilon)$ by default, and s is a final state. Thus $\mathcal{L}(P) = \{\varepsilon\}$. □

In the transition diagram of a PDA, the transitions are picturized as joining arrows from a state to another labeled with an input symbol, followed by "$A \rightarrow \alpha$," when the stack symbol A is replaced by the string α. The state diagram for the PDA of Example 6.1 is given in Fig. 6.2.

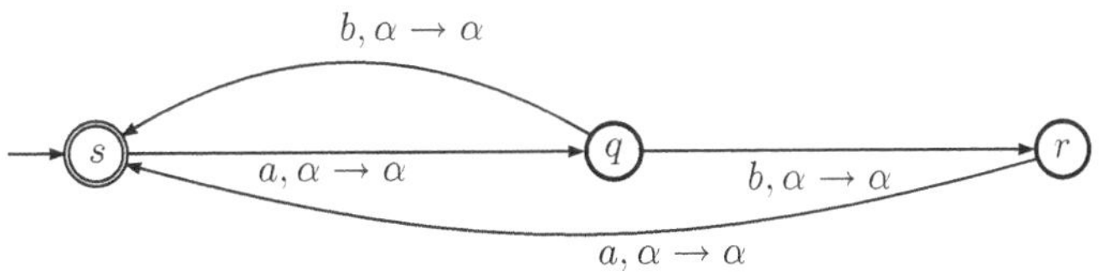

Fig. 6.2. PDA of Example 6.2.

But what is intended in the above example? The transitions in Δ say that the stack symbol is never changed. But somehow this was not expressed correctly. See the following modification:

Example 6.2. What is $\mathcal{L}(P)$ if $P = (\{s, q, r\}, \{a, b\}, \{\alpha, \beta\}, \Delta, s, \{s\})$, where $\Delta = \{(s, a, \varepsilon, q, \varepsilon), (q, b, \varepsilon, s, \varepsilon), (q, b, \varepsilon, r, \varepsilon), (r, a, \varepsilon, s, \varepsilon)\}$?

Solution. We try computations with some strings over $\{a, b\}$. Using the first and the third transitions, we have

$$(s, ab, \varepsilon) \rightsquigarrow (q, b, \varepsilon) \rightsquigarrow (r, \varepsilon, \varepsilon).$$

It is a rejecting computation. However, we cannot just say from this that the string ab is not accepted. There might be another accepting computation. Here is one:

$$(s, ab, \varepsilon) \rightsquigarrow (q, b, \varepsilon) \rightsquigarrow (s, \varepsilon, \varepsilon).$$

Since s is a final state, the string $ab \in \mathcal{L}(P)$. Let us take one more string, say, aba.

$$(s, aba, \varepsilon) \rightsquigarrow (q, ba, \varepsilon) \rightsquigarrow (r, a, \varepsilon) \rightsquigarrow (s, \varepsilon, \varepsilon).$$

It is an accepting computation. So, $aba \in \mathcal{L}(P)$. It is not difficult to show that $\mathcal{L}(P) = (ab \cup aba)^*$. Try it. □

Can you see that the stack is never used in any computation of P of Example 6.2? The corresponding NFA where we do not have a stack can be given by $N = (\{s, q, r\}, \{a, b\}, \Delta', s, \{s\})$ with $\Delta' = \{(s, a, q), (q, b, s), (q, b, r), (r, a, s)\}$. Compare the transitions in P and N; the stack components are simply omitted. Here it is all right since in P the stack is never touched.

Example 6.3. What is $\mathcal{L}(P)$ if $P = (\{s, q\}, \{a, b\}, \{A, B\}, \Delta, s, \{q\})$, where $\Delta = \{(s, a, \varepsilon, s, A), (s, b, \varepsilon, s, B), (s, \varepsilon, \varepsilon, q, \varepsilon), (q, a, A, q, \varepsilon), (q, b, B, q, \varepsilon)\}$?

Solution. Look at the transitions. The automaton reads a in state s and then pushes A to the stack. In the same state, whenever it reads a b, it pushes one B. How is the third transition applied? Does it require the stack to be empty? It is not so, because even if there is a string γ in the stack, $\gamma = \varepsilon\gamma$, and thus, the transition $(s, \varepsilon, \varepsilon, q, \varepsilon)$ is still applicable. Similarly, even if the string is not empty, the transition is still applicable. This transition says that without reading any symbol, without operating on the stack, the automaton may change its state from s to q. The fourth and the fifth transitions pop off or pop out symbols A and B from the stack by reading one a or one b, respectively, provided it is in state q. To see how does it operate, let us try it on the strings ab and $abba$.

$$(s, ab, \varepsilon) \rightsquigarrow (s, b, A) \rightsquigarrow (s, \varepsilon, BA) \rightsquigarrow (q, \varepsilon, BA).$$

Then it halts without accepting the string ab as the stack is not empty. We try another computation with the same string:

$$(s, ab, \varepsilon) \rightsquigarrow (q, ab, \varepsilon).$$

Here also it halts as no transition is applicable. It does not accept the string, as the input string has not been read completely. Let us take one more computation with the same string:

$$(s, ab, \varepsilon) \rightsquigarrow (s, b, A) \rightsquigarrow (q, b, A).$$

Once again it halts without accepting the string. Does there exist an accepting computation at all with the input string ab?

The string $abba$ is accepted by P as the following computation shows:

$$(s, abba, \varepsilon) \rightsquigarrow (s, bba, A) \rightsquigarrow (s, ba, BA) \rightsquigarrow (q, ba, BA) \rightsquigarrow (q, a, A) \rightsquigarrow (q, \varepsilon, \varepsilon).$$

You can find one more computation that would be a rejecting one. But that does not matter. As we have one accepting computation, $abba \in \mathcal{L}(P)$. Look at the above computation once more to understand how it accepts the string $abba$. The key fact in this computation is that it nondeterministically changes state to q and there after it remains in the state q. Show that $\mathcal{L}(P) = \{ww^R : w \in \{a, b\}^*\}$. □

Exercise 6.2. Let $P = (\{s, q\}, \{a, b\}, \{A, B\}, \Delta, s, \{q\})$, with the transition relation $\Delta = \{(s, a, \varepsilon, s, A), (s, b, \varepsilon, s, B), (s, \varepsilon, \varepsilon, q, \varepsilon), (q, a, B, q, \varepsilon), (q, b, A, q, \varepsilon)\}$. Compare this PDA with that in Example 6.3. Does P accept the language $\{w\overline{w} : w \in \{a, b\}^*\}$, where $\overline{w}$ is obtained from w by interchanging a's and b's?

Example 6.4. Construct a PDA to accept any binary number with same number of 0's as 1's, that is, $\{w \in \{0, 1\}^* : \#_0(w) = \#_1(w)\}$.

Solution. Try to solve the problem yourself. Return to this solution after a work out. Let us use the stack to keep note of the excess of either digit found thus far in the input string. To use the stack this way, suppose we push an A to the stack whenever one 0 is read and push one B when one 1 is read. However, when it reads a 0 and the stack symbol on the top is a B, it pops out B instead of putting another A. Similarly, if it reads a 1 while the top symbol on the stack is A, it pops out A instead of putting another B.

With such an operation with the stack, when will a computation finish? Of course, only after the input string is completely read. But then we need the stack to be empty also. To this end, can we include the transition $(q, \varepsilon, \varepsilon, r, \varepsilon)$? However, we do not want such a transition to be applied every where. We can solve the problem by pushing an extra symbol, say, C initially so that only when we find this C on the stack, we will come to know that the stack is, in fact, empty. Then instead of the ε on the stack in the above transition, we simply use this symbol C. Once met, the symbol C will signal that the number of 0's and 1's read thus far is same. So, we try the PDA $P = (\{s, q, r\}, \{0, 1\}, \{A, B, C\}, \Delta, s, \{r\})$, where

$$\Delta = \{(s, \varepsilon, \varepsilon, q, C), (q, 0, C, q, AC), (q, 0, A, q, AA), (q, 0, B, q, \varepsilon),$$
$$(q, 1, C, q, BC), (q, 1, B, q, BB), (q, 1, A, q, \varepsilon), (q, \varepsilon, C, r, \varepsilon)\}.$$

Can you now show that P really does the intended job? □

Exercise 6.3. Find an accepting computation of P in Example 6.4 for the input string 00011001010111. From the working out of P, convince yourself that $\mathcal{L}(P)$ is the set of binary strings with same number of 0's and 1's.

We have used our memory device as a stack that comes with two operations of pushing and popping a string of stack symbols. What happens if we give a PDA more power by enabling it to read also a string from the stack, and not just a symbol? A **generalized PDA** is like a PDA where the transition relation Δ is a finite subset of $(Q \times (\Sigma \cup \{\varepsilon\}) \times \Gamma^*) \times (Q \times \Gamma^*)$. That is, the stack-head can now read and replace strings from Γ^*. However, generalized PDAs are no more powerful than the PDAs. To see this, suppose we have a *generalized* transition (q, a, β, p, γ), where $\beta = B_1 B_2 \cdots B_n$ with $n > 1$. We replace it by adding new transitions

$$(q, \varepsilon, B_1, t_1, \varepsilon), (t_1, \varepsilon, B_2, t_2, \varepsilon), \ldots, (t_{n-2}, \varepsilon, B_{n-1}, t_{n-1}, \varepsilon), (t_{n-1}, a, B_n, p, \gamma),$$

where $t_1, \ldots, t_{n-1}$ are new states. The result is a PDA.

Exercise 6.4. Show that any language that is accepted by a generalized PDA is also accepted by a PDA.

However, a triviality may be bothersome. How to push or pop the empty string? It can be done in two steps: by pushing a new symbol and then popping it out. This is the reason we have allowed the empty string in a stack to be operated with. For example, let us consider a transition of the form $(q, \sigma, \varepsilon, r, \beta)$. This means that without bothering for what is there on the stack, read the symbol σ in state q, and then change state to r while pushing β to the stack.

The same effect can be achieved by having a transition of the form $(q, \sigma, X, r, \beta X)$ for each $X \in \Gamma$. The difference is that in the latter case, the action of the PDA has been more explicit. Such PDAs where these new transitions have been added corresponding to each transition of the form $(q, \sigma, \varepsilon, r, \beta)$ are called **simple pushdown automata**, or SPDA for short.

Formally, an **SPDA** is a PDA $P = (Q, \Sigma, \Gamma, \Delta, a, F)$, where Q, Σ, Γ, s, F are as in a PDA, and Δ satisfies the property that whenever $(q, \sigma, \varepsilon, r, \beta) \in \Delta$, we also have $(q, \sigma, A, r, \beta A) \in \Delta$, for each $A \in \Gamma$.

Let us call this restriction as the **restriction of simplicity**. Our first concern is whether we loose any power by such a restriction on a PDA. The discussion above says that we do not. That means we can always construct an SPDA by introducing new transitions to the old one for accepting the same language. See the following example:

Example 6.5. Consider the PDA in Example 6.3. We add more transitions to it corresponding to the empty-stack-transitions. The corresponding SPDA is $P' = (\{s, q\}, \{a, b\}, \{A, B\}, \Delta', s, \{q\})$, where

$$\begin{aligned}\Delta' = \{&(s, a, \varepsilon, s, A), (s, b, \varepsilon, s, B), (s, \varepsilon, \varepsilon, q, \varepsilon), (q, a, A, q, \varepsilon), (q, b, B, q, \varepsilon),\\ &(s, a, A, s, AA), (s, a, B, s, AB), (s, b, A, s, BA), (s, b, B, s, BB),\\ &(s, \varepsilon, A, q, A), (s, \varepsilon, B, q, B)\}.\end{aligned}$$

□

Exercise 6.5. Show that $\mathcal{L}(P) = \mathcal{L}(P')$, where P is the PDA of Example 6.3 and P' is the PDA of Example 6.5.

Have you completed Exercise 6.5? Well, then you would have realized that addition of new transitions to satisfy the simplicity restriction does not matter for the language it accepts. The construction is as follows:

> For each transition $(q, \sigma, \varepsilon, r, \beta) \in \Delta$ with $\sigma \in \Sigma \cup \{\varepsilon\}$, $\beta \in \Gamma^*$, add to Δ all transitions of the form $(q, \sigma, A, r, \beta A)$ for $A \in \Gamma$.

After the construction, the resulting PDA is an SPDA. On the one hand, if any string is accepted by the PDA without adding the extra ones, then it is also accepted by the SPDA due to nondeterminism in using the transitions. On the other hand, these additions cannot get any new string accepted by the SPDA as application of a transition of the form $(q, \sigma, \varepsilon, r, \beta)$ includes already an application of a transition of the form $(q, \sigma, A, r, \beta A)$ (Why is it so?). So we have proved the following statement:

Lemma 6.1. *Let P be a PDA and P' be the corresponding SPDA as constructed above. Then $\mathcal{L}(P) = \mathcal{L}(P')$.*

Exercise 6.6. Can we afford to delete the transitions $(q, \sigma, \varepsilon, r, \beta)$ from the constructed SPDA?

Problems for Section 6.2

6.1. Let P be the PDA with initial state s, final state q, and transitions $(s, a, \varepsilon, s, A)$, $(s, b, \varepsilon, s, A)$, $(s, a, \varepsilon, q, \varepsilon)$, $(q, a, A, q, \varepsilon)$, $(q, b, A, q, \varepsilon)$. Show that aab, $aba \in \mathcal{L}(P)$ but $bba, bab \notin \mathcal{L}(P)$. Describe $\mathcal{L}(P)$ in English.

6.2. Draw transition diagrams for the PDAs of Examples 6.2 and 6.3.

6.3. What are the languages accepted by the PDAs with initial state p, final states in F, and transitions in Δ as specified below?
(a) $\Delta = \{(p, b, \varepsilon, q, A), (q, b, A, q, AA), (r, a, A, s, \varepsilon), (s, a, A, t, \varepsilon), (t, a, \varepsilon, t, \varepsilon)\}$, $F = \{u\}$.
(b) $\Delta = \{(p, a, \varepsilon, q, A), (p, a, \varepsilon, r, \varepsilon), (q, b, A, q, B), (q, b, B, q, B), (q, a, B, r, \varepsilon)\}$, $F = \{r\}$.
(c) $\Delta = \{(p, \varepsilon, \varepsilon, q, \varepsilon), (p, a, \varepsilon, p, A), (p, b, \varepsilon, p, B), (p, a, A, p, AA), (p, b, A, p, \varepsilon), (p, a, B, p, \varepsilon), (p, b, B, p, BB)\}$, $F = \{p, q\}$.
(d) $\Delta = \{(p, a, \varepsilon, q, A), (p, a, \varepsilon, r, \varepsilon), (q, b, A, q, B), (q, b, B, q, B), (q, a, B, r, \varepsilon)\}$, $F = \{p, q, r\}$.
(e) $\Delta = \{(p, a, \varepsilon, p, A), (p, a, A, p, A), (p, b, A, p, B), (p, a, B, p, B), (p, \varepsilon, B, q, B)\}$, $F = \{q\}$.
(f) $\Delta = \{(p, a, A, q, BA), (p, a, A, s, \varepsilon), (p, \varepsilon, A, s, \varepsilon), (q, a, B, q, BB), (q, b, B, r, \varepsilon), (r, b, B, r, \varepsilon), (q, \varepsilon, A, p, \varepsilon)\}$, $F = \{s\}$.

6.4. Find a PDA that accepts the language of balanced parentheses.

6.5. Construct a PDA for accepting the language generated by the CFG with productions $S \mapsto aB|aaabb$, $A \mapsto a|B$, $B \mapsto bb|AaB$. Does there exist a nonaccepting computation of the string $aaabb$ in your PDA?

6.6. Find CFGs for generating, and PDAs for accepting the languages $\{w \in \{a, b\}^* : P(w)\}$, where $P(w)$ is the property:
(a) w contains at least three b's.
(b) w starts and ends with the same symbol.
(c) $\ell(w)$ is odd.
(d) $\ell(w)$ is odd and the middle symbol of w is b.
(e) w contains more a's than b's.
(f) w is a palindrome.
(g) $w \neq w$.
(h) w has twice as many a's as b's.
(i) $w \neq a^n b^n$ for any $n \in \mathbb{N}$.

6.7. Construct CFGs and PDAs for the following languages over $\{a, b, c\}$:
(a) $\{vcw : v^R$ is a substring of w for $v, w \in \{a, b\}^*\}$.
(b) $\{w_1cw_2c \cdots cw_k : k \geq 1,$ each $w_i \in \{a, b\}^*,$ and for some $m, n, w_m = w_n^R\}$.
(c) $\{ucv : u, v \in \{a, b\}^*, u \neq v\}$.
(d) $\{ucv : u, v \in \{a, b\}^*, u \neq v$ but $\ell(u) = \ell(v)\}$.

6.8. Construct PDAs for accepting the following languages over $\{a, b, c\}$:
(a) $\{ww^R : w \in \{a, b\}^*\}$.
(b) $\{a^nb^n : n \in \mathbb{N}\} \cup \{a\}$.
(c) $\{a^nb^{2n} : n \in \mathbb{N}\}$.
(d) $\{a^mb^nc^{m+n} : m, n \in \mathbb{N}\}$.
(e) $\{a^mb^{m+n}c^n : m \geq 1, n \in \mathbb{N}\}$.
(f) $\{a^mb^n : m \leq n \leq 2n\}$.
(g) $\{a^mb^n : m \leq n \leq 3n\}$.
(h) $\{a^mb^n : m, n \in \mathbb{N}, m \neq n\}$.
(i) $\{ab(ab)^nb(ba)^n : n \in \mathbb{N}\}$.
(j) $\{w \in \{a, b\}^* : \#_b(w) = 2 \cdot \#_a(w)\}$.
(k) $\{w \in \{a, b, c\}^* : \#_a(w) + \#_b(w) = \#_c(w)\}$.
(l) $\{w \in \{a, b\}^* : \#_a(w) < \#_b(w)\}$.
(m) $\{w \in \{a, b, c\}^* : 2 \cdot \#_a(w) \leq \#_b(w) \leq 3 \cdot \#_c(w)\}$.
(n) $\{vcw : v, w \in \{a, b\}^*, v \neq w^R\}$.

6.9. Let $P = (Q, \Sigma, \Gamma, \Delta, s, F)$ be a PDA. Define the *language accepted by P by empty stack* as follows:

$$\mathcal{L}_e(P) = \{w \in \Sigma^* : (s, w, \varepsilon) \rightsquigarrow_P (q, \varepsilon, \varepsilon) \text{ for some } q \in Q\}.$$

Here, we require the stack to be emptied relaxing the condition that the state at the end of the computation can be any state rather than a final state. Similarly, the *language accepted by P by final state* is defined as follows:

$$\mathcal{L}_f(P) = \{w \in \Sigma^* : (s, w, \varepsilon) \rightsquigarrow_P (q, \varepsilon, \alpha) \text{ for some } q \in F, \text{ and some } \alpha \in \Gamma^*\}.$$

Here, we do not bother whether the stack is emptied or not, but only require that at the end, a final state is reached. Note that our definition of $\mathcal{L}(P)$ requires both, that is, $L = \mathcal{L}(P)$ iff $L = \mathcal{L}_e(P) = \mathcal{L}_f(P)$. To prove that all these definitions are equivalent, show the following:
(a) For each PDA P, there is a PDA P_1 such that $\mathcal{L}(P) = \mathcal{L}_e(P_1)$.
(b) For each PDA P, there is a PDA P_2 such that $\mathcal{L}_e(P) = \mathcal{L}(P_2)$.
(c) For each PDA P, there is a PDA P_3 such that $\mathcal{L}(P) = \mathcal{L}_f(P_3)$.
(d) For each PDA P, there is a PDA P_4 such that $\mathcal{L}_f(P) = \mathcal{L}(P_4)$.

6.10. Suppose that we use a start symbol on the stack of a PDA, initially, and allow its temporary removal during computation. That is, such a symbol will always be there on the bottom of the stack; if it is removed by following a transition, then immediately, it has to be pushed back into the stack. This allows transitions of the type $(p, a, \nabla, q, \alpha\nabla)$, but neither of the type $(p, a, \nabla, q, \varepsilon)$ nor of the type $(p, a, \nabla, q, \alpha)$, where ∇ is "bems," the bottom-of-the-stack symbol. At the end of a computation, we consider the stack to be empty when the only symbol in it is ∇. Show that we can always suitably modify the transitions in such a way that, given a PDA, there is another, having a bottom-of-the-stack symbol, accepting the same language. Prove also the converse, that is, given a PDA with a bottom-of-the-stack symbol, there is always a PDA accepting the same language.

6.3 CFG and PDA

Let $G = (N, \Sigma, R, S)$ be a context-free grammar. If $w \in \mathcal{L}(G)$, then by Lemma 5.1, there is a leftmost derivation of w in G. We want to construct a PDA that mimics the leftmost derivation of w while accepting w.

Let us try a PDA only with two states, say, s and q. In order that the stack might keep track of the derivation, we use a stack alphabet that has all the nonterminal symbols of G along with the terminals. But we cannot use the terminals as such, for the stack alphabet must not intersect with the terminals or the states. So, introduce a new symbol corresponding to each terminal symbol, say, by using a prime over them. The new symbols will be in the stack alphabet. Corresponding to any string $u \in \Sigma^*$, we write the string of stack symbols, u', as the string of primed ones. That is,

$$\text{Let } \Sigma' = \{\sigma' : \sigma \in \Sigma\}. \text{ If } u = \sigma_1\sigma_2 \ldots \sigma_n, \text{ then } u' = \sigma_1{}'\sigma_2{}' \ldots \sigma_n{}'.$$

Corresponding to any string of symbols from $N \cup \Sigma'$, we keep the symbols of N as they are and make the symbols in Σ primed. For example, if $u = AbBaa$, then $u' = Ab'Ba'a'$. We also make it a convention to write ε' as ε. Also notice that $(uv)' = u'v'$ for any strings of symbols of $N \cup \Sigma'$. Notice that had we allowed overlapping of symbols in N, Σ, Γ, etc, in grammars or machines, this renaming with "primes" over symbols would not have been needed. However, we continue with the renaming and the nonoverlapping convention of concerned alphabets.

To mimic a leftmost derivation, we will put, initially, S into the stack, and the PDA will enter the state q. On each subsequent step, whenever a production of the form $X \mapsto x$ is applied in the derivation, we must have X on the top of the stack, and then the PDA would replace X by the string x'. To guarantee that whenever a production of the form $X \mapsto x$ is used, we also have X on top of the stack, we have to do something more. What we do is, whenever we see X on the top of the stack where X is not a nonterminal symbol, we simply pop it out. This is continuously monitored.

Formally, given a context-free grammar $G = (N, \Sigma, R, S)$, we construct the PDA $P_G = (\{s, q\}, \Sigma, N \cup \Sigma', \Delta, s, \{q\})$, where

$$\Delta = \{(s, \varepsilon, \varepsilon, q, S)\} \cup \{(q, \varepsilon, X, q, x) : X \mapsto x \in R\} \cup \{(q, \sigma, \sigma', q, \varepsilon) : \sigma \in \Sigma\}.$$

The PDA, while in operation, mimics a leftmost derivation checking a terminal symbol (primed one) on the top of the stack with the currently scanned symbol on the input string. See the following example before arguing abstractly.

Example 6.6. Carry out the above construction of a PDA from the context-free grammar of Example 5.4.

Solution. The grammar G of Example 5.4 has productions

$$S \mapsto aA \mid bB,\ A \mapsto b \mid bS \mid aAA,\ B \mapsto a \mid aS \mid bBB.$$

The corresponding PDA is $P_G = (\{s, q\}, \{a, b\}, \{S, a', b', A, B\}, \Delta, s, \{q\})$, where

$$\begin{aligned}\Delta = \{&(s, \varepsilon, \varepsilon, q, S), (q, \varepsilon, S, q, a'A), (q, \varepsilon, S, q, b'B), (q, \varepsilon, A, q, b'),\\ &(q, \varepsilon, A, q, b'S), (q, \varepsilon, A, q, a'AA), (q, \varepsilon, B, q, a'), (q, \varepsilon, B, q, a'S),\\ &(q, \varepsilon, B, q, b'BB), (q, a, a', q, \varepsilon), (q, b, b', q, \varepsilon)\}.\end{aligned}$$ □

Let us see how an accepting computation mimics a leftmost derivation in Example 6.6. Take a string, say, *aabbba*. It has the leftmost derivation:

$$S \Rightarrow aA \Rightarrow aaAA \Rightarrow aabA \Rightarrow aabbS \Rightarrow aabbbB \Rightarrow aabbba.$$

Correspondingly, P goes on the following computation (Each line below gives one step of the derivation in the grammar):

$$\begin{aligned}&(s, aabbba, \varepsilon) \rightsquigarrow (q, aabbba, S) \rightsquigarrow (q, aabbba, a'A)\\ &\rightsquigarrow (q, abbba, A) \rightsquigarrow (q, abbba, a'AA)\\ &\rightsquigarrow (q, bbba, AA) \rightsquigarrow (q, bbba, b'A) \rightsquigarrow (q, bba, A)\\ &\rightsquigarrow (q, bba, b'S) \rightsquigarrow (q, ba, S)\\ &\rightsquigarrow (q, ba, b'B) \rightsquigarrow (q, a, B)\\ &\rightsquigarrow (q, a, a') \rightsquigarrow (q, \varepsilon, \varepsilon).\end{aligned}$$

It is fairly clear how a string having a leftmost derivation in G is accepted by the PDA P_G. In fact, the mimicking process says a little more.

Lemma 6.2. *Let G be a context-free grammar and P_G be its corresponding PDA as constructed earlier. Let $u \in \Sigma^*$, and either $x = \varepsilon$ or $x \in N(N \cup \Sigma')^*$. If there is a leftmost derivation of the string ux in G, then $(q, u, S) \rightsquigarrow (q, \varepsilon, x')$ in P_G.*

Proof. We use induction on a leftmost derivation of the string ux in G. If the derivation is of length 0, then $ux = S$; consequently, $u = \varepsilon$ and $x = S$. Here we see that $(q, \varepsilon, S) \rightsquigarrow (q, \varepsilon, S)$ as well.

Assume the induction hypothesis that if there is a leftmost derivation of length less than n of the string ux in G, then $(q, u, S) \rightsquigarrow (q, \varepsilon, x')$. Let wy be a string having a leftmost derivation in G, of length n and no fewer. We may view the derivation as

$$S \Rightarrow \ldots \text{ (in } n-1 \text{ steps) } \ldots \Rightarrow v \overset{1}{\Rightarrow} wy.$$

Then v has at least one nonterminal symbol that has been replaced by an allowed string to obtain wy. So, assume that $v = x_1 Z x_2$, for some $x_1, x_2 \in (\Sigma')^*$; the production used was $Z \mapsto z \in R$, and then $wy = x_1 z x_2$. By the induction hypothesis, we have

$$(q, x_1{}', S) \rightsquigarrow (q, \varepsilon, Z x_2{}').$$

The way we have constructed the PDA, corresponding to the production $Z \mapsto z \in R$, we have the transition $(q, \varepsilon, Z, q, z') \in \Delta$. Thus the computation above can be extended to

$$(q, x_1', S) \rightsquigarrow (q, \varepsilon, Zx_2') \rightsquigarrow (q, \varepsilon, z'x_2').$$

As $wy = x_1zx_2$, the proof is complete. Is it? It is not, because we have not yet shown that $(q, w, S) \rightsquigarrow (q, \varepsilon, y')$, in which case the computations can be combined to get the result. To this end, we start from $wy = x_1zx_2$, where $w \in \Sigma^*$, $x_1 \in \Sigma^*$, and either $y = \varepsilon$ or y begins with a nonterminal. Hence $w = x_1\alpha$ for some $\alpha \in \Sigma^*$, so that $\alpha y = zx_2$. Then we have

$$(q, \alpha, z'x_2') \rightsquigarrow (q, \varepsilon, y')$$

due to the transitions of the type $(q, \sigma, \sigma', q, \varepsilon) \in \Delta$. Combining this with the previous computation, we obtain

$$(q, w, S) \rightsquigarrow (q, x_1\alpha, S) \rightsquigarrow (q, \varepsilon\alpha, z'x_2') \rightsquigarrow (q, \alpha, z'x_2') \rightsquigarrow (q, \varepsilon, y).$$ □

Exercise 6.7. In the induction step of the proof of Lemma 6.2, where have you used the "leftmost derivation of wy?" Further, in the last line of the same proof, you have used the statement "If $(q, x_1, S) \rightsquigarrow (q, \varepsilon, z'x_2')$, then $(q, x_1\alpha, S) \rightsquigarrow (q, \varepsilon\alpha, z'x_2')$." Prove this statement formally.

We also prove the converse of the statement in Lemma 6.2.

Lemma 6.3. *Let G be a context-free grammar and P_G be its corresponding PDA as constructed earlier. Let $u \in \Sigma^*$ and $x \in (N \cup \Sigma')^*$. If $(q, u, S) \rightsquigarrow (q, \varepsilon, x)$, then there is a leftmost derivation of the string ux in G.*

Proof. Use induction on the length of a computation of P_G. If $(q, u, S) \rightsquigarrow (q, \varepsilon, x')$ in 0 steps, then $u = \varepsilon, x = S$; consequently, $ux = S$. As $S \Rightarrow S$ in G in 0 steps, the statement is clear. Lay out the induction hypothesis that if $(q, u, S) \rightsquigarrow (q, \varepsilon, x')$ in a computation of length less than n in P_G, then there is a leftmost derivation of the string ux in G.

Suppose $(q, w, S) \rightsquigarrow (q, \varepsilon, y')$ in n steps and in no fewer steps, in P_G. We look at the computation as

$$(q, w, S) \rightsquigarrow \ldots \text{ (in } n-1 \text{ steps) } \ldots \rightsquigarrow (q, v, z') \overset{1}{\rightsquigarrow} (q, \varepsilon, y).$$

The last step says that there is a transition in Δ, which is either in the form $(q, \varepsilon, A, q, \alpha')$ or $(q, \sigma, \sigma', q, \varepsilon)$.

In the former case, $v = \varepsilon$, $z = A\beta'$, $y = \alpha'\beta'$ for some $\beta \in (N \cup \Sigma)^*$. That is, there is a production $A \mapsto \alpha \in R$. Then the computation would look like

$$(q, w, S) \rightsquigarrow \ldots \text{ (in } n-1 \text{ steps) } \ldots \rightsquigarrow (q, \varepsilon, A\beta) \overset{1}{\rightsquigarrow} (q, \varepsilon, y).$$

By the induction hypothesis, there is a leftmost derivation of the form $S \Rightarrow wA\beta$. Because of the production $A \mapsto \alpha$, we have

$$S \Rightarrow aA\beta \Rightarrow w\alpha\beta \Rightarrow wy.$$

In the latter case, when $(q, \sigma, \sigma', q, \varepsilon) \in \Delta$, we must have $z = \sigma' y, v = \sigma$, and $(q, w, S) \rightsquigarrow (q, \sigma, \sigma' y)$ in $n - 1$ steps. As w is the input to P_G, and in $n - 1$ steps the only symbol remaining is σ, we also have $w = v\sigma$ and then $(q, v\sigma, S) \rightsquigarrow (q, \sigma, \sigma' y)$. As the last symbol σ has no effect on the computation of P_G until it has been read, we conclude that $(q, v, S) \rightsquigarrow (q, \varepsilon, \sigma' y)$. By the induction hypothesis, there is a leftmost derivation of $v\sigma' y$ in G. Then $v\sigma y = wy$, and this completes the proof. □

What about mimicking the computation of a PDA by a context-free grammar? For convenience, we will start with an SPDA. Let $P = (Q, \Sigma, \Gamma, \Delta, s, F)$ be an SPDA. Take $N = \{S\} \cup \{[a, X, r] : q, r \in Q \text{ and } X \in \Gamma \cup \{\varepsilon\}\}$. Here we assume that the symbol S has not appeared anywhere in the computation of P. Also read the bracketed triple $[q, X, r]$ as a name of some nonterminal. This naming scheme, in a certain way, remind us how the SPDA handles the input strings. The name $[q, \varepsilon, r]$ stands for the portion of the input string that might have been read during a time period when the SPDA remains in state q without operating with the stack, and after this period, it changes state to r.

Similarly, the name $[q, X, r]$ would remind us the portion of the input string used by P being in state q throughout a time period when there is the string X on the top of the stack, and after this, P changes state to r by removing X from the stack. This is only an intuitive understanding of the naming scheme. With this reading convention of the names, we construct a context-free grammar $G_P = (N, \Sigma, R, S)$, where R has the following four types of productions:

(a) $S \mapsto [s, \varepsilon, q]$ for each final state $q \in F$.
(b) $[q, \varepsilon, q] \mapsto \varepsilon$ for each state $q \in Q$.
(c) $[q, X, p] \mapsto u[r, \varepsilon, p]$ for each $p \in Q$ and for each transition $(q, u, X, r, \varepsilon) \in \Delta$, where $q, r \in Q, u \in \Sigma^*$, and, $X \in \Gamma$ or $X = \varepsilon$.
(d) $[q, X, p] \mapsto u[r, A_1, q_1] \cdots [q_{n-1}, A_n, p]$ for each transition $(q, u, X, r, A_1 A_2 \ldots A_n) \in \Delta$ and for all $p, q_1, \ldots, q_{n-1} \in Q$, where $n > 0, q, r \in Q, u \in \Sigma^*$, $A_i \in \Gamma, X \in \Gamma$, and $X \in \Gamma$ or $X = \varepsilon$.

As P is an SPDA, each of its transitions is in one of the forms of those in (c) or (d) above. A production of the type (a) says that no computation is needed to remain in the same state. Type (b) productions say that if the stack remains unchanged, then P can change its state from the initial to the final. A production of type (c) says that P changes its state from q to r while reading u, and at that moment it either removes X from the stack or it keeps the stack unchanged. Finally, a production of type (d) says that P changes state from q to p while removing X (or ε) from the stack through the n states, which effect the removal of $A_1, \ldots, A_n$ after taking a single move in reading the input u and changing state to r. We now prove the following result:

Lemma 6.4. *Let P be an SPDA and G_P be a context-free grammar as constructed earlier. Let $q, r \in Q, X \in \Gamma \cup \{\varepsilon\}$, and $u \in \Sigma^*$. If $[q, X, r] \Rightarrow u$ in G_P, then $(q, u, X) \rightsquigarrow (r, \varepsilon, \varepsilon)$ in P.*

Proof. If $[q, X, r] \Rightarrow u$ in one step, then a type (a) production has been applied. Then $q = r, u = \varepsilon, X = \varepsilon$, and consequently, $(q, u, X) \rightsquigarrow (r, \varepsilon, \varepsilon)$ trivially. Lay out the induction hypothesis that if $[q, X, r] \Rightarrow u$ in fewer than m steps, then $(q, u, X) \rightsquigarrow (r, \varepsilon, \varepsilon)$. Suppose that $[q, X, r] \Rightarrow u$ in m steps. Then the first step must be an application of a production of one the types (c) or (d).

Let us first consider the type (d) productions. Here, the derivation looks like

$$[q, X, r] \overset{1}{\Rightarrow} v[p, A_1, q_1] \cdots [q_{n-1}, A_n, r] \Rightarrow u.$$

For notational convenience, rename p as q_0 and r as q_n. We have strings $v_i \in \Sigma^*$ such that $[q_{i-1}, A_i, q_i] \Rightarrow v_i$ for each i with $1 \le i \le n$. Notice that each of these derivations was of length fewer than m. Moreover, $u = vv_1 \ldots v_n$. By the induction hypothesis, $(q_{i-1}, v_i, A_i) \rightsquigarrow (q_i, \varepsilon, \varepsilon)$. We also have $(q, v, X) \rightsquigarrow (r, \varepsilon, A_1 \ldots A_n)$. Hence

$$(q, vv_1 \cdots v_n, X) \rightsquigarrow (p, v_1, \cdots v_n, A_1 \cdots A_n) \rightsquigarrow \cdots \rightsquigarrow (r, \varepsilon, \varepsilon).$$

The former case of type (c) productions is similar and a bit easier than this. □

Lemma 6.5. *Let P be an SPDA and G_P be a context-free grammar as constructed earlier. Let $q, r \in Q, X \in \Gamma \cup \{\varepsilon\}$, and $u \in \Sigma^*$. If $(q, u, X) \rightsquigarrow (r, \varepsilon, \varepsilon)$ in P then $[q, X, r] \Rightarrow u$ in G_P.*

Proof. If $(q, u, X) \rightsquigarrow (r, \varepsilon, \varepsilon)$ in 0 step, then $[q, \varepsilon, q] \Rightarrow \varepsilon$ due to type (a) production in G_P. Lay out the induction hypothesis that if $(q, u, X) \rightsquigarrow (r, \varepsilon, \varepsilon)$ in a computation of length fewer than m, then $[q, X, r] \Rightarrow u$. Suppose that $(q, u, X) \rightsquigarrow (r, \varepsilon, \varepsilon)$ in a computation of length m and no fewer. Then such a computation looks like

$$(q, u, X) \overset{1}{\rightsquigarrow} (p, w, A_1 \cdots A_n) \rightsquigarrow (r, \varepsilon, \varepsilon) \quad \text{(in } m - 1 \text{ steps)}.$$

Then there is $v \in \Sigma^*$ such that $u = vw$ and $(q, v, X) \rightsquigarrow (p, \varepsilon, A_1 \cdots A_n)$. That is, there must be a transition in one of the forms

$$(q, v, X, p, A_1 \cdots A_n) \text{ or } (q, v, \varepsilon, p, A_1 A_2 \cdots A_{n-1}).$$

As P is an SPDA, corresponding to each transition of the latter type, there is one of the former type also. Thus, we may consider only transitions of the form $(q, v, X, p, A_1 \cdots A_n)$.

However, we do not yet know whether the machine indeed goes through other states while changing from q to p. Trivially, when $n = 0$, $(p, w, \varepsilon) \rightsquigarrow (r, \varepsilon, \varepsilon)$ by a computation of $m - 1$ steps. By the induction hypothesis, $[p, \varepsilon, r] \Rightarrow w$ in G_P using a type (c) production $[q, X, r] \mapsto [p, \varepsilon, r]$. We see that $[q, X, r] \Rightarrow vw = u$. For the nontrivial case $n > 0$, all A_i's are somehow removed from the stack. The SPDA will then have states $q_1, \ldots, q_{n-1}$ and there are strings $w_1, \ldots w_n$ such that

$$(q_i, w_{i+1} \cdots w_n, A_{i+1} \cdots A_n) \rightsquigarrow (q_{i+1}, w_{i+2} \cdots w_n, a_{i+2} \cdots A_n).$$

As, for $i = 0, 1, \ldots, n-1$ with $q_0 = p$ and $q_n = r$, each computation has fewer than m steps, by the induction hypothesis, $[q_i, A_{i+1}, q_{i+1}] \Rightarrow w_{i+1}$. With the production

$$[q, A, r] \mapsto v[p, A_1, q_1][q_1, A_2, q_2] \cdots [q_{n-1}, A_n, r],$$

we obtain $[q, X, r] \Rightarrow vw_1 \cdots w_n = u$. □

Thus we see that the construction has the desired effect. We summarize the discussion as follows:

Theorem 6.1. *Let L be a language over some alphabet. Then there is a context-free grammar G with $\mathcal{L}(G) = L$ iff there is a PDA P with $\mathcal{L}(P) = L$.*

Proof. Let G be a context free grammar with $\mathcal{L}(G) = L$. Then by Lemma 5.1, each $w \in L$ has a leftmost derivation in G. By Lemmas 6.2 and 6.3 (with $x = \varepsilon, u = w$), there is a PDA, say P such that $(q, w, S) \rightsquigarrow (q, \varepsilon, \varepsilon)$. Moreover, in the construction of P, we have a transition that allows $(s, w, \varepsilon) \rightsquigarrow (q, \varepsilon, \varepsilon)$. This shows that $(s, w, \varepsilon) \rightsquigarrow (q, \varepsilon, \varepsilon)$. That is, $w \in \mathcal{L}(P)$.

Conversely, let $w \in \mathcal{L}(P)$ for a PDA P. By Lemma 6.1, there is an SPDA, say, P' such that $w \in \mathcal{L}(P)$. Then $(s, w, \varepsilon) \rightsquigarrow (f, \varepsilon, \varepsilon)$ for some final state of P'. Using Lemmas 6.4 and 6.5 (with $q = s, X = \varepsilon, r = f$), we obtain a context free grammar G such that $[s, \varepsilon, f] \Rightarrow w$ in G. That is, $w \in \mathcal{L}(G)$. □

Problems for Section 6.3

6.11. Let G be a CFG with productions $S \mapsto c|aSa|bSb$. Take the PDA P with initial state p, final state q, and transitions $(p, \varepsilon, \varepsilon, q, S)$, $(q, \varepsilon, S, q, ASA)$, $(q, \varepsilon, S, q, BSB)$, $(q, \varepsilon, S, q, C)$, $(q, a, A, q, \varepsilon)$, $(q, b, B, q, \varepsilon)$, $(q, c, C, (q, \varepsilon)$. Derive the string $baacaab$ in G and also trace the computation of P on $baacaab$. Relate the two.

6.12. Construct, in each case, a PDA that accepts the language generated by the CFG with productions as given below.
(a) $S \mapsto a|aSA,\ A \mapsto bB,\ B \mapsto b.$
(b) $S \mapsto \varepsilon|SS|aSa.$
(c) $S \mapsto C|aSb,\ C \mapsto \varepsilon|S|bCa.$
(d) $S \mapsto aAA,\ A \mapsto a|aS|bS.$
(e) $S \mapsto aA,\ A \mapsto a|bB|aABC,\ B \mapsto b,\ C \mapsto c.$
(f) $S \mapsto aab|aSbb.$
(g) $S \mapsto ab|aSSS.$
(h) $S \mapsto aAA|aABB,\ A \mapsto a|aBB,\ B \mapsto A|bBB.$
(i) $S \mapsto a|AA,\ A \mapsto b|SA.$

6.13. Find CFGs that generate the languages accepted by the PDAs with initial state p, final state q, and transitions
(a) $(p, a, \varepsilon, p, A), (q, \varepsilon, \varepsilon, p, A), (p, a, A, q, \varepsilon), (p, b, A, r, \varepsilon), (r, \varepsilon, \varepsilon, q, \varepsilon)$.
(b) $(p, a, \varepsilon, p, A), (p, a, A, p, A), (p, b, A, r, \varepsilon), (r, \varepsilon, \varepsilon, q, \varepsilon)$.
(c) $(p, a, \varepsilon, p, A), (p, b, \varepsilon, p, B), (p, \varepsilon, \varepsilon, q, \varepsilon), (q, a, A, q, \varepsilon), (q, b, B, q, \varepsilon)$.
(d) $(p, \varepsilon, \varepsilon, r, C)$, (r, a, C, r, AC), (r, a, A, r, AA), $(r, a, B, r, \varepsilon)$, (r, b, C, r, BC), (r, b, B, r, BB), $(r, b, A, r, \varepsilon)$, $(r, \varepsilon, C, q, \varepsilon)$.
(e) $(p, 1, \varepsilon, p, A), (p, 1, A, p, AA), (p, 0, A, q, A), (p, \varepsilon, A, p, \varepsilon), (q, 1, A, q, \varepsilon)$, $(q, 0, \varepsilon, p, \varepsilon)$, accepted by empty stack.
(f) $(p, 0, \varepsilon, p, A), (p, 0, A, p, AA), (p, 1, A, p, A), (p, \varepsilon, A, q, \varepsilon), (q, \varepsilon, A, q, \varepsilon)$, $(q, 1, A, q, AA), (q, 1, \varepsilon, q, \varepsilon)$, accepted by final states.

6.14. Construct a PDA with two states, one initial and the other final, that accepts the language
(a) $L = \{a^n b^{n+1} : n \in \mathbb{N}\}$.
(b) $L = \{a^n b^{2n} : n \geq 1\}$.
(c) $\{a^m b^n c^k : m = 2n \text{ or } n = 2k;\ m, n, k \in \mathbb{N}\}$.
(d) $\{a^m b^n c^{2(m+n)} : m, n \in \mathbb{N}\}$.

6.15. Is it true that corresponding to each PDA there exists an equivalent PDA with only two states, one of which is the initial state and the other is the final state?

6.16. Consider a PDA M with n states and m number of distinct input symbols. What is the maximum possible number of rejecting computations that M can have on an input of length k?

6.4 Pumping Lemma

There are only a countable number of pushdown automata. Can you see this? In that case, there are only a countable number of context-free languages over any alphabet. But there are uncountable number of languages over the same alphabet. Hence, there are uncountably many languages that are not CFLs. However, this argument neither does give a noncontext-free language nor does it help in proving whether a given language is context-free or not. In this section, we suggest a partial remedy: the pumping lemma. Let us begin with an example.

Example 6.7. Let $G = (\{S\}, \{a, b\}, \{S \mapsto \varepsilon \mid aSb \mid SS\}, S)$. A parse tree in G with yield $aabbab$ is shown in Fig. 6.3.

The leftmost derivation of the same string that has been represented in the parse tree in Fig. 6.3 is

$$S \Rightarrow SS \Rightarrow aSbS \Rightarrow aaSbbS \Rightarrow aa\varepsilon bbS \Rightarrow aa\varepsilon bbaSb \Rightarrow aa\varepsilon bba\varepsilon b = aabbab.$$

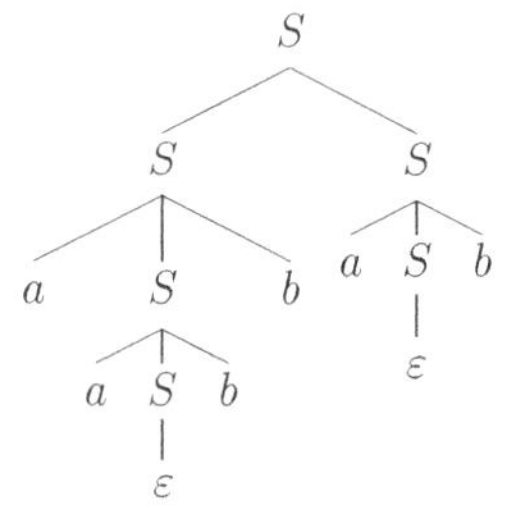

Fig. 6.3. Parse tree for Example 6.7.

Look at the derivation of $aabbab$ and also the parse tree. It is clear that any string of the form $aa^n\varepsilon b^n bab$ can be derived in G. It is, of course, obvious from the grammar itself, as $\mathcal{L}(G)$ is simply the language of matching parentheses [with a as (and b as)]. However, I want you to realize this fact by looking at the parse tree. Can you see it?

In the second step of the derivation, you see that the string (over $N \cup \Sigma$) is $aSbS$. Now, the first S in it can be grown into $a^n Sb^n$ by repeated use of the production $S \mapsto aSb$. But not only that, the same method can be used on any string of length greater than 3. Why?

If we take any string (over Σ) of length greater than 3, there must be a parse tree with height at least 2. This happens because on the right hand side of any production, there are at most three symbols. Then any node in the parse tree can have at most three children. If a parse tree has height 1, then the number of leaves in it will be no more than 3. Thus the height of any parse tree that yields a string of length greater than 3 has to be of height at least 2. In that case, there is a path from the root to a leaf of length (number of edges in the path) at least 2.

That is, such a path has at least three nodes in it. Of these nodes, the last one, the leaf is a terminal symbol from Σ and all the others are from N, nonterminals. As there is only one nonterminal here, namely, S, it must repeat at least twice in the path. That is, while traveling from the root to the leaf along that path, we find S at least twice. For example, in the parse tree of Fig. 6.3, we take the path "S to S to S to S to ε," which gives the tree its height. Now, taking the second and the fourth occurrences of S, we can have a relook at the tree as in Fig. 6.4.

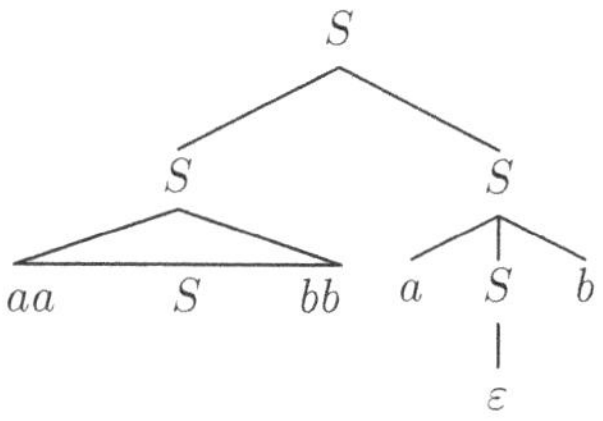

Fig. 6.4. Another parse tree for Example 6.7.

In Fig. 6.4, we have kept the yield aa and bb at the appropriate places by shrinking the tree at the second occurrence of S. As the same node S is repeated, the whole portion shown within a triangle can be inserted below the S again to have another parse tree. Such an insertion will give us the parse tree in Fig. 6.5.

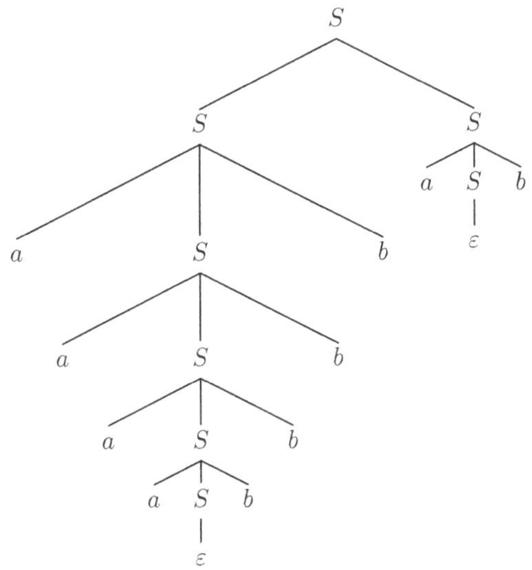

Fig. 6.5. Parse tree for Example 6.7 with inserted nodes.

This corresponds to the derivation of the string $aa(aabb)bbab$. That is, we can always construct a derivation of $aa(aabb)^n bbab$ from that of $aabbab$ by using the insertion of the said portion of the parse tree. Looking at it in a slightly different way, we can get a derivation of $aabbab$ from a derivation of $aa(aabb)^n bbab$, where $n > 1$. Loosely speaking, the latter observation says that for derived strings of certain bigger lengths, we can have derivations of strings of smaller lengths. □

The technique of insertion can be generalized to any context-free grammar.

Lemma 6.6. *Let $G = (N, \Sigma, R, S)$ be a CFG. Let $m = |N|$, the number of nonterminal symbols, and $k = \max\{\ell(\alpha) : X \mapsto \alpha \in R\}$, the maximum number of symbols on the right hand side of any production. Let $w \in \mathcal{L}(G)$ have length more than k^m. Then w has a derivation of the form $S \Rightarrow uAz \Rightarrow uvAyz \Rightarrow uvxyz = w$ for some $A \in N$, $u, v, x, y, z \in \Sigma^*$, where $\ell(vxy) \le k^m$ and at least one of v or y is nonempty.*

Proof. Let T_1 be a parse tree in G with yield $w \in \Sigma^*$, and having the least height among many possible parse trees yielding w. Suppose that $\ell(w) > k^m$, that is, the number of leaves in T_1 is more than k^m. Each node in T_1 can have at the most k children. Hence, height of T_1 is greater than m.

Let P be a path in T_1 from the root to a leaf, which gives T_1 its height. Now, P has at least $m + 1$ edges, and thus has at least $m + 2$ nodes. The last node, the leaf, is a terminal symbol and all the others are nonterminals in N. But N has only m symbols. Thus, by the pigeon hole principle, there is at least one nonterminal that is repeated in P. Choose such a symbol, say, $A \in N$, which is repeated at least twice in P. Take the last two occurrences of A in P.

To understand what is going on, we construct five more parse trees from T_1. Look at the schematic drawing in Fig. 6.6.

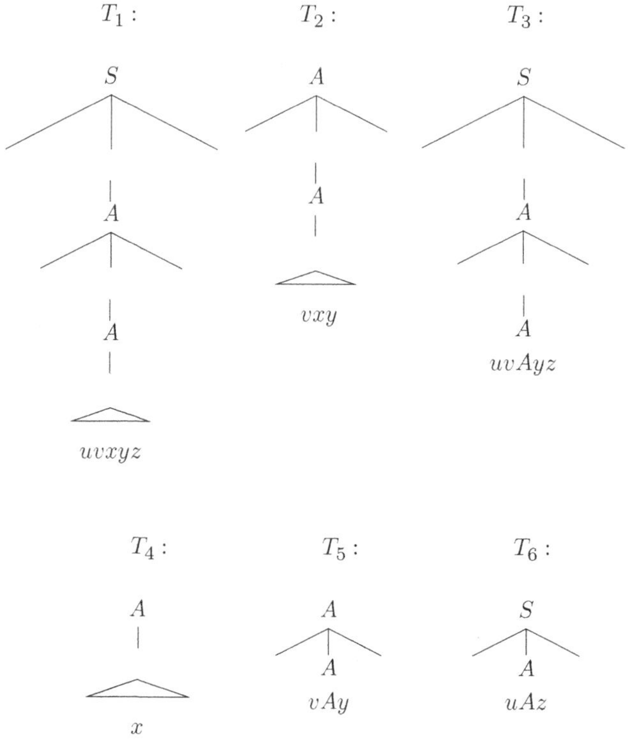

Fig. 6.6. Schematic trees for Lemma 6.6.

Let T_2 be the subtree of T_1 whose root is the last but one occurrence of A in the path P. Let T_3 be the tree obtained from T_1 by cutting down the subtree from the last occurrence of A in P, but keeping this A as a leaf. Let T_4 be the subtree of T_1 whose root is the last occurrence of A in P. That is, you have got T_3 by cutting down T_4 from T_1 and keeping this A. Moreover, T_4 is a subtree of T_2. Let T_5 be the tree obtained from T_2 by cutting down T_4 from it and keeping the last A as a leaf. Let T_6 be the tree obtained from T_3 by cutting down T_5 from it and keeping the A (last but one occurrence in P) as a leaf.

Now, what are the yields of these parse trees? The original parse tree T_1 has the yield w any way. Look at the smaller parse trees first. Suppose that the parse tree T_4 has yield x for some $x \in \Sigma^*$. This means that in G, we have a derivation $A \Rightarrow x$. Similarly, the tree T_5 has the yield vAy and the tree T_6 has a yield of the form uAz for some $u, v, y, z \in \Sigma^*$. Then we have derivations $A \Rightarrow vAy$ and $S \Rightarrow uAz$ in G. Then what are the yields of the other parse trees?

Look at Fig. 6.6 again. The yields are written below the trees. From the yields of T_4 and T_5, we see that the yield of T_2 is vxy. Similarly, from the yields of the trees T_5 and T_6, we obtain the yield of T_3 as $uvAyz$. Finally, from the yields of T_3 and T_4, we get the yield of T_1 as $uvxyz$. The sequence of derivations in terms of growing the parse trees can be given as

$$S \text{ then } T_6 \text{ then } T_3 \text{ then } T_1.$$

In terms of the derivations, we have the corresponding steps as

$$S \Rightarrow uAz \Rightarrow uvAyz \Rightarrow uvxyz = w \text{ for some } u, v, x, y, z \in \Sigma^*.$$

Can both v and y be empty? If so, then the tree T_5 can completely be removed from T_1 without changing the yield. This is done by removing the subtree at the last but one occurrence of A and adding the subtree with the root as the last A at the same node. Notice that the subtree with the last but one A generates vxy, while the subtree at the last A generates x. As $v = y = \varepsilon$, $vxy = x$.

This deletion can be carried out in every such path in T_1 that give the tree its height. This way, the height of the tree T_1 is reduced with the same yield as w. That is, if $v = y = \varepsilon$, then there will exist a parse tree of shorter height than that of T_1, whose yield will be w. This is a contradiction as T_1 is a tree of shortest height with yield w. Therefore, $v \neq \varepsilon$ or $y \neq \varepsilon$.

To see that $\ell(vxy) \leq k^m$, look at the tree T_2, whose yield is vxy. We had chosen the repeated nonterminal A arbitrarily. Now we want to take a special choice so that our job would be done. We choose the repeated nonterminal A in such a way that on the path P from the root of T_2 to the leaf, no other nonterminal is repeated. In turn, this uses an appropriate choice of T_2.

It is obvious that such a nonterminal can always be chosen. As the path P in T_1 was a longest path, the portion of the same path in T_2 is also the longest. However, its length is no more than m as there are only m number of nonterminals in G. Thus, every path from the root of T_2 to any leaf is having length less than or equal to m. Moreover, each node in T_2 has at most k children. Thus, there are at the most k^m leaves in T_2. This shows that $\ell(vxy) \leq k^m$ and completes the proof. □

Lemma 6.6 also shows how certain strings can be pumped into a derived string. This also shows how certain strings can be pumped out! We state and prove a formal version of the Pumping Lemma for context-free languages, from which this *pumping out* will become clearer.

Theorem 6.2 (Pumping Lemma for Context-free Languages). *Let L be an infinite context-free language over an alphabet Σ. Then there is an $n \in \mathbb{Z}_+$ such that any string $w \in L$ with $\ell(w) > n$ can be written as $w = uvxyz$ for strings $u, v, x, y, z \in \Sigma^*$, where at least one of v or y is nonempty, $\ell(vxy) \leq n$, and $uv^jxy^jz \in L$ for every $j \in \mathbb{N}$.*

Proof. As L is a context-free language over Σ, we have a context-free grammar $G = (N, \Sigma, R, S)$ such that $L = \mathcal{L}(G)$. Let $k = \max\{\ell(\alpha) : X \mapsto \alpha \in R\}$, $m = |N|$. Take $n = k^m$. By Lemma 6.6, we have a derivation of w in G of the form

$$S \Rightarrow uAz \Rightarrow uvAyz \Rightarrow uvxyz = w \text{ for some } u, v, x, y, z \in \Sigma^*,$$

where at least one of v or y is nonempty and $\ell(vxy) \le n$. It follows from this (and also from the proof of Lemma 6.6) that $A \Rightarrow x$ and then $S \Rightarrow uAz \Rightarrow uxz$. Moreover, $A \Rightarrow vAy$ shows that for any $j \in \mathbb{Z}_+$,

$$S \Rightarrow uAz \Rightarrow uvAyz \Rightarrow uvvAyyz \Rightarrow \cdots \Rightarrow uv^jAy^jz \Rightarrow uv^jxy^jz. \qquad \square$$

The assumption "infinite" is not really required in the pumping lemma as the conclusion is a conditional statement. However, it guarantees that a string w of length more than n, in fact, exists. Note that the number n is fixed for a language, and the lengths of the pumped strings v and y are also somehow fixed by the language, for the given string w. We have the freedom to choose only the number of pumped strings into an already derived string. As applications of the pumping lemma we will show that some languages are not context-free.

Example 6.8. Show that $L = \{a^m b^m c^m : m \in \mathbb{N}\}$ is not context-free.

Solution. Suppose that L is context-free. Let n be the number that is fixed for L as in the pumping lemma. Take $k = n + 1$. The string $w = a^k b^k c^k \in L$ and it has length more than n. We can thus write $w = uvxyz$, where at least one of v, y is nonempty. Suppose that $v \neq \varepsilon$; the other case that $y \neq \varepsilon$ is similar. What can be this string v? As $\ell(vxy) \le n$, either v has occurrences of only one letter or v has occurrences of two letters.

Suppose v has occurrences of only one letter, say, a's. As $\ell(vxy) \le n$, the string vxy does not contain any c. Now, in the string uv^2xy^2z, there are more a's than c's. The cases that v consists of only b's or of only c's are similar.

On the other hand, suppose v has occurrences of two letters, say, a and b. Then in v all a's must precede all b's. However, in v^2, a b precedes an a. Thus, in the string uv^2xy^2z, a b precedes an a. The cases that v has occurrences of b's and c's, or of c's and a's are similar.

That is, in any case, $uv^2xy^2z \notin L$, contradicting the pumping lemma. Therefore, our assumption that L is a context-free language is wrong. $\square$

Exercise 6.8. Formulate the application of Pumping Lemma for CFLs as a game with the demon.

Example 6.9. Show that $L = \{a^k b^m c^k d^m : k, m \in \mathbb{N}\}$ is not context-free.

Solution. Assume that L is a context-free language. Let $n \in \mathbb{Z}_+$ be the number fixed by the pumping lemma for L. Take $w = a^n b^n c^n d^n \in L$. As $\ell(w) > n$, we can write it as $w = uvxyz$, where at least one of v or y is nonempty, and $\ell(vxy) \le n$. Suppose $v \neq \varepsilon$; the case of $y \neq \varepsilon$ is similar. As $\ell(vxy) \le n$, the string v is in one of the forms

$$a^i,\ b^i,\ c^i,\ d^i,\ a^i b^j,\ b^i c^j,\ \text{or } c^i d^j \text{ for } i, j \ge 1.$$

Suppose $v = a^i$. As $\ell(vxy) \le n$, the string vxy has no occurrence of c. Then $uxz = uv^0xy^0z$ has fewer occurrences of a's than c's. Similar are the cases when v is of the form b^i, c^i, or d^i. In all these cases, $uv^0xy^0z \notin L$.

On the other hand, suppose $v = a^i b^j$. Then in uv^2xy^2z, the symbol b precedes an a. The cases of v being in one of the forms $b^i c^j$ or $c^i d^j$ are similar. In all these cases, $uv^2xy^2z \notin L$.

In either case, pumping lemma is violated. Therefore, L is not context-free. □

Exercise 6.9. Show that $\{a^i b^i c^j : j \geq i \geq 0\}$ is not a context-free language.

Example 6.10. Show that $L = \{a^m b^m a^m : m \in \mathbb{N}\}$ is not context-free.

Solution. Suppose L is context-free. Let $n \in \mathbb{Z}_+$ be the number for L provided by the pumping lemma. Choose $w = a^n b^n a^n$. Pumping Lemma says that you can write $w = uvxyz$ such that $\ell(vxy) \leq n$, $v \neq \varepsilon$ or $y \neq \varepsilon$, and $uv^j xy^j z \in L$ for every $j \in \mathbb{N}$.

In such a rewriting of w, the substring vxy is in one of the forms a^k, b^k, $a^l b^k$ or $b^l a^k$ for some $l, k \in \mathbb{Z}_+$. If $vxy = a^l b^k$, then v cannot be in the form $a^l b^p$, for otherwise, uv^2xy^2z will have blocks of a's, b's, a's, b's, followed by a block of a's in it; which would not be in L. Similarly, $vxy = b^l a^k$ is also impossible.

If $v = a^q$ for some $q \in \mathbb{Z}_+$ (leaving the case $v = \varepsilon$, when you argue with y instead), then uv^2xy^2z will have the first block of a's with length more than the last block (which has still length n); so that it is not in L. Other cases are similar. Thus, L is not a CFL. □

The pumping lemma is, undoubtedly, an important tool in discovering noncontext-free languages. Unfortunately, it does not always succeed in showing that certain languages are not context-free. For an example, try using pumping lemma to prove that the language $\{a^i b^j c^k d^l : \text{ either } i = 0 \text{ or } j = k = l\}$ is not context-free. There are some strengthening of the pumping lemma such as Ogden's Lemma (Problem 6.45) and Parikh's Theorem (Problems 6.67 and 6.68) that help getting the job done.

Proof of the Pumping Lemma gets simplified if we use Chomsky normal form. Some more nontrivial but interesting structural properties of context-free languages can be proved using the normal forms. The best known such properties are the Chomsky–Schützenberger theorem (Problem 6.66) and Parikh's theorem. The first of these results says that each CFL is essentially an intersection of some parentheses language with a regular language; and the second states that the commutative image of a CFL is a semilinear set. To demystify, Parikh's theorem says that if we look at the relative number of occurrences of terminal symbols in the strings of a CFL disregarding their order, then we find only a regular set.

Problems for Section 6.4

6.17. Show how pumping lemma is satisfied for the languages
(a) $\{aa, ba, aba\}$.
(b) $\{a^n b^n : n \in \mathbb{N}\}$.
(c) The set of all palindromes over $\{a, b\}$.

6.18. Show that the following languages are not context-free:
(a) $\{a^{n!} : n \in \mathbb{N}\}$.
(b) $\{a^n : n$ is a prime number$\}$.
(c) $\{a^{n^2} : n \in \mathbb{N}\}$.
(d) $\{a^n b^n : n \in \mathbb{N}\} \cup \{a^n b^{2n} : n \in \mathbb{N}\} \cup \{a^n b^n c^n : n \in \mathbb{N}\}$.
(e) $\{ww : w \in \{a, b\}^*\}$.
(f) $\{www : w \in \{a, b\}^*\}$.
(g) $\{w^R w : w \in \{a, b\}^*\}$.
(h) $\{vcw : v$ is a substring of $w;\ v, w \in \{a, b\}^*\}$.
(i) $\{a^n b^n a^n b^n : n \in \mathbb{N}\}$.
(k) $\{a^n b a^{2n} b a^{3n} : n \in \mathbb{N}\}$.
(l) $\{a^m b^n : m \le n^2;\ m, n \in \mathbb{N}\}$.
(m) $\{a^m b^n : m \ge (n-1)^3;\ m, n \in \mathbb{N}\}$.
(n) $\{baba^2b \cdots ba^{n-1}ba^n b : n \ge 1\}$.
(o) $\{w \in \{a, b, c\}^* : (\#_a(w))^2 + (\#_b(w))^2 = (\#_c(w))^2\}$.
(p) $\{a^m b^n c^k : k = mn;\ m, n \in \mathbb{N}\}$.
(q) $\{a^m b^n c^k : m < n < k;\ m, n \in \mathbb{N}\}$.
(r) $\{a^m b^m c^n : n \le m;\ m, n \in \mathbb{N}\}$.
(s) $\{a^m b^m c^n : m \le n \le 2m;\ m, n \in \mathbb{N}\}$.
(t) $\{a^m b^n c^k : k > m, k > n;\ m, n \in \mathbb{N}\}$.
(u) $\{a^m b^n c^k : m < n, m \le k \le n;\ m, n \in \mathbb{N}\}$.
(v) $\{ww^R w : w \in \{0, 1\}^*\}$.
(w) $\{w_1 c w_2 c \cdots c w_k : k \ge 2,$ each $w_i \in \{a, b\}^*,$ and for some $m \ne n,\ w_m = w_n\}$.

6.19. Following the proofs of Lemma 6.6.and of Theorem 6.2, give another proof of the pumping lemma for regular languages. [Hint: Use regular grammars instead of DFAs]

6.5 Closure Properties of CFLs

You may be wondering why this delay in discussing the closure properties? Indeed, we have gone topsy turvy. For regular languages, we discussed all the machinery almost at the same time. For CFLs, we relied heavily upon the context-free grammars. You will discover the reason shortly; some closure properties can be easily seen via PDAs! Moreover, waiting this far will become fruitful in discovering certain more noncontext-free languages.

You have seen that union, intersection, complement, concatenation, and Kleene star of regular languages are regular. Do they also behave the same way for context-free languages?

Theorem 6.3. *Let L, L_1, L_2 be context-free languages. Then $L^*, L_1L_2, L_1 \cup L_2$ are also context-free.*

Proof. Let $L = \mathcal{L}(G)$, where $G = (N, \Sigma, R, S)$ be a context-free grammar. Define another context-free grammar $G^* = (N, \Sigma, R^*, S)$ with

$$R^* = R \cup \{S \mapsto \varepsilon, S \mapsto SS\}.$$

Is it clear why we are doing this? In G^* you can now derive ε. Also you can derive from the first S in SS any string that can be derived in G and from the second S another such string so that their concatenation can be derived in S. Inductively, that must give you $L^* = \cup_{n \in \mathbb{N}} L^n$.

Let us try to prove that G^* indeed generates L^*. Trivially, $L^0 = \{\varepsilon\} \subseteq \mathcal{L}(G^*)$ due to the production $S \mapsto \varepsilon$. Suppose that $L^m \subseteq \mathcal{L}(G^*)$ for $0 \le m \le k$. Let $w \in L^{k+1}$. Then $w = uv$, where $u \in L^k$ and $v \in L$. That means, we have derivations $S \Rightarrow_{G^*} u$ and $S \Rightarrow_G v$. However, all the productions in G are also in G^*. Thus, $S \Rightarrow_{G^*} v$. A derivation of w in G^* is

$$S \Rightarrow SS \Rightarrow uS \Rightarrow uv = w.$$

By induction, it follows that $L^k \subseteq \mathcal{L}(G^*)$ for every $k \in \mathbb{N}$. Thus, $L^* = \cup_{k \in \mathbb{N}} L^k \subseteq \mathcal{L}(G^*)$. Conversely, whenever a nonempty string is derived in G^*, it has to be a concatenation of strings derived in G. As L^* contains all possible concatenations of strings of L, we have $\mathcal{L}(G^*) \subseteq L^*$. This completes the proof that $\mathcal{L}(G^*) = L^*$.

For the other two operations, let $G_1 = (N_1, \Sigma_1, R_1, S_1)$, $G_2 = (N_2, \Sigma_2, R_2, S_2)$, where $N_1 \cap N_2 \neq \emptyset$, $\mathcal{L}(G_1) = L_1$, and $\mathcal{L}(G_2) = L_2$. Let S be a new symbol not in $N_1 \cup N_2$. Write $N_3 = N_1 \cup N_2 \cup \{S\}$ and $\Sigma_3 = \Sigma_1 \cup \Sigma_2$. Construct

$$G_c = (N_3, \Sigma_3, R_c, S), \text{ where } R_c = R_1 \cup R_2 \cup \{S \mapsto S_1 S_2\}.$$

$$G_\cup = (N_3, \Sigma_3, R_\cup, S), \text{ where } R_\cup = R_1 \cup R_2 \cup \{S \mapsto S_1, S \mapsto S_2\}.$$

You can now verify that $\mathcal{L}(G_c) = L_1 L_2$ and $\mathcal{L}(G_\cup) = L_1 \cup L_2$. □

Exercise 6.10. Show that $\mathcal{L}(G_\cup) = L_1 \cup L_2$ and $\mathcal{L}(G_c) = L_1 L_2$ to complete the proof of Theorem 6.3.

Theorem 6.4. *Intersection of two context-free languages need not be context-free and complement of a context-free language need not be context-free.*

Proof. Let $L_1 = \{a^i b^i c^j : i, j \in \mathbb{N}\}$ and $L_2 = \{a^i b^j c^j : i, j \in \mathbb{N}\}$. These languages are context-free as the context-free grammar with productions

$$S \mapsto AB, A \mapsto \varepsilon \mid aAb, B \mapsto \varepsilon \mid cB$$

generates L_1 and the context-free grammar that has productions

$$S \mapsto AB, A \mapsto \varepsilon \mid aA, B \mapsto \varepsilon \mid bBc$$

generates L_2. But the language $L_1 \cap L_2 = \{a^m b^m c^m : m \in \mathbb{N}\}$ is not context-free as you have seen in Example 6.8. Therefore, intersection of context-free languages need not be context-free.

Suppose that complement of any context-free language is context-free. Then $\overline{L_1}, \overline{L_2}$ are context-free. From Theorem 6.3, $\overline{L_1} \cup \overline{L_2}$ is context-free. Taking complement again, we see that $L_1 \cap L_2 = \overline{\overline{L_1} \cup \overline{L_2}}$ must also be context-free. But it is not. Therefore, complement of a context-free language need not be context-free. □

Exercise 6.11. Show that for a regular language L_1 and a context-free language L_2, the intersection $L_1 \cap L_2$ may not be a regular language. [Hint: What was our first example of a nonregular language?]

Intersection of a context-free language with a regular language does give rise to a context-free language as the following statement shows.

Theorem 6.5. *If L_1 is a context-free language and L_2 is a regular language, then $L_1 \cap L_2$ is a context-free language.*

Proof. Because of the machine characterizations of regular and context-free languages, we start with a PDA P and a DFA D such that $\mathcal{L}(P) = L_1$ and $\mathcal{L}(D) = L_2$. The idea is to construct a PDA that simultaneously simulates P and D and accepts a string when both of them accept one. We name a state in the new PDA by a pair of states where one is from P and the other is from D, following the approach of product automata.

Let $P = (Q_1, \Sigma, \Gamma, \Delta, s_1, F_1)$ be a PDA and $D = (Q_2, \Sigma, \delta, s_2, F_2)$ be a DFA, such that $\mathcal{L}(P) = L_1$ and $\mathcal{L}(D) = L_2$. Define the PDA $P_\cap = (Q, \Sigma, \Gamma, \Delta, s, F)$, where

$$Q = Q_1 \times Q_2, s = (s_1, s_2), F = F_1 \times F_2,$$

$$\Delta = \{((q_1, q_2), \sigma, X, (r_1, r_2), Y) : (q_1, \sigma, X, r_1, Y) \in \Delta_1 \text{ and } (q_2, \sigma) \rightsquigarrow_D (r_2, \varepsilon)\}.$$

See how the idea is at work so that we have $\mathcal{L}(P_\cap) = L_1 \cap L_2$. □

Exercise 6.12. Show that $\mathcal{L}(P_\cap) = L_1 \cap L_2$, as claimed in the proof of Theorem 6.5.

Analogous to the regular languages, you can use the closure properties along with the pumping lemma to prove that certain languages are or are not context-free. See the following examples.

Example 6.11. Show that $L = \{ww : w \in \{a, b\}^*\}$ is not context-free, but $\overline{L}$ is.

Solution. If L is context-free, so is the language $L' = L \cap a^*b^*a^*b^*$, by Theorem 6.5. But $L' = \{a^i b^j a^i b^j : i, j \in \mathbb{N}\}$ is not context-free as we have seen in Example 6.9. (Rename c to a and d to b, or, apply the same argument as in Example 6.9.) Therefore, L is not a context-free language.

However, $\overline{L} = w \in \{a, b\}^* : w \neq uu$ for any $u \in \{a, b\}^*\}$ is a context-free language as it is generated by the CFG with productions:

$$S \mapsto A \mid B \mid AB \mid BA, \;\; A \mapsto a \mid CAC, \;\; B \mapsto b \mid CBC, \;\; C \mapsto a \mid b.$$ □

This language $\overline{L}$ thus provides an example of a CFL whose complement is not a CFL. It seems that these negative results regarding the intersection and complements of CFLs are the effect of nondeterminism. What would happen if we have, instead, the deterministic pushdown automata?

Problems for Section 6.5

6.20. Show that the following languages are context-free:
(a) $\{w \in \{a, b, c\}^* : \#_a(w) = \#_b(w) = \#_c(w)\}$.
(b) $\{w \in \{a, b, c\}^* : \#_a(w) < \#_b(w) < \#_c(w)\}$.
(c) $\{w \in \{a, b, c\}^* : \#_a(w) = \#_b(w) \cdot \#_c(w)\}$.
(d) $\{w \in \{a, b, c\}^* : 2\#_b(w) = \#_a(w) + \#_c(w)\}$.
(e) $\{a^m b^n : m \neq n;\ m, n, \in \mathbb{N}\}$.
(f) $\{a, b\}^* - \{a^n b^n : n \in \mathbb{N}\}$.
(g) $\{w \in \{a, b\}^* : w = w^R\}$.
(h) $\{a^m b^n c^m : m \neq n; m, n \in \mathbb{N}\}$.
(i) $\{a, b\}^* - \{baba^2b \cdots ba^{n-1}ba^nb : n \geq 1\}$.
(j) $\{a^m b^n c^k : m \neq n$ or $m \neq k;\ m, n, k \in \mathbb{N}\}$.
(k) $\{xywy^R : w, x, y \in \{a, b\}^+, \ell(x) = \ell(w) = 2\}$.
(l) $\{vcw : v, w \in \{a, b\}^+, w \neq v^R\}$.
(m) $a^*b^*c^* - \{a^n b^n c^n : n \in \mathbb{N}\}$.

6.21. Are the following languages context-free? Justify.
(a) $\{a^m b^n : m = 2^n,\ m, n \in \mathbb{N}\}$.
(b) $\{a^m b^m : m \in \mathbb{N}, m \neq 2009\}$.
(c) $\{a^n b^n : n \in \mathbb{N}$ and n is not a multiple of 5$\}$.
(d) $\{w \in \{a, b\}^* : \#_a(w) = \#_b(w),\ w$ does not have aab as a substring$\}$.
(e) $\{a^m b^n a^m b^n : m, n \in \mathbb{N}\} \cup \{a^m b^n a^n b^m : m, n \in \mathbb{N}\}$.
(f) $\{a^i b^j a^k b^m : i + j \leq k + m;\ i, j, k, m \in \mathbb{N}\}$.
(g) $\{a^i b^j a^k b^m : i \leq j, k \leq m;\ i, j, k, m \in \mathbb{N}\}$.
(h) $\{a^m ww^R a^m : w \in \{a, b\}^*,\ m \in \mathbb{N}\}$.
(i) $\{a^m b^m c^n : m \leq n;\ m, n \in \mathbb{N}\}$.
(j) $\{a^m b^n c^k : m, n, k \geq 1$ and $(2m = 3k$ or $5k = 7n)\}$.
(k) $\{a^m b^n c^k : m, n, k \geq 1$ and $(2m = 3k$ and $5k = 7n)\}$.
(l) $\{a^m b^n c^k : m, n, k \geq 1$ and $(m \neq 3n$ or $m \neq 5k)\}$.
(m) $\{a^m b^n c^k : m, n, k \geq 1$ and $(m \neq 3n$ and $m \neq 5k)\}$.
(n) $\{a^m b^n c^k : m, n, k \geq 1$ and $m + k = n\}$.
(o) $\{a^m b^n c^k d^r : m \leq k$ or $n = r$ or $m + n = k + r\}$.
(p) $\{a^m b^n c^k d^r : m, n, k, r \geq 1$ and $m = n, k = r\}$.
(q) $\{a^m b^n c^k d^r : m, n, k, r \geq 1$ and $m = k, n = r\}$.
(r) $\{a^m b^n c^k d^r : m, n, k, r \geq 1$ and $m = r, n = k\}$.
(s) $\{ucv : u, v \in \{a, b\}^*,\ u \neq v\}$.
(t) $\{a^{mn} : m, n$ are prime numbers$\}$.
(u) $\{w \in \{a, b, c\}^* : \#_a(w), \#_b(w), \#_c(w)$ are distinct $\}$.
(v) $\overline{L}$, where $L = \{w \in \{a, b, c\}^* : \#_a(w) = \#_b(w) = \#_c(w)\}$.

6.22. Show that the family of CFLs is closed under reversal.

6.23. Classify each of the following languages into (i) regular, (ii) context-free but not regular, or (iii) not context-free:
(a) $\{a^n : n = 2^m$ for some $m \in \mathbb{N}\}$.

(b) $\{w \in \{0, 1\}^* : w$ represents a power of 2 in binary$\}$.
(c) $\{a^m b^n : m \neq n,\ m, n \in \mathbb{N}\}$.
(d) $\{a^m b^n : 5m + 3n = 2009,\ m, n \in \mathbb{N}\}$.
(e) $\{a^m b^n : 5m - 3n = 2009,\ m, n \in \mathbb{N}\}$.
(f) $\{w \in \{a, b\}^* : \#_a(w) < \#_b(w),\ m, n \in \mathbb{N}\}$.
(g) $\{w \in \{a, b, c\}^* : \#_a(w) = \#_b(w) = \#_c(w)\}$.
(h) $\{a^m b^n c^k : m > n > k \geq 0\}$.
(i) $\{a^m b^n c^k : m > n \geq 0$ or $n > k \geq 0\}$.
(j) $\{a^m b^n c^k d^r : 2m = 3n$ and $5k = 7r;\ m, n, k, r \in \mathbb{N}\}$.
(k) $\{a^m b^n c^k d^r : 2m = 3k$ and $5n = 7r;\ m, n, k, r \in \mathbb{N}\}$.
(l) $\{a^m b^n c^k d^r : 2m = 3r$ and $5n = 7k;\ m, n, k, r \in \mathbb{N}\}$.

6.24. What is $\mathcal{L}(G)$, where G has productions $S \mapsto aSb|Aa|bA,\ A \mapsto \varepsilon|aA|Ab$? Give a CFG to generate $\overline{\mathcal{L}}(G)$.

6.6 Deterministic Pushdown Automata

Abstractly speaking, a deterministic pushdown automaton has the same relation with a PDA as a DFA has with an NFA. In a PDA, the machine has the capability to guess a course of action basing on its current state, its currently scanned input symbol, and the stack symbol that is on top of the stack. This guess is nondeterministic in the sense that there may be no possible course of action or there may be many possible actions from which it chooses one; the choice being not specified. In case of determinism, we have either no possible course of action or there is a unique possibility. This was so as far as DFAs are concerned.

To make the matter simple for deterministic pushdown automata in dealing with its stack, we assume that there is always something on the stack. It would behave as a bottom end marker on the stack. Similar to this, we will use a marker for the end of the input string. It will signal when the input is over. Further, determinism demands that at each possible situation there is either no way to proceed, when the machine would halt abruptly, or there is a unique action to be taken. The end marker on the input string will allow the use of ε-transitions due to the extra information on the stack. Further, we will use acceptance by final states only.

Formally, a **deterministic pushdown automaton**, or **DPDA**, for short, is an eight-tuple $P = (Q, \Sigma, \Gamma, \star, \triangledown, \delta, s, F)$, where

Q is a finite set of symbols, the *set of states*,
Σ is a finite set of symbols not containing the blank symbol, the *input alphabet*,
Γ is a finite set of symbols, the *stack alphabet*; the sets Q, Σ, Γ being pairwise disjoint,
$\star \notin \Sigma$ is the end marker on the input string; read $\star$ as "sem" abbreviating string-end-marker,
$\triangledown \notin \Gamma$ is the *bottom end marker* on the stack; read $\triangledown$ as "bems," abbreviating "bottom-end-marker-on-the-stack,"
$s \in Q$ is the *initial state* where P is initially in,

$F \subseteq Q$ is the *set of final states*, and
$\delta : Q \times (\Sigma \cup \{\star, \varepsilon\}) \times \Gamma \to Q \times (\Gamma^* \cup \Gamma^* \triangledown)$ is a partial function, called the *transition function* satisfying:

1. For any $p \in Q$, $\sigma \in \Sigma \cup \{\star, \varepsilon\}$, $A \in \Gamma$, there is at most one transition of the form $\delta(p, \sigma, A) = (q, \beta)$, for $\beta \in \Gamma^*$.
2. Each transition involving $\triangledown$ is of the form $\delta(p, \sigma, \triangledown) = (q, \beta\triangledown)$, for $\sigma \in \Sigma \cup \{\star, \varepsilon\}$ and $\beta \in \Gamma^*$.

The first restriction on the transitions keeps determinism intact, even in the presence of ε-transitions. The second condition says that $\triangledown$ (bems) is always there at the bottom of the stack. The machine may pop it off momentarily, but it must bring it back immediately. We do not allow this symbol to be put on the stack anywhere else.

As earlier, the transition $\delta(q, \sigma, \varepsilon) = (r, \alpha)$ will mean that upon reading the input σ from state q, the machine changes its state from q to r, pushing the string α to the stack so that the α read from left to right matches that on the stack read from top to bottom, irrespective of whatever be there on the stack. Similarly, the transition $\delta(q, \sigma, A) = (r, \varepsilon)$ will simply pop off the stack symbol A, which is currently at the top of the stack, with necessary change of state from q to r upon reading the input symbol σ.

An instantaneous description or a *configuration* of P is any element of $Q \times \Sigma^* \star \times \Gamma^* \triangledown$, which specifies the current state of P, the input string yet to be read (the string to the right of the scanned square and before $\star$), and the string on the stack read from top to bottom (above $\triangledown$). The one-step operation of P is described by the *yield in one step* relation among various possible configurations of P. That is, for $q, r \in Q$, $\sigma \in \Sigma$, $v \in \Sigma^*\star$, $A \in \Gamma$, $\alpha, \beta \in (\Gamma \cup \{\triangledown\})^*$,

$$(q, \sigma v, A\beta) \overset{1}{\rightsquigarrow} (r, v, \alpha\beta) \text{ iff } \delta(q, \sigma, A) = (r, \alpha),$$

As before, we take the transitive and reflexive closure of $\overset{1}{\rightsquigarrow}$ as the *yield relation*, and denote it by $\overset{*}{\rightsquigarrow}$. This means that $C_0 \overset{*}{\rightsquigarrow} C_n$ iff either $n = 0$ or there are configurations $C_1, C_2, \ldots, C_{n-1}$ such that $C_0 \overset{1}{\rightsquigarrow} C_1$, $C_1 \overset{1}{\rightsquigarrow} C_2, \ldots, C_{n-1} \overset{1}{\rightsquigarrow} C_n$. If there is no confusion, we will simply use $\rightsquigarrow$ for $\overset{1}{\rightsquigarrow}$, $\overset{*}{\rightsquigarrow}$, and also for the yield in n steps ($\overset{n}{\rightsquigarrow}$). Similarly, whenever there are many DPDAs involved in a context, we will use a subscript with $\rightsquigarrow$ to denote the computation of that machine. For example, "yield relation of the machine P" will be denoted by $\rightsquigarrow_P$ if we talk about other machines.

We say that a string $u \in \Sigma^*$ is *accepted* by P iff $(s, u\star, \triangledown) \rightsquigarrow (q, \varepsilon, \beta)$ for some $q \in F$ and for some $\beta \in \Gamma^*\triangledown$. That is, u is accepted by the DPDA P if P, upon starting with its initial configuration $(s, u\star, \triangledown)$ (i.e., being in state s, with input u, and having only $\triangledown$ on the stack), is driven to a final state q after reading u completely along with $\star$, no matter whatever is left on the stack. Finally, the *language of* P is the set of all strings accepted by it. That is,

$$\mathcal{L}(P) = \{u \in \Sigma^* : (s, u\star, \triangledown) \rightsquigarrow (q, \varepsilon, \beta) \text{ for some } q \in F \text{ and for some } \beta \in \Gamma^*\triangledown\}.$$

Example 6.12. What is $\mathcal{L}(P)$ if $P = (\{s, q, r\}, \{a, b\}, \{A, B\}, \star, \triangledown, \delta, s, \{r\})$, with $\delta(s, a, A) = (q, A)$, $\delta(q, b, A) = (s, A)$, $\delta(r, a, A) = (s, A)$?

Solution. The initial configuration of P with an input $u \in \{a, b\}^*$ is $(s, u\star, \nabla)$. We see that no transition is applicable on such a configuration, whatever the first symbol of u may be. Further, the initial state s is not a final state. Thus, even the empty string ε is not accepted. The DPDA does not accept any string; thus, $\mathcal{L}(P) = \emptyset$. □

Example 6.13. What is $\mathcal{L}(P)$ if $P = (\{s, q, r\}, \{a, b\}, \{\nabla, A, B\}, \star, \nabla, \delta, s, \{s\})$, with $\delta(s, a, A) = (q, A)$, $\delta(q, b, A) = (s, A)$, $\delta(r, a, A) = (s, A)$?

Solution. The only change we have made to the DPDA in the last example is that s is now the final state instead of r. Then the initial configuration $(s, \star, \nabla)$ becomes a final configuration, and without any nontrivial computation the empty string is accepted. Instead of ε if we have some other input, then the initial configuration would either be $(s, av\star, \nabla)$ or $(s, bv\star, \nabla)$, for some string $v \in \{a, b\}^*$. As no transition is applicable, the string is not accepted. $\mathcal{L}(P) = \{\varepsilon\}$. □

Example 6.14. Construct a DPDA for accepting the language $(ab \cup aba)^*$.

Solution. As the given language is regular, operating with the stack is redundant. So the problem is essentially, constructing a DFA and then modify the transition to take care of the redundant stack. Now, how do we construct a DFA for accepting the language?

First, ε is accepted, so make the initial state s also the final state. Let the machine go to state q after reading an a, and then from q to r after reading one b. Now, r must be a final state. Here, we let the machine go to state p after reading the next a. We also make p a final state so that aba is accepted. If the machine, upon reading another a goes back to state q, then it would accept the string ab thereafter. But one more ab would not then be accepted. We just add another edge from p to r with label b. Check that the DFA in Fig. 6.7 accepts $(ab \cup aba)^*$. [Hint: $ab(aab)^n = (aba)^n ab$.]

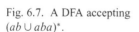

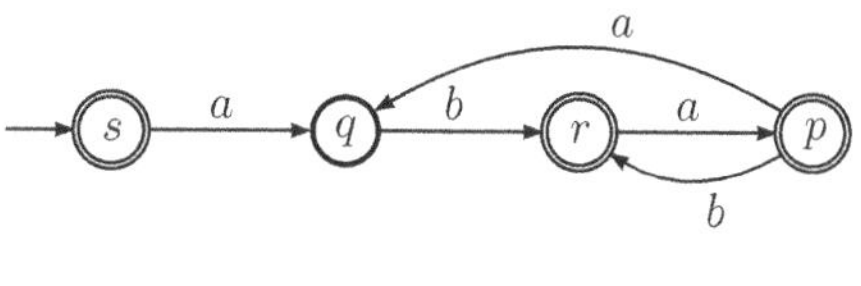

Fig. 6.7. A DFA accepting $(ab \cup aba)^*$.

Once you have checked the claim that the DFA in Fig. 6.7 accepts $(ab \cup aba)^*$, it would be fairly easy to construct a DPDA for accepting the same language. Just use a stack that is not touched at all, that is, the DPDA would pop off ∇ and immediately add it to the stack, with every transition. We keep an unused stack symbol, A, just to make Γ nonempty. So, take $P = (\{s, q, r, p\}, \{a, b\}, \{A\}, \star, \nabla, \delta, s, \{s, r, p\})$, with $\delta(s, a, \nabla) = (q, \nabla)$, $\delta(s, \star, \nabla) = (s, \nabla)$, $\delta(q, b, \nabla) = (r, \nabla)$, $\delta(r, a, \nabla) = (p, \nabla)$, $\delta(r, \star, \nabla) = (r, \nabla)$, $\delta(p, \star, \nabla) = (p, \nabla)$, $\delta(p, a, \nabla) = (q, \nabla)$. □

Exercise 6.13. Check on some inputs that the DPDA of Example 6.14 accepts the language $(ab \cup aba)^*$.

Notice that our DPDAs accept strings by final states. By modifying the restriction (2) of the transitions suitably, it can be shown that acceptance by final states and by empty stack are equivalent, but not as it is!

A **DCFL** or a **deterministic context-free language** is a language accepted by a DPDA. Now you have learnt that each DFA can be simulated by a DPDA. And this is achieved by replacing each transition in the DFA of the form $\delta(q, \sigma) = r$ by $\delta(q, \sigma, \nabla) = (r, \nabla)$, and then adding transitions $\delta(p, \star, \nabla) = (p, \nabla)$ for each final state p. This shows that each regular language is a deterministic context-free language. Is this inclusion proper?

Example 6.15. Show that the language $\{a^m b^m : m \in \mathbb{N}\}$ is a DCFL.

Solution. We keep track of the number of a's and b's by pushing the symbols A and popping it off, respectively. Moreover, we will not define the transitions for other cases when the input is not of the required form, so that the machine will halt abruptly in those cases.

Thus we take $P = (\{s, q, r\}, \{a, b\}, \{A\}, \star, \nabla, \delta, s, \{r\})$, where $\delta(s, a, \varepsilon) = (s, A)$, $\delta(s, b, A) = (q, \varepsilon)$, $\delta(q, b, A) = (q, \varepsilon)$, $\delta(q, \star, \nabla) = (r, \nabla)$, $\delta(s, \star, \nabla) = (r, \nabla)$. Upon reading one b, P cannot read an a. The last two transitions force the stack to be empty (only ∇ on it) upon acceptance of a string. As the stack becomes empty only after the whole input is read, P does not accept any other string than one in the form $a^m b^m$. Verify this. □

This example settles the matter, as $\{a^m b^m : m \in \mathbb{N}\}$ is not a regular language. Thus, the set of regular languages is a proper subset of the set of DCFLs. As each DPDA is, by default, a PDA (You have to modify a PDA taking care of $\star$ and ∇; but that is easy.), each deterministic context-free language is a context-free language. To see whether this inclusion is proper or not, we need to discuss the closure properties of DCFLs.

DCFLs are neither closed under union, concatenation, Kleene star, reversal, nor homomorphism. It can be shown that DCFLs are closed under intersection, inverse homomorphism, and complementation.

For complementation, let P be a DPDA. Introduce a state of no return to make its transition function a total function. (Imitate the same process for DFAs but take care of the stack also.) In the resulting DPDA, we see that each symbol of the input is read during a computation, and there is no abrupt halt during any computation. Then, reverse the final states, that is, make the nonfinal states as final ones and the old final states as nonfinal states.

However, the new DPDA may not accept the complement of the language accepted by the old one! It is because a DPDA can loop by just operating on its stack using ε-transitions. You must do some extra work; see Problem 6.87 in the additional problems to this chapter.

We have seen earlier that CFLs are not closed under complementation. Therefore, the class of DCFLs is not the same as the class of CFLs. To get one such language that is a CFL but not a DCFL, we must look for a language that is context-free but whose complement is not. We have seen in Sect. 6.5 (Example 6.11) that the language $\overline{L} =$

$\{a, b\}^* - \{ww : w \in \{a, b\}^*\}$ is a CFL but its complement $L = \{ww : w \in \{a, b\}^*\}$ is not a CFL. This means, had the language $\overline{L}$ been a DCFL, its complement L would also have been a DCFL, and thus a CFL. Hence, the context-free language $\overline{L}$ is not a DCFL.

Thus, the class of regular languages (over an alphabet) is a proper subclass of that of DCFLs and the class of DCFLs is a proper subclass of CFLs.

Because of the presence of determinism and context-free-ness, all programing languages are DCFLs. It is easier to construct parsers (and thus compilers) for the DCFLs than more general CFLs. But then it makes mandatory to know what kind of grammars generate the DCFLs.

The grammars that correspond to DCFLs are the so-called LR-grammars, which are naturally, restrictions of CFGs. However, the restriction is not too natural or easy conceptually. We will not deal with LR-grammars in this book. We also mention that there are two more operations that preserve DCFLs but not CFLs, in general. They are

$$\min(L) = \{w \in L : \text{no } u \in L \text{ is a proper prefix of } w\}, \text{ and}$$
$$\max(L) = \{w \in L : w \text{ is not a proper prefix of any } u \in L\}.$$

For details regarding the closure properties and LR-grammars, you must consult the references quoted in the summary to this chapter.

Problems for Section 6.6

6.25. Construct a DPDA for accepting each of the following languages:
(a) $\{a^n b^n : n \in \mathbb{N}\}$.
(b) $\{a^m b^n : 0 \le m \le n\}$.
(c) $\{a^m b^n : m \ne n; m, n \in \mathbb{N}\}$.
(d) $\{a^n b^{2n} : n \in \mathbb{N}\}$.
(e) $\{a^n b^n : n \ge 1\} \cup \{a\}$.
(f) $\{a^m b^n : n \ge m + 2;\ m, n \in \mathbb{N}\}$.
(g) $\{w \in \{a, b\}^* : \#_a(w) = \#_b(w)\}$.
(h) $\{w \in \{a, b\}^* : \#_a(w) \ne \#_b(w)\}$.
(i) $\{a^n b^{2n} : n \in \mathbb{N}\}^*$.
(j) $\{a^m b^n a^m : m, n \in \mathbb{N}\}$.
(k) $\{wcw^R : w \in \{a, b\}^*\}$.
(l) $\{ca^m b^n : m \ne n; m, n \in \mathbb{N}\} \cup \{da^m b^{2m} : m \in \mathbb{N}\}$.
(m) $\{a^m cb^n : m \ne n; m, n \in \mathbb{N}\} \cup \{a^m db^{2m} : m \in \mathbb{N}\}$.

6.26. Give reasons why one might conjecture that $\{a^m b^n c^k : m = n \text{ or } n = k\}$ is not a DCFL.

6.27. Is $\{a^m b^n : m = n \text{ or } m = n + 2\}$ a DCFL?

6.28. Show that $\{a^m b^m : n \in \mathbb{N}\} \cup \{a^n b^{2n} : n \in \mathbb{N}\}$ is a CFL but not a DCFL.

6.29. Give an intuitive explanation why $\{wcw^R : w \in \{a, b\}^*\}$ is a DCFL while $\{ww^R : w \in \{a, b\}^*\}$ is not.

6.30. Determine which of the following languages are DCFLs:
(a) $\{ww : w \in \{0, 1\}^*\}$.
(b) $\{w \in \{0, 1\}^* : w = w^R\}$.
(c) $\{a^m b^n : n \leq m \leq 2n\}$.
(d) $\{a^m b^n : 3n \leq m \leq 7n\}$.

6.31. Give a precise algorithm to construct a DPDA from a given DFA for accepting the same language.

6.32. Show that if a language is accepted by a DFA with n states, then it is accepted by a DPDA with n states and just one stack symbol.

6.33. Show that if a language is accepted by a DFA with n states, then it is accepted by a DPDA with two states and n stack symbols. Can you further minimize on the number of states and/or the number of stack symbols of the DPDA?

6.34. Construct a DPDA so that on some input it goes on an infinite computation.

6.35. Show that if L is regular and L' is a DCFL, then $L \cup L'$ and $L \cap L'$ are DCFLs.

6.36. Give an example of a DCFL L such that L^R is not a DCFL.

6.37. Show that each DCFL is accepted by a DPDA that never adds more than one symbol to its stack at a time, that is, each transition of such a DPDA looks like $\delta(p, a, \alpha) = (q, \beta)$, where $\beta \in \{\varepsilon, \triangledown, B, B\triangledown\}$.

6.38. Draw a Venn diagram illustrating the containment relations between the four classes of languages such as regular, deterministic context-free, context-free, and complements of context-free. Supply a language for each nonempty region in your diagram.

6.39. Construct formally a PDA with $\star$ and $\triangledown$ so that each DPDA is now a PDA.

6.7 Summary and Additional Problems

This chapter started with the context-free recognition devices, the PDAs. We have shown equivalence of PDAs with context-free grammars. We have proved a pumping lemma for CFLs and used it to demonstrate some languages that are not context-free. The closure of context-free languages with respect to union, intersection with regular languages, concatenation, and Kleene star have also been proved. We claimed that unlike DFAs and NFAs, the deterministic analogues of PDAs, called DPDAs, are not equivalent to the PDAs.

The pumping lemma for context-free languages was first given in [8] followed by a stronger form by Ogden [97], now called Ogden's lemma. I ask you to prove

Ogden's lemma in one of the problems below. Though PDAs were introduced by Oettinger [96, 115], equivalence of PDAs and CFGs was shown in [16, 30, 115]. Closure properties of CFLs are from [8, 41, 42, 114].

The DPDAs exist not only due to our curiosity, but also due to an important practical concern: parsing in programing languages. Early appearance of DPDAs can be traced back to [31, 39, 49, 115]. The DPDAs are equivalent to LR(k) grammars, or sometimes called LR-grammars. It first appeared in [68]. You can also refer [58] for an exposition of these grammars. For the use of DPDAs and LR-grammars in parsing, you may like to read [68] and some explanatory texts such as [4, 59, 77].

In the following exercises, I ask you to prove Parikh's theorem and Chomsky–Schützenberger theorem. It is a bit too much to expect from you without external help if you are learning the subject for the first time. However, you can refer [16, 18] for Chomsky–Schützenberger theorem and [45, 51, 73, 100] for Parikh's theorem.

Additional Problems for Chapter 6

6.40. Show that $\{a^m b^n : m = n^2;\ n \in \mathbb{N}\}$ is not a CFL.

6.41. Construct a CFG and also a PDA for $\{w \in \{a, b\}^* : \#_a(w) \leq 2 \cdot \#_b(w)\}$.

6.42. Construct a PDA for accepting the language $\{a^m b^n : m \leq n \leq 2m;\ m, n \in \mathbb{N}\}$.

6.43. Show that each CFL is accepted by some PDA with a single state if acceptance is by empty stack.

6.44. Show that $\{a^m b^{2m} c^n : m, n \in \mathbb{N}\}$ and $\{a^m b^n c^{2n} : m, n \in \mathbb{N}\}$ are CFLs. What about their intersection?

6.45. Prove the following stronger versions of the Pumping Lemma for CFLs and then say in what sense they are stronger:

(a) Let L be a CFL over an alphabet Σ containing at least one nonempty string. There is $n \geq 1$ such that if $w \in L$ is of length at least n, then w can be rewritten as $w = uvxyz$ such that $v \neq \varepsilon$, $y \neq \varepsilon$, $\ell(vxy) \leq n$, and for each $k \in \mathbb{N}$, $uv^k xy^k z \in L$.

(b) *Ogden's Lemma*: Let L be a CFL over an alphabet Σ. Let L contain at least one nonempty string. Then there exists an $n \geq 1$ such that the following is satisfied: Let $w \in L$ with $\ell(w) > n$. Mark any n or more occurrences of symbols in w arbitrarily. Then there exists strings $u, v, x, y, z \in \Sigma^*$ such that

1. w can be rewritten as $w = uvxyz$,
2. the string vwx contains n or fewer marked symbols,
3. each of the strings uz, vy, x contains at least one marked symbol, and
4. for all $k \in \mathbb{N}$, the string $uv^k xy^k z \in L$.

6.46. Use Ogden's Lemma to simplify the proof that $\{ww : w \in \{a, b\}^*$ is not a CFL.

6.47. Use Ogden's Lemma to show that the following languages are not CFLs:

(a) $\{a^m b^k a^p : k = \max\{m, p\};\ m, k, p \in \mathbb{N}\}$.

(b) $\{a^m b^m c^p : p \neq m;\ m, p \in \mathbb{N}\}$.

6.48. Let L, L' be CFLs. Show that the following languages are not necessarily CFLs:
(a) $min(L) = \{w \in L :$ no proper prefix of w is in $L\}$.
(b) $max(L) = \{w \in L : wx \notin L$ for $x \neq \varepsilon\}$.
(c) $half(L) = \{w : wx \in L, \ell(w) = \ell(x)$ for some string $x\}$.
(d) $alt(L, L') = \{\sigma_1\tau_1\sigma_2\tau_2 \cdots \sigma_n\tau_n : \sigma_1 \cdots \sigma_n \in L, \tau_1 \cdots \tau_n \in L'\}$.
(e) $shuffle(L, L') = \{v_1w_1v_2w_2 \cdots v_nw_n : v_1v_2 \cdots v_n \in L, w_1w_2 \cdots w_n \in L'\}$.

6.49. Show that $shuffle(L, L')$ is a CFL if L is a CFL and L' is regular. What happens to $shuffle(L, L')$ if L is regular and L' is a CFL?

6.50. For a string x, define $perm(x)$ as any string obtained by reordering the occurrences of symbols in x. For a language L, define $perm(L) = \{perm(x) : x \in L\}$. For example, $perm(\{a^nb^n\})$ is the set of all strings having equal number of a's as b's.
(a) Show that if L is a regular language over $\{a, b\}$, then $perm(L)$ is context-free.
(b) Give an example of a regular language L such that $perm(L)$ is not context-free.

6.51. Let $L = \{a/b : a < b$ and a, b are positive integers written in decimal notation$\}$. Is L a context-free language?

6.52. A CFG is in *two-standard form* if each of its productions is in one of the forms $A \mapsto a$, $A \mapsto aB$, or $A \mapsto aBC$.
(a) Convert the CFG with productions $S \mapsto aSA$, $A \mapsto bABC$, $B \mapsto b$, $C \mapsto aBC$ into a two-standard form.
(b) Prove: If $\varepsilon \notin \mathcal{L}(G)$ for a CFG G, then there is an equivalent two-standard form for G.

6.53. Show that for each CFL, there exists an accepting PDA in which the number of symbols in the stack never exceeds the length of the input string by more than one.

6.54. Show that corresponding to each PDA P, there is an equivalent PDA P' such that each transition in P' is of the form $(p, \sigma, \alpha, q, \beta)$, where $\ell(\alpha) + \ell(\beta) \leq 1$.

6.55. Show that $\{a^n : n \neq m^2$ for any $m \in \mathbb{N}\}$ is not a CFL.

6.56. Show that the family of CFLs is closed under regular difference, that is, if L is a CFL, L' is regular, then $L - L'$ is a CFL. Does it also follow that $L' - L$ is a CFL?

6.57. Show that the family of unambiguous CFLs is neither closed under union nor under intersection.

6.58. Give a noncontext-free language that satisfies the Pumping Lemma for CFLs.

6.59. Show that if L is a CFL and R is a regular language, over an alphabet Σ, then the quotient $L/R = \{w \in \Sigma^* : wx \in L$ for some $x \in R\}$ is a CFL.

6.60. A nonterminal A of a CFG is called a *self-embedding nonterminal* if a string uAv can be derived from A in one or more steps, for some strings u, v of terminals and nonterminals.
(a) Show that if G has no self-embedding nonterminals, then $\mathcal{L}(G)$ is regular.

(b) Suppose G has no useless nonterminals. If $\mathcal{L}(G)$ is regular, is it true that G does not have any self-embedding nonterminal?
(c) Give an algorithm for determining whether a given nonterminal of a CFG is self-embedding.

6.61. Let Σ be an alphabet. A *substitution on* Σ is a map $s : \Sigma \to 2^{\Gamma^*}$; that is, for each symbol $\sigma \in \Sigma$, we have $s(\sigma)$ a language over some alphabet Γ. Extend s to a map $s : \Sigma^* \to 2^{\Gamma^*}$ by defining $s(\sigma_1\sigma_2 \cdots \sigma_n) = s(\sigma_1)s(\sigma_2)\cdots s(\sigma_n)$. Further, if L is a language over Σ, then define $s(L) = \cup_{w \in L} s(w)$, extending s to a map from the power set of Σ^* to the power set of Γ^*. Prove the

Substitution Theorem: Let L be a CFL over Σ. Let s be a substitution on Σ such that $s(\sigma)$ is a CFL for each $\sigma \in \Sigma$. Then $s(L)$ is a CFL.

6.62. Use the Substitution Theorem to show that the class of CFLs (over an alphabet Σ) is closed under union, concatenation, Kleene star, positive closure (+), homomorphism, and inverse homomorphism.

6.63. Show that $\{a^p b^n : p \text{ is a prime number and } n > p\}$ is not a CFL.

6.64. Let PAR_n denote the language of properly nested parentheses of n distinct types, where the n pairs of parentheses are $[^1,]^1, [^2,]^2, \ldots, [^n,]^n$. Show that PAR_n is generated by the CFG with productions $S \mapsto \varepsilon | SS | [^1 S]^1 | [^2 S]^2 | \cdots | [^n S]^n$.

6.65. Let $G = (N, \Sigma, R, S)$ be a CFG in CNF. For a production π in G, define a production π' by

$$\text{if } \pi = A \mapsto a, \text{ then } \pi' = A \mapsto [^1_\pi]^1_\pi \, [^2_\pi]^2_\pi, \text{ and}$$

$$\text{if } \pi = A \mapsto BC, \text{ then } \pi' = A \mapsto [^1_\pi \, B]^1_\pi \, [^2_\pi \, C]^2_\pi .$$

Construct the CFG $G' = (N, \Sigma', R', S)$, where $\Sigma' = \{[^1_\pi,]^1_\pi, [^2_\pi,]^2_\pi : \pi \in R\}$ and $R' = \{\pi' : \pi \in R\}$. See that if there are m number of productions in G, then there are $2m$ distinct pairs of parentheses in Σ'. Let $PAR(\Sigma')$ denote the language of balanced parentheses of distinct types taken from Σ'. Show that, in $\mathcal{L}(G')$, the following happens:
(a) Each $[^1_\pi$ is immediately followed by a $]^1_\pi$.
(b) No $]^2_\pi$ is immediately followed by a left parenthesis.
(c) If $\pi = A \mapsto BC$, then each $[^1_\pi$ is immediately followed by $[^1_\eta$ for some $\eta \in P$ with left-hand side B, and each $[^2_\pi$ is immediately followed by some $[^1_\tau$ for some $\tau \in P$ with left-hand side C.
(d) If $\pi = A \mapsto a$, then each $[^1_\pi$ is immediately followed by $]^1_\pi$ and each $[^2_\pi$ is immediately followed by $]^2_\pi$.
(e) For each string w, if $A \Rightarrow_{G'} w$, then w begins with $[^1_\pi$ for some $\pi \in P$ with left-hand side A.

Then, conclude the following:
(f) $\mathcal{L}(G') \subsetneq PAR(\Sigma')$.
(g) The sets $L_A = \{w \in (\Sigma')^* : A \Rightarrow_{G'} w \text{ and } w \text{ satisfies the properties (a–e) above}\}$ is a regular language.
(h) If $A \Rightarrow_{G'} w$, then $w \in PAR(\Sigma') \cap L_A$.

(i) If $w \in PAR(\Sigma') \cap L_A$, then $A \Rightarrow_{G'} w$.

(j) Define a homomorphism $h : (\Sigma')^* \to \Sigma^*$ by looking at the productions $\pi \in P$ as follows:

$$\text{If } \pi = A \mapsto a, \text{ then } h([^1_\pi) = a,\ h(]^1_\pi) = h([^2_\pi) = h(]^2_\pi) = \varepsilon.$$

$$\text{If } \pi = A \mapsto BC, \text{ then } h([^1_\pi) = h(]^1_\pi) = h([^2_\pi) = h(]^2_\pi) = \varepsilon.$$

Show that $\mathcal{L}(G) = h(\mathcal{L}(G')) = h(PAR(\Sigma') \cap L_S)$.

6.66. *Chomsky–Schützenberger*: Every context-free language is a homomorphic image of the intersection of a parentheses language and a regular language. Prove it.

6.67. Parikh's theorem states that every CFL is letter-equivalent to a regular language. To understand and prove this, we need some definitions.

Let $G = (N, \Sigma, R, S)$ be a CFG in CNF. Let $T, T_1, T_2, \ldots$ denote parse trees of G with a nonterminal at the root, nonterminals as labels of the internal nodes, and terminals or nonterminals at the leaves. Recall that for a parse tree T, $root(T)$ is the nonterminal at the root of T, $yield(T)$ is the string of terminals at the leaves of T, read left to right, and the $depth(T)$ is the maximum number of edges in a path of T connecting a leaf up to the root. Let $N(T)$ denote the set of nonterminals occurring in T.

A parse tree T is called a *pump* if T contains at least two nodes and all leaves of T are labeled with terminal symbols except one, and the nonterminal label of that leaf is the same as the label of the $root(T)$; that is, $yield(T) = u\,root(T)\,v$, for some $u, v \in \Sigma^*$.

For parse trees T_1, T_2, define (a binary relation $\prec$ by) $T_1 \prec T_2$ iff T_2 can be obtained from T_1 by splitting T_1 at a node labeled with some nonterminal A and inserting a pump whose root is labeled A. Though the relation $\prec$ is not a partial order (why?), it has the property that if $T_1 \prec T_2$, then T_1 has less nodes than that of T_2. A pump T is called a *basic pump* if T is $\prec$-minimal among all pumps. That means, a basic pump can never contain another pump that can be cut out of it. It thus follows that if $T' \prec T$ and T is a basic pump, then T' can only be the trivial one-node parse tree labeled with the nonterminal $root(T)$.

A parse tree T_1 is *dominated by* a parse tree T_2, if T_2 can be obtained from T_1 by inserting a finite number of basic pumps into T_1, where these basic pumps use only (not necessarily all) nonterminals that occur in T_1. That is, in T_1, we may choose a nonterminal and insert a basic pump at that occurrence of the nonterminal with the nonterminal as the root of the basic pump. Moreover, the basic pump cannot use any nonterminal that has not already occurred in T_1. This process is repeated a finite number of times to obtain T_2.

Suppose $\Sigma = \{\sigma_1, \sigma_2, \ldots, \sigma_m\}$. The *Parikh map* is the function $\phi : \Sigma^* \to \mathbb{N}^m$ defined by $\phi(w) = (\#_{\sigma_1}(w), \#_{\sigma_2}(w), \ldots, \#_{\sigma_m}(w))$. The Parikh map records the number of occurrences of symbols of Σ in the string w. For a language L over Σ, we define the *commutative image* of L as the image of L under ϕ, which equals $\phi(A) = \{\phi(w) : w \in L\}$. Note that the commutative image of L is a subset of $\mathbb{N}^m$. For a string α of terminals and/or nonterminals, denote by x_α, the string obtained from α by deleting all occurrences of nonterminals. Then, define $\phi(\alpha) = \phi(x_\alpha)$.

For $u = (n_1, n_2, \ldots, n_m) \in \mathbb{N}^m$ and $v = (r_1, r_2, \ldots, r_m) \in \mathbb{N}^m$, we define their addition as the component-wise addition, that is, $u+v = (n_1+r_1, n_2+r_2, \ldots, n_m+r_m)$. For finite subsets $B, C \subseteq \mathbb{N}^m$, we take (the affine combination).

$$\mathcal{A}(B, C) = \{b + c_1 + c_2 + \cdots + c_p : b \in B,\ c_1, c_2, \ldots, c_p \in C,\ p \in \mathbb{N}\}.$$

A subset $D \subseteq \mathbb{N}^m$ is called a *linear set* iff $D = \mathcal{A}(B, C)$ for some $B, C \subseteq \mathbb{N}^m$. For example, the set $\{(m, n) : m \le n\}$ is linear as it equals $\mathcal{A}(B, C)$, where $B = \{(0, 0)\}$ and $C = \{(0, 1), (1, 1)\}$; it is a linear subset of $\mathbb{N}^2$. Similarly, $E = \phi(\{a^n b^n : n \in \mathbb{N}\})$ is linear as

$$E = \phi(\{w \in \{a, b\}^* : \#_a(w) = \#_b(w)\}) = \{(n, n) : n \in \mathbb{N}\} = \mathcal{A}(\{(0, 0)\}, \{(1, 1)\}).$$

However, the set $\{(n^2, n) : n \in \mathbb{N}\}$ is not linear. A subset of $\mathbb{N}^m$ is called *semilinear* iff it is a finite union of linear sets. See that $\{(n^2, n) : n \in \mathbb{N}\}$ is not a semilinear set.

Show the following:

(a) Let n be the number of nonterminals in N. If T is a basic pump, then $depth(T) \le 2n$. Therefore, there are a finite number of basic pumps.
(b) Let T be a parse tree with $yield(T) \in \Sigma^*$. If T is not $\prec$-minimal, then T contains a basic pump.
(c) Let T, T' be parse trees. The set $\{\phi(yield(T)) : T'$ is dominated by $T\}$ is linear.
(d) Let n be the number of nonterminals in N and p be the number of distinct basic pumps. If T is any parse tree not dominated by any other parse tree, then $depth(T) \le (n+1)(p+1)$.
(e) Let M be the set of all parse trees having root as S, yield in Σ^*, and that are not dominated by any other parse tree. Then

$$\phi(\mathcal{L}(G)) = \cup_{T \in M}\{\phi(T') : T \text{ is dominated by } T'\}.$$

6.68. Prove *Parikh's Theorem*: If L is context-free, then $\phi(L)$ is semilinear.

6.69. What is the semilinear set $\phi(L)$ if L is
(a) $\{a^m b^n : n \le m \le 2n\}$?
(b) $\{ww^R : w \in \{a, b\}^*\}$?
(c) $\{a, b\}^* - \{baba^2ba^3b \cdots ba^{n-1}ba^n : n \ge 1\}$?

6.70. Show that the complement of a semilinear set is semilinear.

6.71. Let $\mathcal{S}$ be the set of all semilinear subsets of $\mathbb{N}^k$. Prove or disprove:
(a) $\emptyset \in \mathcal{S}$.
(b) $\mathbb{N}^k \in \mathcal{S}$.
(c) If $L, K \in \mathcal{S}$, then $L \cup K \in \mathcal{S}$.
(d) If $L, K \in \mathcal{S}$, then $L \cap K \in \mathcal{S}$.
(e) If $L \in \mathcal{S}$, then $\mathbb{N}^k - L \in \mathcal{S}$.
(f) If $L, K \in \mathcal{S}$, then $L \times K \in \mathcal{S}$.

6.72. Using Parikh's theorem, show that the following languages are not context-free:
(a) $\{a^m b^n : m > n$ or (m is prime and $m \le n$)$\}$.
(b) $\{a, b\}^* - \{a^m b^n : n = m^2,\ m \in \mathbb{N}\}$.

6.73. Show that if $K \subseteq \mathbb{N}$ is semilinear, then $\{a^n : n \in K\}$ is regular.

6.74. Let $K \subseteq N^k$ be a semilinear set. Prove that there is a DFA M such that $K = \phi(\mathcal{L}(M))$.

6.75. Give an example of a language L which is not context-free but $\phi(L)$ is semilinear.

6.76. Show that the family of CFLs is closed under homomorphic images and homomorphic preimages. We also say that CFLs are closed under homomorphisms and inverse homomorphisms. That is, if h is a homomorphism, and L is a CFL, then show that $h(L)$ and $h^{-1}(L)$ are also CFLs.

6.77. *Ginsburg and Rice*: Any CFL over a one-symbol alphabet is regular. Prove it.

6.78. Show that $head(L) = \{w : wu \in L \text{ for some } u \in \Sigma^*\}$ is a CFL if L is a CFL.

6.79. Recall that for a string w, $perm(w)$ is a permuted string where the occurrences of symbols in w are reordered. For a language L, $perm(L)$ is the set of all permutated strings of L. In each case below, decide whether $perm(L)$ is regular, context-free, or neither:
(a) $L = (ab)^*$.
(b) $L = a^* \cup b^*$.
(c) $L = (abc)^*$.
(d) $L = (ab \cup bc)^*$.

6.80. Let L be a CFL and R be a regular language. Which of the following languages are CFLs and which is not?
(a) $cycle(L) = \{wx : xw \in L\}$.
(b) $perm(L) = \{w : \text{ for some } x \in L, \text{ for all } \sigma \in \Sigma, \#_\sigma(x) = \#_\sigma(w)\}$.
(c) $shuffle(L_1, L_2) = \{w_1v_1w_2v_2 \cdots w_nv_n : w_1w_2 \cdots w_n \in L_1, v_1v_2 \cdots v_n \in L_2\}$.

6.81. Let L be any language over $\{a, b, c\}$. Define $L_1 = \{ww : w \in L\}$ and $L_2 = \{w : ww \in L\}$. Which one of the following is true, and which three are false? Give reasons.
(a) If L is regular then so is L_1.
(b) If L is regular then so is L_2.
(c) If L is context-free then so is L_1.
(d) If L is context-free then so is L_2.

6.82. Let $L_1 = \{a^nb^n : n \in \mathbb{N}\}$, $L_2 = \{a^nb^{2n} : n \in \mathbb{N}\}$, and $L_3 = \{a^nb^nc^n : n \in \mathbb{N}\}$. Show that
(a) $L_1 \cup L_2$ is a CFL.
(b) Given a DPDA for accepting $L_1 \cup L_2$, construct a PDA for $L_1 \cup L_2 \cup L_3$.
(c) Conclude that there is no DPDA accepting $L_1 \cup L_2$.

6.83. Let $b(n)$ denote the binary representation of n, where leading zeros are omitted, for example, $b(5) = 101$ and $b(12) = 1100$. Show that
(a) $\{b(n)2b(n+1) : n \geq 1\}$ is not a CFL.
(b) $\{(b(n))^R2b(n+1) : n \geq 1\}$ is a CFL.

6.84. Show that there is no DPDA that accepts $\{ww^R : w \in \{a, b\}^*\}$.

6.85. Let L be a CFL over an alphabet Σ. Define $L_1 = \{w : wx \in L,$ for some $x \in \Sigma^*\}$ and $L_2 = \{w : wx \in L,$ for all $x \in \Sigma^*\}$. Exactly one of L_1, L_2 is context-free; which one is, and which one is not? Justify.

6.86. Show that each DCFL is generated by some unambiguous CFG. Conclude that no DCFL is inherently ambiguous.

6.87. Here is an outline of the proof that complements of DCFLs are DCFLs. Let $M = (Q, \Sigma, \Gamma, \delta, \star, \nabla, s, F)$ be a DPDA. Construct the DPDA M' from M by $M' = (Q \cup Q' \cup \{h, \hbar\}, \Sigma, \Gamma, \delta', \star, \nabla, s, F')$, where $h, \hbar \notin Q \cup Q'$, $Q' = \{q' : q \in Q\}$, $F' = \{q' : q \in F\}$, and δ is extended to δ' by performing the following in that sequence:

1. For each transition of the form $\delta(p, \sigma, A) = (q, \beta)$, include $\delta'(p', \sigma, A) = (q', \beta)$.
2. Replace each transition of the form $\delta(p, \star, A) = (q, \beta)$ with $\delta'(p, \star, A) = (q', \beta)$.
3. Replace each transition of the form $\delta'(p', \varepsilon, A) = (q', \beta)$ with $\delta'(p', \varepsilon, A) = (p', A)$, for $p' \in F$
4. Include the transitions $\delta'(h, \sigma, A) = (h, A)$, $\delta'(h, \star, A) = (\hbar, A)$, and $\delta'(\hbar, \varepsilon, A) = (\hbar, A)$, for each $\sigma \in \Sigma$, $A \in \Gamma$.
5. Replace each transition of the form $\delta(p, \varepsilon, A) = (q, \beta)$ with $\delta'(p, \varepsilon, A) = (h, A)$, if $p \in Q$, or with $\delta'(p, \varepsilon, A) = (\hbar, A)$, if $p \in Q'$.

Prove the following:

(a) After the modifications (1–3), $\mathcal{L}(M = \mathcal{L}(M')$.

(b) Let w be any input. Because of determinism, there exists a unique infinite sequence of configurations of the new machine obtained after the modifications (1–4) on input w. Let γ_i denote the stack contents of the modified machine at time $i \in \mathbb{N}$. Then there exists an infinite sequence of times $i_0 < i_1 < i_2 < \cdots$ such that, for each $i_k \leq i$, $\ell(\gamma_{ik}) \leq \ell(\gamma_k)$.

(c) In the modified machine, after (1–4) have been applied, there exists a transition $\delta'(p, \varepsilon, A) = (q, \beta)$ such that, it has been applied infinitely often, that is, at times $j_0, j_1, j_2, \ldots$.

(d) With M' as the modified machine where all of (1–5) have been applied, we have $\mathcal{L}(M) = \mathcal{L}(M')$.

(e) Let M'' be a DPDA obtained from M' by taking $F'' = \{\hbar\}$ and keeping all others as they were in M'. Then $\mathcal{L}(M'') = \overline{\mathcal{L}}(M') = \overline{\mathcal{L}}(M)$.

6.88. Show that the family of DCFLs is closed under regular difference, and hence, under complementation.

6.89. Show that the family of DCFLs is closed under inverse homomorphism.

6.90. Let L be a DCFL over Σ. Define languages $L_1 = \{w : w\sigma \in L$ for some $\sigma \in \Sigma\}$ and $L_2 = \{w : \sigma w \in L$ for some $\sigma \in \Sigma\}$. Show that one of L_1, L_2 is a DCFL, but the other is not necessarily a DCFL.

6.91. Using Problem 6.90, show that the class of DCFLs is not closed under reversal.

6.92. If the DPDAs are required to accept by empty stack (not by final states), then the recognizing capability is limited. A language L has *prefix property* iff there are no two distinct strings $x, y \in L$ such that x is a prefix of y. Prove that

(a) If a language L is accepted by some DPDA by empty stack, then L has prefix property.

(b) If a language L is accepted by some DPDA by empty stack, then L is accepted by some DPDA by final states.

(c) If L has the prefix property and L is accepted by a DPDA by final states, then there is some DPDA that accepts L by empty stack.

(d) There is a regular language that is not accepted by any DPDA by empty stack.

6.93. A *simplified* DPDA is one that does not use ε-transitions, and has neither $\star$ nor ∇. Such a DPDA is a PDA with its transition relation as a partial function. Define formally a simplified DPDA. Show that any language accepted by a simplified DPDA is a DCFL. Is it true that each DCFL is accepted by some simplified DPDA?

6.94. A *restricted* PDA is one in which each transition is of the form (p, a, A, q, α), where $\ell(\alpha) \leq 2$. That means, at any single move, the stack of a restricted PDA can increase the height of the stack by at most one. Show that corresponding to each PDA, there is an equivalent restricted PDA.

6.95. A *generalized* PDA is one in which each transition is of the form (p, a, α, q, β), where $\alpha \in \Gamma^*$ is any string of stack symbols, instead of a single symbol. That means, at any single move, the stack of a generalized PDA can read and pop off a string from the stack. Show that corresponding to each generalized PDA, there is an equivalent restricted PDA.

7 Computably Enumerable Languages

7.1 Introduction

In the last chapter, you have tried but failed to construct a context-free grammar for the language $\{a^n b^n c^n : n \in \mathbb{N}\}$. Later you could also prove that no such grammar might be able to generate this language. Obviously, the restrictions on the productions in a context-free grammar have some roles to play in this regard. The productions in a context-free grammar look like $A \mapsto u$, where A is a nonterminal and u can be any string of terminals and nonterminals.

What happens if we generalize the productions, for example, by allowing any string of terminals and nonterminals in the left side of a production? In such a generalization, how do we apply a production? Suppose we have a production of the form $aAbBbc \mapsto aabbcc$ and that we have already generated the string $abcaAbBccab$. By replacing the substring $aAbBbc$ with $aabbcc$, we obtain the string $abcaabbccab$. But, neither A will be replaced by a nor B will be replaced by b in isolation. This is not really hopeless. Let us give a try and see what generality is achieved.

7.2 Unrestricted Grammars

An **unrestricted grammar** is a quadruple $G = (N, \Sigma, R, S)$, where

N is an alphabet, called the set of nonterminals,
Σ is an alphabet, called the set of terminals (terminal symbols) with $\Sigma \cap N = \emptyset$,
$S \in N$ is a designated nonterminal symbol, called the start symbol, and
R is a set of productions of the form $u \mapsto v$ with $u, v \in (N \cup \Sigma)^*$, where the string u contains at least one nonterminal symbol.

The unrestricted grammars are called *grammars* without any adjectives, and also *rewriting systems*.

As usual, the productions can be given as ordered pairs of strings u, v instead of using the symbol $\mapsto$. But you are now matured enough to express facts formally!

A. Singh, *Elements of Computation Theory*, Texts in Computer Science,

That is, R is a finite relation from $(N \cup \Sigma)^* N (N \cup \Sigma)^*$ to $(N \cup \Sigma)^*$. For any strings $x, y \in (N \cup \Sigma)^*$ and for any production $u \mapsto v$, we say that xuy **derives in one step** xvy, or that xvy is **derived from** xuy in one step, and we write it as $xuy \stackrel{1}{\Rightarrow} xvy$. This constitutes an application of the production $u \mapsto v$. We use the symbol $\Rightarrow$ (derives) for the reflexive and transitive closure of the relation of $\stackrel{1}{\Rightarrow}$. That is, $w_0 \Rightarrow w_n$ iff either $w_n = w_0$ or there is a sequence of derivations

$$w_0 \stackrel{1}{\Rightarrow} w_1,\ w_1 \stackrel{1}{\Rightarrow} w_2 \ldots w_{n-1} \stackrel{1}{\Rightarrow} w_n \text{ for } n \geq 1.$$

We write $\stackrel{n}{\Rightarrow}$ to denote "derives in n steps." If more than one grammar occur in a certain context, then we will write $\Rightarrow_G$ to denote the derivation in the grammar G. Similarly, if no confusion arises, we may also write $\Rightarrow$ even for a one-step or an n-step derivation.

For a grammar G, we define the **language of** G as $\mathcal{L}(G) = \{w \in \Sigma^* : S \Rightarrow w\}$. We will also read $\mathcal{L}(G)$ as the language *generated by* the grammar G; it is the set of all strings over the alphabet Σ that are derived from the start symbol by following the production rules of G. A language is called a **computably enumerable language** if it is generated by some (unrestricted) grammar. Computably enumerable languages were previously (before 1999) called as *recursively enumerable* languages. See the following examples.

Example 7.1. Are the languages $L_1 = \{a^n : n \in \mathbb{N}\}$ and $L_2 = \{a^n b^n : n \in \mathbb{N}\}$ computably enumerable?

Solution. With $G_1 = (\{S\}, \{a\}, \{S \mapsto \varepsilon, S \mapsto aS\}, S)$ and $G_2 = (\{S\}, \{a, b\}, \{S \mapsto \varepsilon, S \mapsto aSb\}, S)$, you find that $\mathcal{L}(G_1) = \{a^n : n \in \mathbb{N}\}$ and $\mathcal{L}(G_2) = \{a^n b^n : n \in \mathbb{N}\}$. Hence both L_1 and L_2 are computably enumerable. □

In fact, every context-free (and hence, regular) language is computably enumerable because each context-free grammar is also an unrestricted grammar; see it formally from the shape of productions. Is this generalization of a grammar nontrivial? That is, do we really have computably enumerable languages that are not context-free (therefore, not regular)?

Example 7.2. Is the language $\{a^n b^n c^n : n \in \mathbb{N}\}$ computably enumerable?

Solution. Let $G = (\{S, A, B, C, \alpha, \beta, \gamma\}, \{a, b, c\}, R, S)$, where

$$R = \{S \mapsto \varepsilon, S \mapsto ABCS, BA \mapsto AB, CB \mapsto BC, CA \mapsto AC, C\gamma \mapsto \gamma c, C\gamma \mapsto \beta c, B\beta \mapsto \beta b, B\beta \mapsto \alpha b, A\alpha \mapsto \alpha a, \alpha \mapsto \varepsilon\}.$$

In this grammar, $S \Rightarrow \varepsilon$ as $S \mapsto \varepsilon \in R$. Also,

$$S \Rightarrow ABC\underline{S} \Rightarrow AB\underline{C\gamma} \Rightarrow A\underline{B\beta}c \Rightarrow \underline{A\alpha}bc \Rightarrow \underline{\alpha}abc \Rightarrow abc.$$

In the derivation of abc, I have used underlines to tell you exactly which substring has been matched for obtaining the next string. Now you know how to generate

$a^n b^n c^n$. First generate $(ABC)^n \gamma$, then sort out A, B, C to derive $A^n B^n C^n \gamma$. Next, γ slowly goes to the left by changing each C to c and becoming β; β then goes to the left changing B into b and becoming α. Finally, α changes A into a and becomes ε. Argue that the only way the nonterminals would vanish is this way and then convince yourself that nothing else is generated in G. □

Example 7.3. Construct a grammar for $\{w \in \{a, b, c\}^* : \#_a(w) = \#_b(w) = \#_c(w)\}$.

Solution. We will use a construction similar to that in Example 7.2. That is, our plan is to first generate $(ABC)^n$ and then somehow permute the A's, B's, and C's, and finally replace A, B, C with a, b, c, respectively. Take $G = (\{S, A, B, C\}, \{a, b, c\}, R, S)$ with

$$R = \{S \mapsto \varepsilon, S \mapsto ABCS, AB \mapsto BA, BA \mapsto AB, BC \mapsto CB, CB \mapsto BC, CA \mapsto AC, AC \mapsto CA, A \mapsto a, B \mapsto b, C \mapsto c\}.$$

To know exactly what is going on, derive the strings $abbacc$ and $cabacb$ in this grammar. Can you show that the grammar G generates the language? □

Example 7.4. Is there a grammar G such that $\mathcal{L}(G) = \{a^{2^n} : n \in \mathbb{N}\}$?

Solution. Yes. Consider the grammar $G = (\{S, A, B, C, D\}, \{a\}, R, S)$ with

$$R = \{S \mapsto CAD, C \mapsto CB, BD \mapsto D, BA \mapsto AAB, C \mapsto \varepsilon, D \mapsto \varepsilon, A \mapsto a\}.$$

A derivation of the string a^4 in G goes as follows:

$$S \Rightarrow \underline{C}AD \Rightarrow \underline{C}BAD \Rightarrow CB\underline{BA}D \Rightarrow C\underline{BA}ABD \Rightarrow CAA\underline{BA}BD$$
$$\Rightarrow CAAAAB\underline{BD} \Rightarrow CAAAA\underline{BD} \Rightarrow \underline{C}AAAAD \Rightarrow AAAA \Rightarrow aaaa.$$

How does the grammar work? First, S gives CAD, then C gives CB^n having the derivation by now of CB^nAD. The B's make the A's double while going to the right, thus generating $CA^{2^n}B^nD$. Then B^nD gets absorbed into D. Finally, C, D both become ε, A becomes a, and the string a^{2^n} is generated. □

It looks as though we have generalized the concept of a grammar too wildly. Almost any language can be generated by such grammars! But we are going too fast in throwing conjectures. First, we must see what kind of automaton accepts computably enumerable languages. If an unrestricted grammar is the ultimate type of a grammar, then we must have a corresponding automaton with the same property of being the ultimatum in some sense.

Problems for Section 7.2

7.1. Decide whether the string $baabbabaaabbaba$ is generated by the grammar with productions $S \mapsto aABa$, $A \mapsto baABb$, $B \mapsto Aab$, $aA \mapsto baa$, $bBb \mapsto abab$.

7.2. Let G be the grammar with productions $S \mapsto aSb|aA|bB|a|b$, $A \mapsto aA|a$, $B \mapsto bB|b$. Decide whether $ababa$, $aababa$, $aabbaa \in \mathcal{L}(G)$.

7.3. Show that the grammar with productions $S \mapsto AAB$, $Aa \mapsto SaB$, $Ab \mapsto SBb$, $B \mapsto SA|ab$ generates no string.

7.4. Let Σ be an alphabet. Construct a grammar for generating all and only the strings of length at most 2009 over Σ.

7.5. Construct a grammar that generates all (and only) even natural numbers below 2009, written in usual decimal notation.

7.6. Find a grammar that generates all strings of the form $a^m b^m c^m$, where $m \geq 1$.

7.7. Let G be the grammar with productions $S \mapsto \varepsilon|aa|bb|aSa|bSb$. Show that $\mathcal{L}(G)$ has no string of odd length, and also that the number of strings in $\mathcal{L}(G)$ of even length is 2^n.

7.8. What is $\mathcal{L}(G)$ if G has the productions $S \mapsto AB$, $A \mapsto Abba$, $B \mapsto cBbd$? Is $\mathcal{L}(G)$ regular, context-free, or neither?

7.9. For each of the following languages, find a grammar that generates it:
(a) $L_1 = \{a^n b^{2n} : n \geq 0\}$.
(b) $L_2 = \{a^{n+2} b^n : n \geq 1\}$.
(c) $L_3 = \{a^n b^{n-3} : n \geq 3\}$.
(d) $L_4 = \{a^m b^n : n > m \geq 0\}$.
(e) $L_4 L_2$.
(f) $L_4 \cup L_2$.
(g) L_4^3.
(h) L_4^*.
(i) $L_4 - \overline{L_3}$.

7.10. Find grammars to generate the following languages over the alphabet $\{a, b\}$:
(a) $\{ww : w \in \{a, b\}^*\}$.
(b) $\{a^m b^n a^n b^m : m, n \geq 1\}$.
(c) $\{w : \#_a(w) = \#_b(w)\}$.
(d) $\{w : \#_a(w) = 1 + \#_b(w)\}$.
(e) $\{w \in \{a, b\}^* : \#_a(w) > \#_b(w)\}$.
(f) $\{w \in \{a, b\}^* : \#_a(w) - \#_b(w) = \pm 1\}$.

7.11. Find grammars to generate the following languages over the alphabet $\{a, b, c\}$:
(a) $\{a^{2^n} : n \in \mathbb{N}\}$.
(b) $\{w : \#_a(w) = 1 + \#_b(w)\}$.
(c) $\{w : \#_a(w) = 2 \cdot \#_b(w)\}$.
(d) $\{a^m b^m c^n : m \geq 1, n \geq 0\}$.

7.12. What difficulties would arise if we allow ε on the left side of a production in an unrestricted grammar?

7.13. What is the language generated by the grammar with productions
(a) $S \mapsto aAB,\ Aa \mapsto SaB,\ Ab \mapsto SBb,\ B \mapsto SA|ab$?
(b) $S \mapsto aSa|bSb|aa|bb|\varepsilon$?
(c) $S \mapsto aS|A,\ A \mapsto bc|bAc$?
(d) $S \mapsto AB,\ A \mapsto Ab|a,\ B \mapsto cB|d$?
(e) $S \mapsto a|CD,\ C \mapsto ACB|AB,\ AB \mapsto aBA,\ Aa \mapsto aA,\ Ba \mapsto aB,\ AD \mapsto Da,\ BD \mapsto Ea,\ BE \mapsto Ea,\ E \mapsto a$?
(f) $S \mapsto a|CD,\ C \mapsto ACB|AB,\ AB \mapsto aBA,\ Aa \mapsto aA,\ Ba \mapsto aB,\ AD \mapsto Da,\ BD \mapsto Ea,\ BE \mapsto Ea,\ E \mapsto a$?
(g) $S \mapsto aSBA|abA,\ AB \mapsto BA,\ bB \mapsto bb,\ bA \mapsto ba,\ aA \mapsto aa$?
(h) $S \mapsto ABC,\ AB \mapsto aAD|bAE,\ DC \mapsto BaC,\ DC \mapsto BbC,\ Da \mapsto aD,\ Db \mapsto bD,\ Ea \mapsto aE,\ Eb \mapsto bE,\ aB \mapsto Ba,\ bB \mapsto Bb,\ AB \mapsto \varepsilon,\ C \mapsto \varepsilon$?
(i) $S \mapsto ABCS|Z,\ CA \mapsto AC,\ BA \mapsto AB,\ CB \mapsto BC,\ CZ \mapsto Yc|Zc,\ BY \mapsto Xb|Yb,\ AX \mapsto Xa,\ X \mapsto \varepsilon$?
(j) $S \mapsto AB,\ A \mapsto aAb,\ bB \mapsto bbbB,\ aAb \mapsto aa,\ B \mapsto \varepsilon$?

7.3 Turing Machines

You have worked with regular grammars and finite automata, and context-free grammars along with pushdown automata. Now you are working with unrestricted grammars. The definition of an unrestricted grammar looks simpler than others, as there are no restrictions on the productions. But then it allows so much of freedom that life becomes difficult with too many choices. Freedom brings forth responsibility as it is said. You will see, similarly, that Turing machines will be simple by definition, but it may become difficult to work with ample opportunities and manipulating the unrestricted power. However, it will be certainly enjoyable.

Like any other automaton, a Turing machine also has states, a reading head, and an input tape. But it has no memory, no stack, or no pushdown storage. Then where from the power comes? Well, instead of the reading head, a Turing machine uses a read–write head, that is, it can use the same device to read symbols from the tape and also write symbols on the tape. Thus the input tape is also used as an output tape and hence it would work as a temporary memory.

Moreover, it can now move both left and right instead of the only-right movement of the reading heads in earlier automata. The set of states, also called the control unit, works in discrete steps as earlier. We will also have final states, which we rather call as halt states. There will be exactly two halt states with each Turing Machine. Once a machine enters a halt state, it will stop operating thereafter. There will also be an initial state which is, of course, different from the halt states. From the initial state the machine starts operating.

Unlike other automata, a Turing machine can move to the left. It then comes with the danger of moving out of the tape. This is avoided by an extension of the tape itself. The input tape is extended to both left and right, that is, there is neither a left end nor a right end of the tape unlike the tapes of earlier machines.

Once the input is written on the tape, all squares to the left of the input string are left blank. We write the blank space as $\not b$. All squares to the right of the input string are also left blank, that is, every square preceding and following the input string contains the blank symbol $\not b$. The machine will start with its read–write head placed on the 0th square; (In fact, it is only conceptual; we do not number the squares.) it is the square or cell just to the left of the input string, which initially contains the blank symbol $\not b$. See Fig. 7.1.

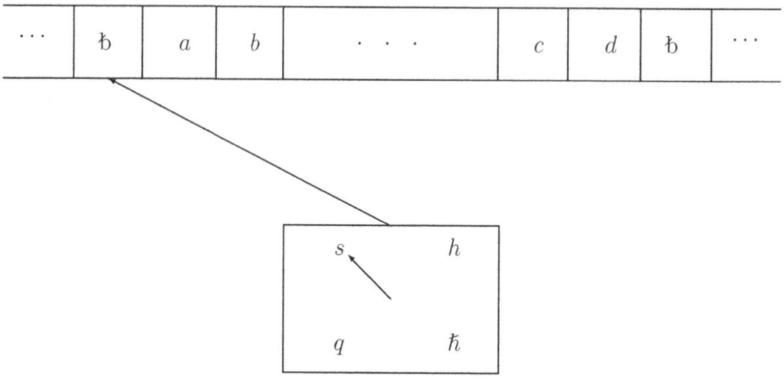

Fig. 7.1. A Turing machine.

We equip every Turing machine with a partial function. For example, if the partial function is f with $f(q, \not b) = (r, \not L)$, then it would mean that "if the machine, being in state q, reads $\not b$, then it would go to state r and its read–write head will move one square to the left of the scanned square." The symbols $\not b$, $\not L$, $\not R$ are common to every Turing machine. The symbol $\not b$ can be used to write the blank space (for erasing nonblank symbols), $\not L$ and $\not R$ signify the movement of the read–write head one square to the left and one square to the right, respectively. A formal description follows.

A **Turing machine** or a **TM** is a seventuple $M = (Q, \Sigma, \Gamma, \delta, s, h, \hbar)$, where

Q is a finite set, called the set of *states,*
Σ is an alphabet, called the *input alphabet,* satisfying $Q \cap \Sigma = \emptyset$ and $\not b \notin \Sigma$,
$\Gamma \supseteq \Sigma \cup \{\not b\}$ is a finite set, called the *tape alphabet* with $\not L, \not R \notin \Gamma$, so that these special symbols are also not in Σ.
$s \in Q$ is called the *start state* or the *initial state,*
$h \in Q$ is called the *accepting halt state,*
$\hbar \in Q$ is called the *rejecting halt state*, and
δ is a partial function from $(Q - \{h, \hbar\}) \times \Gamma$ to $Q \times (\Gamma \cup \{\not L, \not R\})$, called the *transition function.*

Both the states h and $\hbar$ are called the *halt states*. We further assume that the symbols $s, h, \hbar$ are all different and that $Q \cap (\Gamma \cup \{\not L, \not R\}) = \emptyset$, that is, the states and tape symbols are different and the special symbols do not occur anywhere. The transitions of the form $\delta(q, \beta) = (p, \gamma)$ are also called *instructions* of the machine M. Moreover, the input alphabet Σ is also called the *output alphabet* of M.

We must describe how the machine M works. Suppose the machine is currently in the state q scanning (its read–write head is on a square containing) the symbol a. If the partial function δ is not defined for the pair (q, a), then the machine stops working. In particular, the machine halts whenever it enters one of the halt states h or $\hbar$ as δ is never defined for these states, whatever be the scanned symbol. We will use the difference between the nonfunctioning of a Turing machine with a halt state and with a nonhalt state in a very meaningful way later.

If $\delta(q, a) = (h, \not R)$, then the read–write head of the machine moves one square to the right and then halts. Similarly, when $\delta(q, a) = (h, \not L)$, the machine goes one square to the left and then halts. Similar interpretation is given to the transitions $\delta(q, a) = (\hbar, \not L)$ and $\delta(q, a) = (\hbar, \not R)$ when the machine first moves to the left or right accordingly, and then halts in the state $\hbar$.

At any instant of time, the machine M is in some state and must be scanning some symbol. Also, there must be some string on the tape to the left of the currently scanned square and one to its right. These four pieces of information of the state, the left-string, currently scanned symbol, and the right-string describe the current position of the machine. Further, we will represent the infinite sequence of $\not b$'s (blanks) on both sides of the input string (or even on both sides of whatever string is there currently) by the empty string ε. An *instantaneous description* or a *configuration* of any machine will have these four components.

The work-out of the machine can be described by specifying how the machine passes from one configuration to another. In passing from one configuration to another, it uses a transition as given by δ. Suppose M is currently in state q scanning the symbol a to the left of which is the string ab on the tape, and on the right hand side of the currently scanned symbol is the string $abab$. Then we write the current configuration of M as a quadruple $(q, ab, a, abab)$.

If s is the initial state and if we place the read–write head on the square preceding the input u, then the initial configuration will be $(s, \varepsilon, \not b, u)$. Notice that the infinite $\not b$'s on the right are simply taken as ε and thus it is merged with the input u. The configuration $(q, \varepsilon, \not b, \varepsilon)$ means that currently the read–write head of the machine is on the 0th square scanning the symbol $\not b$ and to its left is the string ε and to its right is also the string ε. That means empty string ε is the current input to the machine.

The configuration $(q, ab, a, abab)$ can be abbreviated to $qab\underline{a}abab$. The configuration $(q, \varepsilon, \not b, \varepsilon)$ can then be abbreviated to $q\underline{\not b}$. We mention the convention again that the infinite number of $\not b$'s to the right or left are represented by ε and thus not explicitly mentioned in abbreviated configurations. Now from such a configuration, if the read–write head moves one square to the left, then it will scan the $\not b$ on the left, and again the $\not b$'s on the right will merge giving the new configuration as $(q, \varepsilon, \not b, \varepsilon)$ or as $q\underline{\not b}$ again.

Formally, a **configuration** is an element of $Q \times \Gamma^* \times \Gamma \times \Gamma^*$. We use the *abbreviated configurations* written as $qu\underline{a}v$ in place of (q, u, a, v). This includes the abbreviations of (q, ε, a, v) as $q\underline{a}v$, of $(q, \not b, a, \varepsilon)$ as $q\not b\underline{a}$, and of $(q, \varepsilon, \not b, \varepsilon)$ as $q\underline{\not b}$.

We can dispense with the underline in the currently scanned symbol by writing configurations in yet another way. In this notation, the configuration $(q, ab, a, abab)$ is written as $abqaabab$. In general, if q is the current state, σ is the currently scanned symbol, u is the string to the left of the currently scanned symbol, and v is the string right to the currently scanned symbol, then the configuration (q, u, σ, v) is abbreviated to the *string* $uq\sigma v$. When need arises to represent a configuration as a string over the alphabet $Q \cup \Gamma$, we will use this abbreviation. Otherwise, we continue with the abbreviated configurations written in the form $qu\underline{\sigma}v$.

The working out of the machine M is described by how a configuration **yields in one step** another configuration. Obviously, the transitions will have some role to play in this yield of configurations. Though this can be done abstractly for any pair of configurations, we will go the other way around. We will define the yield in one step relation of configurations starting from the transitions. A formal definition of the "yield in one step" written as $\stackrel{1}{\rightsquigarrow}$ can be given as in the following:

Let $M = (Q, \Sigma, \Gamma, \delta, s, h, \hbar)$ be a Turing machine. Let p, q be generic states. Let σ, τ be generic symbols from Γ, and u, v be generic strings over Γ.

(a) If $\delta(p, \sigma) = (q, \tau)$, then $pu\underline{\sigma}v \stackrel{1}{\rightsquigarrow} qu\underline{\tau}v$.
(b) If $\delta(p, \sigma) = (q, \not{L})$, then $pu\tau\underline{\sigma}v \stackrel{1}{\rightsquigarrow} qu\underline{\tau}\sigma v$.

In particular, $pu\sigma\underline{\not{b}} \stackrel{1}{\rightsquigarrow} qu\underline{\sigma}$ as $\not{b}$ merges with ε on the right. Also, $p\underline{\sigma}v \stackrel{1}{\rightsquigarrow} q\underline{\not{b}}\sigma v$ as $\not{b}$ appears from the ε on the left.

(c) If $\delta(p, \sigma) = (q, \not{R})$, then $pu\underline{\sigma}\tau v \stackrel{1}{\rightsquigarrow} qu\sigma\underline{\tau}v$.

In particular, If $\delta(p, \not{b}) = (q, \not{R})$, then $pu\underline{\not{b}}\sigma v \stackrel{1}{\rightsquigarrow} pu\not{b}\underline{\sigma}v$. This includes the cases: $p\underline{\not{b}} \stackrel{1}{\rightsquigarrow} q\not{b}\underline{\not{b}}$ as a $\not{b}$ can appear from ε at the right, and $p\underline{\not{b}}\sigma v \stackrel{1}{\rightsquigarrow} p\underline{\sigma}v$ as $\not{b}$ merges with ε on the left.

(d) A configuration of the form $hu\underline{\sigma}v$ or $\hbar u\underline{\sigma}v$ does not yield any configuration.
(e) If $\delta(p, \sigma)$ is not defined, then $pu\underline{\sigma}v$ does not yield any configuration.

The relation of **yield**, written as $\rightsquigarrow$, between configurations is defined as the reflexive and transitive closure of the one-step yield relation $\stackrel{1}{\rightsquigarrow}$. That is, if C, C' are configurations, then $C \rightsquigarrow C$ and also we have $C \rightsquigarrow C'$ iff there is a sequence of configurations $C_1, C_2, \ldots, C_n$ such that

$$C \stackrel{1}{\rightsquigarrow} C_1, C_1 \stackrel{1}{\rightsquigarrow} C_2, \cdots, C_{n-1} \stackrel{1}{\rightsquigarrow} C_n, C_n \stackrel{1}{\rightsquigarrow} C'.$$

In such a case, we also write $C \stackrel{n+1}{\rightsquigarrow} C'$ mentioning in how many steps C yields C'. If there is no confusion, we will simply write $\rightsquigarrow$ instead of $\stackrel{1}{\rightsquigarrow}$ or $\stackrel{n}{\rightsquigarrow}$. The sequence of yields in one step that shows how C yields C' is called a **computation** of M starting from the configuration C.

We also specify the configuration from which a machine starts its work. When we use the phrase "M starts with input u," we will assume that the initial configuration of the machine is $s\underline{\not b}u$, that is, the read–write head is kept on the blank square immediately to the left of the input string u and the machine is in its initial state s. See the following example.

Example 7.5. Let $M = (\{s, q, h, \hbar\}, \{a\}, \{a, \not b\}, \delta, s, h, \hbar)$ be the Turing machine with $\delta(s, \not b) = (q, \not R)$, $\delta(q, a) = (q, \not R)$, $\delta(q, \not b) = (h, \not b)$. What is the computation of M with input aaa?

Solution. The initial configuration of M with input aaa is $s\underline{\not b}aaa$. As $\delta(s, \not b) = (q, \not R)$, the read–write head moves one square to the right while M changes its state to q. Thus, the next configuration is $q\underline{a}aa$. In terms of the yield (in one step) relation, we write it as

$$s\underline{\not b}aaa \overset{1}{\rightsquigarrow} q\underline{a}aa \text{ or simply as } s\underline{\not b}aaa \rightsquigarrow q\underline{a}aa.$$

Now, the transition or instruction $\delta(q, a) = (q, \not R)$ becomes applicable. Thus in the next step of computation, M goes one square to the right while remaining in the same state q. This is recorded as the yield $q\underline{a}aa \rightsquigarrow qa\underline{a}a$. You go three steps further to get the computation as

$$s\underline{\not b}aaa \rightsquigarrow q\underline{a}aa \rightsquigarrow qa\underline{a}a \rightsquigarrow qaa\underline{a} \rightsquigarrow qaaa\underline{\not b} \rightsquigarrow haaa\underline{\not b}.$$

Here M halts with the accepting halt state h. □

You can also represent a Turing machine by its *transition diagram* as you were doing for the DFAs. A transition diagram of a Turing machine has states and transitions as usual. The transition diagram of the machine in Example 7.5 is given in Fig. 7.2.

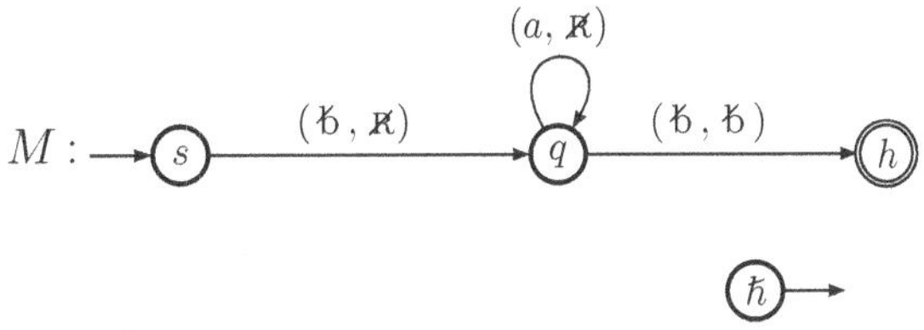

Fig. 7.2. Transition diagram For Example 7.5.

The transitions now go from one state to another but are labeled with a pair, the first component is what it reads on the scanned square and the second component specifies its action. Thus a transition from p to q labeled (a, b) would mean:

> if the machine is currently scanning the symbol a being in state p, then it changes state to q and replaces a with b.

This is so if b is a tape symbol including $\not b$. If $b = \not L$, then the machine changes its state to q and goes to the left square from the currently scanned one. Similarly, if $b = \not R$, then the machine goes one square right instead.

We must also have some way of telling which state is our accepting state and which one is the rejecting state. We will have two circles around the accepting halt state as we were doing for the final states in a DFA. We will put an outgoing arrow from the rejecting state. The isolated node with $\hbar$ along with the out-going arrow is sometimes omitted.

Problems for Section 7.3

7.14. Let M be a Turing machine with transitions $\delta(s, \not{b}) = (p, \not{R}), \delta(p, \not{b}) = (h, \not{b}), \delta(p, a) = (q, \not{b}), \delta(q, \not{b}) = (r, \not{R}), \delta(r, a) = \delta(r, b) = (r, \not{R}), \delta(r, \not{b}) = (t, \not{L}), \delta(t, b) = (t, \not{b}), \delta(t, \not{b}) = (u, \not{L}), \delta(u, a) = \delta(u, b) = (u, \not{L}), \delta(u, \not{b}) = (p, \not{R})$, with s as the initial state. Describe the computation of M with the input aba and $aaabbbb$. Is there any input that drives M to do an infinite computation?

7.15. Design a Turing machine with least number of transitions that goes into an infinite loop.

7.16. Design a Turing machine that started on a square not containing a, somewhere in the middle of an input string not containing $\not{b}$, searches for an occurrence of a to its right. If it finds an a, then it halts, else, it gets a $\not{b}$ before getting an a, and in that case, it should enter an infinite computation.

7.17. Let $M = (\{s, p, q\}, \{a\}, \{a, \not{b}\}, s, \delta, h, \hbar)$ be the Turing machine with δ given by $\delta(s, \not{b}) = (s, \not{b}), \delta(s, a) = (p, \not{L}), \delta(p, \not{b}) = (h, \not{b}), \delta(p, a) = (q, \not{b}), \delta(q, \not{b}) = (s, \not{L})$, and $\delta(q, a) = (q, a)$. Let $n \in \mathbb{N}$. Describe what M does when started in the configuration $s \not{b} a^n \underline{a}$.

7.18. What is the maximum number of transitions in a Turing machine?

7.4 Acceptance and Rejection

A Turing machine accepts or rejects languages depending upon its last configuration. A halted configuration of a Turing machine is a configuration whose state component is one of the halt states. An accepting configuration is a halted configuration whose state component is the accepting halt state. A rejecting configuration is a halted configuration with the state component as the rejecting halt state. The machine accepts a string when it finally halts in an accepting configuration. Similarly, it rejects a string when it eventually halts in a rejecting configuration.

Formally, let $M = (Q, \Sigma, \Gamma, \delta, s, h, \hbar)$ be a Turing machine. The two types of configurations whose state components are one of h or $\hbar$ are called the **halted configurations**. A halted configuration of the type $hu\underline{\sigma}v$ is called an *accepting* *configuration*. Similarly, a halted configuration of the type $\hbar u\underline{\sigma}v$ is called a *rejecting configuration*. Let $w \in \Sigma^*$ be a string. We say that M **accepts** w iff $s\underline{\not{b}}w \rightsquigarrow hu\underline{\sigma}v$ for some $\sigma \in \Gamma$, $u, v \in \Gamma^*$. The **language accepted by** M is defined as the set of all strings accepted by M. That is,

$$\begin{aligned}\mathcal{L}(M) &= \{w \in \Sigma^* : M \text{ accepts } w\}\\ &= \{w : s\underline{\not{b}}w \rightsquigarrow hu\underline{\sigma}v \text{ for some } u, v \in \Gamma^*,\ \sigma \in \Gamma\}.\end{aligned}$$

Similarly, the **language rejected by** the machine is the set of all strings from Σ^* that are rejected by it. That is,

$$\begin{aligned}\mathcal{L}_R(M) &= \{w \in \Sigma^* : M \text{ rejects } w\}\\ &= \{w : s\underline{\not{b}}w \rightsquigarrow \hbar u\underline{\sigma}v \text{ for some } u, v \in \Gamma^*,\ \sigma \in \Gamma\}.\end{aligned}$$

The machine can, however, stop operating in the middle due to the simple reason that the transition function is not defined for the current state–symbol pair. The configurations whose state component is neither h nor $\hbar$ are called *nonhalted configurations*. If the machine stops operating in a nonhalted configuration, then we will refer to this situation by telling that the machine has come to an *abrupt halt*. A Turing machine that neither comes to an abrupt halt nor goes on an infinite computation must halt for each input from Σ^*. Such a machine is called a **total Turing machine**. Each input string from Σ^* drives a total Turing machine to one of the halt states h or $\hbar$ eventually.

Exercise 7.1. Show that if M is a total Turing machine, then $\mathcal{L}(M)$ and $\mathcal{L}_R(M)$ are complements of each other. Does the converse hold?

We also say that the machine M accepts $\mathcal{L}(M)$ and it rejects $\mathcal{L}_R(M)$. The language of the machine (i.e., the language accepted by the machine) in Example 7.5 is, obviously, a^* and the language rejected by it is $\emptyset$. Notice that we do not bother to take care what happens if an input is given over a different alphabet to the machine. Because, the machine already *knows* what its input alphabet is. Informally, the case of irrelevant input strings is referred to as *garbage in and garbage out*. This also applies to all the earlier machines such as finite automata and pushdown automata. For example, exactly the same transitions in Example 7.5 can be used for constructing a TM for accepting a^* over the alphabet $\{a, b\}$.

Example 7.6. Construct a total TM for accepting the language $a^* \subseteq \{a, b\}^*$.

Solution. The way it differs from the earlier machine of Example 7.5 is that the strings having at least one b must now be rejected, as it is total. So, we modify the earlier machine by changing its transitions taking into account the bigger input alphabet. Take $M = (\{s, q, h, \hbar\}, \{a, b\}, \{a, b, \not{b}\}, \delta, s, h, \hbar)$ with $\delta(s, \not{b}) = (q, \not{R})$, $\delta(q, a) = (q, \not{R})$, $\delta(q, \not{b}) = (h, \not{b})$, $\delta(q, b) = (\hbar, b)$. The transition diagram of this machine is given in Fig. 7.3. Check that it does its job. □

Notice that in case of Turing machines, it is not required that the input string is read completely for acceptance or rejection.

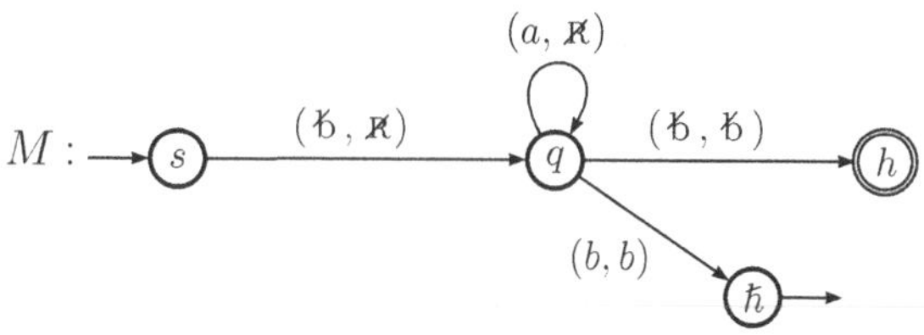

Fig. 7.3. Transition diagram for Example 7.6.

Example 7.7. Is the language $L = \{a^n b^n : n \in \mathbb{N}\}$ accepted by any Turing machine?

Solution. Of course, you are hopeful that a Turing machine M can be constructed to accept L. It must be able to match the number of a's and b's and only in that particular form, where no b can precede any a. When M meets the first a, it should look for a b and then it should come back to the next a, etc. To fix the idea, suppose that M meets the first a from the left, it matches the last b with this a and then it comes back to the second a, goes for matching with it the last but one b, and so on. But how does it come back to the second a after the first matching? To be able to sense how far it has matched, let us change the already met a to another symbol, say, c. In that case, we will simply go on accumulating c's!

Well, something can be done. We will rewrite the a it meets first, and after the matching with a b, we will erase this c and then start the next matching process, again rewriting the next a as a c, and so on. It looks that this much extra symbol will do. Let us try. At this point I would suggest you to take a piece of paper and a pen and build up the transition function δ as we go along.

First, our set of states must contain the initial state s. As the initial configuration is of the form $s\underline{\not{b}}w$ for the input w, the machine must go to the right from this. That means we must have $\delta(s, \not{b}) = (p, \not{R})$ for some state p. Here, start building the function δ as a table. Draw a horizontal line on the top of the piece of paper. Draw a vertical line crossing the horizontal line towards the left hand side of the paper, put δ on the top left corner of the table. Above the horizontal line keep the first symbol as $\not{b}$, second symbol as a, third as b, and fourth as c. To the left of the vertical line write the first symbol as s, and second as p. We will write the states one below the other and the tape symbols on the horizontal one next to the other. Now fill in the first entry corresponding to the pair $(s, \not{b})$ as $(p, \not{R})$. The table now looks like:

δ	$\not{b}$	a	b	c
s	$(p, \not{R})$			
p				

The first input we should think about is ε. In such a case, the initial configuration is $s\underline{\not{b}}$. As $\delta(s, \not{b}) = (p, \not{R})$, this configuration would yield $p\underline{\not{b}}$, which must yield an accepting configuration. So, we must have $\delta(p, \not{b}) = (h, \not{b})$. Here, $(h, \not{b})$ may also be replaced by any other ordered pair whose first component is h. Let us stick to this pair. At this point, update your transition function (table) to

δ	$\not{b}$	a	b	c
s	$(p, \not{R})$			
p	$(h, \not{b})$			

Suppose the input string is starting with an a, that is, it is in the form au. Now how should our machine behave? As earlier, it reaches the configuration $p\underline{a}u$. We want to replace a by c being in the same state. So add $\delta(p, a) = (p, c)$. Had it been b instead of a, our machine should not have accepted it; so we leave $\delta(p, b)$ undefined. With input a, the machine is now having the configuration $p\underline{c}u$. It must go to the right and change state to, say, q. Then we add $\delta(p, c) = (q, \not{R})$. Now it is reading the first symbol of u. Whatever this may be, it should go to the right till it gets a $\not{b}$, just to the right of u. So, we add $\delta(q, a) = \delta(q, b) = (q, \not{R})$. The updated table now looks like

δ	$\not{b}$	a	b	c
s	$(p, \not{R})$			
p	$(h, \not{b})$	(p, c)		$(q, \not{R})$
q		$(q, \not{R})$	$(q, \not{R})$	

The next configuration is $qcu\underline{\not{b}}$. From this place it should go left by changing its state. So, add $\delta(q, \not{b}) = (r, \not{L})$, where r is another state. We have replaced one a by one c and now we want to erase one b from the right end of the input, where the read–write head is currently scanning the b. So, we must have $\delta(r, b) = (r, \not{b})$.

What if the last symbol was not a b? If the last symbol is an a, the string should not be accepted; we leave $\delta(r, a)$ undefined. If it is a c, then we have an excess a in the input, thus $\delta(r, c)$ must not be defined. Now, the square that was having a b contains a $\not{b}$ and the machine is in state r. So, what should the machine do? It must go to the left searching for the c. Can it be in the same state r while going left?

We have already decided to leave $\delta(r, a)$ undefined. If there is one more a to the left, somewhere else, then we will be in soup. So, let us introduce one more state, say, t and take $\delta(r, \not{b}) = (t, \not{L})$. Our plan is to move left until we get the c, being in state t. So, add $\delta(t, a) = (t, \not{L})$, $\delta(t, b) = (t, \not{L})$. When the read–write head reaches c, it must erase it, as agreed upon. Let us then add $\delta(t, c) = (t, \not{b})$. After this, the machine should start afresh as it had started initially from state s. So, have $\delta(t, \not{b}) = (p, \not{R})$. The updated table of instructions is shown in Table 7.1.

Table 7.1. Transition function for Example 7.7

δ	$\not{b}$	a	b	c
s	$(p, \not{R})$			
p	$(h, \not{b})$	(p, c)		$(q, \not{R})$
q	$(r, \not{L})$	$(q, \not{R})$	$(q, \not{R})$	
r	$(t, \not{L})$		$(r, \not{b})$	
t	$(p, \not{R})$	$(t, \not{L})$	$(t, \not{L})$	$(t, \not{b})$

The constructed machine is $M=(\{s, p, q, r, t, h, \hbar\}, \{a, b\}, \{a, b, \not{b}, c\}, \delta, s, h, \hbar)$. The transition diagram is given in Fig. 7.4, where we have omitted the node $\hbar$. You can very well have this node as in Fig. 7.2.

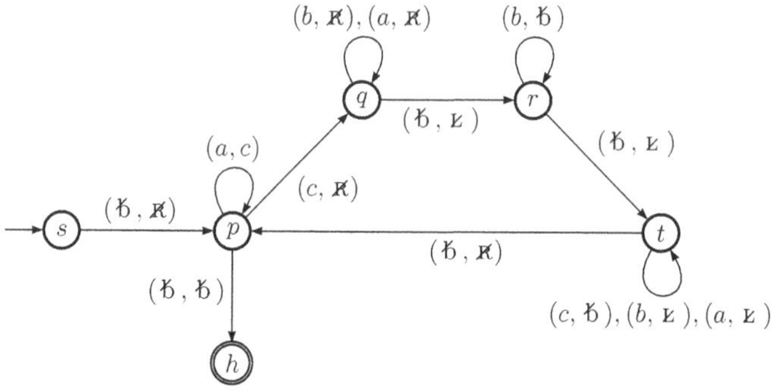

Fig. 7.4. Transition diagram for Example 7.7.

The following is a verification that the machine works as intended for the inputs $\varepsilon, a, b, ab, aba, abb, abab, aabb$. Read each $\rightsquigarrow$ as $\overset{1}{\rightsquigarrow}$. Verify with other inputs.

1. $s\underline{\not{b}} \rightsquigarrow p\underline{\not{b}} \rightsquigarrow h\underline{\not{b}}$.
2. $s\underline{\not{b}}a \rightsquigarrow p\underline{a} \rightsquigarrow p\underline{c} \rightsquigarrow q\underline{\not{b}} \rightsquigarrow r\underline{c}$, an abrupt halt.
3. $s\underline{\not{b}}b \rightsquigarrow p\underline{b}$, an abrupt halt.
4. $s\underline{\not{b}}ab \rightsquigarrow p\underline{a}b \rightsquigarrow p\underline{c}b \rightsquigarrow qc\underline{b} \rightsquigarrow qcb\underline{\not{b}} \rightsquigarrow rc\underline{b} \rightsquigarrow rc\underline{\not{b}} \rightsquigarrow t\underline{c} \rightsquigarrow t\underline{\not{b}}ylp\underline{\not{b}} \rightsquigarrow h\underline{\not{b}}$.
5. $s\underline{\not{b}}aba \rightsquigarrow p\underline{a}ba \rightsquigarrow p\underline{c}ba \rightsquigarrow qc\underline{b}a \rightsquigarrow qcb\underline{a} \rightsquigarrow qcba\underline{\not{b}} \rightsquigarrow rcb\underline{a}$, an abrupt halt.
6. $s\underline{\not{b}}abb \rightsquigarrow p\underline{a}bb \rightsquigarrow p\underline{c}bb \rightsquigarrow qc\underline{b}b \rightsquigarrow qcb\underline{b} \rightsquigarrow qcbb\underline{\not{b}} \rightsquigarrow rcb\underline{b} \rightsquigarrow rcb\underline{\not{b}} \rightsquigarrow tc\underline{b}$ $\rightsquigarrow t\underline{c}b \rightsquigarrow t\underline{\not{b}}b \rightsquigarrow p\underline{b}$, an abrupt halt.
7. $s\underline{\not{b}}abab \rightsquigarrow p\underline{a}bab \rightsquigarrow p\underline{c}bab \rightsquigarrow qc\underline{b}ab \rightsquigarrow qcbab\underline{\not{b}} \rightsquigarrow rcba\underline{b} \rightsquigarrow rcba\underline{\not{b}}$ $\rightsquigarrow tcb\underline{a} \rightsquigarrow t\underline{c}ba \rightsquigarrow t\underline{\not{b}}ba \rightsquigarrow p\underline{b}a$, an abrupt halt.
8. $s\underline{\not{b}}aabb \rightsquigarrow p\underline{c}abb \rightsquigarrow qcabb\underline{\not{b}} \rightsquigarrow rcab\underline{b} \rightsquigarrow rcab\underline{\not{b}} \rightsquigarrow tca\underline{b} \rightsquigarrow t\underline{c}ab$ $\rightsquigarrow t\underline{\not{b}}ab \rightsquigarrow p\underline{a}b \rightsquigarrow h\underline{\not{b}}$, as in 4.

From the above, you see that the computations in 1, 4, 8 are accepting computations, while those in 2, 3, 5, 6, 7 do not end in a halted configuration; thus the strings $a, b, aba, abb, abab$ are not accepted by M. Now you can convince yourself how the machine M works by arguing abstractly with what the states do and why $\mathcal{L}(M) = L$. □

It is possible to construct a Turing machine to accept the language $\mathcal{L}(M) = \{a^n b^n : n \in \mathbb{N}\}$ without using the extra symbol c; you may have to introduce more states. Why don't you try it?

What is the language rejected by the machine constructed in Example 7.7? In that machine there is no transition that drives the machine to the rejecting state $\hbar$. This means that the machine rejects no string, that is, $\mathcal{L}_R(M) = \emptyset$. Can you construct a

TM that accepts $a^n b^n$ for each $n \in \mathbb{N}$ and rejects every other string over the alphabet $\{a, b\}$? Start your work from Table 7.1.

All the transitions that were left undefined was so intended just to ensure that the strings that are not in the form $a^n b^n$ are never accepted; in such cases, the machine halts abruptly. Is it always the case that if the transition function of a machine is a total transition function, then the machine is a total Turing machine? In this particular case, if we modify the transition function by filling in those places driving the machine to the rejecting state $\hbar$, the machine would reject the unwanted strings. The updated transition function is now a total function as given in Table 7.2.

Table 7.2. Modified transition function

δ	$\not b$	a	b	c
s	$(p, \not R)$	$(\hbar, \not b)$	$(\hbar, \not b)$	$(\hbar, \not b)$
p	$(h, \not b)$	(p, c)	$(\hbar, \not b)$	$(q, \not R)$
q	$(r, \not L)$	$(q, \not R)$	$(q, \not R)$	$(\hbar, \not b)$
r	$(t, \not L)$	$(\hbar, \not b)$	$(r, \not b)$	$(\hbar, \not b)$
t	$(p, \not R)$	$(t, \not L)$	$(t, \not L)$	$(t, \not b)$

The updated machine is $M' = (\{s, p, q, r, t, h, \hbar\}, \{a, b\}, \{a, b, \not b, c\}, \delta, s, h, \hbar)$. The transition diagram of Fig. 7.4 can be modified accordingly. Verify for some inputs whether it really does the job. Here are some of them:

$$s\underline{\not b}a \rightsquigarrow p\underline{a} \rightsquigarrow p\underline{c} \rightsquigarrow q\underline{\not b} \rightsquigarrow r\underline{c} \rightsquigarrow \hbar\underline{\not b}.$$

$$s\underline{\not b}b \rightsquigarrow p\underline{b} \rightsquigarrow \hbar\underline{\not b}.$$

$$s\underline{\not b}ba \rightsquigarrow p\underline{b}a \rightsquigarrow \hbar\underline{\not b}a.$$

$$s\underline{\not b}aba \rightsquigarrow p\underline{a}ba \rightsquigarrow p\underline{c}ba \rightsquigarrow qc\underline{b}a \rightsquigarrow qcb\underline{a} \rightsquigarrow qcba\underline{\not b} \rightsquigarrow rcb\underline{a} \rightsquigarrow \hbar cb\underline{\not b}.$$

$$s\underline{\not b}abab \rightsquigarrow p\underline{a}bab \rightsquigarrow p\underline{c}bab \rightsquigarrow qc\underline{b}ab \rightsquigarrow qcbab\underline{\not b} \rightsquigarrow rcba\underline{b} \rightsquigarrow rcba\underline{\not b}$$
$$\rightsquigarrow tcb\underline{a} \rightsquigarrow t\underline{c}ba \rightsquigarrow t\underline{\not b}ba \rightsquigarrow p\underline{b}a \rightsquigarrow \hbar\underline{\not b}a.$$

The above computations show that the strings $a, b, ba, aba, abab$ are all rejected. A moment's thought will convince you that the modified machine M' accepts the language $\{a^n b^n \in \{a, b\}^* : n \in \mathbb{N}\}$ and rejects the language

$$\{w \in \{a, b\}^* : w \neq a^n b^n \text{ for any } n \in \mathbb{N}\}.$$

Here, $\mathcal{L}_R(M') = \overline{\mathcal{L}}(M')$. Recall that we use $\overline{\mathcal{L}}(M')$ instead of the precise $\overline{\mathcal{L}(M')}$. Will this strategy of modifying an already existing machine work in all cases? To be specific, suppose M accepts $L \subseteq \Sigma^*$ for some alphabet Σ. Construct a machine M' by redefining the missing transitions to $(\hbar, \not b)$. Clearly, $\mathcal{L}(M') = \mathcal{L}(M)$. Will it be the case that $\mathcal{L}_R(M') = \overline{\mathcal{L}}(M')$?

The strategy would not work when the machine M *does not accept* a string, by entering an infinite computation. Recollect that unlike finite automata, a Turing machine can go on infinite computations. Moreover, a Turing machine accepts a string by eventually reaching its accepting state; it does not accept a string either by rejecting it, or coming to an abrupt halt, or by entering an infinite computation. If M enters an infinite computation when it starts with an input u, so will M'. In that case, M' will never reject the input u, which is required to satisfy the condition $\mathcal{L}_R(M') = \overline{\mathcal{L}}(M')$. For example, the machine

$M = (\{s, p, h, \hbar\}, \{a, b\}, \{a, b, \not{b}\}, \delta, s, h, \hbar)$ with $\delta(s, \not{b}) = (s, \not{R})$, $\delta(s, a) = (h, a)$, $\delta(s, b) = (p, \not{R})$, $\delta(p, \not{b}) = (p, \not{R})$, $\delta(p, a) = (p, \not{R})$, $\delta(p, b) = (p, \not{R})$

accepts the language $\mathcal{L}(M) = \{aw : w \in \{a, b\}^*\}$ and rejects $\emptyset$. The strings in $\{a, b\}^* - \mathcal{L}(M)$ cause M to enter an infinite computation by moving to the right. The transition function is a total function having nothing more to be filled in. The strategy of redefining the undefined transitions gives $M' = M$. Thus it fails. However, you can have another machine for accepting the same language where there is no such transition that might cause the machine to enter an infinite computation. In that case, you can use the strategy for rejecting strings from $\overline{\mathcal{L}}(M)$. Can we always do that?

The question is a bit deeper now. To be precise, suppose we have a Turing machine that accepts a language L. Can we have a Turing machine that accepts L but for no strings in its input alphabet it enters an infinite computation? Equivalently, if L is a language accepted by a Turing machine, can we have a machine that accepts L and rejects $\overline{L}$? Notice that this question asks about the closure properties of the class of languages accepted by Turing machines. We will need some more experience with Turing machines to answer this question.

Problems for Section 7.4

7.19. Consider $M = (Q, \{a, b\}, \{a, b, \not{b}\}, \delta, s, h, \hbar)$, where $Q = \{s, p, q, r, t, u, v, h, \hbar\}$, and $\delta(s, \not{b}) = (p, \not{R})$, $\delta(p, \not{b}) = (h, \not{b})$, $\delta(p, a) = (q, \not{b})$, $\delta(q, \not{b}) = (r, \not{R})$, $\delta(r, \not{b}) = (t, \not{L})$, $\delta(r, a) = (r, \not{R})$, $\delta(r, b) = (r, \not{R})$, $\delta(t, b) = (u, \not{b})$, $\delta(u, \not{b}) = (v, \not{L})$, $\delta(v, \not{b}) = (p, \not{R})$, $\delta(v, a) = (v, \not{L})$, $\delta(v, b) = (v, \not{L})$. Determine $\mathcal{L}(M)$ and $\mathcal{L}_R(M)$.

7.20. Let M be a Turing machine with transitions $\delta(s, \not{b}) = (p, \not{R})$, $\delta(p, \not{b}) = (h, \not{b})$, $\delta(p, a) = (q, \not{b})$, $\delta(q, \not{b}) = (r, \not{R})$, $\delta(r, a) = \delta(r, b) = (r, \not{R})$, $\delta(r, \not{b}) = (t, \not{L})$, $\delta(t, b) = (t, \not{b})$, $\delta(t, \not{b}) = (u, \not{L})$, $\delta(u, a) = \delta(u, b) = (u, \not{L})$, $\delta(u, \not{b}) = (p, \not{R})$, with s as the initial state. What are $\mathcal{L}(M)$ and $\mathcal{L}_R(M)$?

7.21. Design total Turing machines that accept the following languages over $\{a, b\}$:
(a) $\emptyset$.
(b) $\{\varepsilon\}$.
(c) $\{a\}$.
(d) a^*ba^*b.

7.22. Design Turing machines for accepting the following languages over $\{a, b\}$:
(a) ab^*ab^*a.
(b) $\{a^nb^n : n \geq 1\}$.
(c) $a(a \cup b)^*$.
(d) bab^*a.

(e) $\{w : \ell(w) \text{ is even}\}$.
(f) $\{w : \ell(w) \text{ is a multiple of } 3\}$.
(g) $\{a^m b^n : m \geq 1,\ m \neq n\}$.
(h) $\{w : \#_a(w) = \#_b(w)\}$.
(i) $\{a^m b^n a^{m+n} : m \geq 0,\ n \geq 1\}$.
(j) $\{(a^m b^m)^2 : m \in \mathbb{N}\}$.
(k) $\{a^n b^{2n} : n \geq 1\}$.
(l) $\{ww : w \in \{a, b\}^*,\ \ell(w) \geq 1\}$.
(m) $\{ww^R : w \in \{a, b\}^*\}$.

7.23. Give an example of a Turing machine M for each of the following cases:
(a) M has the single halt state h and M does not accept any language.
(b) M has both the halt states h and $\hbar$ and M does not accept any language.

7.24. Turing machines can be defined with more final states than two as for other automata. Define such a machine and show that for each such machine there exists a Turing machine with only two final states (as we defined) accepting the same language as the original.

7.25. What happens if we put forth the convention of starting a Turing machine from scanning the first input symbol instead of starting it from the $\not{b}$ preceding the input string?

7.26. What is $\mathcal{L}(M)$, if $M = (\{p, q, r, s\}, \{a, b\}, \{a, b, \not{b}\}, \delta, p, h, \hbar)$ with δ given by:
(a) $\delta(p, \not{b}) = (q, \not{R})$, $\delta(q, a) = (r, \not{R})$, $\delta(r, b) = (q, \not{R})$, $\delta(r, \not{b}) = (h, \not{R})$?
(b) $\delta(p, \not{b}) = (p, \not{R})$, $\delta(p, a) = (q, \not{b})$, $\delta(q, \not{b}) = (r, \not{R})$, $\delta(r, b) = (q, \not{b})$, $\delta(r, \not{b}) = (h, \not{R})$?
(c) $\delta(p, \not{b}) = (p, \not{R})$, $\delta(p, a) = (q, b)$, $\delta(q, b) = (r, \not{R})$, $\delta(r, b) = (s, a)$, $\delta(s, a) = (t, \not{L})$, $\delta(t, b) = (p, \not{R})$, $\delta(r, \not{b}) = (h, \not{R})$.?

7.5 Using Old Machines

Our experience shows that it is not easy to construct Turing machines that accept a given language. We will, in fact, develop better mechanism for constructing Turing machines instead of always trying from the first principle.

Look back on the solution of Example 7.7. You first had an informal way of spelling out how the machine would work. The informal way took care of small jobs such as "go to left unless one $\not{b}$ is met," then "turn right," etc. Imagine that there are machines to do these small jobs. Then can you make bigger machines to do the job at hand using these small ones?

Suppose you have a machine, say, M_L that goes to the left only one square, and a machine M_R that goes to one square on the right. You can think of combining these machines so that they work one after the other. Say, somewhere on the middle of the tape, the read–write head is already there. Now if M_L works on it, the read–write head goes to the square to the left of the original. If M_R works immediately on the current one, then it would take the read–write head back to the original square.

That is, the combined machine $M_L \to M_R$ does not alter the input at all, leaving the read–write head as it was, that is, it does not alter the current configuration. Similarly, $M_L \to M_L \to M_R$ will have the net effect of M_L.

Notice that these machines need not start from the blank square preceding the input. In fact, this holds for all machines. Only when we require a machine to accept a language, it has to start from the blank square just to the left of the input.

However, all machines will start from their initial states and if nothing is defined, they would halt abruptly. An abrupt halt would signal that the string is not accepted. In combining machines, we will ensure that the first machine comes to its accepting halt state, and then the second machine takes over. Though this is not necessary, it will be easy, conceptually leading to a clear presentation.

To fix the idea, if we have two machines to be operated one after the other, then the first machine starts from its initial state, it should come to its halt state after some computation, and then the second machine takes over starting from its own initial state. The first machine can start from anywhere on the tape as per our requirements, but the second machine must start from wherever the first one has left the tape. Think of taking out the control unit of the first machine (Fig. 7.1) and plugging in the control unit of the second machine leaving the tape and the read–write head as they were. How do we formally describe such a combined machine? Well, with $\Gamma = \Sigma \cup \{b\!\!\!/\}$, take

$$M_L = (\{s_1, h, \hbar\}, \Sigma, \Gamma, \delta_L, s_1, h_1, \hbar_1) \text{ and } \delta_L(s_1, \gamma) = (h_1, L\!\!\!/), \text{ for each } \gamma \in \Sigma \cup \{b\!\!\!/\}.$$

$$M_R = (\{s_2, h, \hbar\}, \Sigma, \Gamma, \delta_R, s_2, h_2, \hbar_2) \text{ and } \delta_R(s_2, \gamma) = (h_2, R\!\!\!/), \text{ for each } \gamma \in \Gamma.$$

The machine $M_L \to M_R$ will have the initial state as that of M_L, that is, s_1. On any given input when the new machine works, the machine M_L will be initiated. Upon halting of M_L, the new machine will go to the initial state of the machine M_R and then M_R must work. Thus, no halt states of M_L should be a halt state of the new machine. The only accepting halt state of the new machine is the accepting halt state of M_R. In addition, we must have a transition to drive the new machine from the halt states of M_L to the initial state of M_R. That is,

> $M_L \to M_R$ is the TM $(\{s_1, h_1, \hbar_1, s_2\}, \Sigma, \Gamma, \delta, s_1, h_2, \hbar_2)$, where for each $\gamma \in \Gamma$, $\delta(s_1, \gamma) = (h_1, L\!\!\!/)$, $\delta(h_1, \gamma) = (s_2, \gamma)$, $\delta(s_2, \gamma) = (h_2, R\!\!\!/)$, $\delta(\hbar, g) = (s_2, \gamma)$.

In general, if we have two machines M_1 and M_2 and we want to combine them, we can think of a conditional combination. That is, suppose M_1 has halted after its work, leaving the read–write head wherever it was at the time of its halting. If the symbol it was scanning is σ, then we want M_2 to work. Otherwise, we may want another machine, say, M_3 to start work. Or, more generally, we may add another condition for M_3 to take over, say, if the scanned square upon the halting of M_1 was τ, then only M_3 would work. In such a case, the new machine, call it M, can be shown as in Fig. 7.5.

As you have noticed, the diagrams are just like those for finite automata, the only difference being that we do not have final *states* here. The states in this diagram are, of course, the old machines. We will read them as machines and not states. The initial machine is M_1 as represented above by an arrow coming to it from nowhere. The diagram in Fig. 7.5 can be read as

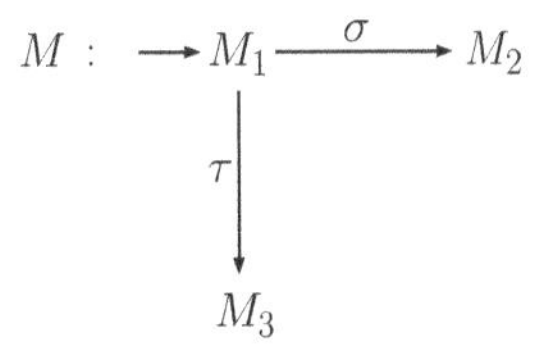

Fig. 7.5. A combined machine.

Start in the initial state of M_1 and operate as M_1 does. Upon the halting of M_1, check the currently scanned symbol. Then do one of the following:

(a) If the currently scanned symbol is σ, then change state to the initial state of M_2 and operate as M_2 does.
(b) If the currently scanned square is τ, then change state to the initial state of M_3 and operate as M_3 does.

Formally, let $M_i = (Q_i, \Sigma, \Gamma, \delta_i, s_i, h_i, \hbar_i)$, for $i = 1, 2, 3$ be Turing machines. Assume that the states in Q_1, Q_2, Q_3 and the halt states h_i and $\hbar_i$ are all distinct. Then the combined machine of Fig. 7.5 is defined as $M = (Q, \Sigma, \Gamma, \delta, s_1, h, \hbar)$, where

$Q = Q_1 \cup Q_2 \cup Q_3 \cup \{h, \hbar\}$, with $h, \hbar$ as new states, and for each $\gamma \in \Gamma$, $q \in Q$, $\delta(q, \gamma)$ is defined as in the following:

(a) $\delta(q, \gamma) = \delta_i(q, \gamma)$ for $i = 1, 2, 3$, and $q \notin \{h_1, h_2, h_3, \hbar_1, \hbar_2, \hbar_3\}$;
(b) $\delta(h_1, \sigma) = (s_2, \sigma)$, $\delta(h_1, \tau) = (s_3, \tau)$, $\delta(\hbar_1, \sigma) = (s_2, \sigma)$, $\delta(\hbar_1, \tau) = (s_3, \tau)$;
(c) $\delta(h_2, \gamma) = \delta(h_3, \gamma) = (h, \gamma)$, $\delta(\hbar_2, \gamma) = \delta(\hbar_3, \gamma) = (\hbar, \gamma)$.

For rejection of strings, only the accepting and rejecting states of M_2 and M_3 matter and not the rejecting state of M_1. When the scanned square is neither σ nor τ, the combined machine must halt abruptly due to the reason that no transition would be defined in that case. Similarly, for acceptance of a string, the accepting states of M_2 and M_3 matter and not that of M_1.

If in one of the machines M_i, the transition function δ_i is not defined for $(\hbar_i, \sigma)$, then the transition function δ in the combined machine is left undefined for $(\hbar_i, \sigma)$. For example, the transition $\delta(\hbar_i, \sigma)$ in the combined machine $M_1 \rightarrow M_2$ may be well defined provided $\delta_1(\hbar_1, \sigma)$ has been defined in M_1. In the combined machine $M_L \rightarrow M_R$, we have not included such a transition as in M_L the transition $\delta_L(\hbar_1, \sigma)$ has not been defined.

We can have much more complex combinations of machines, examples of which you will see later. A simple modification of the above machine is where the new machine simulates M_3 whenever the currently scanned symbol is any thing other than τ. The arrow from M_1 to M_3 can, in such a case, be written simply as $\overline{\tau}$ or as $\gamma \neq \tau$. This can be depicted by any of the diagrams in Fig. 7.6. We use the former when we do not require the currently scanned symbol anymore; and we use the latter when we require to use the information that the currently scanned symbol is γ.

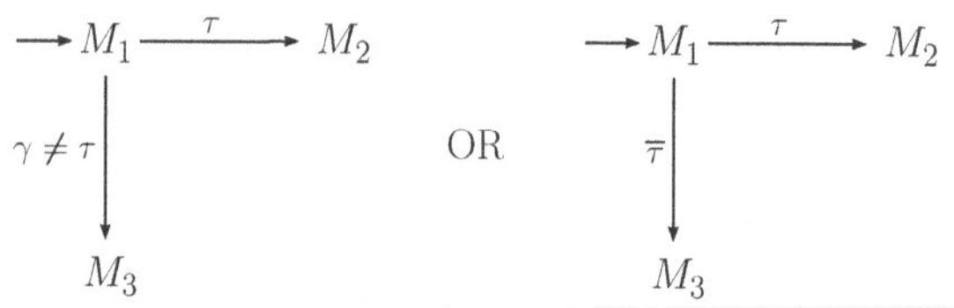

Fig. 7.6. Depicting machine diagrams.

The combined machine of Fig. 7.6 works as in the following:

> If M_1 has halted on a square containing the symbol τ, then M_2 would take over, else, if M_1 has halted on a square containing the symbol not equal to τ, then M_3 would take over.

We will refer to these diagrams as *machine diagrams*. Moreover, when machines are combined unconditionally, we may also omit the arrow. For example, the combined machine $M_1 \rightarrow M_2$ can also be written as $M_1\ M_2$ just by juxtaposing the machines one after the other. By induction, many machines can be combined.

It would be nice to have some simple machines and then from those we would like to build complex ones. Some such simple machines are as follows:

Symbol Writing Machines: Let Σ be an alphabet. Let $\sigma \in \Gamma \supseteq \Sigma \cup \{\not{b}\}$. Define the machine $M_\sigma = (\{s, h, \hbar\}, \Sigma, \Gamma, \delta, s, h, \hbar)$, where $\delta(s, \tau) = (h, \sigma)$ for each $\tau \in \Gamma$. This machine simply writes the symbol σ on the currently scanned square and then halts. The machine $M_{\not{b}}$ is called the *erasing machine* as it writes the symbol $\not{b}$ (i.e., erases the symbol) on the currently scanned square.

Head Moving Machines: Let Σ be an alphabet and let $\Gamma \supseteq \Sigma \cup \{\not{b}\}$. Define the machine $M_{\not{L}} = (\{s, h, \hbar\}, \Sigma, \Gamma, \delta, s, h, \hbar)$, where $\delta(s, \tau) = (h, \not{L})$ for each $\tau \in \Gamma$. This machine simply moves the read–write head to one square left of the currently scanned square and then halts. This is called the *left moving machine.*

Similarly, we will also have a *right moving machine*, which would move one square to the right of the currently scanned square and then halt. That is, with $\Gamma \supseteq \Sigma \cup \{\not{b}\}$, $M_{\not{R}} = (\{s, h, \hbar\}, \Sigma, \Gamma, \delta, s, h, \hbar)$, where $\delta(s, \tau) = (h, \not{R})$ for each $\tau \in \Gamma$.

Halting Machines: Let Σ be an alphabet and let $\Gamma \supseteq \Sigma \cup \{\not{b}\}$. Define the machine $M_h = (\{s, h, \hbar\}, \Sigma, \Gamma, \delta, s, h, \hbar)$ with $\delta(s, \tau) = (h, \tau)$ for each $\tau \in \Gamma$. This machine is called the **accepting machine**, which halts in the accepting state h without doing any thing else. Similarly, we have the **rejecting machine** $M_\hbar = (\{s, h, \hbar\}, \Sigma, \Gamma, \delta, s, h, \hbar)$ with $\Gamma \supseteq \Sigma \cup \{\not{b}\}$ and $\delta(s, \tau) = (\hbar, \tau)$ for each $\tau \in \Gamma$.

A Convention: These machines will be used so often that we may require to use simpler notation. We write the symbol writing machines M_σ simply as σ, and $M_{\not{b}}$ as $\not{b}$. Similarly, we write $\not{L}$ instead of $M_{\not{L}}$ for the left moving machine, and $\not{R}$ for the right moving machine $M_{\not{R}}$. We also abbreviate the halting machines M_h and $M_\hbar$ to h and $\hbar$, respectively. From the context, you should be able to know whether the symbols σ, $\not{b}$, $\not{L}$, $\not{R}$, h, $\hbar$ are used as symbols or as machines. Further, in the machine diagrams for combining the machines, we will drop using multiple arrows from one machine to the other; instead, we will use a single arrow and label it with multiple symbols separated by commas. For example, suppose there are three arrows from the machine M_1 to the machine M_2, the first labeled σ, second labeled τ, and the third labeled $\gamma \neq \not{b}$, then we will replace all the three by a single arrow with the label $\sigma, \tau, \gamma \neq \not{b}$. However, an arrow in the other direction, that is, from M_2 to M_1 cannot be mixed with the previous one. Recollect also the other abbreviations: $M_1 \to M_2$ is written as $M_1 M_2$, and an arrow labeled $\tau \neq \sigma$ as $\overline{\sigma}$, when the particular symbol τ different from σ is not used latter.

A combined machine is also called a **machine scheme**, and a **machine diagram** when depicted in a figure using labeled graphs. See the following examples.

Example 7.8. What does the following machine do?

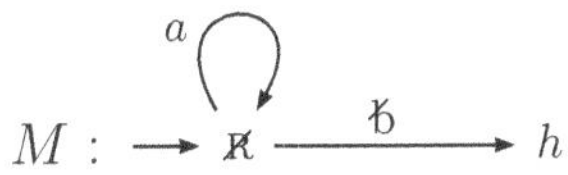

Solution. As usual, the input to a combined machine must be in the specified form $\underline{\not{b}}u$. Suppose we give the input aaa to M. This means that the initial configuration of M is $s\underline{\not{b}}aaa$. The initial machine $\not{R}$ takes over and the read–write head moves to the first input symbol. We can no more write the configurations as there is no trace of the states of the combined machine. Moreover, the machines in a combined machine are just copies of the usual head moving or symbol writing machines. Thus, their exact states are not known. However, we can write the tape appearances. For example, initially, the tape looks like $\underline{\not{b}}aaa$, and then after the $\not{R}$ machine has worked, the tape looks like $\underline{a}aa$, etc.

At this point, we make a convention in referring to the tape appearances in a combined machine as the *configurations* of it, overusing the word "configuration." We also use the same symbol $\rightsquigarrow$ between successive tape appearances for writing down a computation in a combined machine.

The current configuration of M is $\underline{a}aa$. The machine is now scanning a and the machine diagram of M says that if it is scanning an a, it must simulate $\not{R}$. (The loop around $\not{R}$ is labeled a.) Thus the next configuration of M is $a\underline{a}a$. Twice the process is repeated and the configuration $aaa\underline{\not{b}}$ is reached. The currently scanned symbol is $\not{b}$ and the machine diagram of M says that it must simulate the accepting machine h, which then forces M to halt in an accepting state. The computation of M on the input aaa is as follows:

$$s\underline{\not{b}}aaa \rightsquigarrow \underline{a}aa \rightsquigarrow a\underline{a}a \rightsquigarrow aa\underline{a} \rightsquigarrow aaa\underline{\not{b}} \rightsquigarrow haaa\underline{\not{b}}.$$

Notice the state components in the first and last of the above tape appearances. We will make it a practice to write the s in the beginning to mark the initial configuration. The string following this $s\underline{\not{b}}$ is the input string. Similarly, the state of the final computation step is also marked with h or $\hbar$ so that it will be easy to read it through for acceptance or rejection of strings.

It is thus obvious that any string a^n is accepted by M. The empty string ε is clearly accepted by M. If there is any other symbol in the input string, then the machine would halt abruptly upon reaching that symbol. That is, $\mathcal{L}(M) = \{a^n : n \in \mathbb{N}\}$. □

Exercise 7.2. Do the combined machines $\not{L}\,\not{R}$ and $\not{R}\,\not{L}$ do the same thing?

You can also construct a total TM for accepting the same language a^* modifying the machine diagram of Example 7.8. What you need is that the machine must execute the rejecting machine $\hbar$ when it meets some symbol other than a. Look at Fig. 7.7.

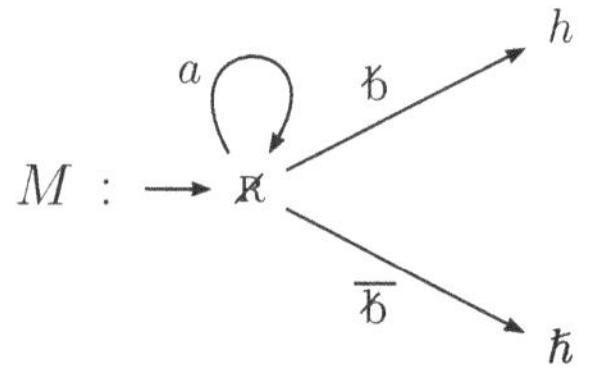

Fig. 7.7. A total TM accepting a^*.

You see that there is no need to think about the states of the combined machine. Those can be found out by following the combination rules. This is possible because we assume that all the machines used in a combined machine have distinct states. If a machine is used many times in a combined machine, then the many occurrences of the former machine are not same but identical copies.

In Example 7.8, the machine $\not{R}$ has been simulated four times and each time a different copy of $\not{R}$ has been used. For example, a copy of the machine $\not{R}$ is $M_{\not{R}} = (\{s_i, h_i, \hbar_i\}, \Sigma, \Gamma, \delta_i, s_i, h_i, \hbar_i)$, where $\Gamma \supseteq \Sigma \cup \{\not{b}\}$ and $\delta_i(s_i, \tau) = (h_i, \not{R})$ for each $\tau \in \Gamma$.

Example 7.9. What does the following machine do?

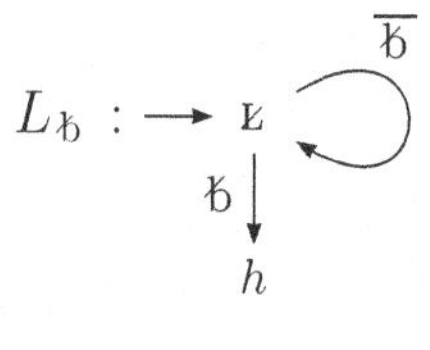

Solution. The machine $L_{\not{b}}$ first goes one square to the left. Next, if the square it is currently in (that is, one left to the original) contains a nonblank symbol, then it moves another square to the left. It repeats the steps and stops when it finds a $\not{b}$. It does not really matter what the underlying input alphabet is; the label $\overline{\not{b}}$ on the loop takes care. If at the start the machine is scanning the blank symbol $\not{b}$, then it moves one square to the left anyway and then depending upon whether the next left square is a $\not{b}$ or not, it proceeds as earlier. Upon finding the left $\not{b}$, the machine $L_{\not{b}}$ halts in the accepting halt state h. □

Look at the combined machines in Fig. 7.8.

The machine $L_{\overline{\not{b}}}$ moves its read–write head to the left until it reaches a square containing a nonblank symbol. Once it gets a nonblank symbol, it halts then and there in its accepting state. The machine $R_{\not{b}}$ goes on moving its read–write head to the right and halts in its accepting state upon reaching a square containing the blank symbol $\not{b}$. Similarly, The machine $R_{\overline{\not{b}}}$ halts at the first nonblank symbol to the right of the currently scanned square.

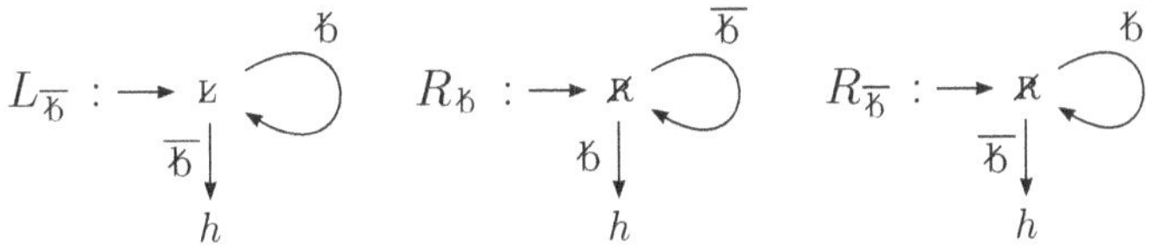

Fig. 7.8. Some combined machines.

In Fig. 7.9, the machines L_σ and $L_{\bar{\sigma}}$ halt after finding σ and a symbol other than σ to the left, respectively. The machines R_σ and $R_{\bar{\sigma}}$ find the symbols σ and one other than σ to the right, respectively. We can use these machines to build other machines; see the following example.

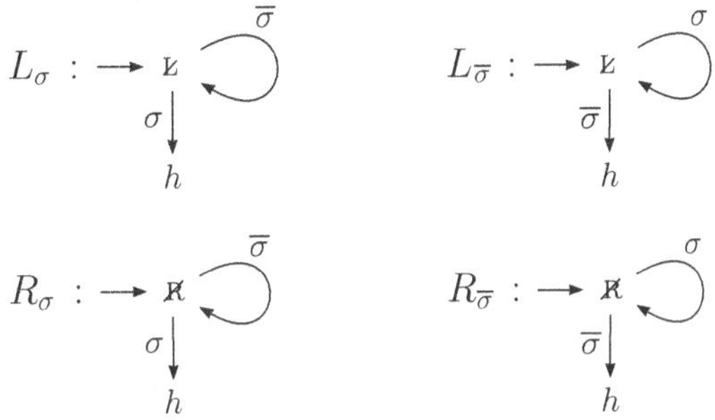

Fig. 7.9. More combined machines.

Example 7.10. What is the language accepted by the combined machine M of Fig. 7.10?

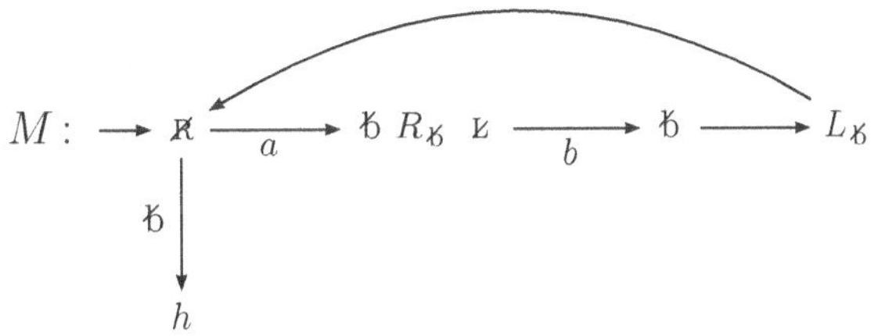

Fig. 7.10. The combined machine for Example 7.10.

Solution. First, the read–write head of M goes to the right. If that symbol happens to be an a, then it erases that a and then it moves to the right until a blank symbol is met. Next, it turns left and if that symbol is b, then this b is erased. It then comes back to the blank on the left, which must be on the square where it erased an a. Then the control is transferred to the initial machine $R\!\!\!/$. Till now, the net effect is erasing an a from the start-side and erasing a b from the end-side of the input string. It continues to operate this way until the initial machine gets a blank, when it halts accepting the string. This happens if the input string has the same number of a from the start-side as the number of b's from the end-side, and there should not be anything else in the middle. Thus, $\mathcal{L}(M) = \{a^n b^n : n \in \mathbb{N}\}$. □

Computation of M of Fig. 7.10 with the input $aabb$ can now be shown as in the following. While reading or writing such a computation sequence by a combined machine, you must keep one finger on the current machine that is operating and then proceed.

$$s\underline{b\!\!\!/}aabb \rightsquigarrow \underline{a}abb \rightsquigarrow \underline{b\!\!\!/}abb \rightsquigarrow abb\underline{b\!\!\!/} \rightsquigarrow ab\underline{b} \rightsquigarrow ab\underline{b\!\!\!/} \rightsquigarrow \underline{b\!\!\!/}ab \rightsquigarrow \underline{a}b \rightsquigarrow \underline{b\!\!\!/}b$$
$$\rightsquigarrow b\underline{b\!\!\!/} \rightsquigarrow \underline{b} \rightsquigarrow \underline{b\!\!\!/} \rightsquigarrow \underline{b\!\!\!/} \rightsquigarrow h\underline{b\!\!\!/}.$$

Exercise 7.3. Show computations of the machine M of Example 7.10 (Fig. 7.10) on the inputs $aabab$ and $aabba$.

Exercise 7.4. Follow the idea used in Example 7.10 to construct a Turing machine with states and transitions for accepting $\{a^n b^n : n \in \mathbb{N}\}$.

You can now easily construct a machine diagram for accepting the language $\{a^n b^n : n \in \mathbb{N}\}$ and rejecting any string in $\{a, b\}^* - \{a^n b^n : n \in \mathbb{N}\}$. One such modification of the diagram of Fig. 7.10 is shown in Fig. 7.11; it has to be a total Turing machine.

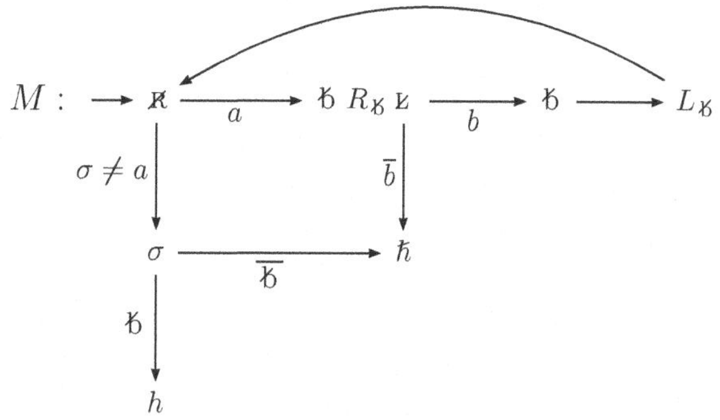

Fig. 7.11. A total TM accepting $\{a^n b^n : n \in \mathbb{N}\}$.

Example 7.11. Construct a machine to accept $\{a^n b^n c^n : n \in \mathbb{N}\}$.

Solution. How do you proceed? You are now thinking algorithmically. First give a try before reading further. Well, for the language at hand, our approach should be analogous to that in Example 7.10. You now know how to match a's with b's. Match them by replacing b's with d's instead of erasing the b's altogether. Then you are left with d's and c's. This can be tackled as in Example 7.10.

However, we will use another idea to solve this problem. Suppose you have erased one a from the left-end, one c from the right-end, and one b from somewhere in the middle. Then you are left with a string of the form

$$a^{n-1} b^k b\!\!\!/ b^{n-1-k} c^{n-1}.$$

You are not able to use the same procedure as you might have followed earlier to erase a, b, c, one each, because this is not in the form $a^m b^m c^m$. If somehow, you can transform this string to the initial form, then things will be easier. Say, we will transform $a^{n-1}b^k \underline{b}b^{n-1-k}c^{n-1}$ to $\underline{b}a^{n-1}b^k b^{n-1-k}c^{n-1}$. Can you have a machine to do it? What you require is that if the machine is given an input in the form $u\underline{b}v$, then upon halting, it would leave the string $\underline{b}uv$ on the tape, where u contains no blanks and v may contain blanks. That is, the machine is bringing the currently scanned b to the square just left of the string u which contains no b's.

The machine S_R of Fig. 7.12 is such a machine, called the **right-shifting machine**, as it just shifts the left string containing no blanks one square to the right. S_R copies the string u symbol by symbol.

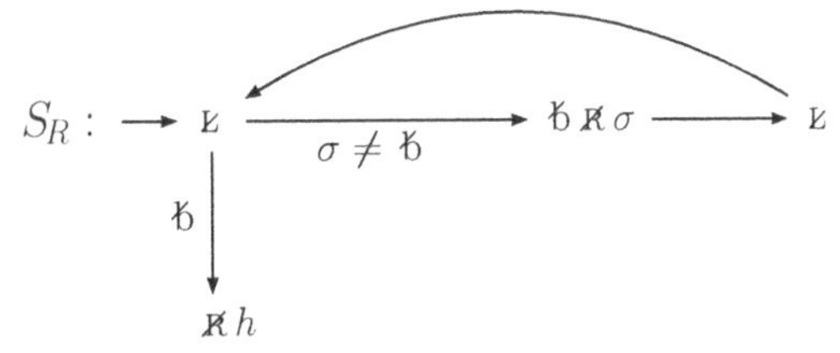

Fig. 7.12. Right shifting machine.

S_R starts from the b following the string, comes left one square, remembers that symbol (which is not equal to b), erases the symbol, goes one square to the right, writes the symbol there, turns left two squares. By now it is at the square just to the left of the original blank square. It repeats the loop so that the string is copied symbol by symbol from the right side. The read–write head is also moving one square to the left each time the loop is executed. Finally, it reaches the square where the first symbol of the original string was written, now containing b. It goes left and checks whether the input string is over or not as usual. If the left one is a b, then the string is over. To come to the square containing the moved b, a right move is taken. S_R thus transforms $wbu\underline{b}v$ to $wb\underline{b}uv$, where u does not contain b.

Exercise 7.5. Use the notation of $\rightsquigarrow$ for the combined machines to write a computation of S_R on the input $abbbaabab\underline{b}babaa$. (Strictly speaking, it is a tape appearance rather than an input; we agree to overuse the term "input.")

We can now use the right-shifting machine S_R for constructing a machine that accepts $\{a^n b^n c^n : n \in \mathbb{N}\}$. If the input is ε, then the string is to be accepted. This is done by turning right, and if found a b, then we simply halt. Otherwise, we will erase the first a, the last c, the last b (the last b is the first b coming from the right side), and then right-shift the remaining a's and b's. We then come back to the b on the left, the first square to the left of anything left over from the input. The process is repeated till we end up with the empty string on the tape; this happens when the input was really $a^n b^n c^n$. If the input string is not of this form, then our machine halts abruptly somewhere in the middle. At this point you must yourself complete the construction. See Fig. 7.13.

See that this machine really accepts the language $\{a^n b^n c^n : n \in \mathbb{N}\}$. Try designing another machine to accept this language that does not use S_R; see Problem 7.31. □

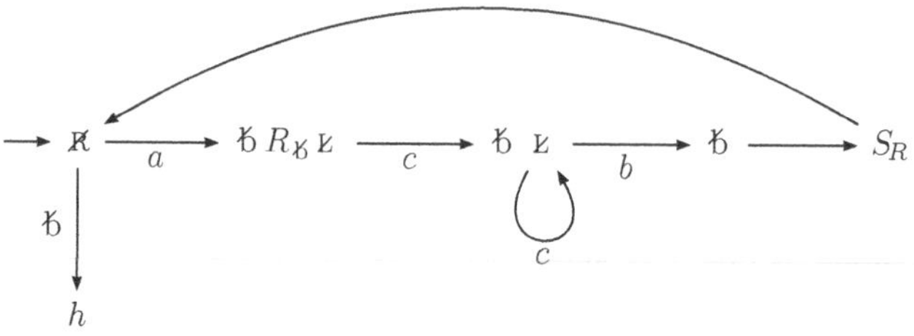

Fig. 7.13. TM accepting $\{a^n b^n c^n : n \in \mathbb{N}\}$.

Notice that the language accepted by the machine in Fig. 7.13 is not a context-free language.

Example 7.12. Construct a machine that shifts a string containing no blanks to one square left, assuming that the square just left to the string contains a blank. That is, it must transform $v\underline{ƀ}uƀw$ to $vu\underline{ƀ}ƀw$, where u does not contain a ƀ.

Solution. The idea is similar to S_R. We shift the string symbol by symbol from the left. See the machine diagram for S_L in Fig. 7.14.

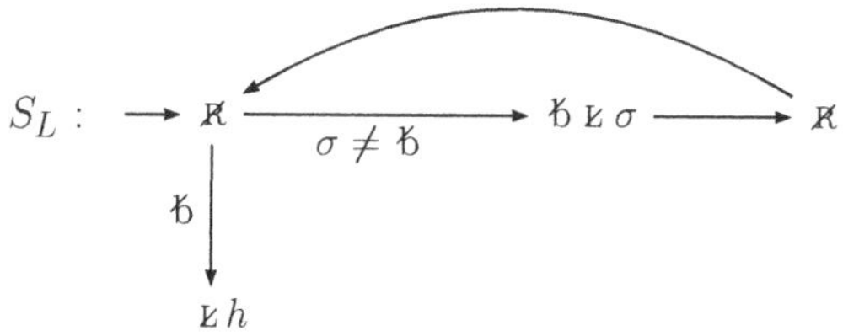

Fig. 7.14. Left-shifting machine.

Let us see how S_L operates. It starts with the blank just to the left of the string and then ends at the blank just to the right of the left-shifted string. Here is its operation on the tape appearing as $\underline{ƀ}ab$:

$$\underline{ƀ}ab \rightsquigarrow \underline{a}b \rightsquigarrow \underline{ƀ}b \rightsquigarrow \underline{ƀ}ƀb \rightsquigarrow \underline{a}ƀb \rightsquigarrow a\underline{ƀ}b \rightsquigarrow aƀ\underline{b} \rightsquigarrow aƀ\underline{ƀ} \rightsquigarrow a\underline{ƀ} \rightsquigarrow a\underline{b}$$
$$\rightsquigarrow ab\underline{ƀ} \rightsquigarrow abƀ\underline{ƀ} \rightsquigarrow ab\underline{ƀ}.$$ □

Example 7.13. Construct a machine that copies a string to its right, leaving a blank in between the string and its copy. Further, at the end, the machine must be scanning this middle blank. That is, the machine should transform $v\underline{ƀ}u$ to $vƀu\underline{ƀ}u$, where u does not contain a blank and v may be any string. As usual, we assume that there are infinite ƀ's to the right of u.

Solution. Verify that the machine C in Fig. 7.15 does the work. □

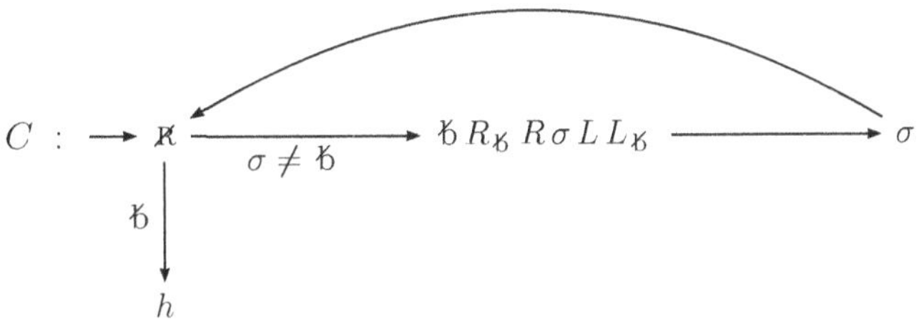

Fig. 7.15. Machine for Example 7.13.

Problems for Section 7.5

7.27. Design a Turing machine that replaces the symbol following the leftmost a by a ƀ. Moreover, if the input contains no a, it replaces the rightmost nonblank symbol by a b.

7.28. Design a Turing machine (combined machines allowed) with $\Sigma = \{0, 1, 2\}$ that, when started on any square containing a 0 or a 1, will halt iff its tape has a 2 somewhere on it. Note that there can be ƀ's in between strings over Σ.

7.29. Construct another copying machine that copies from the right side unlike the one in Example 7.13 that copies from the left; the copy must be on the right as earlier.

7.30. Design a Turing machine that outputs ww when given input w. Check your design on the input $aabba$.

7.31. Design a machine diagram for accepting the language $\{a^n b^n c^n : n \in \mathbb{N}\}$ that matches an a from the beginning with a c from the end, deletes this a and writes that c as a d. Next, it matches a b from the left with a d from the right, recursively.

7.32. Design machine schemes for accepting the following languages:
(a) a^*ba^*b.
(b) $\{ww^R : w \in \{a, b\}^*\}$.
(c) $\{ww : w \in \{a, b\}^*\}$.
(d) $\{w \in \{a, b\}^* : w = w^R\}$.
(e) $\{a, b\}^* - \{ww^R : w \in \{a, b\}^*\}$.
(f) $\{a^m b^n : n = m^2,\ n \in \mathbb{N}\}$.
(g) $\{w \in \{a, b\}^* : w = a^p,$ where p is a prime number$\}$.
(h) $\{w \in \{a, b\}^* : \#_a(w) > \ell(w)/2\}$.

7.33. Can you modify all your combined machines for the languages in the preceding problem so that they accept the given language and reject the complement of the corresponding language?

7.34. Describe combined machines for accepting the languages:
(a) $\{a^n : n = 2^m \text{ for some } m \in \mathbb{N}\}$.
(b) $\{wcw : w \in \{a, b\}^*\}$.
(c) $\{w \in \{a, b, c\}^* : \#_a(w) = \#_b(w) = \#_c(w)\}$.
(d) $\{a^m b^n c^k : m \times n = k,\ m.n.k \geq 1\}$.
(e) $\{x2y2z : x, y, z \in \{0, 1\}^+, \text{ and as binary numbers, } x \times y = z\}$.

7.35. Describe combined machines that accept the given language and reject the complement of the corresponding language in the previous problem.

7.36. Design a combined machine that accepts the language $\{u2v2w : u, v, w \in \{0, 1\}^* \text{ and are distinct}\}$.

7.6 Multitape TMs

You have sufficient experience in working with Turing machines as language acceptors. We can now try to answer the question whether these machines accept all computably enumerable languages or something else. As you guess, the answer is in the affirmative. But then we have to do some more work to see that it is indeed so. We have to have a way to simulate the working of each grammar by a Turing machine and vice versa. We will do it, albeit, indirectly. Our plan is to first try to extend the notion of Turing machines a little bit and then bring in the question of estimating its power.

In this scheme, first we think of equipping a Turing machine with more number of input tapes. Imagine a machine having, say, three tapes and three read/write heads. We mark which head works on which tape; the first head works on the first tape, the second head on the second tape, and the third head on the third tape. As earlier, we have the same control unit with possibly many states, one of which is marked as the initial state. We also have the two special states h and $\hbar$ possibly different for different machines.

We give input to this three-tape machine only on the first tape, and the other tapes are initially blank. The machine may use them whenever required. As earlier, the machine has a transition function now specifying its action on all the three tapes. That is, at certain instant of time, if the machine is in a certain state, the three heads scanning a symbol each on their respective heads, the transition function must give the information as to what to do next. The next action of the machine will be possibly a change of state, and also doing one of the three actions of writing symbols on each of the scanned squares, or moving to the left or moving to the right by the respective heads on their respective tapes.

Formally, a **three-tape machine** is a seventuple $M = (Q, \Sigma, \Gamma, \delta, s, h, \hbar)$, where $Q, \Sigma, \Gamma, s, h, \hbar$ are as they were in a Turing machine, and δ is a partial function from $(Q - \{h, \hbar\}) \times \Gamma \times \Gamma \times \Gamma$ to $Q \times (\Gamma \cup \{\not{L}, \not{R}\}) \times (\Gamma \cup \{\not{L}, \not{R}\}) \times (\Gamma \cup \{\not{L}, \not{R}\})$.

The transition function δ tells what action is taken on each of the three tapes while changing a state. We define an instantaneous description of the machine or a *configuration* as a quadruple $(q, u_1\underline{a_1}v_1, u_2\underline{a_2}v_2, u_3\underline{a_3}v_3)$. This configuration is to be understood as

The current state of the machine is q, on its ith tape, the currently scanned symbol is a_i, to the left of it is the string u_i, and to the right of it is the string v_i, for $i = 1, 2, 3$.

The yield relation of the configurations is then defined to describe any computation of the machine, as earlier, but taking care of the three tapes. We say that such a three-tape machine M accepts a string $u \in \Sigma^*$ iff the initial configuration $(s, \underline{b}u, \underline{b}, \underline{b})$ yields a configuration whose state component is h. And then the language accepted by M is defined as the set of all strings from Σ^* accepted by M. Similarly, a string is rejected by the three-tape machine is one that drives the machine to the rejecting state $\hbar$ and the language rejected by the machine is the set of all strings rejected by it.

Obviously, any work that is done a by a Turing machine (standard, or one-tape) can also be done a by a three-tape machine. For example, the three-tape machine that operates on its first tape just as the original standard machine works, and which does not operate with the second and third tapes at all, would achieve this. That is, any language accepted by a standard Turing machine is also accepted by a three-tape machine. We intend to show the converse also. This is a bit harder. The idea is to simulate the three tapes on a single tape.

Suppose M is a three-tape machine. It works with the three tapes simultaneously, but step by step. This means that in a single step it changes its three head positions, and there is only one change of states. Imagine a standard Turing machine M', which is to simulate faithfully one step of a computation of M. How will it work? We may give M' the same states as that of M (a renaming of states, of course).

What about the working out of the three heads? At the first instant, we may try to have a copy of the first tape of M on the tape of M', then put a copy of the second tape keeping some blanks in-between, and then some blanks, followed by a copy of the third tape. But this will not work, for we do not know how many squares will be used from the first tape to its right. If we need more than the number of separating blanks, then contents of the second tape on M' will get unduly rewritten. So this method of keeping copies will fail. There is, of course, a way out. The way is while simulating M on M' we can always make a space and then follow M. But we have an easier alternative.

Think of the first squares on all the three tapes of M as a triple, and write them in three consecutive squares on the tape of M'. For example, the blanks preceding the strings on the three tapes are taken as a block of three blanks on the single tape of M'. This triple represents the 0th squares of M. The first b is for the b of the first tape of M, the second b for the b on the second tape of M, and the third b for the b on the third tape of M. In such a case, one step of computation of M will be carried out by three steps of M' doing the works of the three heads of M in succession.

For example, a right movement of the first tape of M would be accomplished by three right movements of M' starting from the corresponding square. Think of a situation when the first head of M is on the second square (on the first tape) and the second head of it is on the first square (on the second tape).

While simulating M, the head of M' must be on one of the squares corresponding to one of these heads. Then how does it keep track of different positions of different heads? We will use a trick here. Instead of using a triple for the three symbols on the corresponding squares on the tape, we will use a six-tuple.

Suppose that we have a, b, c on the second squares of each of the tapes first to third, respectively. And the first head is scanning a, second b, and third c. Then, the corresponding six-tuple will be $(a, 1, b, 1, c, 1)$. Suppose, for example, the second head is not scanning the symbol b right now, but first is scanning a and third is scanning c. Then we would represent it as the six-tuple $(a, 1, b, 0, c, 1)$. Is the scheme clear now? The next right square to the symbol will contain 1 if the corresponding head is scanning it, else, it contains 0. Of course, the symbols 0 and 1 are new to Γ.

We reword the above scheme of *simulation of a three-tape machine by a standard machine*. At some instant of time, suppose the nth squares of the three tapes of M contain the symbols $\sigma_1, \sigma_2, \sigma_3$, respectively. Then form the nth six-tuple as $(\sigma_1, b_1, \sigma_2, b_2, \sigma_3, b_3)$, where $b_1, b_2, b_3 \in \{0, 1\}$. If $b_i = 1$, then it means that the ith head is scanning σ_i and if $b_i = 0$, then the ith head is not scanning the symbol σ_i right now. The first squares on the tapes of M are written on the first six squares on the tape of M', the second squares of tapes of M are written on the next six squares of the tape of M' and so on.

For example, initially, the first six squares of the tape of M' will look like $\flat 1 \flat 1 \flat 1$. We put an extra symbol, say, 2 to distinguish between the six-tuples; this is not needed, but will be handy. The next six squares will look like $a0\flat 0\flat 0$ if the first input symbol is a, which is written on the first tape of M, while all other tapes of M are blank. The point is, this is only a conceptual device so far. The following is a description of M', which accomplishes this representation and then simulation of M. (We assume that 0, 1, 2 have not been used in M.) Given an input u to the three-tape machine M, the machine M' operates as follows:

1. The input u is given to M' on its tape. M' goes to the blank on the right ($R_{\flat}$) and writes $\flat 1 \flat 1 \flat 1$, using the six squares in succession, goes right and writes 2, comes back to the beginning of the tape ($L_{\flat}\ L_{\flat}\ L_{\flat}$), goes right, remembers the symbol, say, σ (the first input symbol), replaces it with τ, a new symbol, goes to the blank (after 1), writes there in succession $\sigma, 0, \flat, 0, \flat, 0, 2$, comes back to τ, erases it, goes right, scanning a square containing a $\flat$. This is done iteratively until the input to M is over. The tape after the completion of this step is as per the above description of input to M' representing the input to M.
2. Now M' simulates M step by step by doing one of the following for each $i \in \{1, 2, 3\}$:
   (a) If the ith head of M goes to the left, then M' finds a 1 as the $(2i)$th symbol after a 2, and then changes that 1 to 0, comes left seven squares, changes that 0 to 1.
   (b) If the ith head of M goes to the right, then M' finds a 1 as the $(2i)$th symbol after a 2 (or the $\flat$ where M starts), and then changes that 1 to 0, goes right seven squares, changes that symbol α to 1. Here, the symbol α need not be always 0, as the final blank positions of the tapes in M can be accessed in a right movement. In that case, when $\alpha \neq 0$, M is using a $\flat$ on the ith tape while other tapes corresponding to that square are also containing $\flat$'s. Here, M' replaces α as 1 and replaces the $\flat$'s on the other 0–1 positions as 0. That

is, if $i = 1$, then after replacing α as 1, it moves right twice, writes 0, moves right twice and writes 0. Similarly, it takes appropriate action for $i = 2, 3$.

(c) If the ith head of M replaces a symbol σ by β on the ith tape, then M' finds the 1 on the $(2i)$th square to the right of a 1 and (re)writes there β.

3. When M halts on its input, M' erases all symbols from its tape, except the square containing 1 and the second square of every seven squares starting from the first square onwards. Then it shifts all the symbols taking away all ƀ's to the right. Finally, it erases the 1, takes that ƀ to the right and comes back to the symbol that followed the 1, and halts. This will leave the head of M' on the same symbol that is scanned by M upon halting.

Now it is clear that M' simulates faithfully M on a single tape. It is also obvious that any string that is accepted (or rejected) by M is also accepted (or rejected) by M' and nothing else. Generalization of a three-tape machine to a k-tape machine is similar and so also its simulation by a standard Turing machine. Thus, the following theorem is proved:

Theorem 7.1. *For every k-tape machine M, there exists a standard Turing machine* M' *such that* $\mathcal{L}(M) = \mathcal{L}(M')$ *and* $\mathcal{L}_R(M) = \mathcal{L}_R(M')$.

In what follows, we will use a kind of restriction on Turing machines. The restricted machines will be having some extra power. The extra power I am talking of is that a Turing machine *can know* the square on the tape to the left of which there are only ƀ's, that is, the left end of the string on the tape, after some computations. Note that this is a restriction, as in an arbitrary Turing machine, we do not have any way of determining the position on the tape from where there exists nothing interesting, but only the infinite number of ƀ's. This knowledge can be exploited to erase all that is left on the tape if required.

For this purpose, we will equip the Turing machines with an extra symbol, say, $*$, which will always be there on the tape marking the left end of the interesting portion. To the left of $*$, there will only be infinite number of ƀ's. Initially, such a machine will be given input in the form of $*\underline{b\!\!\!/}u$, the machine scanning the ƀ as the underlining shows. Whenever the machine tries to go to the left of $*$, it will first move that $*$ one square to the left, and then continue its operation. Thus the net effect will be that the $*$ behaves like a ƀ but remains on the left edge of the used portion of the tape. Specifically, when the machine reaches the symbol $*$, say, in state p, it will erase this $*$, go one square to the left, write the symbol $*$, and then comes one square to the right, coming back to the state p. This can be achieved by adding two extra states, say, e and f to the old machine and adding the transitions

$$\delta(p, *) = (e, b\!\!\!/), \delta(e, b\!\!\!/) = (f, L\!\!\!/), \delta(f, b\!\!\!/) = (e, *), \delta(e, *) = (p, R\!\!\!/),$$
$$\text{for each state } p,$$

to the transitions of the old machine. Thus, the new machine will always get a ƀ to the left of the interesting portion on the tape as intended, simultaneously keeping the symbol $*$ to the left of which there is only the infinite number of ƀ's. A similar modification can be done on the right side also by adding another extra symbol, say,

$\star$ on the right and adding two extra states, say, e' and f' to the machine. We then add the following transitions:

$$\delta(p, \star) = (e', \not{b}),\ \delta(e', \not{b}) = (f', \not{R}),\ \delta(f', \not{b}) = (e', \star),\ \delta(e', \star) = (p, \not{L}),$$
$$\text{for each state } p \text{ which is neither } e \text{ nor } f.$$

The modified machine with two extra symbols $*$, $\star$ and four new states e, e', f, f' now always have $*$ on the left and $\star$ on the right, keeping an extra $\not{b}$ in between if need arises for going to the left or right, one square. This way, the modified machine always works within the variable portion of the tape marked by $*$ on the left and $\star$ on the right. What is its impact on acceptance and rejection of a language?

Let M' be the modified machine that works as M but always working in between $*$ and $\star$. We want another modification of M'. Add four more states $m, m', h', \hbar'$ to M', and call the new machine M''. Let h' be the accepting state of M'' and $\hbar'$ be its rejecting state. The accepting state h of M' is no more an accepting state of M''. Similarly, the rejecting state of M' is not the rejecting state of M''. The transitions of M'' are those of M' plus

$$\delta(h, \tau) = (m, \not{R}),\ \delta(h, *) = (m, \not{R}),\ \ \delta(\hbar, \tau) = (m', \not{R}),\ \delta(\hbar, *) = (m', \not{R}),$$

$$\delta(m, \star) = (m, \not{b}),\ \delta(m, \tau) = (m, \not{b}),\ \delta(m, \not{b}) = (m, \not{L}),\ \delta(m, *) = (h', \not{b}),$$
$$\delta(m', \star) = (m', \not{b}),\ \delta(m', \tau) = (m', \not{b}),\ \delta(m', \not{b}) = (m', \not{L}),\ \delta(m', *) = (\hbar', \not{b}),$$
$$\text{for every } \tau \neq *, \tau \neq \star.$$

The new machine M'' does everything that M' does. After this, instead of halting, M'' goes to the right until it finds $\star$. It erases $\star$, comes left erasing whatever it gets on the way. It also erases $*$, and then halts in h', provided M' would have halted in h, otherwise, it halts in $\hbar'$ had M' halted in $\hbar$. It is clear that M'' accepts the same language as M', and it rejects the same language as M'. In turn, M'' accepts the same language as M and rejects the same language as M. Thus we have shown that

Theorem 7.2. *Corresponding to each standard Turing machine, there is at least one standard Turing machine that accepts or rejects the same language leaving the tape empty.*

Problems for Section 7.6

7.37. Describe how a two-tape TM might accept $\{a^n b^n : n \in \mathbb{N}\}$ without going back and forth.

7.38. Consider a Turing machine that cannot write a $\not{b}$, that is, if $\delta(p, a) = (q, b)$, then $b \neq \not{b}$. Show how such a machine can simulate a standard TM.

7.39. Give a formal definition of a TM with a single tape, many control units each with a read–write head. Show how such a machine can be simulated by a multi-tape TM.

7.40. Suppose we restrict a Turing machine by requiring that it cannot write the same symbol on the same square that it reads, that is, if $\delta(p, a) = (q, b)$, then $a \neq b$. Does it restrict the power of TMs?

7.41. Consider a TM with a different decision process in which transitions are made if the currently scanned symbol is not an element of a given set. For instance, $\delta(p, \{a, b\}) = (q, \not R)$ would mean that if the currently scanned symbol is neither an a nor a b, and the machine is in state p, then change state to q and go a square right. Show that such machines can be simulated by standard TMs.

7.42. Consider a variation of TMs, where transitions depend not only on the symbol currently scanned, but also on the symbols on the squares immediately to the left and/or right of the currently scanned square. Show that such a TM can be simulated by a standard TM.

7.43. A multihead TM has a single tape but more than one read–write head working on the same tape. Give a formal definition of such a machine and describe how it can be simulated by a standard TM.

7.7 Nondeterministic TMs and Grammars

We will extend Turing machines to accommodate nondeterminism. The extension is similar to one of the NFAs from DFAs. To begin with, we will have only one tape and only one read/write head. In place of the transition function, we will have a transition relation so that when the machine is in a state scanning a particular symbol, it might have one or many possible ways of changing to another state and performing one of the left–right movements or writing a symbol. Finally, it would accept a string if *there exists at least one computation* that halts in an accepting state starting from the string as an input.

Formally, a **nondeterministic Turing machine** or an **NTM** is a seventuple $M = (Q, \Sigma, \Gamma, \Delta, s, h, \hbar)$, where Q, Σ, Γ, s, F are as they were for a (standard) Turing machine, and Δ is a finite relation from $(Q - \{h, \hbar\}) \times \Gamma$ to $Q \times (\Gamma \cup \{\not L, \not R\})$, called the *transition relation.*

The configurations and yield relation between them are defined as usual. However, the transitions are in the form of ordered pairs of states and symbols; earlier, we had $\delta(q, a) = (p, b)$, but now we have $((q, a), (p, b)) \in \Delta$ instead. We will abbreviate the ordered pair $((q, a), (p, b))$ to (q, a, p, b) as usual. Because of nondeterminism, one configuration may now yield more than one configurations. At this point you must define formally the relation of yield (symbolized as $\leadsto$).

A **halted configuration** is one whose state component is either h or $\hbar$. The halted configurations with state component h are called **accepting configurations** and those with state component as $\hbar$ are called **rejecting configurations** as earlier. A string u is **accepted** by M iff *there is at least one computation* that starts from the initial configuration $s\underline{\not b}u$ and ends in an accepting configuration. Similarly, a string u is

rejected by M iff *all computations* starting from the initial configuration $s\,\underline{\text{ƀ}}u$ end in a rejecting configuration. The language accepted by M and those rejected by M are defined as

$$\mathcal{L}(M) = \{u \in \Sigma^* : (s, \underline{\text{ƀ}}u) \text{ yields an accepting configuration in some computation of } M\}.$$

$$\mathcal{L}_R(M) = \{u \in \Sigma^* : (s, \underline{\text{ƀ}}u) \text{ yields a rejecting configuration in every computation of } M\}.$$

Notice that though the definition of $\mathcal{L}(M)$ is almost the same as that for a Turing machine, the yield here tells something more. It says that the language accepted by M is simply the set of all strings accepted by it; and a string is accepted by M whenever there is a way to drive the machine from the initial state to the accepting state by reading the input.

If there is a computation of M that does not halt on input u or one that rejects the string u, then that does not mean that u is never accepted by M, for, there might be another computation that accepts u! Thus, in a way, a nondeterministic machine does not know that it accepts a string, but we decide from its behavior whether it accepts the string or not. Strictly speaking, the same is the case for a standard Turing machine; do you see why?

In a nondeterministic machine, usually there are more than one way a computation can proceed from a given input. You can, conceptually, draw a tree of all possible computations starting from an input, whereas in deterministic machines this tree is necessarily a path. In such a tree you take the nodes as the configurations of the machines. The children of a node keep track of all possible one-step computations. The leaves in such a computation tree are then the configurations from where no more computation is possible.

The situation corresponds to either an abrupt halt, or an accepting configuration with state component h, or a rejecting configuration with state component as $\hbar$. On the other hand, an infinite path in the computation tree says that the machine goes on computing for ever, if at all it follows that path.

A string is accepted if one of the leaves in the computation tree is an accepting configuration and it is rejected if all the leaves (so there is no infinite path) are rejecting configurations. (See, how biased we are towards acceptance.) It is thus obvious that the language rejected by a nondeterministic machine need not be the complement of the language accepted by it. A **total** nondeterministic Turing machine is one where this condition is fulfilled, that is, there is no infinite computation and each string over its alphabet is either accepted or rejected.

Example 7.14. What is the language accepted by the nondeterministic machine $M = (\{s, p, q, h, \hbar\}, \{a, b\}, \{a, b, \text{ƀ}\}, \Delta, s, h, \hbar)$, where

$$\Delta = \{(s, \text{ƀ}, p, \not{R}), (p, \text{ƀ}, q, \not{R}), (p, b, h, \text{ƀ}), (p, a, p, \not{R}), (p, \text{ƀ}, h, \text{ƀ})\}?$$

Solution. We will follow the same shorthand for writing the configurations, that is, the configuration (s, u, σ, v) will be written as $su\underline{\sigma}v$. Let us see what happens for the empty input. Computation might proceed as in the following:

$$s\underline{\text{ƀ}} \rightsquigarrow p\underline{\text{ƀ}} \rightsquigarrow q\underline{\text{ƀ}}.$$

This does not accept the empty string. However, we must see all possible computations to decide whether M accepts ε or not. Here is another computation:

$$s\underline{\not{b}} \rightsquigarrow p\underline{\not{b}} \rightsquigarrow h\underline{\not{b}}.$$

This is an accepting computation. We need only one such computation to conclude that M accepts ε. Therefore, $\varepsilon \in \mathcal{L}(M)$.

What about the string b? Here follows a computation:

$$s\underline{\not{b}}b \rightsquigarrow p\underline{b} \rightsquigarrow h\underline{\not{b}}.$$

This is also an accepting computation. So $b \in \mathcal{L}(M)$.

Similarly, a is also accepted. But not only one a, a sequence of a's is accepted. If one b follows such a sequence of a's, then also from state p the machine can go to h and accept the input. Thus it accepts any string in the language $a^* \cup a^*b$. If there is another symbol after the b in $a^n b$, say, an a, then this a is not at all read. As M is already in the accepting halt state h, the string $a^n ba$ is also accepted. That is, M accepts the language $a^* \cup a^*b(a \cup b)^*$. This means that each and every string in $(a \cup b)^*$ is accepted by M, that is, $\mathcal{L}(M) = (a \cup b)^*$. □

Exercise 7.6. Construct an NTM (not deterministic) for accepting $\{\varepsilon\}$. Also construct an NTM for accepting $a^* \cup a^*b$.

As the partial function of any Turing machine is also a relation, every Turing machine is, by default, a nondeterministic Turing machine. The combinations of machines (when we use machine diagrams) can also be extended for nondeterministic machines. To start with, we may take the basic machines as earlier, which are deterministic. (Henceforth, when both types of machines occur in a context, we will refer to the standard Turing machines as deterministic machines (**DTM**). However, deterministic machines also include k-tape Turing machines.)

Recollect that the basic machines were the head moving machines $\not{L}$ and $\not{R}$, the symbol writing machines σ, for every $\sigma \in \Gamma$, the tape alphabet, and the two halting machines h and $\hbar$. In earlier diagrams, when we have an arrow from M_1 to M_2 labeled τ, we would never have an arrow from M_1 to another machine M_3 labeled τ. Now we allow that to create nondeterministic machine combinations. Thus in a nondeterministic machine combination (its diagram), we can have multiple arrows emanating from the same machine with same labels, going to different machines.

For example, the machine diagram of Fig. 7.16 represents a nondeterministic combination of machines. The nondeterministic Turing machine M of Fig. 7.16 first

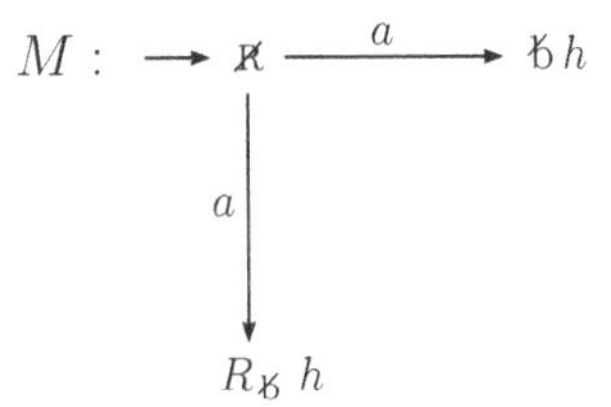

Fig. 7.16. A nondeterministic combined machine.

goes to the right. If that symbol happens to be an a, then it either erases it and halts, or it moves to the right until it finds a ƀ, when it halts.

Example 7.15. Construct a nondeterministic Turing machine to accept the language $((a \cup b)^* b(a \cup b)^*) \cup \{\varepsilon\}$.

Solution. You can construct one for yourself, even a deterministic one. But try constructing one nondeterministic machine diagram that is not a deterministic diagram before reading further, and then come back to compare your solution with the one given in Fig. 7.17.

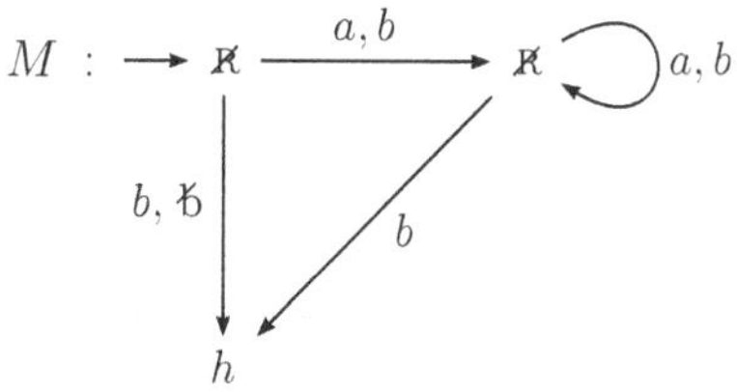

Fig. 7.17. An NTM for Example 7.15.

Here, M is not a deterministic machine, because from the first $R\!\!\!/$, the machine can either go to h or $R\!\!\!/$ upon getting one b. □

Exercise 7.7. Show why a simpler machine as drawn in the following diagram does not accept $L = ((a \cup b)^* b(a \cup b)^*) \cup \{\varepsilon\}$. Can you make it still simpler to accept L?

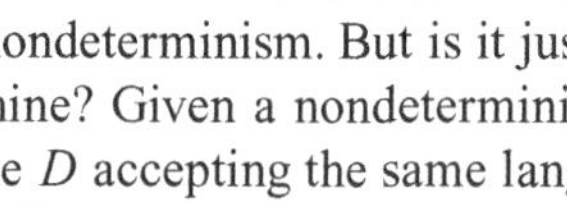

It is now clear that such machine diagrams do indeed give rise to nondeterministic machines. We will use such diagrams for constructing nondeterministic machines instead of writing the states and transition relations in detail. Further we will refer to such machine diagrams as **nondeterministic machine diagrams**.

It is convenient to have nondeterminism. But is it just convenient, or it increases the power of a Turing machine? Given a nondeterministic machine M, does there exist a deterministic machine D accepting the same language as M? Again, we will see that the answer is in the affirmative. The idea is, D can be constructed by looking through all possible computations of M on any given input, as at any one particular step, there are only a finite number of choices for M.

Suppose $M = (Q, \Sigma, \Gamma, \Delta, s, h, \hbar)$ is a nondeterministic Turing machine. Let C and C' be two configurations such that $C \overset{1}{\rightsquigarrow} C'$ in M. For a given C, there can be many C's. However, if C has the state component q and the symbol being scanned is σ, then the number of such C's is exactly the same as the number of pairs of the form $((q, \sigma), (p, \tau))$ in Δ. Obviously, this number is finite and is bounded by $|Q| \cdot (|\Gamma|+2)$, where $|A|$ denotes the number of elements in the set A. The "plus 2" comes from the extra symbols $\not{L}$ and $\not{R}$. Now, if r is the maximum number of transitions applicable at any C, then $r \leq |Q| \cdot (|\Gamma| + 2)$. This r can, of course, be found out by looking at Δ. With such an r, the machine M starts from the initial configuration and then follows up one of the possible r (or less) choices at every step.

If C is the current configuration of M, then it may follow one of the r or less paths of the computations in the next step:

$$C \rightsquigarrow C_1, \text{ or } C \rightsquigarrow C_2, \text{ or } \ldots \text{ or } C \rightsquigarrow C_r.$$

Similarly, from any C_i, it may follow one of the paths:

$$C_i \rightsquigarrow C_{i1}, \text{ or } C_i \rightsquigarrow C_{i2}, \text{ or } \ldots \text{ or } C_i \rightsquigarrow C_{ir}.$$

This r here may be different from the previous r. Taking $r = |Q| \cdot (|\Gamma| + 2)$, we have at the maximum r choices at every step of any computation. That is, at the most there are r^2 possible two-steps computations starting from any configuration. In general, there are r^m possible m-step computations at the most starting from any configuration.

The standard Turing machine D that would simulate M checks all possible one-step computations first, and then accepts a string whenever there was a one-step computation starting from C, which yields an accepting configuration (with state component h) of M in one step. When it fails, it goes for checking all two-steps computations for acceptance, and so on. D does not accept a string that did not yield an accepting configuration of M in those finite-steps computations. If we can have such a machine D, it would definitely accept $\mathcal{L}(M)$.

We take the Turing machine D as a three-tape deterministic machine. The first tape of D is the tape of M. Initially D copies the contents of the first tape to the second tape. Throughout the operation, D never changes its first tape. D writes an integer n on its third tape and then generates a string of n integers (not necessarily distinct) from among the numbers $1, 2, \ldots, r$, separating them by a $\not{b}$ and keeps this string next to n separating it by a $\not{b}$. D works on the second and third tapes towards simulating M.

For simulating the first step of computation of M, D chooses one of the r (at the most) possible choices. It keeps its second head on the $\not{b}$ following the first integer on the second tape. If that integer is m, it means that out of the r choices, it has chosen the mth choice. This means that if the current state–symbol pair is (q, σ), and if we have the applicable transitions in the forms $((q, \sigma), (q_1, \sigma_1)), \ldots ((q, \sigma), (q_r, \sigma_r))$, then it has decided to follow the mth transition $((q, \sigma), (q_m, \sigma_m))$. So, we assume an ordering of transitions of M.

It begins its work by placing its third head on the first integer after n. On the second tape it does the work by following the chosen transition; the mth applicable one from the current configuration of M if the first integer is m. Once this computation of one step is over, it puts its second head on the next integer. It follows the same strategy again and continues until it halts or until it finishes the nth integer on the third tape (It leaves the n in the beginning of the tape any way.) If by this time D has not halted, then it erases its second and third tapes, chooses n as $n + 1$ and starts work afresh. Notice that D simply simulates M by exploring all possible n-step computations of M at each phase. By fixing $n = 1$ in the beginning, all possible computations are eventually explored.

If M accepts an input u, say, in n steps, where n is the least such integer, then D would enter a halted configuration in the nth phase, after $n - 1$ unsuccessful phases of work. That is, u is accepted by D any how. Conversely, if D enters an accepting configuration in the nth phase starting with input u, then n is the least integer such that M accepts u in an n-steps computation.

Similarly, equipping a nondeterministic machine with more than one tape makes no difference. Thus we have proved the following result:

Theorem 7.3. *A language is accepted by a Turing machine iff it is accepted by a nondeterministic Turing machine iff it is accepted by a k-tape nondeterministic Turing machine.*

Exercise 7.8. Show that in the above simulation of a nondeterministic Turing machine M by a deterministic Turing machine D, if M accepts u in n steps then D accepts u in at most $r + r^2 + \cdots + r^n$ steps.

Though nondeterministic machines are no more powerful than the standard Turing machines in accepting languages, it is often convenient to construct nondeterministic machines for a given job at hand. For example, you can construct a nondeterministic machine to accept the set $\{4, 6, 8, 9, 10, 12, 14, 15, \ldots\}$ of all composite numbers relatively easily. The idea is that given a number n, the machine would generate two numbers p, q (Here nondeterminism is used.), multiply the numbers to get $p \cdot q$ and then compares $p \cdot q$ with n.

Exercise 7.9. What about the languages rejected by deterministic and nondeterministic Turing machines? Does a statement similar to Theorem 7.3 hold?

We want to see that a more unified statement holds. That is, each computably enumerable language ought to be accepted by a Turing machine. To this end, we plan to simulate a grammatical derivation in a nondeterministic Turing machine and try to simulate the work of any Turing machine by a grammar. This will accomplish our job due to Theorems 7.1–7.3.

Theorem 7.4. *A language is accepted by a Turing machine iff it is computably enumerable.*

Proof. Let G be a grammar. We will describe a three-tape nondeterministic machine M to simulate derivations in G. The machine M generates a string u on the first tape. It then writes all the rules of the grammar G on the second tape. Next, it nondeterministically chooses a rule on the second tape and chooses a square on the third tape where it writes the start symbol S of the grammar G.

If the rule is $u \mapsto v$, then it tries to match u on the third tape starting from the next square to the one it is scanning on the third tape. If it matches, then it rewrites that u as v. Here it may need to shift some squares on the third tape for making space for v, etc. This constitutes simulating one step of a derivation in the grammar G.

After the end of each such step, M matches whether the input string u on the first tape is the same as the generated string w on the third tape. If yes, it stops in the accepting halt state h, else it enters the rejecting halt state $\hbar$. Notice that the $\hbar$ here does not mean that the string on the first tape is rejected. Because of nondeterminism, a string will be rejected if all possible computations on the input result in the state $\hbar$.

It is now clear that M halts on an input u only when it is generated by the grammar G. This proves the *if* part of the theorem that every computably enumerable language is accepted by a Turing machine due to Theorems 7.1–7.3.

Conversely, suppose $D = (Q, \Sigma, \Gamma, \delta, s, h, \hbar)$ is a Turing machine that accepts a language L. We want to construct a grammar G that generates L. We take $G = (N, \Gamma, R, S)$, where $N = Q \cup \Gamma \cup \{X, S\}$ with two new symbols as X and S. The set of productions, R, contains the rules as given in the following:

1. For any $p, q \in Q$, $\sigma, \tau \in \Gamma$, if $\delta(p, \sigma) = (q, \tau)$ for some, then $\tau q \mapsto \sigma p \in R$.
2. For any $p, q \in Q$, $\tau \in \Gamma$, if $\delta(p, \tau) = (q, \not{R})$, then $\tau q \mapsto \sigma q \tau \in R$ for all $\sigma \in \Gamma$, and $\tau \text{ƀ} p X \mapsto \tau q X \in R$.
3. For any $p, q \in Q$, $\tau \in \Gamma - \{\text{ƀ}\}$, if $\delta(p, \tau) = (q, \not{L})$, then $q\tau \mapsto \tau p \in R$.
4. For any $p, q \in Q$, if $\delta(p, \text{ƀ}) = (q, \not{L})$, then $q \text{ƀ} \tau \mapsto \text{ƀ} p \tau \in R$ for every $\tau \in \Gamma$, and $qX \mapsto \text{ƀ} p X \in R$.
5. $S \mapsto hA$, $S \mapsto \hbar A \in R$; and $s \mapsto \varepsilon$, $A \mapsto \varepsilon \in R$.

Our plan is to simulate the working of the machine by the derivations in the reverse order. The extra rule as in (2) takes care of the extension of the tape to the right by a new blank; the extra rule in (4) similarly accommodates the erasing of extra blanks on the left. The rules in (5) erase the part of the final string to the left of the input and erases the extra symbol A. It is clear from the description above that

$$(p, u\underline{s}v) \rightsquigarrow_D (q, x\underline{\tau}y) \text{ iff } x\tau q y A \Rightarrow_G u\sigma p v A. \tag{7.1}$$

Now, since the only rules involving S, s are as in (5), and A can give a terminal only by the rule $A \mapsto \varepsilon$ as in (5), we have

$$w \in \mathcal{L}(G) \text{ iff } S \Rightarrow_G \text{ƀ} h A \Rightarrow_G \text{ƀ} s w A \Rightarrow_G wA \Rightarrow_G w.$$

From (7.1) above, we conclude that $w \in \mathcal{L}(G)$ iff D halts on the input w accepting it. This completes the proof. □

Problems for Section 7.7

7.44. Construct an NTM that accepts the language $\{u2v : u, v \in \{0, 1\}^*,\ u \neq v\}$.

7.45. Show that for each NTM, there exists an equivalent NTM that has a unique accepting configuration for each accepted input. [Can you prove a theorem analogous to Theorem 7.2?]

7.46. Construct a Turing machine to accept the language $(ab)^+$ and then construct a grammar equivalent to it. Derive the string $abab$ in this grammar.

7.47. Design NTMs for accepting the following languages:
(a) $a^+b^+ba^+$.
(b) $\{w \in \{a, b\}^* : abaab$ is a substring of $w\}$.
(c) $\{ww : w \in \{a, b\}^+\}$.
(d) $\{vv^R ww^R : v, w \in \{a, b\}^*\}$.
(e) $\{ww^R w : w \in \{a, b\}^+\}$.
(f) $\{uww^R v : u, v, w \in \{a, b\}^+,\ \ell(u) \geq \ell(v)\}$.
(g) $\{0^n : n = m^2 + k^2$ for some $m, k \in \mathbb{N}\}$.
(h) $\{a^n : n \in \mathbb{N}$ is not a prime number$\}$.

7.48. Let N be the NTM with input alphabet $\{a, =\}$, initial state s, and transitions $(s, =, p, a)$, $(s, =, p, =)$, $(p, =, p, =)$, (p, a, s, R), (p, a, h, R). Starting from the initial configuration $s\underline{=}$, find all possible computations of length less than five. Explain what M eventually does starting from this initial configuration.

7.49. Suppose the tape of a TM has one nonblank symbol, say, 0, at one square only; all other squares contain ƀ. Design an NTM that starts from some square containing a ƀ and finds out the 0 and then halts. Design a DTM for doing the same job.

7.50. Design NTMs (not deterministic) that decide the following languages over $\{a, b\}$, and then construct equivalent grammars from the NTMs.
(a) $\emptyset$.
(b) $\{\varepsilon\}$.
(c) $\{a\}$.
(d) a^*ba^*b.

7.51. Prove: for each NTM M, we can construct a standard TM D such that for any string not containing ƀ,

(a) If M accepts the string w, then D halts in its accepting state in some configuration in which M would have halted.
(b) If M does not accept the string w, then D does not accept w.

7.52. What is the maximum number of transitions in a nondeterministic Turing machine?

7.8 Summary and Additional Problems

As a way of generalizing grammars, we introduced unrestricted grammars. The versatility of these grammars lead us to think that probably we have reached the limit; we may look for no more generalizations. We then introduced the deterministic Turing machines as language recognizers that would probably correspond to the unrestricted grammars. As it seemed to stay as the most generalized version of automata, we tried to make shortcuts in terms of modularity. The assembling of old Turing machines could be used to make new combined Turing machines in a manner very similar to high level programing languages.

Then we considered nondeterministic Turing machines and multitape Turing machines that turned out to be equivalent to the deterministic ones. In turn, we have shown the equivalence of all these mechanisms with the unrestricted grammars. The Turing machines are capable of doing an important job of rejecting languages, which we have not discussed so far for the other types of automata. This means that total TMs not only recognize a string but also decide whether a given string is in a given language or not. We have not discussed so far whether this can always be done.

The Turing machines were first introduced by Turing in [131] for solving the decision problem of Hilbert. In fact, this started the whole branch of Theory of computation. Later, the other types of automata were introduced for different purposes. For a good account of the related topics you can see [54, 67]. Equivalence of unrestricted grammars (also called Type-0 grammars) to Turing machines was shown in [15].

For a biography of Turing and the background of problems he tried to solve, see [56]. For a brief history of computability and why the term "computably enumerable" is more appropriate than "recursively enumerable," see [124, 125]. The multi-stack and counter machines and the restrictions of TMs that I ask you to tackle in the additional problems below are from [32, 89]. For more references on the theme, see the last section of the next chapter. For an introduction to TMs at the level of this text, you can refer [59, 71, 78, 79, 81, 90, 122].

Additional Problems for Chapter 7

7.53. Find grammars for the following languages over $\Sigma = \{a\}$:
(a) $\{w : \ell(w) \mod 3 = 0\}$.
(b) $\{w : \ell(w) \mod 3 > 0\}$.
(c) $\{w : \ell(w) \mod 3 \neq \ell(w) \mod 2\}$.
(d) $\{w : \ell(w) \mod 3 \geq \ell(w) \mod 2\}$.

7.54. Find grammars to generate the following languages over $\{a, b\}$:
(a) $L = \{w : \#_a(w) = 2 \cdot \#_b(w)\}$.
(b) $\{a, b\}^* - \{ww^R : w \in \{a, b\}^*\}$.

7.55. Consider the grammar G with productions $S \mapsto ABaC, Ba \mapsto aaB, BC \mapsto DC|E, aD \mapsto Da, AD \mapsto AB, aE \mapsto Ea, AE \mapsto \varepsilon$. Derive three different strings in $\mathcal{L}(G)$. Give a conjecture about how to describe $\mathcal{L}(G)$.

7.56. Show that for each unrestricted grammar, there exists an equivalent grammar in which all productions not of the form $A \mapsto \varepsilon$ are of the form $u \mapsto v$, where the strings u, v of terminals and nonterminals satisfy $1 \leq \ell(u) \leq \ell(v)$.

7.57. Show that the conclusion of the preceding problem holds even if we add further conditions that $1 \leq \ell(u) \leq \ell(v) \leq 2$.

7.58. Show that any grammar can be converted into one in which each rule is of the form $uAv \mapsto uwv$, where A is a nonterminal and u, v, w are strings of terminals and nonterminals.

7.59. Show that each unrestricted grammar is equivalent to one in which no terminal symbol occurs on the left hand side of any production rule.

7.60. Consider a variation of an unrestricted grammar in which the starting point of a derivation is a string rather than the start symbol S. Are these grammars more powerful than the usual unrestricted grammars?

7.61. A *normal system* is an unrestricted grammar in which a derivation is defined in a different way. In a normal system, we say that $x \Rightarrow y$ in one step iff there is a string w of terminals and nonterminals and a production rule $u \mapsto v$ such that $x = uw$ and $y = wv$. Derivations in zero or more steps, and then the language of the normal system are defined in the usual manner. Show that each computably enumerable language is the language of some normal system.

7.62. Describe a Turing machine that accepts the following language:
$\{cw_1cw_2c \cdots cw_m : \text{ each } w_i \in \{a, b\}^* \text{ and } w_i \neq w_j \text{ for each } i \neq j\}$.

7.63. Show that every computation that can be done by a standard TM can be done by a two-tape TM with at most two states (besides h and $\hbar$).

7.64. Show that for each TM, there exists an equivalent standard TM with at most six states. Can you reduce the number of states further, say, to three?

7.65. Let M be a TM. Assume, without loss of generality (why?), that each computation of M is of even length. For any such computation $q_0\underline{b}w \rightsquigarrow w_1 \rightsquigarrow \cdots \rightsquigarrow w_n$, construct the *string* $q_0\underline{b}w \rightsquigarrow w_1^R \rightsquigarrow w_2 \rightsquigarrow w_3^R \rightsquigarrow \cdots \rightsquigarrow w_{n-1}^R \rightsquigarrow w_n$. This is called a *valid computation* of M on w. Show that for every M, three CFGs G_1, G_2, and G_3 can be constructed such that
(a) The set of all valid computations is $\mathcal{L}(G_1) \cap \mathcal{L}(G_2)$.
(b) The complement of the set of all valid computations is $\mathcal{L}(G_3)$.

7.66. Consider a model of Turing machine in which each move permits the read–write head to travel more than one square on either direction, the number of squares it can travel being the third argument of δ. Give a precise definition of such a machine, and then show that such a machine can be simulated by a standard Turing machine.

7.67. A *nonerasing Turing machine* is one which cannot replace a nonblank symbol by a b. This means, if $\delta(p, a) = (q, b)$, then $a = b$. Show that nonerasing machines are no less powerful than standard TMs.

7.68. A *write-once Turing machine* is a standard TM, where the read–write head is allowed to replace a symbol on a cell at most once. Show that this variant of TM is equivalent to the standard TM.

7.69. Show that single-tape TMs that are not allowed to write on the portion of the tape containing the input accept only regular languages.

7.70. A *stay-put Turing machine* is a TM with a left end beyond which the read–write head cannot move. Further, these machines do not have one-step left-movements; the

Ł signifies jumping back to the left-most cell. Show that this variant is not equivalent to the standard TMs. What class of languages are accepted by these machines?

7.71. A *queue automaton* is an automaton in which the temporary storage is a queue, instead of a stack in a PDA. Show that a queue automaton can simulate a standard Turing machine.

7.72. A *counter automaton* is a deterministic automaton with counters. Each counter is a stack with the bottom-of-the-stack marker and just one more symbol. This other symbol can be pushed onto the stack or be removed from it, from the top of the stack. Show that a standard TM can be simulated by a counter automaton with four counters. What about machines with three counters? And two counters?

7.73. A *two-stack automaton* is a PDA with two stacks. Formally define a two-stack automaton and then prove that two-stack automata are equivalent to Turing machines.

7.74. A *deterministic two-dimensional Turing machine* is like a Turing machine where instead of a tape we have a two-dimensional tape, a chess board extended to infinite in all the four directions. Cells are numbered (m, n) for integers m and n. Initially, the read–write head is put on the cell $(0, 0)$ and input string of length n is written on the cells numbered $(1, 0), (2, 0), \ldots, (n, 0)$, other cells contain the ƀ symbol as usual. Being in a state, scanning a symbol, it either writes a symbol on the scanned square, or, moves in one of the four directions: top, bottom, left, or right, according to its transition function δ. It accepts an input by entering into its accepting halt state, thereby it erases all the symbols from the tape, and halts. Define such a machine formally. Prove that these machines are equivalent to standard Turing machines, in the sense that a machine of one type can be simulated by some machine of the other type.

7.75. A *Post System* P is a quadruple (C, V, A, R), where $C = N \cup \Sigma$ is a finite set with $N \cap \Sigma = \emptyset$, N being the set of nonterminals and Σ being the set of terminals, V is a finite set of variables, A is a finite subset of C^*, called the axioms, and R is a finite set of production rules. The productions in R are of the form

$$x_1 V_1 x_2 \cdots V_n x_{n+1} \mapsto y_1 W_1 y_2 \cdots W_m y_{m+1},$$

where $x_i, y_i \in C^*$, and $V_i, W_i \in V$ are such that any variable can occur at most once on the left, and that each variable on the right must occur on the left, that is, $V_i \neq V_j$ for $i \neq j$, and $\{W_j\} \subseteq \{V_i\}$. Suppose we have a string of terminals of the form $x_1 w_1 x_2 w_2 \cdots w_n x_{n+1}$. We can then match such a string with the production of the form given above identifying V_1 as w_1, V_2 as w_2, etc, and then we substitute these values for the W_j's on the right of the production. We thus get the new string of y's and w's by the derivation $x_1 w_1 x_2 w_2 \cdots w_n x_{n+1} \Rightarrow y_1 w_1 y_2 w_2 \cdots y_{m+1}$. The language of the Post System P is defined by $\mathcal{L}(P) = \{w \in \Sigma^* : w_0 \Rightarrow w \text{ for some } w_0 \in A\}$.

(a) Give derivations of ab and $aabb$ from ε in the Post System $P = (N \cup \Sigma, \{V_1\}, \{\varepsilon\}, \{V_1 \mapsto a V_1 b\})$, where $N = \emptyset$ and $\Sigma = \{a, b\}$. What is $\mathcal{L}(P)$?

(b) What is $\mathcal{L}(P)$, where P is the Post System $(N \cup \Sigma, \{V_1, V_2, V_3\}, \{1+1=11\}, R)$ with $N = \emptyset$, $\Sigma = \{1, +, =\}$, and $R = \{V_1 + V_2 = V_3 \mapsto V_1 1 + V_2 = V_3 1,\ V_1 + V_2 = V_3 \mapsto V_1 + V_2 1 = V_3 1\}$?

(c) Find a Post System to generate the language $\{ww : w \in \{a, b, c\}^*\}$.

(d) What is the language of the Post System that has $\Sigma = \{a\}$, $A = \{a\}$, and the only production $V_1 \mapsto V_1 V_1$?

(e) Prove that a language is computably enumerable iff it is the language of some Post System.

(f) In a restricted Post System, we further require that $m = n$, that is, the number of occurrences of variables on the left and on the right sides of a production is the same. Show that restricted Post Systems generate the same class of languages as the Post Systems.

8 A Noncomputably Enumerable Language

8.1 Introduction

In Sect. 7.7, I gave a passing comment that the set of composite numbers is computably enumerable. There is a minor huddle in using the terminology to arbitrary sets, for example, a set of numbers, rather than to languages. That is not a big huddle, for we just represent a countable set by a language and try to solve the problem about the language. Agreed that we can use the adjective "computably enumerable" for sets, how do we proceed to show that the set of composite numbers is computably enumerable? Its solution, as suggested there, requires a Turing machine capable of multiplying two numbers p and q.

Now this is something unnatural to the way we were using all our automata. Why? We have used all varieties of automata as language acceptors, which, given a string, would signal to us whether it accepts the string or not. Here, when we say that we want multiplication of p and q, our requirement is a bit more demanding. We want a Turing machine not only to operate and halt on an input, but also it should give us an *output* upon halting.

To add to the problem, when a machine halts after giving an output, another machine may also work on the same tape, taking the content of the tape as its input tape. That is, the first machine must leave the tape in such a way that it can be used as an initial tape for another machine. As our inputs are written as $\underline{ƀ}\, u$, where the read/write head is placed initially on the ƀ preceding the input, we may like our output to be in the same form. This means that the first machine must scan the symbol ƀ preceding the output while halting so that the second machine will start working.

This requirement amounts to imagining the machine as the control unit only, whereas the tape and the read/write head can be detached from it and attached to another machine that might work after the first had completed its job. It goes along well with our machine combinations also. We will put it as a convention while computing any function.

A. Singh, *Elements of Computation Theory*, Texts in Computer Science,

8.2 Turing Machines as Computers

Computing a function $f(\cdot)$ means that given an input x to the machine, it must output the value $f(x)$ upon halting. As we have adopted our data structure as strings, we will use them for computing functions mapping strings to strings. Shortly, we will see how to compute arbitrary functions via representations of other types of data as strings.

Formally, let $f : \Sigma^* \to \Sigma_1^*$ be a partial function. We take $\Gamma \supseteq \Sigma \cup \Sigma_1$ and think of f as a partial function from Σ^* to Γ^*. A Turing machine $M = (Q, \Sigma, \Gamma, \delta, s, h, \hbar)$ is said to **compute** the partial function f iff for every $u \in \Sigma^*$, whenever $f(u) = v \in \Sigma_1^* \subseteq \Gamma^*$, we have $s\underline{b}u \rightsquigarrow h\underline{b}v$ in M. Moreover, if $f(u)$ is undefined, then M either does not halt on the input u going on an infinite computation or halts in the state $\hbar$. Notice that we exclude the case of an abrupt halt here. A partial function $f : \Sigma^* \to \Sigma_1^*$ is said to be **computable** iff there exists a Turing machine that computes f.

If f is a total function from Σ^*, then the machine M that computes it must halt on every input from Σ^*. It may perhaps go on an infinite computation on inputs over other alphabets, but never so when the input is a string over Σ.

Exercise 8.1. How does a nondeterministic Turing machine compute a function?

We should first address the issue of representing other types of inputs as strings. If the inputs are, say, numbers, then how do we go about representing them as strings? You may do it in many ways. As a start, we use **unary representation** of natural numbers. Here, 0 is represented as 0 itself, 1 is represented as 00, and so on. In general, the natural number n is represented as 0^{n+1}. If $f : \mathbb{N} \to \mathbb{N}$ is a partial function, then we **represent** f as $f' : 0^* \to 0^*$, where if $f(m) = n$, then $f'(0^{m+1}) = 0^{n+1}$. For example, the square function $f : \mathbb{N} \to \mathbb{N}$ with $f(n) = n^2$ is represented as $f' : 0^* \to 0^*$ with $f'(0^{m+1}) = 0^{m^2+1}$.

What about functions from $\mathbb{N} \times \mathbb{N}$ to $\mathbb{N}$, such as addition or multiplication of natural numbers? Here again we follow the same procedure, but we must take care as to how to represent an ordered pair (m, n) as an input. We just separate the numbers m and n or rather their representations by a 1. The ordered pair (m, n) is represented as $0^{m+1}10^{n+1}$. Then as an input, the initial tape appears as $\underline{b}\,0^{m+1}10^{n+1}$. That is, the symbol 1 represents the comma (,) in between the pair of numbers.

We assume that the tape alphabet of the machine that would compute such a function will have 1 in it and this symbol 1 is never used for any other purpose. Thus the operation of addition of two natural numbers is represented as a (partial) function $+' : 0^*10^* \to 0^*$, where if the initial configuration of the machine that computes it is $s\underline{b}0^{m+1}10^{n+1}$, the final configuration must be $h\underline{b}0^{m+n+1}$.

In general, any partial function $f : \mathbb{N}^m \to \mathbb{N}^n$, called an **arithmetic function**, is represented as the partial function $f' : 0^*10^*1 \cdots 10^* \to 0^*10^*1 \cdots 10^*$ defined by (Number of 1's on the left is $m - 1$, and on the right is $n - 1$.)

If $f(k_1, k_2, \ldots, k_m) = (p_1, p_2, \ldots, p_n)$, then
$f'(0^{k_1+1}10^{k_2+1}1 \cdots 10^{k_m+1}) = 0^{p_1+1}10^{p_2+1}1 \cdots 10^{p_n+1}$.

That is, the input tape of the machine that would compute f' appears initially as

$$\underline{\not{b}}\,0^{k_1+1}10^{k_2+1}1\cdots 10^{k_m+1} \quad (m-1 \text{ number of 1's}).$$

The output tape of the machine must appear as

$$\underline{\not{b}}\,0^{p_1+1}10^{p_2+1}1\cdots 10^{p_n+1} \quad (n-1 \text{ number of 1's}).$$

In what follows, we will not distinguish between the numbers (or functions) and their corresponding representations. We say that f is computable whenever the corresponding f' is computable. For computing a function f, we simply discuss how to compute f'. We will loosely use the symbol f, although what we mean is its representation f'.

Example 8.1. Which arithmetic function is computed by the machine

$$M: \ \longrightarrow 0\,\not{L}\,h.$$

Solution. The machine M is designed to compute a function from $\mathbb{N}$ to $\mathbb{N}$. If the input is given as (input tape appears as) $\underline{\not{b}}\,0^{n+1}$, then M writes 0 in place of the first blank and then turns left. The left square to the initial blank is assumed to contain $\not{b}$ as all the squares to the left of the input are left blank initially. Hence the output contains one more 0 than the input. M thus computes the **successor function** $f(n) = n + 1$, and then it shows that the successor function is computable. Notice that we are not at all worried what M does if the input is not in the form 0^{n+1}. □

Example 8.2. Show that the addition function is computable.

Solution. Initially, the tape appears as $\underline{\not{b}}\,0^{m+1}10^{n+1}$. We rewrite that 1 as a 0. The tape contains the string 0^{m+n+3}. We delete two 0's so that the output will be 0^{m+n+1} as required. This is accomplished by the following machine:

$$\longrightarrow R_1 0 L_{\not{b}}\,\not{R}\,\not{b}\,\not{R}\,\not{b}\,h.$$

□

Example 8.3. Let Σ be any alphabet. For each string $u \in \Sigma^*$, define $f(u) = u^R$, the reversal of the string u. This defines f to be the *reversal function*. Is it computable?

Solution. Our plan is to use the copying machine and then read a symbol from the right end of the copied string, erase it, and write it on the left end of the original string. When the copied string is over, we must have the reversal of the string on the tape, and finally, we must come back to the blank square preceding the reversed string. Recollect that after a string is copied, the copying machine C scans the blank square in between the string and its copy, that is, C transforms the tape appearance $\underline{\not{b}}\,u$ to $u\underline{\not{b}}\,u$. The machine diagram for reversing a string is given in Fig. 8.1.

You can also construct a machine to reverse a string without using the copying machine. Try it! You can also simplify the machine in Fig. 8.1. □

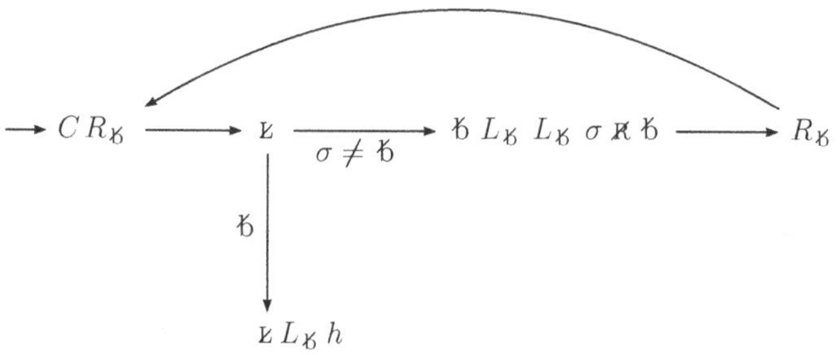

Fig. 8.1. Computing reversal of a string.

Exercise 8.2. Show that multiplication of two natural numbers is a computable function. [Hint: Use repeated copying of the string 0^n exactly m times and then add one more 0 to transform the tape from $\underline{\not b}\, 0^{m+1}10^{n+1}$ to $\underline{\not b}\, 0^{m \cdot n+1}$. You must take care of the cases $m = 0$ and/or $n = 0$.]

Exercise 8.3. Use Exercise 8.2 to show that the set of all composite numbers is computably enumerable (as a language over $\{0\}$). What about the set of all prime numbers?

Suppose M is a Turing machine with an input alphabet Σ that accepts a language $L \subseteq \Sigma^*$. We construct another three-tape Turing machine M' using M. The machine M' writes a symbol, say, # on the first tape, to the left of which it generates two natural numbers m, n. It then generates all strings over Σ of length m on its second tape following some order, say, lexicographic ordering.

On its third tape, M' simulates M carrying this simulation up to computations of length at most n. If M halts on the currently generated string (on the second tape), then M' writes this string to the right of the rightmost # and writes one more # on the square to the right of the string. Then M' erases its third tape and goes on processing the next string of length m that is on the second tape. Once all strings on the second tape have been processed this way, M' erases the earlier pair of numbers. This constitutes a step of M' simulating M. In the next step, it generates another pair of numbers m, n and writes this pair to the left of the first #, and then proceeds as earlier. The generation of the pairs of numbers m, n can also be ordered.

As M' goes on simulating M this way, each pair of natural numbers m, n is eventually generated, and each string from Σ^* is eventually processed. Only the strings that are accepted by M are written on the first tape with a # separating them. That is, M' thus *enumerates* all the strings that are accepted by M.

Conversely, suppose M_1 is a Turing machine that enumerates strings of a language $L \subseteq \Sigma^*$, one by one. It is not necessary that this machine be a three-tape machine. A standard Turing machine can also do the same job. Construct a three-tape Turing machine M_1' that uses M_1. The machine M_1' generates strings from Σ^* on its first tape one by one, say, in lexicographic order. It also generates a natural number k on its second tape. It then simulates M_1 on its third tape up to k steps of computations of M_1. During this simulation, M_1 goes on generating all strings that are enumerated by it within k steps of its computation. When this simulation step is over, M_1' searches the contents of third tape, that is, the enumerated strings for the string on the first tape. If matching is found, then, M_1' halts in its accepting state. You see that any string that can be generated by M_1 is eventually accepted by M_1'.

Let us say that a Turing machine M **enumerates** a language $L \subseteq \Sigma^*$ if M lists all elements of L (and none of $\overline{L}$), one by one, as output. Such a machine is called an *enumerator*. The above discussion can be summarized as follows:

Theorem 8.1. *A language is computably enumerable iff it is enumerated by some Turing machine.*

In fact, this is the reason why computably enumerable languages are called *computably enumerable*!

Problems for Section 8.2

8.1. What function from $\mathbb{N}$ to $\mathbb{N}$ is computed by the TM with initial state s and transitions given by $\delta(s, \not b) = (p, \not L)$, $\delta(p, \not b) = (h, \not R)$, $\delta(p, 0) = (q, \not b)$, $\delta(q, \not b) = (r, \not L)$, $\delta(r, \not b) = (h, \not R)$, $\delta(r, 0) = (h, \not R)$?

8.2. Which function is computed by the TM with initial state s and transitions $\delta(s, \not b) = (p, \not L), \delta(p, a) = (p, \not L), \delta(p, b) = (p, \not L)\delta(p, \not b) = (q, b), \delta(q, a) = (q, \not R), \delta(q, b) = (q, \not R), \delta(q, \not b) = (h, \not b)$?

8.3. Design Turing machines with states and transitions for doing the following jobs:

(a) Addition of two natural numbers.

(b) Given a string of a's, it leaves a blank to the right of the given string and then copies the string.

(c) Given two strings of a's with a blank in between, it outputs 1 if first string is of length at least that of the second; else, it outputs 0.

(d) Given a string $w \in \{0, 1\}^+$ as input, it outputs w^R.

8.4. Give a combined machine to compute the function $f : \{a, b\}^* \to \{a, b\}^*$ defined by $f(w) = ww^R$.

8.5. Design a combined machine to compute the function $f : \mathbb{N}^3 \to \mathbb{N}$, where $f(m, n, k)$ is the maximum of the numbers m, n, k, provided none of them equals another, else, $f(m, n, k)$ equals 0.

8.6. Design a combined machine to compute the function $f : (a^*)^3 \to \{0, 1, 2, 3\}$, where $f(w_1, w_2, w_3)$ is the i such that $\ell(w_i) = \max(\ell(w_1), \ell(w_2), \ell(w_3))$, provided w_1, w_2, w_3 are each of different lengths; else, $f(w_1, w_2, w_3)$ is 0.

8.7. Design a combined machine to compute the function $g : (a^*)^3 \to \{0, 1, 2, 3\}$, where $g(w_1, w_2, w_3)$ equals 0 if $\ell(w_1) = \ell(w_2) = \ell(w_3)$; otherwise, $g(w_1, w_2, w_3)$ is the least i such that $\ell(w_i) = \max(\ell(w_1), \ell(w_2), \ell(w_3))$.

8.8. Design a combined machine which, given a string $a_1 \cdots a_n a_{n+1} \cdots a_{2n}$ of even length, outputs $a_1 \cdots a_n c a_{n+1} \cdots a_{2n}$, where $c \notin \Sigma$. That is, it finds the middle of the string.

8.9. Design combined machines for computing
(a) $f : \mathbb{N} \to \mathbb{N}$, where $f(n) = 3n$.
(b) $f : \mathbb{N} \to \mathbb{N}$, where $f(n) = n \mod 5$.
(c) $f : \mathbb{N} \to \mathbb{N}$, where $f(n) = n^2$.
(d) $f : \mathbb{N} \times \mathbb{N} \to \mathbb{N}$, where $f(m, n) = m - n$ if $m > n$, else, $f(m, n) = 0$.
(e) $f : \mathbb{N} \times \mathbb{N} \to \mathbb{N}$, where $f(m, n) = 2m + 3n$.
(f) $f : \mathbb{N} \to \mathbb{N}$, where $f(n) = \lceil n/2 \rceil$.
(g) $f : \mathbb{N} \to \mathbb{N}$, where $f(n) = \lceil \log_2 n \rceil$.
(h) $f : \mathbb{N} \times \mathbb{N} \to \mathbb{N}$, where $f(n) = \lceil n/m \rceil$, if $m > 0$, else, $f(n) = 0$, if $m = 0$.
(i) $f : \mathbb{N} \times \mathbb{N} \to \mathbb{N}$, where $f(m, n) = \gcd(m, n)$ using the *Euclidean Algorithm*.

8.10. In computing numerical functions suppose we use binary representation of numbers instead of the unary notation. Describe how the successor function is computed by a machine.

8.11. Design a Turing machine to compute multiplication of two positive integers. In addition, if one of the inputs is 0, then the machine should not halt at all.

8.12. Represent a negative integer by putting a 2 at the beginning of a string of 0s. For example, -3 is represented as 20000. Design Turing machines to compute
(a) Addition of two integers.
(b) Subtraction of two integers.
(c) Multiplication of two integers.
(d) Comparison of two integers. (Output 1 if $x \geq y$, else, 0.)

8.13. Using the TMs for addition, subtraction, multiplication, comparison of two natural numbers, and the copying machine, design TMs for computing the following functions $f : \mathbb{N} \to \mathbb{N}$, where
(a) $f(n) = n(n + 1)$.
(b) $f(n) = n^3$.
(c) $f(n) = 2^n$.
(d) $f(n) = n!$.
(e) $f(n) = n^{n!}$.

8.14. Construct a TM that computes concatenation of two strings, that is, it must compute the function $f : \Sigma^* \times \Sigma^* \to \Sigma^*$ defined by $f(u, v) = uv$.

8.15. Describe a Turing machine that converts a given positive integer written in decimal notation to one in binary.

8.16. Suggest a method to represent rational numbers as strings and then give a description of a Turing machine for adding and subtracting rational numbers.

8.17. Describe Turing machines for adding and subtracting natural numbers given in decimal notation.

8.18. Let $f : \mathbb{N} \to \mathbb{N}$ be a total function. The *graph of* f is the set of ordered pairs $(x, f(x))$. Answer the following with informal but clear constructions:
(a) How to construct a TM that accepts the graph of f from one that computes f?
(b) How to construct a TM that computes f from one that accepts the graph of f?
(c) If f is a partial function, then a TM that computes it need not halt on the inputs for which f is undefined. Answer (a and b) for a partial function $f : \mathbb{N} \to \mathbb{N}$.

8.19. Let M be a TM with initial state s. The language $L = \{w : s\underline{\not b} \rightsquigarrow_M h\underline{\not b}w\}$ is called an *output language* of M. Show that a language L is computably enumerable iff it is the output language of a TM.

8.20. Prove that a language (or a set) A is computably enumerable iff it is the range of a computable function.

8.21. Show that if a language is computably enumerable, then there is a TM that enumerates it without ever repeating its strings.

8.3 TMs as Language Deciders

A particular way Turing machines can be used is to solve an important problem called the *decision problem*. A decision problem is in the form of a question such as "Whether this road goes to the airport or it does not?," "Whether 5 is a prime number or it is not?." This is important because of the presence of nontrivial questions such as "Whether every even number greater than two is a sum of two prime numbers or not?."

You can visualize a decision problem as determining whether a given element of a set satisfies certain property. As a unary property is again thought of as a subset of the given set, the decision problem can be rephrased, in an abstract way, as "Given a set and a unary property or a subset of it, whether a given element of the set is a member of the subset or not."

However, a subset P of a given set A can always be determined by its *characteristic* or *indical function* χ_P of P. Whenever, $x \in P$, we have $\chi_P(x) = 1$ and when $x \notin P$, we have $\chi_P(x) = 0$. It is clear that the answer to the question "whether $x \in P$ or not" is obtained by finding the value of $\chi_P(x)$. As we are using 0, 1 elsewhere for other purposes, we may use α and ν instead.

Turing machines can now be used to compute the respective indical functions for solving any particular decision problem. It is not, however, clear why every decision problem should have an answer. For example, we do not know yet whether every even number greater than two is a sum of two primes. But the question is, whether we will have any answer to such a question or not. The ones that can be answered are called *decidable problems*.

Formally, let $L \subseteq \Sigma^*$ be any language over an alphabet Σ. We say that L is **decidable** iff its indical function χ_L is a computable function. The indical function of $L \subseteq \Sigma^*$ is defined by

$$\text{for every } u \in \Sigma^*, \text{ if } u \in L, \text{ then } \chi_L(u) = \alpha, \text{ else, } \chi_L(u) = \nu.$$

Decidable languages were also called *recursive* languages, though decidability and recursiveness have different intensional meanings. Usually the terms "acceptable" and "decidable" are used together and similarly the terms "computably enumerable" and "recursive." Here, the word "recursive" has nothing to do with algorithms that call themselves. To avoid this confusion in terminology, we will use the word "decidable" instead.

If M is a Turing machine that computes χ_L, then upon input u to it, M outputs α if $u \in L$, and M outputs ν if $u \notin L$. In such a case, we say that M **decides** the language L.

Compare the definitions of accepting a language and deciding one. If M accepts a language L, then on inputs u from L, it halts (in the accepting state h), and for other inputs not from L, the machine either enters an infinite computation, or eventually enters a nonhalt state, or it halts in the rejecting state $\hbar$. While M decides L, it halts on every input from Σ^* and outputs different symbols depending upon whether the input is from L or not. Such a TM is called a **decider** of the language L. A decider is necessarily a total TM, be it deterministic or nondeterministic.

Decidable languages can also be defined without using the notion of computable functions. It is because we have two halting states, h for accepting and $\hbar$ for rejecting strings. If $L \subseteq \Sigma^*$ is decidable, then given an input string $u \in \Sigma^*$, a machine M halts with output α or ν depending on whether $u \in L$ or not. We can construct another machine that works after M has worked in such a way that it halts in state h starting from α and it halts in $\hbar$ starting from ν. Look at the machine M' in Fig. 8.2. If M computes the characteristic function χ_L, of $L \subseteq \Sigma^*$, then M' clearly accepts L and rejects $\overline{L}$.

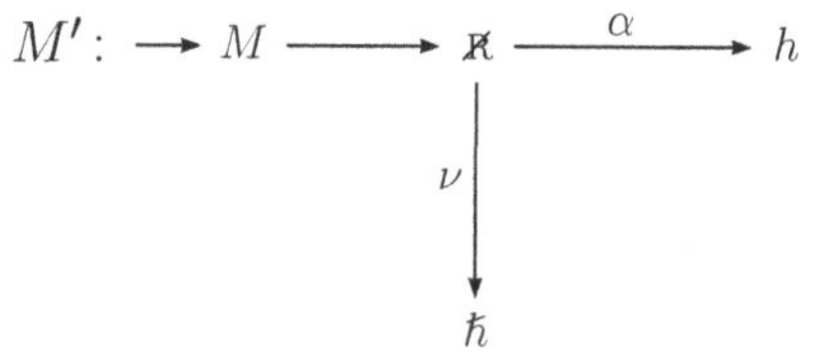

Fig. 8.2. From computing a characteristic function to acceptance and rejection.

Conversely let M be a machine that halts in state h for inputs from L and it halts in state $\hbar$ for inputs from $\overline{L}$. Assume, without loss of generality, that it has erased every symbol on the tape before halting (due to Theorem 7.2). Suppose $M = (Q, \Sigma, \Gamma, \delta, s, h, \hbar)$ is such a TM. Take four new states h_1, h_2, H, H' and construct the TM $M'' = (Q \cup \{h_1, h_2, H, H'\}, \Sigma, \Gamma, \delta', s, H, H')$, where δ' is an extension of δ with additional transitions as

$$\delta'(h, \not{b}) = (h_1, \not{R}),\ \delta'(h_1, \not{b}) = (h_1, \alpha),\ \delta'(h_1, \alpha) = (H, \not{L}),$$
$$\delta'(\hbar, \not{b}) = (h_2, \not{R}),\ \delta'(h_2, \not{b}) = (h_2, \nu),\ \delta'(h_2, \nu) = (H, \not{L}).$$

Now M'' works as M does, except instead of halting in h or $\hbar$ as in earlier computations of M, it writes α or ν being in the states h or $\hbar$, comes to the $\not{b}$ preceding the α or ν accordingly, and then halts in the new accepting halt state H. Thus, M''

computes the characteristic function of L. Essentially it amounts to telling that M'' *simulates* M, then it erases all symbols from the tape and writes α or ν according as M halted in h or $\hbar$. We mention this result in the following:

Theorem 8.2. *A language* $L \subseteq \Sigma^*$ *is decidable iff there is a Turing machine* M *such that* $\mathcal{L}(M) = L$ *and* $\mathcal{L}_R(M) = \overline{L}$ *iff there is a total Turing machine (even nondeterministic) that accepts* L.

Exercise 8.4. Instead of taking four new states in M'' in the above discussion, take two new states H and H' and replace each transition of the form $\delta(q, \sigma) = (h, \not b)$ and $\delta(q, \sigma) = (\hbar, \not b)$ by $\delta(q, \sigma) = (H, \not b)$ and $\delta(q, \sigma) = (H', \not b)$, respectively. Add transitions of the form $\delta(H, \not b) = (H, \alpha)$, $\delta(H, \alpha) = (h, \not L)$, $\delta(H', \not b) = (H', \nu)$ and $\delta(H', \nu) = (h, \not L)$. Does this new TM do the same work as M'' constructed there?

Let M be a Turing machine, be it deterministic or nondeterministic. In view of Theorem 8.2, we say that M **decides** a language L whenever M is a total Turing machine and $L = \mathcal{L}(M)$. Alternatively, a language L is decidable iff we have a machine M that accepts L and rejects $\overline{L}$, and in such a case, M decides the language L. A decider is necessarily a total TM, be it deterministic or nondeterministic. In this context, computably enumerable languages are also called *semi-decidable languages.* A machine that accepts a semi-decidable language is said to *semi-decide* the language.

For example, is the singleton $\{\varepsilon\}$ a decidable language? As a problem, the question is: "Given a string, whether it is empty or not?" To see that $\{\varepsilon\}$ is decidable, you construct a Turing machine, which upon input ε, outputs α, and on any other input it outputs ν. Look at Fig. 8.3. It computes the function $\chi_\varepsilon : \Sigma^* \to \{\alpha, \nu\}$ defined by $\chi_\varepsilon(\varepsilon) = \alpha$ and $\chi_\varepsilon(u) = \nu$ for every $u \neq \varepsilon$.

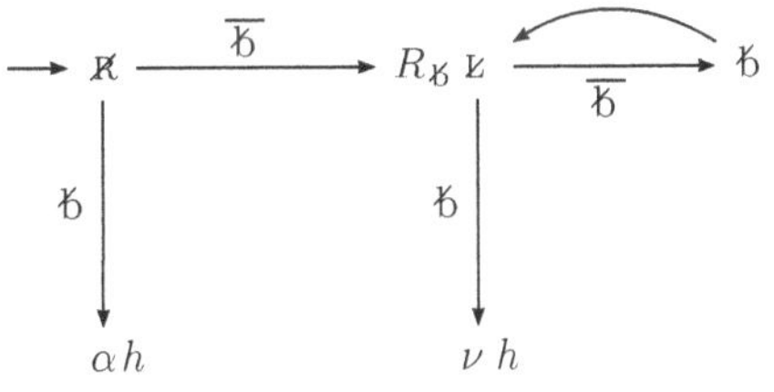

Fig. 8.3. First TM deciding $\{\varepsilon\}$.

When the input is ε, the tape will appear as $\underline{\not b}$, otherwise it would look like $\underline{\not b}u$ for some nonempty string u. So, the read–write head of the TM in Fig. 8.3 just goes once to the right. If the symbol there is $\not b$, we have ε as the input, otherwise, it is not ε. After determining this, the machine should erase the symbols if required, and then output α or ν accordingly.

Alternatively, you can design a machine that enters h if it finds no symbol on the right, else it enters $\hbar$. This is, perhaps, easier to do. See Fig. 8.4.

Exercise 8.5. Construct a TM with states and transitions that decides $\{\varepsilon\}$.

Example 8.4. Design a nondeterministic TM to decide $L = \{a, b\}^* a \{a, b\}^*$.

Fig. 8.4. Second TM deciding $\{\varepsilon\}$.

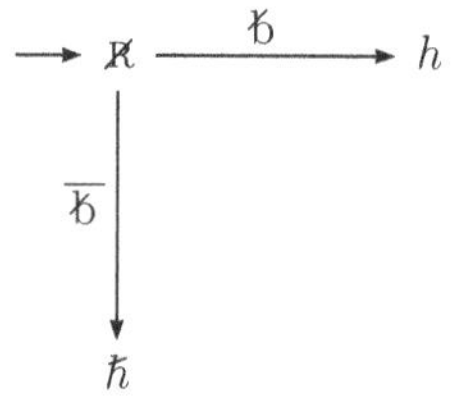

Solution. L is a regular language; it should be easy to design a DTM for deciding it. Given an input, all that we check is whether there is an occurrence of an a in it or not. For example, let the read–write head of the TM move right if it gets a b, and halts in h if it gets an a. Of course, this TM is also an NTM. But to see a truly nondeterministic machine, let us ask the machine to guess a position on the string where it expects an a. If it is an a, accept, else, reject. See Fig. 8.5 for an implementation of this idea.

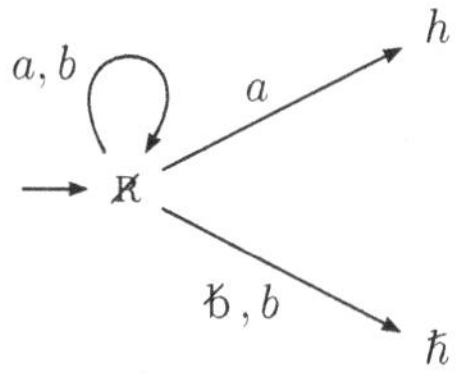

Fig. 8.5. A nondeterministic decider for $\{a, b\}^*a\{a, b\}^*$.

The machine just takes some right movements to guess a position on the input string, and then it checks whether the scanned symbol is an a. If it is, it accepts. If it is not, it does not accept. But, input strings that do not contain an a are always rejected. Hence, it decides the language L. □

What does the phrase "if it is an a, accept, else reject" mean? Does it mean that if at the guessed position, an a is not there, it will reject the string? No. Because, for rejection of the string, we must ascertain that every computation on the string leads to the rejecting state. In such a case, it simply tries another computation. The nondeterministic acceptance and decision process is encapsulated by:

> Choose a value of some parameter, possibly, depending on the input. Do computation on the input using this parameter. If a condition is satisfied, accept; else, reject.

The parameter could be a substring of the input, a symbol at some particular position of the string, or even, any string or number not related to the input at all. The last sentence above is heavily loaded. It is somewhat more favorable towards acceptance than rejection. If the specified condition is satisfied, then it accepts, of course. But if the condition is not satisfied, then the word "reject" does not mean the input string is rejected. It really means that the choice of the parameter is rejected. The NTM goes on guessing another value of the parameter and repeats the computation again, setting the time spent till now to zero. The time spent on such rejections are not accounted for. We will see later how this matters in a significant way.

From the definition of a decidable language, it is clear that it has to be computably enumerable. Are there computably enumerable languages that fail to be decidable? The following statement gives a partial answer, or rather a feeling towards a possible partial answer.

Theorem 8.3. *A language L over an alphabet Σ is decidable iff both L and its complement $\overline{L}$ are computably enumerable.*

Proof. Let $L \subseteq \Sigma^*$. $\overline{L} = \Sigma^* - L$. Assume that L is decidable. Then there is a Turing machine M which, given any arbitrary string u from Σ^* as an input, outputs α when $u \in L$, and outputs ν when $u \in \overline{L}$. We construct two machines M_1 and M_2, which accept the languages L and $\overline{L}$, respectively. They are drawn in Fig. 8.6.

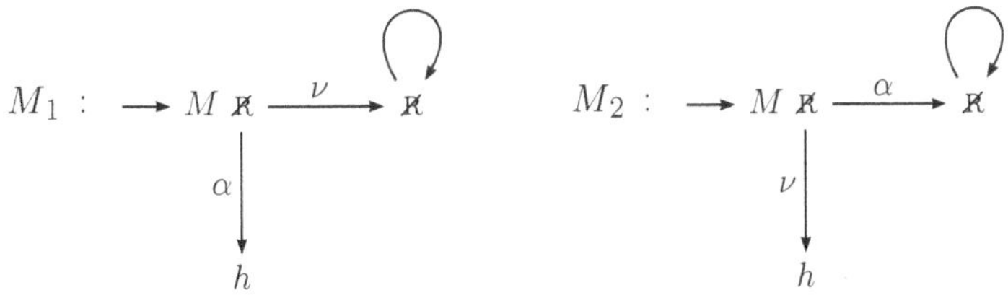

Fig. 8.6. TMs for accepting L and $\Sigma^* - L$.

Alternatively, if L is decidable, by Theorem 8.2, there is a total TM M such that $L = \mathcal{L}(M)$ (and $\overline{L} = \mathcal{L}_R(M)$). Then M also accepts L. Moreover, you can construct M' from M by renaming h as $\hbar$ and $\hbar$ as h, that is, by interchanging the accepting and rejecting states. Then M' would accept $\overline{L}$.

Conversely, suppose both L and $\overline{L}$ are computably enumerable. Let M_3, M_4 be the machines that accept the languages $L, \overline{L}$, respectively. We want to construct a machine M that would decide L, of course, by simulating M_3 and M_4 somehow. We will describe a three-tape total TM M for this purpose.

On the first tape of M is given the input string u. M copies u to both the second and the third tapes. Next, it simulates M_3 on the second tape and M_4 on the third tape, step by step. That is, M uses only one transition of M_3 on the second tape, then one transition of M_4 on the third tape. It repeats this *simultaneous simulation* of M_3 and M_4 on the second and third tapes until one of the machines M_3 or M_4 halts. M now acts in the following way:

If M_3 halts in its accepting halt state, then M halts in h.
If M_3 halts in its rejecting halt state, then M halts in $\hbar$
If M_4 halts in its accepting halt state, then M halts in $\hbar$.
If M_4 halts in its rejecting halt state, then M halts in h.

As it is assumed that L is accepted by M_3 and $\overline{L}$ is accepted by M_4, one of them, and only one of them is guaranteed to halt. Thus, M either ends up with halting in h or in $\hbar$. In the first case, the input is in L, and in the second case, the input is in $\overline{L}$. Thus, $\mathcal{L}(M) = L$ and $\mathcal{L}_R(M) = \overline{\mathcal{L}}(M)$. By Theorem 8.2, L is decidable. □

Exercise 8.6. It follows from Theorem 8.3 that "if a language L is decidable then so is its complement." Give a direct proof of this statement.

For deciding a language L, Theorem 8.3 gives you the freedom to design TMs for accepting L and $\overline{L}$ separately instead of constructing one total TM that would accept L and reject $\overline{L}$ simultaneously.

Exercise 8.7. Design machines that accept the languages $\{a^n : n \text{ is odd}\}$ and $\{a^n : n \text{ is even}\}$. Use these to construct a machine that decides $\{a^n : n \text{ is odd}\}$.

You have seen that Turing machines accept the same languages as generated by unrestricted grammars. It then seems reasonable to believe that we are tackling the most general type of automata. That there cannot be any automata (or even any computational device) strictly more general than Turing machines is to assert that, whatever ingenious computational device we may think of, all that it can do can also be done by a Turing machine. If such a computational device can accept a language, we can have a Turing machine to accept the same language. If such a device can compute a function, we can have a Turing machine to compute the same function.

In fact, many such devices are already in existence, for example, λ-calculus, μ-recursive functions, Post's machine, unlimited register machines, Markov's normal algorithms, while-programs, and many other grammatical models of computation such as P-Systems with suitable constraints, membrane computing, genetic algorithms, DNA-computing, and quantum computing. All these devices have been shown to compute the same class of functions as standard Turing machines.

This gives evidence to the belief that a total Turing machine is the correct formalization of the notion of an "algorithm." This belief, commonly known as **Church-Turing thesis**, amounts to asserting that any class of problems that cannot be solved by total Turing machines is, in fact, *algorithmically unsolvable*. Nevertheless, particular problems of the same class can very well be solved by algorithms (total Turing machines), but there cannot be a single algorithm for the whole class.

Problems for Section 8.3

8.22. Let L be a language. What is the difference between a TM accepting L and a total TM accepting L?

8.23. Let Σ, Γ be two alphabets not containing $\flat$. Suppose $A, B \subseteq \Sigma^*$, $C \subseteq \Gamma^*$, and $f : \Sigma^* \to \Gamma^*$ is a partial function. Give detailed proofs of the following:

(a) $\emptyset$ and Σ^* are computably enumerable.

(b) If A, B are computably enumerable, then so are A^*, A^R, $A \cup B$, $A \cap B$, and AB.

(c) If A is finite, then both A, $\overline{A}$, and A^+ are computably enumerable.

(d) If f is computable, then the domain and range of f are computably enumerable.

(e) If f is computable and C is computably enumerable, then $f^{-1}(C) = \{u \in \Sigma^* : f(u) \in C\}$ is computably enumerable.

(f) If f is one-to-one and computable, then $f^{-1} : \Gamma^* \to \Sigma^*$ is a computable partial function.

8.24. Show that the family of decidable languages is closed under Kleene star, reversal, union, intersection, concatenation, and complementation. What about the family of computably enumerable languages?

8.25. Construct a TM to determine whether a string is a substring of another.

8.26. Describe how an NTM may compute a function. Give a function where it is more natural to compute by an NTM than by a DTM.

8.27. Give a language that an NTM decides more naturally than a DTM.

8.28. Let $C = \{L_1, L_2, \ldots, L_k\}$ be a collection of languages over an alphabet Σ such that no string is in any two languages, and each string is in one of the languages. If each language in C is computably enumerable, show that each is decidable.

8.29. Prove that the complement of a CFL is decidable. Conclude that decidable languages need not be CFLs.

8.30. Use NTMs to show that both the classes of decidable languages and of computably enumerable languages are closed under union, concatenation, and Kleene star.

8.4 How Many Machines?

Theorem 8.3 does not answer the question whether decidability is any different from acceptability. It only says that decidable languages are closed under complementation. It does not say anything similar about computably enumerable languages. It is still possible that each computably enumerable language is decidable. It is also possible that there is a computably enumerable language that is not decidable. All that Theorem 8.3 says is that in the latter case, the complement of such a language is not computably enumerable.

So, our question now takes the form: "Whether there exists a computably enumerable language whose complement is not computably enumerable?" We ask a weaker question first: whether there are languages at all that are not computably enumerable? It leads to an easy attempt, the counting argument. Such an attempt might succeed if there are more languages than machines. But then how many languages and how many machines are there? How do we determine how many Turing machines are there?

Each Turing machine is a seven-tuple that uses finite number of states, a finite set for its tape alphabet, of which a subset is the input alphabet, and a finite relation, the transition relation (or a function). We can thus safely assume that the set of states for each machine is drawn from a countable set

$$Q_\infty = \{q_0, q_1, q_2, \ldots\}.$$

For different machines, we can use altogether different states by renaming the states if necessary. The idea is that each machine has a set of states, none of the states being used in any other machine and each such set is a finite subset of Q_∞. Notice that h and $\hbar$ are also some states in Q_∞; they, of course, differ from machine to machine.

Similarly, we can assume that the tape alphabet Γ of each machine contains $\flat$ and some finite number of symbols drawn from a countable set

$$\Gamma_\infty = \{\sigma_0, \sigma_1, \sigma_2, \ldots\}.$$

Naturally, we have the special symbols $\flat$, $\rlap{/}L$, $\rlap{/}R$ used in every machine. (For the machines that would decide languages, you may also require the special symbols α and ν. But you can avoid them by using the halting states h and $\hbar$ for accepting, rejecting, and deciding languages.) Our plan is to represent all these symbols as strings of 0's. We do this with the help of a map, call it ρ, which is defined as follows:

$$\rho(\flat) = 0, \quad \rho(\rlap{/}L) = 0^2, \quad \rho(\rlap{/}R) = 0^3, \text{ and}$$

$$\rho(\sigma_m) = 0^{m+4}, \quad \rho(q_m) = 0^{m+1}, \text{ for every } m \in \mathbb{N}.$$

The function $\rho : \{\flat , \rlap{/}L , \rlap{/}R\} \cup Q_\infty \cup \Gamma_\infty \to 0^*$ is not a one-to-one function. But ρ from Q_∞ to 0^* is one-to-one. Similarly, ρ from $\{\flat , \rlap{/}L , \rlap{/}R\} \cup \Gamma_\infty$ to 0^* is also one-to-one. For example, the string 000 represents both the state q_2 and also the symbol $\rlap{/}R$. Once we know whether the string 000 is a state or a symbol, we will be able determine what it is. Our main goal is to represent any Turing machine by a string of some symbols.

A Turing machine is a seven-tuple $M = (Q, \Sigma, \Gamma, \delta, s, h, \hbar)$. It is completely determined from its transition function δ, the initial state s, and the halting states $h, \hbar$. Notice that any transition can be written as a quadruple, for example, $\delta(p, a) = (q, b)$ can simply be written as (p, a, q, b). Thus, δ can be written as a set of such quadruples.

We must somehow represent the initial state, the halting states, and the quadruples (for transitions) so that M is faithfully coded. Further, if we want to write them one after another, there must be a way how to separate them from each other. For example, we must say when the representation of the initial state is complete. Then will follow the set of halting states, a separator, and the quadruples with separators in between.

Each state q will be written as $1\,\rho(q)1$ instead of just $\rho(q)$. To say that it is the initial state, we will have another 1 on the right. That is, if q_0 is the initial state of a machine M, we will write 1011 in the beginning of the string that might represent M. Then will follow the halting states. As each state is now written with a 1 on either side of ρ-of the state, the two consecutive 1's will tell us when that state has finished and the other has started. For example, if M has the halting states as $h = q_2, \hbar = q_3$, then we will represent it as 10001100001. The first halting state, that is, the accepting state q_2 is represented as 10001 and the second state q_3, that is, the rejecting state is represented as 100001, which are then just concatenated. Thus the string

$$101110^3110^41 \text{ stands for the sequence } q_0, 1, q_2, q_3.$$

The extra 1's in between say that q_0 is the initial state, and q_2 is the accepting state, and q_3 is the rejecting state. Notice the occurrence of 111; it says that the description of the initial state is over and the description of the halting states is starting. The 11 after that tells that the accepting state is over and the rejecting state is starting.

We will represent the quadruples (p, σ, q, τ) again the same way by separating each of the components by a 1, that is, as $1\,\rho(p)1\,\rho(\sigma)1\,\rho(q)1\,\rho(\tau)1$. For example, the string

$$1010^5 10^3 10^2 1 \text{ stands for the quadruple } (q_0, \sigma_1, q_2, \not L).$$

A set of quadruples will simply be written by concatenating the strings of the above type. For example, the string

$$1010^5 10^3 10^2 110^2 10^2 1010^3 1 \text{ stands for the set } \{(q_0, \sigma_1, q_2, \not L), (q_1, \not b, q_0, \not R)\}.$$

Look at the 11 (Read it as "one one," not as "eleven.") in the above string. It signals the end of one transition and the beginning of the other. There will be no confusion in knowing from such a binary number which one is a state and which one represents a nonstate. The first and the third sequences of 0's are states, and the second and the fourth sequences of 0's are symbols from $\{\not b, \not L, \not R\} \cup \Gamma_\infty$.

Now when we write the sequence of states and then the sequence (or set) of quadruples, we will have again two 1's. We will insert one more 1 as a separator. That is, if we have in M, the initial state q_0, the halting states q_1, q_2, and the transitions $(q_0, \sigma_1, q_2, \not L), (q_1, \not b, q_0, \not R)$, then M will be represented as

$$101110^2 110^3 111010^5 10^3 10^2 110^2 101010^3 1.$$

Again look at the occurrences of 1's carefully. The first three 1's say that the initial state is over and the halting states are starting. The second occurrence of 111 says that the halting states are over and the transitions are starting. We will write such a sequence of 0's and 1's for representing a machine M as $\psi(M)$. It is clear that for each M, there is a unique binary number $\psi(M)$. Moreover, from $\psi(M)$, the machine M can be constructed back easily.

Example 8.5. Let $M = (\{q_0, q_1, q_2, q_3\}, \{\sigma_1, \sigma_2\}, \{\sigma_1, \sigma_2, \not b\}, \delta, q_0, q_1, q_2\}$, where $\delta(q_0, \not b) = (q_0, \not R)$, $\delta(q_0, \sigma_1) = (q_2, \not L)$, $\delta(q_1, \not b) = (q_0, \not R)$. What is $\psi(M)$?

Solution. The first transition $\delta(q_0, \not b) = (q_0, \not R)$ is rewritten as the quadruple $(q_0, \not b, q_0, \not R)$. As $\rho(q_0) = 0$, $\rho(\not b) = 0$, $\rho(\not R) = 0^3$, the quadruple is represented as the string $10101010^3 1$. Other representations are mentioned above. We then get

$$\psi(M) = 101110^2 110^3 1110101010^3 11010^5 10^3 10^2 110^2 101010^3 1.$$

Going in the reverse direction, you just scan the above binary number from left to right. The first sequence of 0's gives you the initial state q_0, then there are three 1's. Next starts the sequence of halting states, separated by 1's. They are 0^2, 0^3 giving the halting states as q_1 and q_2. Then starts the transitions (after the three 1's, out of which the last 1 is in the first transition). They are $10101010^3 1$, $1010^5 10^3 10^2 1$, and $10^2 101010^3 1$, which correspond to the transitions $\delta(q_0, \not b) = (q_0, \not R)$, $\delta(q_0, \sigma_1) = (q_2, \not L)$, and $\delta(q_1, \not b) = (q_0, \not R)$, respectively. Thus M is reconstructed. □

Note that the extra symbols and states that are not at all used in the transitions will not appear in the reconstruction of M from $\psi(M)$. But by that we loose nothing, it is the same machine M anyway. Thus, our implicit assumption with any machine M here is that there are no unnecessary states and symbols mentioned in Q or Γ that are not used in the transitions.

Formally, let $M = (Q, \Sigma, \Gamma, \delta, s, h, \hbar)$ be a Turing machine. Scan all the transitions in M. If a nonhalt state $q \in Q$ is never used in any transition, delete it from Q. If any symbol $\sigma \in \Gamma$ is never used in any transition, then delete σ from Γ (also from Σ). It is clear that the new machine with updated Q and updated Γ does the same work as M. Call this updated machine as the **least form** of M. We use only the machines in their least form. We then find out $\psi(M)$ as described earlier.

Further, there is a possible confusion of self-reference regarding the coding of tape symbols in terms of 0's and 1's. Suppose that 0 is a symbol already used in a machine such as one that computes an arithmetic function. Are we rewriting this 0 or keeping it as just a 0? Is the question clear? We have assumed that our set of symbols Γ_∞ contains all possible symbols that any Turing machine might use.

Is 0 in Γ_∞ or not? In general, it will be there. Suppose $\sigma_m = 0$ for some m. Then, in the above scheme, $\rho(0) = \rho(\sigma_m) = 0^{m+4}$ and not just 0. We can, of course, put down a convention that 0 will be written as 0, but we will not do that. We will go in conformity with our rewriting scheme, and represent 0, in this case, by 0^{m+4}. Similar comments for the symbol 1 also hold. In this sense, the 0 and 1 used in the representation are really different from the 0 and 1 as symbols in any input or tape alphabet. We now make our assumptions explicit to disambiguate possible confusions.

Assumption on Representation: All Turing machines are in least form, unless otherwise mentioned. The 0 and 1 in Γ_∞ are treated as any other symbols, whose representations $\rho(0)$ and $\rho(1)$ are possibly not just 0 and 1.

With this assumption, we have thus proved that

Theorem 8.4. *There exists a one–one function ψ from the set of all Turing machines to* $\{0, 1\}^*$.

Notice that $\psi(M)$, as constructed earlier, always has exactly two occurrences of three consecutive 1's; also, it begins with a 1 and ends with a 1. Thus we do not get all strings of 0's and 1's in the set of all $\psi(M)$'s. That is, ψ is not an onto function.

Moreover, for every Turing machine M, the string $\psi(M)$ is a binary number. We will not distinguish between $\psi(M)$ as a string over $\{0, 1\}$ and as the number of which it is the binary representation. That is, we take $\psi(M) = n$, where n has the binary representation $\psi(M)$. These numbers $\psi(M)$ give an enumeration of all Turing machines.

Let $\mathcal{M}$ be the set of all Turing machines. For each machine $M \in \mathcal{M}$, we have $\psi(M)$, a binary number. That is, $\psi : \mathcal{M} \to \mathbb{N}$ is a one-to-one mapping. Hence $\mathcal{M}$ is countable and each machine $M \in \mathcal{M}$ has now a unique number, $\psi(M)$. We have an enumeration of $\mathcal{M}$. It thus makes sense to talk of the ***m*th machine**. In fact, for all the machines, we arrange their corresponding ψ's in ascending order. If $\psi(M)$ comes as the mth in this list, then M is the mth machine. (This m is not necessarily equal to $\psi(M)$.)

Let Σ be an alphabet. There are uncountable number of languages over Σ. But altogether there are only a countable number of Turing machines, and thus, only a countable number of decidable languages. Similarly, there are only a countable number of computably enumerable languages. That means there must be an uncountable number of languages that are not decidable, and an uncountable number of languages that are not computably enumerable.

Each machine computes a single function, if at all it computes any. Therefore, there are countably many computable functions. But corresponding to each language, there is its indical function. There are, of course, functions that are not indical functions. Nonetheless, there are uncountably many functions from Σ^* to Σ^*. Hence there are uncountably many functions that are not computable. (Contrast this with Example 1.10.) We summarize these important observations in the following theorem.

Theorem 8.5. *Over any alphabet Σ, there are uncountable number of languages which are not decidable; there are uncountable number of languages which are not computably enumerable; and there are uncountable number of functions on Σ^* which are not computable. In particular, there are uncountable number of arithmetic functions (functions from $\mathbb{N}^m$ to $\mathbb{N}^n$) which are not computable.*

Notice that even without encoding various parts of an arbitrary TM, we could have shown that there are only a countable number of TMs. We have followed this path due to another reason, which will be clarified shortly.

Problems for Section 8.4

8.31. Describe how to determine whether or not a string from $\{0, 1\}^*$ is an encoding of a Turing machine.

8.32. Encode the TM having the transitions $\delta(q_1, \sigma_1) = (q_2, \not{R})$, $\delta(q_1, \sigma_1) = (q_4, \sigma_1)$, $\delta(q_4, \sigma_1) = (q_3, \not{L})$, $\delta(q_3, \sigma_1) = (q_5, \sigma_2)$, $\delta(q_5, \sigma_2) = (q_2, \not{L})$.

8.33. Describe a Turing machine that enumerates the set $\{a, b\}^+$ in dictionary order. Find its encoding in binary.

8.5 Acceptance Problem

However, Theorem 8.5 does not settle the problem whether there are recursively enumerable languages that are not decidable. What we require is at least one computably enumerable language that is not decidable, or we must show that such a language does not exist. To this end, our representation of any machine as a binary string will come of help. A bit of lateral thinking is of value here. We have a scheme to represent any machine as a binary string. Can we do that for any input string?

In general, an input to any Turing machine is given as a string over its input alphabet. For example, integers m and n are rewritten in unary notation as 0^{m+1} and 0^{n+1}, respectively; the ordered pair (m, n) is rewritten as $0^{m+1}10^{n+1}$ whenever they are used as inputs to Turing machines. The 1 used here is not from 0^*. However, the string $0^{m+1}10^{n+1}$ is from $\{0, 1\}^*$, enlarging our input alphabet. Thus, any machine that computes an arithmetic function has the input alphabet $\{0, 1\}$. What about any other functions, not necessarily from $\mathbb{N}$ to $\mathbb{N}$?

Any general input can be thought of as an ordered k-tuple $(u_1, u_2, \ldots, u_k)$ of strings over an alphabet $\Sigma \subseteq \Gamma_\infty$. Each u_i is represented as $\rho(u_i)$, which is a string of 0's and 1's. The 1's separate ρ of each symbol occurring in u_i, and also 1's appear in the beginning and the end once extra, as earlier. For example, the string $u = \sigma_0\sigma_1\sigma_2$ is represented by $\rho(u) = 10^4 10^5 10^6 1$. Then we separate each such string of 0's and 1's by another 1 to write the k-tuple. For example, the ordered pair $(\sigma_0\sigma_1\sigma_2, \sigma_3\sigma_0)$ is represented as $10^4 10^5 10^6 110^7 10^4 1$.

Name this representation of an input as $\psi(u)$, where u is an input, a k-tuple of strings from an alphabet Σ. As earlier, ψ is a one–one map from the set of inputs to $\{0, 1\}^*$. It is so, because from a string over $\{0, 1\}$, we can first find out occurrences of double 1's, then separate out the components of the k-tuple input (if there are $k - 1$ number of such double 1's in $\psi(u)$). Then, we separate out the symbols in each such component by looking at the occurrence of 1's.

To demystify, we are aiming at representing any Turing machine and a possible input to it by a binary string. That is, not only the machine M and the input u, but also we must be able to represent the phrase "the machine M is supplied with the input u." It is fairly easy to do this. We have $\psi(M)$, we also have $\psi(u)$; then we just concatenate them with an extra 1 in-between. That is, the phrase is represented by $\psi(M)1\psi(u)$. Is this representation faithful? Can we reconstruct the machine M and the input u from $\psi(M)1\psi(u)$ uniquely?

In $\psi(M)$, there are just two occurrences of three consecutive 1s. After that the next such occurrence (of three consecutive 1's) is at the end of $\psi(M)$ and at the beginning of $\psi(u)$. Once $\psi(u)$ is separated, u is reconstructed from it uniquely. Given a specific Turing machine M and a specific input u to it, we can now represent faithfully the pair (M, u) so that this can be given as an input to other machines. We first define a specific language from the pair (M, u).

Let M be any arbitrary Turing machine and u be an input string to M. We construct the **acceptance language**

$$L_A = \{\psi(M)1\psi(u) : \text{ the Turing machine } M \text{ accepts the input } u\}.$$

Notice that the language L_A does not depend upon specific machine M or any specific input u to it. The M and u are simply used as notation (variables indeed).

What is the meaning of decidability of L_A? If it is decidable, then we must have a Turing machine, say, U that decides it. As $L_A \subseteq \{0, 1\}^*$, the input alphabet of U is $\{0, 1\}$. Thus, given any binary string w (a string of 0's and 1's), U must halt in a halting state. Moreover, if $w = \psi(M)1\psi(u)$ for some Turing machine M and some input u to M, then U halts in state h iff M accepts u. Further, U halts in state $\hbar$ iff either M rejects u or M enters an infinite computation. Notice that we are using

$h, \hbar$ as the halting states of U. (Alternatively, you can formulate U to output α and ν instead of halting in states h and $\hbar$, respectively.) In summary,

> U decides L_{A} means that "Given any Turing machine M, any input u to it, U decides whether M accepts u or not."

Because of this reason, the language L_{A} is called the *acceptance language*. The corresponding decision problem:

> given any machine M and any input u to it, whether M accepts u

is named as the **acceptance problem**. We ask: "Is L_{A} computably enumerable? Is L_{A} decidable?"

Theorem 8.6. *The acceptance language L_{A} is computably enumerable.*

Proof. We design a (deterministic) Turing machine U with three tapes for accepting L_{A}. On the first tape is given the binary string w as an input to U. Then U copies this input to the second tape. It first checks whether the input w is in the form $\psi(M)1\psi(v)$ for some machine M and some string v. If w is not in this form, then U halts in state $\hbar$. That is, in this case, U does not accept w. If w is in the specified form, then U first copies $\psi(v)$ to the third tape, erasing it from the second. Now U simulates M (which it gets from the second tape) on the input v, which is on the third tape in the form $\psi(v)$. If M accepts v, then U halts in state h. If M does not accept v, then U also does not accept its own input as it simulates M step by step. Clearly, U accepts w iff M accepts v, where $w = \psi(M)1\psi(v)$ for the Turing machine M and input v to M. Therefore, U accepts L_{A}. □

The machine U as constructed in the proof of Theorem 8.6 is called a **universal Turing machine** that accepts L_{A}. In general, a universal Turing machine is one that can simulate any other machine on any given input. The acceptance language L_{A} is also called the *universal language*; it is canonical of all computably enumerable languages in the sense that its decidability characterizes the decidability of all computably enumerable languages at a time. See the following statement:

Theorem 8.7. *The acceptance language L_{A} is decidable iff every computably enumerable language is decidable.*

Proof. If every computably enumerable language is decidable, then from Theorem 8.6, L_{A} is decidable.

Conversely, let L_{A} be decidable. Let L be any computably enumerable language over some alphabet Σ. Suppose M is a Turing machine that accepts L. Let U be a machine that decides L_{A}. We design a machine D that would decide L. Obviously, we will use the machines M and U for designing D. As D is required to decide L, a typical input w to D will be a string from Σ^*.

Our machine D is a three-tape machine on the first tape of which is given the input w. D first copies w to the second tape. Next, D erases the first tape and generates $\psi(M)$ on it by following the procedure discussed earlier. (M is hard-wired into D.) Then D constructs $\psi(w)$ on its third tape by reading w from its second tape.

(Our scheme for representing states, symbols, strings, etc. is also hard-wired into D.) It writes a 1 after the $\psi(M)$ on the first tape and then copies $\psi(w)$ from the third tape to the first tape to the right of this 1. Then D erases the second and the third tapes. Now, the first tape contains the string $\psi(M)1\psi(w)$, and second and the third tapes are blank. Next, D comes back to the $\not{b}$ preceding $\psi(M)1\psi(w)$ on the first tape and simulates U on it.

From the construction of D, it is now obvious that the result of the working out of D on w is the same as that of U on $\psi(M)1\psi(w)$. That is, D halts in its accepting state if U halts in state h, and D halts in its rejecting state if U halts in state $\hbar$. However, U decides L_{A}. Thus, U always halts, either in state h or in state $\hbar$ depending upon whether M accepts w or M does not. This means that D halts in its accepting state on input w if M accepts w, and D halts in its rejecting state whenever M does not accept w. But M accepts L. Therefore, D halts in its accepting state whenever $w \in L$ and D halts in its rejecting state whenever $w \notin L$. In other words, D decides L; L is decidable. □

This is the reason L_{A} is an important language. Instead of searching for a computably enumerable language that is not decidable, it is enough to work with L_{A}. In fact, our construction does not end here. We go for constructing another interesting language using the ideas involved in L_{A}. It is

$$L_{\text{SA}} = \{\psi(M) : \text{ the Turing machine } M \text{ accepts the input } \psi(M)\}.$$

This is called the **self-acceptance language**. The reason for such a name is that when L_{SA} is decidable by a machine U_S, it would mean that U_S gives an answer to the question: "Whether any arbitrary machine accepts itself as its own input?."

Can you see it how? The machine U_S must first determine whether its input string from $\{0, 1\}^*$ represents a machine M, and then it goes for deciding whether that M accepts the binary string $\psi(M)$. Similar to the acceptance language, we have the corresponding problem for the language L_{SA}. The **self-acceptance problem** is the problem of deciding

whether a machine accepts the input which is its own representation.

There is a close connection between L_{A} and L_{SA}; see the following statement.

Theorem 8.8. *If L_{A} is decidable, then L_{SA} is also decidable.*

Proof. In the light of Theorem 8.7, it is enough to show that L_{SA} is computably enumerable. But here is a more direct proof.

Suppose U decides L_{A}. We design a three-tape machine D that would decide L_{SA}. First, D copies its input w from the first tape to the second. Next, it computes $\psi(w)$ from w and writes it on the third tape. It then writes 1 on the square next to w on the first tape and then next to it, keeps a copy of $\psi(w)$ from the third tape. It then erases the second and the third tapes, comes back to the $\not{b}$ preceding w on the first tape, and transfers control to U.

Now, U works on the input $w1\psi(w)$ on its first tape. As U decides L_{A}, it halts in state h if $w = \psi(M)$ for some Turing machine M and M accepts the input w, else, U

halts in state $\hbar$. That is, U halts in state h if M accepts $\psi(M)$, otherwise it halts in state $\hbar$. Accordingly, D halts in its accepting state or its rejecting state. Therefore, D decides L_{SA}. □

L_{SA} is not a canonical language as L_{A}. How then does it help in determining whether there exists an undecidable language or not? The reason is the following result:

Theorem 8.9. *L_{SA} is not a decidable language.*

Proof. On the contrary, suppose L_{SA} is decidable. Then, so is its complement $\overline{L}_{\mathrm{SA}} = \{0,1\}^* - L_{\mathrm{SA}}$. We then have two Turing machines U_S and $\overline{U}_S$ such that U_S accepts L_{SA} and $\overline{U}_S$ accepts $\overline{L}_{\mathrm{SA}}$. The machine $\overline{U}_S$ can be represented by $\psi(\overline{U}_S)$. This particular string $\psi(\overline{U}_S)$ is a string over the alphabet $\{0,1\}$. We ask:

Whether $\psi(\overline{U}_S) \in L_{\mathrm{SA}}$ or $\psi(\overline{U}_S) \in \overline{L}_{\mathrm{SA}}$?

If $\psi(\overline{U}_S) \in L_{\mathrm{SA}}$, then by the definition of the language L_{SA}, the Turing machine $\overline{U}_S$ accepts the input $\psi(\overline{U}_S)$. But $\overline{U}_S$ accepts the language $\overline{L}_{\mathrm{SA}}$. Hence $\psi(\overline{U}_S) \in \overline{L}_{\mathrm{SA}}$.

If $\psi(\overline{U}_S) \in \overline{L}_{\mathrm{SA}}$, as $\overline{U}_S$ accepts $\overline{L}_{\mathrm{SA}}$, we see that $\overline{U}_S$ accepts the input $\psi(\overline{U}_S)$. By the construction of L_{SA}, the string $\psi(\overline{U}_S) \in L_{\mathrm{SA}}$.

Hence, $\psi(\overline{U}_S) \in L_{\mathrm{SA}}$ iff $\psi(\overline{U}_S) \in \overline{L}_{\mathrm{SA}}$. This is a contradiction. Therefore, L_{SA} is not decidable. □

Using Theorems 8.3, 8.6, 8.8, and 8.9, you conclude that

Theorem 8.10. *L_{A} is not decidable and $\overline{L}_{\mathrm{A}}$ is not computably enumerable.*

Exercise 8.8. Show directly that the language $\overline{L}_{\mathrm{SA}}$ is not computably enumerable but L_{SA} is computably enumerable. This will also prove Theorem 8.9.

The languages L_{A} and L_{SA} are computably enumerable but not decidable. Moreover, the languages $\overline{L}_{\mathrm{A}}$ and $\overline{L}_{\mathrm{SA}}$ are not computably enumerable. As L_{A} is computably enumerable, there is a (one-tape) Turing machine M that accepts it. The machine M has states and a transition function. Given any input w to M, it starts referring to its transition function and enters a computation. Now, undecidability of L_{A} means that we cannot have any Turing machine to determine whether M accepts an arbitrarily supplied input or not.

Whenever M accepts w, clearly it can be so determined. For example, by a machine that simulates M; even M itself does the job. But if M does not accept w, then it may either halt abruptly, or rejects w, or goes on an infinite computation. If the last case happens, that is, when M in fact, goes on an infinite computation on the input w, then no Turing machine can find out that the machine M in fact goes on an infinite computation on the input w. This is the reason we had to take recourse in constructing the self-acceptance language.

Any singleton language $\{w\}$ is a decidable language (Why?). But there may not be any Turing machine to decide a countably infinite union of decidable languages. However, finite union of decidable languages is decidable.

Exercise 8.9. Show that every finite language is decidable.

One final comment about Theorem 8.10. This theorem is often referred to as telling that

> Acceptance problem is unsolvable.

Here, the terminology is that a problem is considered **solvable** iff the corresponding language is decidable. Unsolvability of the acceptance problem means that

> It is not the case that there exists a Turing machine U such that given any Turing machine M and any string w as an input to M, the machine U would be able to decide whether M accepts w or not.

It does not say that "For a given Turing machine M, and a given string w, there cannot exist a machine that would determine whether M accepts w or not." The latter is very much possible for certain Turing machines, especially when a machine decides a language. However, for the particular machine U that accepts the acceptance language L_{A}, this stronger statement also holds. That is, there cannot be any machine to decide whether U accepts a given input or not. The same holds for any machine that accepts an undecidable language. Why?

Recall that at the end of Sect. 8.4, I have asked whether corresponding to each TM, there exists a total TM accepting the same language. Does Theorem 8.10 answer it?

Problems for Section 8.5

8.34. Let L be decidable and L' be computably enumerable. Show that $L' - L$ is computably enumerable. Is $L - L'$ decidable?

8.35. Suppose that L is decidable. Is L^+ decidable?

8.36. Determine which of the following sets are computably enumerable and which are not computably enumerable:

(a) $\{(\psi(M1\psi(N) : M$ takes fewer steps than N on input $\varepsilon\}$.
(b) $\{\psi(M) : M$ takes fewer than 2009^{2009} steps on some input$\}$.
(c) $\{\psi(M) : M$ takes fewer than 2009^{2009} steps on at least 2009^{2009} different inputs$\}$.
(d) $\{\psi(M) : M$ takes fewer than 2009^{2009} steps on all inputs$\}$.
(e) $\{\psi(M) : M$ accepts at least 2009 strings$\}$.
(f) $\{\psi(M) : M$ accepts at most 2009 strings$\}$.
(g) Complement of $\{\psi(M) : M$ accepts at least 2009 strings$\}$.
(h) Complement of $\{\psi(M) : M$ accepts at most 2009 strings$\}$.
(i) $\{\psi(M) : \mathcal{L}(M)$ contains at least 2009 strings$\}$.
(j) $\{\psi(M) : \mathcal{L}(M)$ contains at most 2009 strings$\}$.
(k) $\{\psi(M) : M$ halts on all inputs of length less than 2009$\}$.
(l) Complement of $\{\psi(M) : M$ halts on all inputs of length less than 2009$\}$.

8.37. Like a universal TM, a universal finite automaton would accept all strings $\psi(M)1\psi(w)$, where M is a finite automaton accepting the input w. Explain why a universal finite automaton cannot exist.

8.38. Show that computably enumerable languages are closed under homomorphism but decidable languages are not.

8.39. Let Σ be an alphabet not containing #. Suppose $L \subseteq \Sigma^*\{\#\}\Sigma^*$ and $L' = \{w \in \Sigma^* : w\#v \in L \text{ for some } v \in \Sigma^*\}$.
(a) Show that if L is computably enumerable, then so is L'.
(b) Is L' decidable if L is decidable?

8.6 Chomsky Hierarchy

We have tried extensions of Turing machines by equipping it with more tapes and the power of nondeterministic choice. However, we see that no more power is added by doing so. Now, we want to see how a nondeterministic machine can be restricted (not extended) in a useful manner.

The restriction is on the use of the tape, which is potentially infinite. Instead of using any finite portion of the tape, the restriction says that only a predefined portion of the tape has to be used throughout computation. The input alphabet of such a machine includes two special symbols, say, the left end-marker $*$ and the right end-marker $\star$ on the tape. It never moves to the left of $*$ and also it never moves to the right of $\star$. Moreover, it never rewrites these two special symbols. Read $*$ as lem and $\star$ as sem.

A **linear bounded automaton (LBA)** is an eight-tuple $B = (Q, \Sigma, \Gamma, \Delta, s, k, h, \hbar)$, where $\not{L}, \not{R}, *, \star \notin \Gamma$, and k is a given positive integer, and all else are as for a nondeterministic Turing machine.

The symbols $*$ and $\star$ never occur in any transition so that when the read–write head encounters these symbols, the machine halts abruptly. The integer k fixes the portion of the tape that may be used by the machine throughout any computation. Specifically, the portion is fixed as $kn + 4$, where n is the length of the input string. Thus depending upon the input string, a liner size on the tape is fixed as its space resource. We will have in the beginning this resource bound in the form of an input.

An input u will always appear as $*\underline{b\!\!/}\,u\,b\!\!/^{\,n(k-1)}\,b\!\!/\,\star$, where n is the length of the string u. The machine has to operate without going beyond the portion of the tape between $*$ and $\star$. However, we will refer to this appearance on the tape as by telling that "u is the input to the machine."

A configuration is written as $q * u\underline{\sigma}v\star$, as earlier but with the additional $*$ and $\star$ on the tape. The relation of yield among configurations is defined in the usual manner. B accepts a string u iff there is a computation in which $(s, *\underline{b\!\!/}\,u\,b\!\!/^{\,n(k-1)}\,b\!\!/\,\star) \rightsquigarrow (h, v)$ for some $v \in \Gamma^*$. The language accepted by B is

$$\mathcal{L}(B) = \{u \in \Sigma^* : B \text{ accepts u}\}.$$

Exercise 8.10. Define the language rejected by a linear bounded automaton.

Corresponding to a linear bounded automaton, the generative device of a language is called a context sensitive grammar. A **context sensitive grammar** is any (unrestricted) grammar with productions of the form $S \mapsto \varepsilon$ or $uAv \mapsto uwv$, where S is the start symbol, u, v are any strings of terminals and nonterminals, A is a nonterminal, and $w \neq \varepsilon$ is a string of terminals and nonterminals. Moreover, if the production $S \mapsto \varepsilon$ is present in a context sensitive grammar, then S never occurs on the right side of any other production.

Notice that the length of a string on the right side of any rule is never less than that on the left side of the rule, the only exception being the rule $S \mapsto \varepsilon$. We allow this special rule to generate the empty string if required in such a language.

An application of a rule of the form $uAv \mapsto uwv$ always looks for the context: u on the left and v on the right, and then replaces the nonterminal A by w. As $w \neq \varepsilon$, length of w is never less than that of A. Thus every application of a production rule will increase the length of a string except when $S \mapsto \varepsilon$ is applied. Also, when another nonterminal enters a derivation, the rule $S \mapsto \varepsilon$ cannot be applied after that. This is so, because there will not be an S on the derived string any more as S cannot occur on the right side of any other rule. Therefore, during a derivation, the length of the currently derived string can never decrease.

A language generated by a context sensitive grammar is called a **context sensitive language** due to its obvious connotation.

It can be shown that a language is context sensitive iff it is accepted by a linear bounded automaton. It can also be shown that each context-free language is context sensitive. However, each linear bounded automaton, by definition, is a nondeterministic Turing machine. Thus, from Theorem 7.3, it follows that each context-sensitive language is accepted by a Turing machine. However, a stronger result holds. Each context-sensitive language is decidable.

Exercise 8.11. Show that there are only a finite number of possible distinct configurations of a linear bounded automaton with an input of length n. Use it to prove that each context-sensitive language is decidable.

There is an extension of regular languages. In regular grammars, a nonterminal symbol can occur as the right-most symbol of the string on the right side of any production. Such grammars are also called **right linear grammars**. When each production in a grammar has at most one nonterminal on its right side and that nonterminal occurs as the left most symbol on the right side, the grammar is called a **left linear grammar**. In a **linear grammar**, each production has at most one nonterminal on the right side, without any restriction on its position.

It can be shown that **linear languages**, that is, languages that are generated by linear grammars, form a subset of context-free languages. Obviously, each regular language is linear. The class of linear languages is different from that of deterministic context-free languages. These results are summarized by the so-called **Chomsky's Hierarchy** for formal languages. It states that

Each regular language is linear.
Each regular language is deterministic context-free.

Each linear language is context-free.
Each deterministic context-free language is context-free.
Each context-free language is context-sensitive.
Each context-sensitive language is decidable.
Each decidable language is computably enumerable.

Moreover, the containment relations between language classes as stated earlier are proper. That means

There are computably enumerable languages which are not decidable.
There are decidable languages which are not context-sensitive.
There are context-sensitive languages which are not context-free.
There are context-free languages which are not deterministic context-free.
There are context-free languages which are not linear.
There are deterministic context-free languages which are not regular.
There are linear languages which are not regular.
There are deterministic context-free languages which are not linear.
There are linear languages which are not deterministic context-free.

Linear bounded automata are restrictions of nondeterministic Turing machines in terms of the linear space bounds on the tape. Analogous restriction on the deterministic Turing machines ensue the so-called *deterministic linear bounded automata*. It is not yet known whether the languages accepted by deterministic linear bounded automata form a proper subset of those accepted by (nondeterministic) linear bounded automata.

Fix an alphabet Σ. Let $RE, REC, CS, CF, LIN, DCF, REG$ denote, respectively, the classes of languages containing all computably enumerable (recursively enumerable), decidable (recursive), context sensitive, context-free, linear, deterministic context-free, and regular languages over Σ. The Chomsky hierarchy can be written down as

$$REG \subsetneq LIN \cap DCF, \;\; LIN \subsetneq CF, \;\; DCF \subsetneq CF, \;\; CF \subsetneq CS \subsetneq REC \subsetneq RE.$$

Further, regular languages are those that are accepted by finite automata; Context-free languages are those that are accepted by push down automata; Context sensitive languages are those that are accepted by linear bounded automata; and computably enumerable languages are those that are accepted by Turing machines.

Because of this hierarchy, computably enumerable languages are named as **Type-0** languages, context sensitive languages as **Type-1** languages, context-free languages as **Type-2** languages, and regular languages as **Type-3** languages. There are lower type languages that are not of the higher type and each higher type language is, by default, of the lower type. This is referred to as the *Chomsky Hierarchy* and the results stated above constitute the extended Chomsky hierarchy.

Problems for Section 8.6

8.40. Construct context-sensitive grammars for generating the languages:
(a) $\{a^n b^n c^n : n \geq 1\}$.
(b) $\{a^m b^n c^m d^n : m, n \geq 1\}$.

8.41. Construct a left-linear and a right-linear grammar for the language $\{a^m b^n : m \geq 2, n \geq 3\}$.

8.42. Let $G_1 = (N_1, \Sigma, S_1, R_1)$ be a left-linear grammar and $G_2 = (N_2, \Sigma, S_2, R_2)$ be a right-linear grammar with $N_1 \cap N_2 = \emptyset$. Let $S \notin N_1 \cup N_2$. Construct a grammar $G = (N, \Sigma, S, R)$, where $N = N_1 \cup N_2 \cup \{S\}$, $R = R_1 \cup R_2 \cup \{S \mapsto S_1 | S_2\}$. Is $\mathcal{L}(G)$ regular?

8.43. Show that the language $\{a^m b^m c^n : m, n \in \mathbb{N}\}$ is linear.

8.44. Prove the *Pumping Lemma for Linear Languages*: Let L be an infinite linear language. Then there exists a positive integer n such that any $w \in L$, with $\ell(w) \geq n$, can be rewritten as $w = uvxyz$ such that $\ell(vy) \geq 1$, $\ell(uvyz) \leq n$, and $uv^i xy^i z \in L$, for each $i \in \mathbb{N}$.

8.45. Show that the following languages are not linear:
(a) $\{a^m b^m a^n b^n : m, n \in \mathbb{N}\}$.
(b) $\{w \in \{a, b\}^* : \#_a(w) = \#_b(w)\}$.
(c) $\{w \in \{a, b\}^* : \#_a(w) \geq \#_b(w)\}$.
(d) $\{w \in \{a, b, c\}^* : \#_a(w) + \#_b(w) = \#_c(w)\}$.
(e) The language generated by the CFG with productions $S \mapsto A | S{+}A$, $A \mapsto B | A \times B$, $C \mapsto D | (S)$, $D \mapsto a | b | c$.

8.46. Construct context-sensitive grammars for the following languages:
(a) $\{a^{n+1} b^n c^{n-1} : n \geq 1\}$.
(b) $\{a^n b^n c^{2n} : n \geq 1\}$.
(c) $\{ww : w \in \{a, b\}^+\}$.
(d) $\{w \in \{a, b\}^+ : \#_a(w) = \#_b(w) = \#_c(w)\}$.
(e) $\{w \in \{a, b\}^+ : \#_a(w) = \#_b(w) < \#_c(w)\}$.

8.47. Design linear bounded automata for the following languages:
(a) $\{a^{n^2} : n \geq 1\}$.
(b) $\{a^{n!} : n \in \mathbb{N}\}$.
(c) $\{a^p : p \text{ is a prime number}\}$.
(d) $\{a^m : m \text{ is not a prime number}\}$.
(e) $\{ww : w \in \{0, 1\}^+\}$.
(f) $\{w^n : w \in \{0, 1\}^+, n \geq 1\}$.
(g) $\{ww^R w : w \in \{0, 1\}^+\}$.
(h) $\{a, b\}^* - \{a^{n!} : n \in \mathbb{N}\}$.

8.48. Find a linear grammar for generating the language $\{a^m b^n : m \neq n; m, n \in \mathbb{N}\}$.

8.49. Construct grammars G_1, G_2, G_3, G_4, G_5 to generate the language $\{a^n : n \geq 1\}$, where
(a) G_1 is regular.
(b) G_2 is linear but not regular.
(c) G_3 is context-free but not linear.
(d) G_4 is context-sensitive but not context-free.
(e) G_5 is unrestricted but not context-sensitive.

8.50. In the Chomsky hierarchy, where do you put the class of languages whose complements are computably enumerable? This class is named as *co-computably enumerable languages.*

8.51. Let G be the grammar with productions $S \mapsto A,\ A \mapsto aB|\varepsilon,\ B \mapsto Ab$. What is $\mathcal{L}(G)$? Is it regular?

8.52. Let G be the grammar with productions $S \mapsto Aab,\ A \mapsto Aab|B,\ B \mapsto a$. Is $\mathcal{L}(G)$ regular?

8.53. Let G be the grammar with productions $S \mapsto abc|aSAc, bA \mapsto bb, cA \mapsto Ac$, and G' be the grammar having productions $S \mapsto aSBC|aBC,\ CB \mapsto BC,\ aB \mapsto ab$, $bB \mapsto bb,\ bC \mapsto bc,\ cC \mapsto cc$. Are G and G' equivalent?

8.54. Show that the language $\{vwv : v, w \in \{a, b\}^+\}$ is context-sensitive.

8.55. If two grammars generate the same language, is it that they are of the same type? Justify.

8.56. You can prove that context-sensitive languages are decidable without using LBAs. For this, use the *sentential forms,* that is, strings of terminals and nonterminals those can be derived from the start symbol. Fix a context-sensitive grammar with start symbol S. Write S_i as the set of all sentential forms that are derived from S in ith step. Show that for each $i \in \mathbb{N}$, $S_i \subseteq S_{i+1}$ and that there exists a $k \in \mathbb{N}$ such that $S_k = S_{k+1}$. Next, show that a string of terminals is generated by the grammar iff it is in S_k. Now, write the algorithm to check whether a given string is at all generated by the grammar or not.

8.57. Let G be the context-sensitive grammar with productions $S \mapsto aSa|AcA$, $Aa \mapsto A|a,\ aA \mapsto A|a$ and G' be one with productions $S \mapsto aSa|AcA,\ Aa \mapsto A|a,\ aA \mapsto a$. Which one of $\mathcal{L}(G)$ and $\mathcal{L}(G')$ is regular and which one is context-free but not regular?

8.58. For each statement below, give one example of a language L that satisfies it.
(a) L is computably enumerable but not decidable.
(b) L is decidable but not context-sensitive.
(c) L is context-sensitive but L is not context-free.
(d) L is context-free but not deterministic context-free.
(e) L is linear but not deterministic context-free.
(f) L is deterministic context-free but not linear.
(g) L is linear but not regular.
(h) L is deterministic context-free but not regular.

8.59. Give a language $L \subseteq a^*$ for each of the following classes of languages, or explain why such an L cannot exist:
(a) Regular.
(b) Linear but not regular.
(c) DCFL but not linear.
(d) CFL but not DCFL.
(e) Decidable but not CFL.
(f) Recursively enumerable but not decidable.
(g) Not computably enumerable.

8.60. Let r be an irrational number between 0 and 1. Write r in decimal notation and remove the beginning decimal point from it; call the *infinite string* of digits so obtained as s. Which language class(es) in the Chomsky hierarchy the set of prefixes of s belong to?

8.7 Summary and Additional Problems

The versatility of Turing machines have been demonstrated by using them as computers of numerical functions, and then as language deciders. Though the Turing machines are believed to be most powerful automata, a simple counting argument shows that there are more languages that are not recognized by them than otherwise. Our quest has been to construct one such language. This has been done by using the diagonalization technique of Georg Cantor via encoding of TMs as binary strings. This also demonstrated the usefulness of our chosen data structure as strings. We have shown that the acceptance problem of TMs leading to a noncomputably enumerable language, and in turn, a problem that cannot be solved by TMs. We then discussed the hierarchy of languages, called the extended Chomsky hierarchy.

The diagonalization technique was invented by Cantor [12] to prove that there are fewer algebraic real numbers than the real numbers. A. M. Turing [131] has used it to show that the halting problem for TMs (given an arbitrary TM and an arbitrary string, whether the TM halts on input as that string) is unsolvable. We have used his technique in showing that the language L_{A} is not decidable; and will see, in the next chapter, that the language of the halting problem is undecidable.

Church's thesis has been first formulated in [19], observing that the μ-recursive functions of Gödel and Herbrand [44], and the λ-calculus of Kleene and Church talked about the same class of functions, unaware of Turing's work [131], where Turing's thesis appeared. For a distinction between Church's thesis and Turing's thesis see [124]. Turing has shown that his machines compute the same class of functions as defined by the λ-calculus; moreover, his machines present a more compelling definition of the notion of computability. A good reference on λ-calculus is [9]. Other systems mentioned in the text that have been shown to be equivalent to the Turing machine model of computation are Post systems [103, 104], combinatorial logic [24], Markov's normal algorithms [82], the unlimited register machines [117], and the while-programs.

You can find introductory material on Post systems, μ-recursive functions, λ-calculus, combinatorial logic, and while programs in [27, 58, 71, 78, 79, 118, 122]. You can see the references cited in these books for advanced texts on computability. Cutland [25] works with unlimited register machines as we work with Turing machines, and it also deals with logical theories, in detail.

Ackermann's function used in one of the exercises below is from [2]; discovery of this led Gödel to extend primitive recursive functions to μ-recursive functions. The relationship between the for-programs and the primitive recursive functions was discussed in [87]. Subsequently, the while-programs were found to be equivalent to the Turing machine model. For linear bounded automata and context sensitive grammars, you can see [74, 75, 93].

Additional Problems for Chapter 8

8.61. Let L be a computably enumerable but an undecidable language over an alphabet Σ not containing 0 and 1. Define $L_1 = \{0w : w \in L\} \cup \{1w : w \notin L\}$ and $L_2 = \Sigma^* - L_1$. Are L_1 and L_2 computably enumerable? What happens if $0, 1 \in \Sigma$?

8.62. Let $f : \Sigma_0^* \to \Sigma_1^*$ be a total onto computable function. Show that there is a total computable function $g : \Sigma_1^* \to \Sigma_0^*$ such that $f(g(w)) = w$, for each $w \in \Sigma_0^*$.

8.63. Here is a construction of an uncomputable total function. (Again, contrast it with Example 1.10.) The *busy beaver function* is the function $\beta : \mathbb{N} \to \mathbb{N}$ defined by

> $\beta(n)$ is the largest natural number m such that there is a TM $(Q, \{0, 1\}, \{b\!\!\!/, 0, 1\}, \delta, h, \hbar)$, where Q has exactly n states, and when the TM is started on the blank tape, it eventually halts at the configuration $h\underline{b\!\!\!/}\,0^m$.

Prove the following:

(a) If f is any total computable function, then there exists an integer k_f such that $\beta(n + k_f) \geq f(n)$.

(b) β is an uncomputable total function.

8.64. Let $G = (N, \Sigma, R, S)$ be an unrestricted grammar. We say that G computes the function $f : \Sigma^* \to \Sigma^*$ if and only if, for all $v, w \in \Sigma^*$, we have "$SwS \Rightarrow_G v$ iff $v = f(w)$." Such a function f is then called *grammatically computable*. Show that f is grammatically computable iff f is computable (by a TM).

8.65. Let G be the grammar with productions $S \mapsto X|Z$, $XZ \mapsto \varepsilon$, $Xa \mapsto XA$, $Xb \mapsto XB$, $Aa \mapsto aA$, $Ab \mapsto bA$, $Ba \mapsto aB$, $Bb \mapsto bB$, $AZ \mapsto Za$, $BZ \mapsto Zb$. What is the function computed grammatically by G?

8.66. Show by construction of appropriate grammars that if $f : \Sigma_0^* \to \Sigma_1^*$ is grammatically computable, then so are the functions g and h, where $g(w) = f(w)^R$ and $h(w) = f(w)f(w)$.

8.67. When grammars are used to compute functions, the order in which rules are applied is indeterminate. This indeterminacy is avoided by a Markov system due to

A.A. Markov. A *Markov System* is a quadruple $G = (N, \Sigma, R, R')$, where N is the set of nonterminals, Σ is the set of terminals, R is a sequence (not just a set) of production rules $(u_1 \mapsto v_1, \ldots, u_k \mapsto v_k)$, and R' is a set of rules from R. Let w be a string. If there is an $i \geq 1$ such that u_i is a substring of w, then let i be the smallest such number. Let w_1 be the shortest string such that $w = w_1 u_i w_2$. Then $w \Rightarrow_G w_1 v_i w_2$. Thus, if a rule is at all applicable, then there is a unique rule that is to be applied, and it is applied exactly in one position. We say that G *computes* a function $f : \Sigma^* \to \Sigma^*$ if for all $x \in \Sigma^*$, $x = x_0 \Rightarrow_G x_1 \Rightarrow_G \cdots \Rightarrow_G x_{n-1} \Rightarrow_G x_n = f(x)$ and provided that the rules that have been used for deriving $x_0 \Rightarrow_G x_1$ and $x_{n-1} \Rightarrow_G x_n$ are the same rule and it is from R'. Show that a function is computable by a Markov System iff it is computable (by a TM).

8.68. Prove that left-linear grammars generate all and only the regular languages.

8.69. Suggest a method to obtain a left-linear grammar from an NFA directly. Use this to construct a left-linear grammar that generates the language $(aab^*ab^*)^*$.

8.70. Let L be a linear language not containing ε. Show that there is a linear grammar having productions in the forms $A \mapsto a$, $A \mapsto aB$, or $A \mapsto Ba$ that generates L. [Here is a linear grammar without ε-productions and without unit productions.]

8.71. You can say that a CFG is a linear grammar if no production in it has its right hand side string with more than one nonterminal. A PDA is similarly called *single-turn* whenever once its stack decreases in height, it never increases thereafter. Formalize such a PDA. Show that a language is linear iff it is accepted by a single-turn PDA.

8.72. Let L be a language over an alphabet Σ. Assume that $\varepsilon \notin L$. Prove that L is context sensitive iff $L = \mathcal{L}(M)$ for some LBA M.

8.73. Show that the family of linear languages is closed under union and reversal, but neither under concatenation nor under intersection. What about complementation?

8.74. Show that if L is linear and L' is regular, then LL' is linear.

8.75. Show that the family of linear languages is closed under homomorphism.

8.76. Prove that every infinite regular language has a subset that is not computably enumerable.

8.77. Does there exist a TM that accepts an infinite language, which does not contain an infinite regular language?

8.78. A *directed graph* is an ordered pair (V, E), where V is a nonempty finite set, called the set of vertices, and $E \subseteq V \times V$, called the set of edges. A *path* in a directed graph is a sequence of distinct vertices $v_1, v_2, \ldots, v_n$ such that there is an edge from each v_i to v_{i+1}. Give an encoding of vertices, edges, and finally, a directed graph as bits, that is, if x is a vertex, e is an edge, G is a directed graph, then $\rho(x)$, $\rho(e)$, $\rho(G) \in \{0, 1\}^*$. Design an NTM that accepts the language $\{\rho(G)2\rho(u)2\rho(v) :$ there is a path from u to v in $G\}$.

8.79. In an *undirected graph*, both the edges (u, v) and (v, u) are considered the same edge. In fact, the ordered pair (u, v) is replaced by the two-elements-set $\{u, v\}$ of vertices. In an undirected graph, a *clique* is a subset of vertices such that each vertex in this subset is connected to each other vertex in this subset, by an edge. The *size of a clique* is the number of vertices in a clique. Show that the language $\{\rho(G)2\rho(k) : G$ has a clique of size $k\}$ is computably enumerable.

8.80. A *Hamiltonian path* in an undirected graph is a path that contains each vertex of the graph exactly once. Show that the language $\{\rho(G) : G$ contains a Hamiltonian path$\}$ is computably enumerable.

8.81. Describe NTMs for accepting the following languages. Take advantage of the nondeterminism available, by preferring more branches of computations while each branch is short.

(a) $\{uvwvx : \ell(v) = 99,\ u, v, w, x \in \{a, b\}^*\}$.

(b) $\{w_1 2 w_2 2 \cdots 2 w_n : n \geq 1$, each $w_i \in \{0, 1\}^*$, and for at least one j, w_j is the binary representation of $j\}$.

(c) $\{w_1 2 w_2 2 \cdots 2 w_n : n \geq 1$, each $w_i \in \{0, 1\}^*$, and for at least two indices j, k, w_j is the binary representation of j and w_k is the binary representation of $k\}$.

8.82. A TM *lexicographically enumerates* a language iff the following holds: "If $q\underline{\not b}v \rightsquigarrow q\underline{\not b}w$, then w comes lexicographically after v" for some fixed state q. Show that a language is decidable iff it is lexicographically enumerated by some TM.

8.83. Prove that every infinite computably enumerable language has an infinite subset which is decidable.

8.84. There is a variation of Turing machines, where a machine when writes a symbol, also moves, either to the left or to the right. A typical transition that looks like $\delta(p, a) = (q, b, \not L)$ means that if the machine is in state p, scans a, then in one move, it goes to state q replaces that a by a b, and its read–write head is positioned on the square left to the current one. Similarly, the transitions in an NTM are taken as quintuples like $(p, a, q, b, \not L)$ instead of quadruples.

(a) What is $\mathcal{L}(M)$ if $M = (\{p, q, r, s\}, \{0, 1\}, \{0, 1, 2, 3, \not b\}, \delta, p, h, \hbar)$ with $\delta(p, \not b) = (p, \not b, \not R)$, $\delta(p, 0) = (q, 2, \not R)$, $\delta(p, 3) = (s, 3, \not R)$, $\delta(q, 0) = (q, 0, \not R)$, $\delta(q, 1) = (r, 3, \not L)$, $\delta(q, 3) = (q, 3, \not R)$, $\delta(r, 0) = (r, 0, \not L)$, $\delta(r, 2) = (p, 2, \not R)$, $\delta(r, 3) = (r, 3, \not L)$, $\delta(s, 3) = (s, 3, \not R)$, $\delta(s, \not b) = (h, \not b, \not R)$?

(b) Suppose we represent numbers in binary. Then which numerical function is computed by $M = (\{p, q, r, s, t, u\}, \{0, 1\}, \{0, 1, \not b\}, \delta, p, h, \hbar)$, where $\delta(p, \not b) = (p, \not b, \not R)$, $\delta(p, 0) = (q, \not b, \not R)$, $\delta(p, 1) = (u, \not b, \not R)$, $\delta(q, 0) = (q, 0, \not R)$, $\delta(q, 1) = (r, 1, \not R)$, $\delta(r, 0) = (s, 1, \not L)$, $\delta(r, 1) = (r, 1, \not R)$, $\delta(r, \not b) = (t, \not b, \not L)$, $\delta(s, 0) = (s, 0, \not L)$, $\delta(s, 1) = (s, 1, \not L)$, $\delta(s, \not b) = (p, \not b, rt)$, $\delta(t, 0) = (t, 0, \not L)$, $\delta(t, 1) = (t, \not b, \not L)$, $\delta(t, \not b) = (h, 0, \not R)$, $\delta(u, 0) = (u, \not b, \not R)$, $\delta(u, 1) = (u, \not b, \not R)$, $\delta(u, \not b) = (h, \not b, \not R)$?

(c) What is $\mathcal{L}(M)$ if $M = (\{p, q, r\}, \{0, 1\}, \{0, 1, \not{b}\}, \Delta, h, \hbar)$ is the NTM with transitions $(p, \not{b}, p, \not{b}, \not{R})$, $(p, 0, p, 1, \not{R})$, $(p, 0, q, 1, \not{R})$, $(q, 1, r, 0, \not{L})$, $(r, 1, p, 1, \not{R})$, and $(q, \not{b}, h, \not{b}, \not{R})$?

(d) Show that these machines accept the same class of languages as the standard machines. Further, show that these machines compute the same functions as the standard ones.

8.85. In another variation, a Turing machine has a tape having a left end but no right end unlike ours, which have both the ends extending to infinity. During computations of such a machine, if the read–write head attempts to move off the left end of the tape, then the machine halts abruptly. Show that these machines compute the same class of functions as ours.

8.86. An *off-line Turing machine* is a DTM having two tapes, one of them is the input tape, the other is both a working tape and the output tape. From the input tape, the first reading head reads symbols, while everything else is done by the second read–write head on the second tape. Show that off-line machines are equivalent to standard Turing machines.

8.87. Consider an off-line TM with two tapes, in which the input can be read only once, moving left to right, and not rewritten. On its work tape, it can use at most n extra squares for work space, where n is fixed for the machine, not depending upon inputs. Show that such a machine is equivalent to a finite automaton.

8.88. Give a direct proof of the fact that each language accepted by a PDA is also accepted by some linear bounded automaton.

8.89. Define a *deterministic linear bounded automaton* as an LBA where the TM is deterministic. Show that corresponding to each deterministic LBA, there exists a deterministic LBA that halts on every input.

8.90. A *deterministic context sensitive language* is a language accepted by a deterministic LBA. Show that the class of deterministic context-sensitive languages is closed under complementation.

8.91. In what way the class of deterministic context-sensitive languages and the class of context-free languages are related?

8.92. We define a certain class of k-ary numerical functions, the functions from $\mathbb{N}^k$ to $\mathbb{N}$, for various values of $k \in \mathbb{N}$. Notice that for $k = 0$, the 0-ary functions are simply natural numbers, as such a function has nothing to depend upon. First, we define the so-called *basic functions.* These are the following:

1. The *k-ary zero function* is defined by $zr_k(n_1, \ldots n_k) = 0$ for all $n_1, \ldots, n_k \in \mathbb{N}$.
2. The jth *k-ary identity function* is defined by $id_{k,j}(n_1, \ldots, n_k) = n_j$, for all $n_1, \ldots, n_k \in \mathbb{N}$ and for all j with $0 < j \leq k$. These functions are also called *projection functions.*
3. The *successor function* is the same as earlier, $succ(n) = n + 1$.

New functions are built from the basic functions by applying the following two operations:

(i) Let $g : \mathbb{N}^t \to \mathbb{N}$ be a t-ary function, and let $h_1, h_2, \ldots, h_t$ be m-ary functions. The *composition of g with* $h_1, \ldots, h_t$ is the m-ary function f defined by $f(n_1, \ldots, n_m) = g(h_1(n_1 \ldots, n_m), \ldots, h_t(n_1, \ldots, n_m))$.

(ii) Let g be a k-ary function and h be a $(k+2)$-ary function. The *function defined recursively* by g and h is the $(k+1)$-ary function given by:

$$f(n_1, \ldots, n_k, 0) = g(n_1, \ldots, n_k),$$
$$f(n_1, \ldots, n_k, m+1) = h(n_1, \ldots, n_k, m, f(n_1, \ldots, n_k, m)),$$

for all $n_1, \ldots, n_k, m \in \mathbb{N}$.

The *primitive recursive functions* are functions from $\mathbb{N}^k$ to $\mathbb{N}$ obtained from the basic functions by any (finite) number of successive applications of composition and recursive definition. Show that the following functions are primitive recursive:

(a) $p2(n) = n + 2$.
(b) $plus(m, n) = m + n$.
(c) $mult(m, n) = mn$.
(d) $exp(m, n) = m^n$.
(e) Predecessor function: $pred(0) = 0$ and $pred(n+1) = n$.
(f) Monus function: $m \sim n = \max(m - n, 0)$.
(g) Constant functions: $f(n_1, \ldots, n_k) = n$, for a fixed $n \in \mathbb{N}$.
(h) Sign function: $sgn(n) = 0$ for $n = 0$, and $sgn(n) = 1$ for $n > 0$.
(i) All polynomials in one or many variables.

8.93. A *primitive recursive predicate* is a primitive recursive function that only takes values 0 or 1. When its value is 1, we say that the predicate holds; and when its value is 0, we say that the predicate does not hold. For instance, the relation $m > n$ is captured by a primitive recursive predicate, which is evaluated to 1 when $m > n$, else, it is evaluated to 0. Similarly, $equal(m, n)$ is a primitive recursive predicate, which compares m and n for equality. This predicate holds, that is, it is evaluated to 1, when $m = n$, else, $equal(m, n)$ is evaluated to 0, in which case, $m \neq n$.

Show that if g and h are primitive recursive functions of arity k, and p is a primitive recursive predicate with the same arity k, then the *function* $f : \mathbb{N}^k \to \mathbb{N}^k$ *defined by cases* as

$f(n_1, \ldots, n_k) = g(n_1, \ldots, n_k)$ if $p(n_1, \ldots, n_k)$ holds, and
$f(n_1, \ldots, n_k) = h(n_1, \ldots, n_k)$ if $p(n_1, \ldots, n_k)$ does not hold,

is also a primitive recursive function.

8.94. Show that the following functions are primitive recursive:

(a) $rem(0, n) = 0$;
$rem(m+1, n) = 0$, if $equal(rem(m, n), pred(n))$ holds;
$rem(m+1, n) = rem(m, n) + 1$ if $equal(rem(m, n), pred(n))$ does not hold.
(b) $div(0, n) = 0$;
$div(m+1, n) = div(m+1, n) + 1$ if $equal(rem(m, n), pred(n))$ holds;
$div(m+1, n) = div(m, n)$, if $equal(rem(m, n), pred(n))$ does not hold.
(c) $digit(m, n, p) =$ the m-th least significant digit of the base-p representation of n.
(d) $sum_f(m, n) = f(m, 0) + \cdots + f(m, n)$, provided $f(m, n)$ is primitive recursive.

8.95. *Ackermann's function* is the function $A : \mathbb{N} \to \mathbb{N}$ defined by

$A(0, n) = n + 1;$
$A(m, 0) = A(m - 1, 1)$ for $m \geq 1;$
$A(m, n + 1) = A(m - 1, A(m, n))$, for $m \geq 1, n \geq 0.$

Prove that

(a) $A(1, n) = n + 2,\ A(2, n) = 2n + 3$, and $A(3, n) = 2^{n+3} - 3$.
(b) Ackermann's function is a total and computable function.
(c) For each primitive recursive function $f : \mathbb{N} \to \mathbb{N}$, there exists a natural number k such that $f(j) < A(k, j)$ holds for all $j \geq k$.
(d) Ackermann's function is not primitive recursive.

8.96. Can you find a numerical function, other than Ackermann's function, which is a computable total function but not primitive recursive? Does there exist a numerical partial function, which is computable but not primitive recursive?

8.97. Let g be a $(k + 1)$-ary numerical function. The *minimalization* of g is the k-ary function f defined by $f(n_1, \ldots, n_k) =$ the least m such that $g(n_1, \ldots, n_k, m) = 1$ if such an m exists, else, $f(n_1, \ldots, n_k) = 0$.

The minimalization of g is denoted by $\mu m[g(n_1, \ldots, n_k, m) = 1]$. Regard μm as a quantifier such as $\forall x$ or $\sum_i$.

The function g is called *minimalizable* if for each $n_1, \ldots, n_k \in \mathbb{N}$, there exists an $m \in \mathbb{N}$ such that $g(n_1, \ldots, n_k, m) = 1$. In such a case, the least such m is given by the following algorithm:

$m := 0$; `while` $g(n_1, \ldots, n_k, m) \neq 1$ `do` $m := m + 1$ `od`; `print` m

We call a numerical function *μ-recursive* if it can be obtained from the basic functions by the operations of composition, recursive definition, and minimalization of minimalizable functions. Show that the following functions are μ-recursive:

(a) $\log(m, n)$, which is defined as the smallest power to which $m + 2$ must be raised to get an integer at least as big as $n + 1$
(b) $lastpos(n, b)$, the last (rightmost, least significant) position in the string encoding n in base b.
(c) $f^{-1} : \mathbb{N} \to \mathbb{N}$ provided $f : \mathbb{N} \to \mathbb{N}$ is a μ-recursive bijection.

8.98. Show that a numerical function is μ-recursive iff it is computable (by a TM).

8.99. Prove that no computably enumerable set of total μ-recursive functions can contain all total computable functions.

8.100. Consider a simple programing language with variables $Var = \{x, y, \ldots\}$ ranging over $\mathbb{N}$ containing the following constructs:

1. Simple Assignments: $x := 0 \quad x := y + 1 \quad x := y$
2. Sequential Composition: $p\, ;\, q$
3. Conditional: `if` $x < y$ `then` p `else` q `fi`
4. For Loop: `for` y `do` p `od`
5. While Loop: `while` $x < y$ `do` p `od`
6. Print: `print` x

In (3 and 4), the relation $<$ can be replaced by any one of $>$, $\leq$, $\geq$, $=$, or $\neq$. Programs built inductively from these constructs without using (5) are called *for-programs* and those from all of these constructs are called *while-programs*.

The meanings are as usual in programing languages. For instance, entering the for-loop: `for` y `do` p `od`, the current value of the variable y is determined, and the program p is executed that many times. Similarly, upon entering the while-loop: `while` $x < y$ `do` p `od`, the condition $x < y$ is tested with the current values of x and y. If the condition $x < y$ is false, then it is considered that the loop has been executed. On the other hand, if the condition $x < y$ is satisfied, then p is executed once, and then, the complete loop is executed again. We say that a program computes a numerical function f if given an input m, it prints $f(m)$.

Let g be a numerical function. Prove that
(a) g is primitive recursive iff there is a for-program to compute it.
(b) g is μ-recursive iff there is a while-program to compute it.

8.101. An *unlimited register machine* or a *URM* has an infinite number of registers labeled $R_1, R_2, \ldots$ each of which, at any instant of time, contains a natural number. Imagine a tape with a left end and squares on it as usual, but now each square can contain a whole number and not just a symbol. The contents of the registers can change by the URM in response to certain *instructions* that it can recognize. A finite set of instructions is a *program*. The instructions are of four kinds, as given below.

1. *Zero Instructions*: For each $n \geq 1$, there is a zero instruction $Z(n)$. The response of a URM to $Z(n)$ is to change the content of R_n to 0, leaving all other registers unaltered.
2. *Successor Instructions*: For each $n \geq 1$, there is a successor instruction $S(n)$. The response of a URM to $S(n)$ is to increase the content of R_n by 1, leaving all other registers unaltered.
3. *Transfer Instructions*: For each $m \geq 1, n \geq 1$, there is a transfer instruction $T(m, n)$. The response of a URM to $T(m, n)$ is to replace the content of R_n by the number r_m that is in R_m, that is, transfer r_m to R_n, leaving all other registers (including R_m) unaltered.
4. *Jump Instructions*: Each instruction in a program is given a number, for example, as is done by the line numbers in the BASIC language. For each $m \geq 1$, $n \geq 1, k \geq 1$, there is a jump instruction $J(m, n, k)$. The response of a URM to $J(m, n, k)$ is as follows. Suppose $J(m, n, k)$ is encountered in a program P. The contents of r_m of R_m and r_n of R_n are compared. If $r_m = r_n$, then the URM proceeds to the instruction numbered k in P. If $r_m \neq r_n$, then the URM proceeds to the next instruction in P. (That is, next to where this $J(m, n, k)$ is; not necessarily to the $(k+1)$th.)

A URM is given the numbers $a_1, a_2, \ldots$ in its registers, initially; this is its input. It is also given a program. Then it computes responding to the instructions in the program sequentially, and thereby changing the contents of the registers or jumping to various instructions in the program. The URM continues its computation until the last instruction in the program is executed. If, after the last instruction is executed, the URM halts, then the contents of its registers, as a sequence of natural numbers, is taken as its output. To compute a function $f : \mathbb{N} \to \mathbb{N}$, we give input to a URM in the form of m in the first register and keeping the contents of all other registers as 0, If the URM with a suitable program halts with $f(m)$ in its first register and 0 elsewhere, we say that it computes f. Similarly, numerical functions $g : \mathbb{N}^m \to \mathbb{N}^n$ can be computed by URMs.

Prove that a function $f : \mathbb{N} \to \mathbb{N}$ is μ-recursive iff it is computed by a URM with a suitable program. [Note: The so called *random access Turing machines* are a kind of URM.]

8.102. Write a program in C that simulates a Turing machine. Such a simulator program should accept the initial configuration, the description of a Turing machine, and give the output as the output of the machine.

8.103. How can a computer be simulated by a Turing machine?

9 Algorithmic Solvability

9.1 Introduction

You have come a long way starting from regular grammars to unrestricted grammars and from deterministic finite automata to Turing machines. You have seen how versatile the Turing machines are. They seem to be the most general kind of computing devices in the sense that all known algorithms can be realized as Turing machines. Quoting various approaches to computing models such as unrestricted grammars, λ-calculus, μ-recursive functions, Post's machines, Markov's normal algorithms, unrestricted memory machines, and while-programs, I asked you to believe in the fact that all these models characterize the same set of languages. The search is for a correct formalization of the informal notion of an algorithm.

Informally, an algorithm is a step-by-step procedure that uses a finite amount of space and time to do a bit of work, where each step is unambiguous (effectiveness). Any formalization of this notion must also have the capability to express any effective procedure in a step-by-step manner, while it being such a procedure itself. That is, any object that is formally defined as an algorithm must be an algorithm informally. This means that the formally defined entities that are claimed to be algorithms construe a subset of all informal algorithms.

On the other hand, any object that has been accepted as an algorithm informally has been shown to be an algorithm formally, using any of the known formalizations. Surprisingly, all the formalizations created so far agree, in the sense that all of them have been found to describe the same subset of the set of all informal algorithms. Thus we come to the conclusion that probably we have captured the notion of an algorithm; the subset that any or all of these formal devices define is possibly the whole set of informal algorithms. This is the content of Church–Turing thesis. It says that an algorithm is nothing but a total Turing machine. As of any scientific theory, this thesis cannot be proved, but can only be verified, and possibly, be falsified.

If in future, someone defines another formalization of the notion of an algorithm, which do all works that total Turing machines do and which do some works that a total Turing machine cannot do, then Church–Turing thesis will be falsified. This of course is a remote possibility, a possibility nonetheless. As to the verification part,

A. Singh, *Elements of Computation Theory*, Texts in Computer Science,

you have seen in the last chapter how many odd works such as copying, reversing, and shifting of strings could be done by designing simple Turing machines (diagrams).

The purpose of working with machine diagrams were twofold. The first was to convince you that most of the usual works for which we may think of an easy step-by-step procedure could be done by total Turing machines. The second goal was to see that machine diagrams are nothing but flow-chart-like translations of procedural English, which is commonly used to express algorithms informally.

Turing machines, when written down with all its details of states and transitions, are comparable to the machine language of a modern computer. The machine diagrams built from the basic macros such as the symbol writing machines, the head moving machines, and the halting machines are analogous to macros in an assembly language. The informal procedural English used for writing an algorithm is comparable with a high level language such as C or Java. Once you realize this point, we will express our procedures or Turing machines in simple procedural English. In each case of an algorithm written this way, you must convince yourself that a machine diagram can be constructed so that we are, infact, talking of a Turing machine.

9.2 Problem Reduction

As an evidence of the prowess of Turing machines, we had demonstrated how certain machines could take other machines as inputs via encoding of the latter. In doing so, we have used the notation $\psi(M)$ for the string of 0's and 1's that encodes the Turing machine M. Similarly, we have also denoted the encoding of an input string u by $\psi(u)$. When both M and u are given to another machine as inputs, we use the notation $\psi(M)1\psi(u)$. We will continue the same notation. However, you may abstract away the strict encoding of M and u with binary strings. It is enough to remember that $\psi(M)$ and $\psi(u)$ are the encodings of M and u in some alphabet.

Sometimes, we may use a different symbol such as # instead of 1, separating the encodings $\psi(M)$ and $\psi(u)$ when the input is the pair (M, u). The first instance where an encoding has been used is for showing that the acceptance problem is unsolvable. To refresh the result, let me quote a paragraph from the last chapter. Acceptance problem is unsolvable means that

> It is not the case that there exists a Turing machine D such that given any Turing machine M and any string u as an input to M, the machine D would be able to decide whether M accepts u or not.

Implicitly it is assumed that D is a total TM (Turing machine) and when M accepts u, D signals it by giving 1 as the output, else, D outputs 0. (We write 1 for α and 0 for ν.) The same fact is expressed by telling that the acceptance language L_{A} is not decidable. With regard to Church–Turing thesis, a problem is called solvable when there is an algorithm that gives a "yes" or "no" answer to each instance of the problem. Thus, solvability here means "algorithmic solvability." For example, if number theory is decidable, it would mean that there is an algorithm, a total TM, which would take any possible sentence of number theory as an input and would answer "yes" (output 1) if the sentence is true, else, the algorithm answers "no" (output 0).

In the rest of this chapter, you will work through many such problems from language theory and will come to know many interesting problems from other branches of mathematics towards their algorithmic solvability. The main technique in proving the results on solvability has already been illustrated in showing that the acceptance problem is unsolvable.

First, we have formulated a suitable language corresponding to the problem; the acceptance language L_{A} for the acceptance problem. Second, we have constructed another language L_{SA}, the self-acceptance language. Specifically, we have shown that if L_{A} is decidable, then so is L_{SA}. That is, we have reduced L_{SA} to L_{A}. Third, we have used Cantor's diagonalization technique to show that L_{SA} is not decidable.

Notice the direction of reduction. We say that L_{SA} has been reduced to L_{A}, not otherwise. It is because if we have an algorithm to decide L_{A}, then we can construct an algorithm to solve L_{SA}. That is, if we can solve the acceptance problem, then we can solve the self-acceptance problem, but not otherwise. In this section, we concentrate on the second step of problem reduction. Once we have an unsolvable problem, we can reduce this to another problem so that the latter becomes unsolvable.

Suppose B and C are two problems. First we construct the corresponding languages L_B and L_C over some appropriate alphabets Σ_B and Σ_C, respectively. To reduce B to C means that when we have an algorithm for solving C; and we can somehow use this algorithm to solve B. As languages, L_B is to be transformed to L_C. As a decision procedure for L_C is expected to work for L_B, via this transformation, the transformation of L_B to L_C must be captured by another algorithm.

That is, we should have a map from Σ_B^* to Σ_C^*, which takes $L_B \subseteq \Sigma_B^*$ to exactly $L_C \subseteq \Sigma_C^*$. Consequently, $\Sigma_B^* - L_B$ is taken to $\Sigma_C^* - L_C$ by the transformation. Moreover, as the languages L_B and L_C are suitable representations of the problems B and C, respectively, we can have a uniform alphabet for both of them. Thus, without loss of generality, we assume that $\Sigma_B = \Sigma_C$. Problem reduction via a transformation or a map such as this is called a *map reduction* or a mapping reduction, and is formally defined as follows.

Let L and L' be languages over an alphabet Σ. A total function $f : \Sigma^* \to \Sigma^*$ is called a **map reduction of L to L'** if f is computable, and provided for each $w \in \Sigma^*$, we have $w \in L$ iff $f(w) \in L'$. In such a case, we say that L *is reduced to* L' *by the map* f, and write it as $L \overset{f}{\leq_m} L'$. When the map f is not so important to be mentioned, we will write $L \leq_m L'$ and read it as L is **map-reduced to** L'.

The subscript m in $L \leq_m L'$ indicates that this reduction is via a map, which is, in general, many-one. The goal of such a reduction is to be able to decide L by using a decision procedure for L'.

Theorem 9.1. *If $L \leq_m L'$ and L' is decidable, then L is decidable.*

Proof. Let $L,\ L' \subseteq \Sigma^*$ and let $L \leq_m L'$ by the map f. If L' is decidable, then its characteristic function $\chi_{L'}$ is a computable function. As f is computable, the composition map $\chi_{L'} \circ f$ is computable. We show that $\chi_{L'} \circ f = \chi_L$. To this end, let $w \in \Sigma^*$. Either $w \in L$ or $w \notin L$. If $w \in L$, then $f(w) \in L'$, and then $\chi_{L'}(f(w)) = 1 = \chi_L(w)$. On the other hand, if $w \notin L$, then $f(w) \notin L'$, and then $\chi_{L'}(f(w)) = 0 = \chi_L(w)$. Therefore, $\chi_{L'} \circ f = \chi_L$. And hence χ_L is computable. That is, L is decidable. □

Exercise 9.1. Let $f_1 : \Sigma_1^* \to \Sigma_2^*$, $f_2 : \Sigma_2^* \to \Sigma_3^*$ be computable functions. Show that $f_2 \circ f_1 : \Sigma_1^* \to \Sigma_3^*$ is computable.

To see the above proof more directly, let M' be a total TM that decides L'. Construct a TM M as follows:

On input w, M computes $f(w)$ and then simulates M' on $f(w)$.

As f is a computable function, $f(w)$ can be computed. As M' is a total TM, M halts on w and answers 1 when M' halts on $f(w)$ and answers 1; else, M halts on w and answers 0 as M' halts on $f(w)$ and answers 0. Therefore, M decides L.

The same proof can be put formally and more easily by using combined machines. Let M' decide L' and let M'' compute f. Then the combined machine $M'' \to M'$ decides L. Show it.

Exercise 9.2. Define the **partial indical function** ξ_L of a language $L \subseteq \Sigma^*$ by "for every $u \in \Sigma^*$, $\xi_L(u) = 1$ if $u \in L$, else, $\xi_L(u)$ is left undefined." Show that a language $L \subseteq \Sigma^*$ is computably enumerable iff its partial indical function ξ_L is computable.

Theorem 9.2. *If $L \leq_m L'$ and L' is computably enumerable, then L is computably enumerable.*

Proof. The same proof as that of Theorem 9.1 that uses $\xi_{L'}$ instead of $\chi_{L'}$ would do. However, you can have an alternate proof as in the following:

Let M' be a TM that accepts L' and let M'' be a TM that computes f, where L is reduced to L' by the map f. Let M be the combined machine $M'' \to M'$. It takes input w from Σ^*, computes $f(w)$ by simulating M'', and then accepts $f(w)$ by simulating M'. As $f(w)$ is accepted only when $f(w) \in L'$ and that happens only when $w \in L$, we see that M accepts w when $w \in L$. Therefore, M accepts L. □

By combining Theorems 9.1 and 9.2 and using contraposition, we have the following result that justifies the implicit direction in the symbol $\leq_m$ of a map reduction:

Theorem 9.3. *Let $L \leq_m L'$. Then*

1. *If L is not computably enumerable, then L' is not computably enumerable;*
2. *If L is not decidable, then L' is not decidable.*

Exercise 9.3. Using computability of the composition of two computable functions show that map reduction is transitive.

To see an application of map reduction, we take the **halting problem**. The problem is, "Does there exist an algorithm to decide whether an arbitrary TM halts on a given input?" Similarly, the **self-halting problem** is posed as: "Does there exist an algorithm to decide whether an arbitrary TM halts on itself?" The problems differ from acceptance problem and the self-acceptance problem by asking a question about halting in any of the states h or $\hbar$ rather than in the accepting state h. The suitable languages are

Halting language $L_{\text{H}} = \{\psi(M)1\psi(u) :$ the TM M halts on the input $u\}$.

Self-halting language $L_{\text{SH}} = \{\psi(M) :$ the TM M halts on the input $\psi(M)\}$.

To reduce the corresponding acceptance problems to the halting problems (watch the direction of reduction), we must define maps $f, g : \{0, 1\}^* \to \{0, 1\}^*$ satisfying

$$w \in L_{\text{A}} \text{ iff } f(w) \in L_{\text{H}}, \text{ and } w \in L_{SA} \text{ iff } g(w) \in L_{\text{SH}}, \text{ for each } w \in \{0, 1\}^*.$$

As f, g are required to be computable functions, we use TMs to define them. We construct TMs M_f and M_g and then declare that whatever functions they compute are, respectively, f and g. This would define the functions and also prove that they are computable.

Let M_f be the TM that checks whether the input is in the form $\psi(M)1\psi(u)$ for some TM M and some string u. If not, then it enters an infinite loop. If the input is in the correct form, then M_f constructs another TM M' as follows:

M' on input $\psi(M)1\psi(u)$, simulates M on u.
If M accepts u, M' enters the accepting state h, else M' enters an infinite loop.
Finally, M_f outputs $\psi(M')1\psi(u)$.

Clearly, the function f that is computed by M_f has the property

$$w \in L_{\text{A}} \text{ iff } f(w) = \psi(M')1\psi(u) \in L_{\text{H}}.$$

Similarly, M_g is constructed by taking the input as $\psi(M)$ and giving its output as $\psi(M')$. As $\overline{L}_{\text{A}}, \overline{L}_{\text{SA}}$ are not computably enumerable, Theorem 9.3 proves the following:

Theorem 9.4. *L_{H} is not decidable and $\overline{L}_{\text{H}}$ is not computably enumerable. Moreover, L_{SH} is not decidable and $\overline{L}_{\text{SH}}$ is not computably enumerable.*

Exercise 9.4. Are L_{H} and L_{SH} computably enumerable?

Therefore, both the halting problem and the self-halting problem are (algorithmically) unsolvable problems. You will meet many more unsolvable problems in the rest of this chapter. The proofs that they are unsolvable use reduction in some form or other. However, map reduction is not the most general type of reduction. Intuitively, if a problem B is reduced to another problem C, then an algorithm that might solve C can be used to solve B. Map reduction is not capable of doing this in its full generality. For example, an algorithm for deciding L_{A} can be used to decide $\overline{L}_{\text{A}}$ by just interchanging the outputs 0 and 1. But there is no map reduction of $\overline{L}_{\text{A}}$ to L_{A} as $\overline{L}_{\text{A}}$ is not computably enumerable while L_{A} is computably enumerable.

The requirement that an algorithm for deciding a language L' can be used for deciding a language L is called a **Turing reduction** of L to L'. We write $L \leq_T L'$ for "L is Turing reducible to L'." Thus both $\overline{L}_{\text{A}}$ and $\overline{L}_{\text{SA}}$ are Turing reducible to L_{A}. You will see the use of Turing reduction in the rest of the chapter.

Problems for Section 9.2

9.1. The *state-entry problem* is: Given a Turing machine M, a state q of M, and an input string $w \neq \varepsilon$, whether or not M enters q computing on w. Show that the state-entry problem is unsolvable by giving a reduction of halting problem to the state-entry problem.

9.2. Can you use the unsolvability of the state-entry problem to show that the problem raised by the question "Given a TM M and a state q of M, does M enter the state q on some input?" is also unsolvable?

9.3. The *blank-tape problem* is: Given a Turing machine M, whether it halts when started with the blank-tape. Give a reduction of the halting problem to the blank-tape problem.

9.4. Let $f : \mathbb{N} \to \mathbb{N}$ be the function where $f(n)$ equals the maximum number of moves that can be made by any n-state TM having tape alphabet $\{0, 1, ƀ\}$, that halts when started with a blank tape. First, show that f is well-defined. Show that f is not a computable function by reducing the blank-tape problem to the problem of computability of f.

9.5. Reduce the halting problem to the problem: Given any TM M, a symbol $\sigma \in \Sigma$, and a nonempty string $w \in \Sigma^*$, determine whether or not the symbol σ is ever written when M works on the input w.

9.6. The *restricted acceptance problem* is: Given any string w, whether there exists a TM M such that M accepts w. Give a reduction of the acceptance problem to the restricted acceptance problem. Note that in the acceptance problem, we seek an algorithm that would work for each pair of a machine M and a string w, while in the restricted acceptance problem, we only ask for an algorithm that would work for each w.

9.7. Determine which of the following problems about an arbitrary TM M are decidable, either by giving a decision procedure or by reducing the acceptance problem to the problem at hand:
(a) Does M have at least 2009 states?
(b) Does M take more than 2009 steps on input ε?
(c) Does M take more than 2009 steps on some input?
(d) Does M take more than 2009 steps on every input?
(e) Does M move to some square at least 2009 squares away from the square containing the initial ƀ ?
(f) Does M accept any string at all?
(g) Is $\mathcal{L}(M)$ finite?

9.8. Show that there exists a fixed TM M_0 and a fixed unrestricted grammar G_0 such that the following problems about them are unsolvable:
(a) Given a string w whether $w \in \mathcal{L}(M_0)$.
(b) Given a string w whether $w \in \mathcal{L}(G_0)$.

9.9. Suppose we are given a DFA for accepting L and a TM for accepting L'. Show that we cannot necessarily construct a DFA for L/L' effectively. Note that if $L \subseteq \Sigma^*$ is regular, and $L' \subseteq \Sigma^*$, then $L/L' = \{w \in \Sigma^* : wx \in L \text{ for some } x \in L'\}$ is regular. Similarly, formulate and prove an unsolvability result for $L \backslash L' = \{w \in \Sigma^* : wx \in L \text{ for all } x \in L'\}$.

9.10. Show that the Turing reduction relation $\leq_T$ is reflexive and transitive, and that the map reduction $\leq_m$ refines $\leq_T$.

9.11. Suppose $L \leq_m L'$ and that L' is a regular language. Does that imply L is also regular?

9.12. Recall that $L_{\text{A}} = \{\psi(M)\# \psi(w) : \text{the TM } M \text{ accepts } w\}$. Let $L_{\emptyset} = \{\psi(M) : M$ is a TM and $\mathcal{L}(M) = \emptyset\}$. Is $L_{\text{A}} \leq_m L_{\emptyset}$? Is $L_{\emptyset} \leq_m L_{\text{A}}$?

9.13. Suppose that L is computably enumerable and $L \leq_m \overline{L}$. Is L decidable?

9.14. Is it that each computably enumerable language is map-reducible to L_{A}?

9.15. Give an example of an undecidable language L such that $L \leq_m \overline{L}$.

9.16. Show that $\overline{L_{\text{A}}}$ is Turing reducible to L_{A} but $\overline{L_{\text{A}}}$ is not map-reducible to L_{A}.

9.17. (See Example 1.9) For solving the halting problem, consider two TMs as in the following:

M_1 : Take input as TM M and a string w. Output "M halts on w."

M_2 : Take input as TM M and a string w. Output "M does not halt on w."

As either M halts on w or not, one of M_1, M_2 reports correctly whatever be the case. Therefore, halting problem is solvable. What is wrong with this argument?

9.18. Revisit the proof of Theorem 8.8. Is the reduction of L_{SA} to L_{A} there a map reduction? If not, does there exist such a map reduction?

9.19. In the light of Theorem 8.8, give a reduction of L_{SH} to L_{H}.

9.20. Prove directly that L_{H} is computably enumerable but its complement is not.

9.21. Prove directly that L_{SH} is computably enumerable but its complement is not.

9.3 Rice's Theorem

To see how Turing reducibility is at work, we prove a nontrivial result about nontrivial properties of computably enumerable languages. Unless otherwise stated, a property is a unary relation on a set. For example, in the domain of natural numbers, primeness is a property. A natural number may or may not have this property, that is, it may or may not be a prime number. Thus the property of primeness can be identified with the subset of prime numbers.

When we speak of a property of computably enumerable languages, we fix an alphabet, say, Σ, and then consider the set RE of all computably enumerable languages over Σ. A **property of computably enumerable languages** (over Σ) is then a subset of RE. We say that such a property is **nontrivial** if the subset is neither the empty set $\emptyset$ nor the whole set RE, that is, a proper nonempty subset of RE. Moreover, we want to talk about decidability of such subsets. That means we must first represent such subsets as languages over some alphabet.

As RE is the set of all computably enumerable languages over Σ, it can be characterized by the set of all TMs that accept languages over Σ. However, our aim is to have a property of computably enumerable languages and not of TMs; that is, such a property may or may not hold for a computably enumerable language, but not necessarily for a Turing machine. For example, if M is a TM, the property "$\mathcal{L}(M)$ is regular" is a property of $\mathcal{L}(M)$ and not of M; for if M' is another TM that accepts the same language as M, we have "$\mathcal{L}(M')$ is regular iff $\mathcal{L}(M)$ is regular." On the other hand, the property such as "TMs with at least five states" is dependent upon TMs, for there can be another TM with less than five states accepting the same language as one from the other variety.

Such a property should be independent of particular TMs used. Let P be a nontrivial property of computably enumerable languages over an alphabet Σ. We can characterize P by the set of all TMs that accept some language in P. That is, we may take P as

$$\{M : M \text{ is a TM with input alphabet } \Sigma \text{ and } \mathcal{L}(M) \in P\}.$$

However, we need to see that P is represented as a language and not just a set of TMs. This is tackled easily by using the encodings of TMs as binary strings. Thus, we represent P as the language

$$L_P = \{\psi(M) : \mathcal{L}(M) \in P \text{ where } M \text{ is a TM with input alphabet } \Sigma\}.$$

Notice that we should write $L_P(\Sigma)$ instead of L_P. To reduce heavy symbolism, we make it a convention to suppress the alphabet Σ. Similarly, we identify RE (in fact, $RE(\Sigma)$), the set of all computably enumerable languages over Σ by the set

$$RE = \{M : M \text{ is a TM with input alphabet } \Sigma\}.$$

The set RE is again represented as a language over $\{0, 1\}$ by

$$L_{RE} = \{\psi(M) : M \text{ is a TM with input alphabet } \Sigma\}.$$

The language representations of RE and of P only show that it is meaningful to talk of the decidability of a property of computably enumerable languages. Decidability of the property P of computably enumerable languages is then posed as the decidability of L_P as a subset of L_{RE}. Instead, we will continue with the informal set forms of both RE and P. The corresponding problem is:

> Given a computably enumerable language L over Σ does L have the property P, that is, does L belong to P?

We show that this problem, called **the decision problem for** P, is unsolvable by reducing the self-acceptance problem to this problem, that is, formally, you would reduce L_{SA} to L_P. Notice that the phrase "a *property* P of computably enumerable languages" is represented here as a language $L_P \subseteq \{0, 1\}^*$; and L_P satisfies the property:

$$\text{If } \psi(M) \in L_P \text{ and } \mathcal{L}(M) = \mathcal{L}(M') \text{ for TMs } M, M', \text{ then } \psi(M') \in L_P.$$

Similarly, P is nontrivial amounts to asserting that there are TMs M, M' such that $\psi(M) \in L_P$ but $\psi(M') \notin L_P$.

Theorem 9.5 (Rice's theorem). *Every nontrivial property of computably enumerable languages is undecidable.*

Proof. We continue with the informal P rather than the formal L_P. We would then show that the decision problem for P is unsolvable. Let P be a nontrivial property of computably enumerable languages. For the time being assume that P does not contain the empty language $\emptyset$. As P is a nontrivial property, it contains a nonempty computably enumerable language, say, L. Let M_L be a Turing machine that accepts L. We use M_L for reducing the self-acceptance problem to decision problem for P.

Recall that L_{SA} consists of strings of the form $\psi(M)$, where M accepts $\psi(M)$. We give an algorithm to construct a TM M' given a string $\psi(M)$, by specifying what M' does on an input w. It is as follows:

M' simulates M on $\psi(M)$. If M accepts $\psi(M)$, then M' simulates M_L on w.

Thus, the TM M_L is hard-wired into M' this way. Now, either M accepts $\psi(M)$ or not. In the first case, M' does as M_L would do to the string w. Here, M' accepts all and only strings of L. That is, when M accepts $\psi(M)$, $\mathcal{L}(M') = L$. On the other hand, when M does not accept $\psi(M)$, M' also does not accept w. In fact, no w is accepted, in this case. That is, when M does not accept $\psi(M)$, $\mathcal{L}(M') = \emptyset$. Since $L \in P$ and $\emptyset \notin P$, we see that

(a) if M accepts $\psi(M)$, then $\mathcal{L}(M') \in P$, and
(b) if M does not accept $\psi(M)$, then $\mathcal{L}(M') \notin P$.

Here ends the reduction of the self-acceptance problem to the decision problem for P. How?

Suppose there is an algorithm to determine whether any given computably enumerable language is in P. Feed to this algorithm the language $\mathcal{L}(M')$. The algorithm then determines whether $\mathcal{L}(M')$ is in P or not. Writing the contrapositive statements of (a and b) above, we see that the algorithm then determines whether M accepts $\psi(M)$ or not. As the self-acceptance problem is not solvable (Theorem 8.9), the decision problem for P is not solvable.

The proof is not yet complete. We must get rid of the extra assumption that $\emptyset \notin P$. If $\emptyset \in P$, consider $\overline{P} = RE - P$. The above yields a proof for $\overline{P}$ instead of P. That is, $L_{\overline{P}}$ is undecidable. Then, so is L_P. □

Rice's theorem says that almost whatever interesting we speak about computably enumerable sets is undecidable. But what about acceptability, if not decidability? It can be shown that P, as a subset of RE, is computably enumerable iff P satisfies all of the following:

monotonicity: If $L \in P$, $L \subseteq L'$, and $L' \in RE$, then $L' \in P$.
compactness: If $L \in P$ is infinite, then $L' \in P$ for some finite subset L' of L.
finite enumerability: All finite languages in P can be enumerated by a TM.

We discuss many more unsolvable problems about TMs as applications of Rice's theorem.

Theorem 9.6. *The following problems about* TMs, *and hence, about unrestricted grammars, are unsolvable:*

1. Emptiness Problem: *Does an arbitrary* TM *accept the empty language* $\emptyset$*?*
2. All-TM Problem: *Does an arbitrary TM with input alphabet* Σ *accept* Σ^**?*
3. ε-TM Problem: *Does an arbitrary* TM *accept the empty string* ε*?*
4. Eq-TM Problem: *Do two arbitrary* TMs *accept the same language?*
5. Reg-TM Problem: *Is the language accepted by an arbitrary* TM *regular?*
6. CF-TM Problem: *Is the language accepted by an arbitrary* TM *context-free?*
7. CS-TM Problem: *Is the language accepted by an arbitrary* TM *context-sensitive?*
8. D-TM Problem: *Is the language accepted by an arbitrary* TM *decidable?*

Proof. We fix an alphabet Σ. We encode machines M and strings u as binary strings $\psi(M)$ and $\psi(u)$, respectively. The technique is to realize each of the problems above as a nontrivial property of computably enumerable languages.

1. The suitable language representing the emptiness problem is

$$L_E = \{\psi(M) : M \text{ is a TM and } \mathcal{L}(M) = \emptyset\}.$$

Clearly, $\emptyset \subsetneqq L_E \subsetneqq RE$. This means that $\{\emptyset\}$ is a nontrivial property of computably enumerable languages. By Rice's Theorem L_E is not decidable. Thus, the emptiness problem for TMs is unsolvable.

The argument is similar in all other cases; we just give their language representations. You can find out the corresponding nontrivial properties that Rice requires. For example, corresponding to (2) below, the property is $\{\Sigma^*\}$.

2. $L_{ALL} = \{\psi(M) : M \text{ is a TM and } \mathcal{L}(M) = \Sigma^*\}$.
3. $L_\varepsilon = \{\psi(M) : M \text{ is a TM and } \varepsilon \in \mathcal{L}(M)\}$.
4. $L_{EQ} = \{\psi(M)\#\psi(M') : M \text{ and } M' \text{ are TMs with } \mathcal{L}(M) = \mathcal{L}(M')\}$.
5. $L_{RTM} = \{\psi(M) : M \text{ is a TM and } \mathcal{L}(M) \text{ is regular}\}$.
6. $L_{CFTM} = \{\psi(M) : M \text{ is a TM and } \mathcal{L}(M) \text{ is context-free}\}$.
7. $L_{CSTM} = \{\psi(M) : M \text{ is a TM and } \mathcal{L}(M) \text{ is context-sensitive}\}$.
8. $L_{DTM} = \{\psi(M) : M \text{ is a TM and } \mathcal{L}(M) \text{ is decidable}\}$.

□

Exercise 9.5. Write the nontrivial properties corresponding to each language in Theorem 9.6 (3–8) and then complete the proof. Also try direct proofs of all of them without using Rice's Theorem, but by reducing a known unsolvable problem to the one at hand.

You should be able to translate the statements in Theorem 9.6 to unsolvable problems about unrestricted grammars. For example, unsolvability of the fourth statement there reads as: "It is unsolvable whether two arbitrary unrestricted grammars generate the same language or not." Some more problems about grammars and automata will be discussed in the following sections.

Problems for Section 9.3

9.22. Show that there is no algorithm to determine whether or not a given TM eventually halts with an empty tape given any input.

9.23. Is the problem of determining whether or not an arbitrary TM revisits its initial square (the cell with ƀ which is followed by the input) solvable?

9.24. Using reduction, prove that $L_{\emptyset} = \{\psi(M) : \mathcal{L}(M) = \emptyset\}$ is not computably enumerable. Also, prove that $L_{not\,\emptyset} = \{\psi(M) : \mathcal{L}(M) \neq \emptyset\}$ is computably enumerable.

9.25. Let C, C' be two possible configurations of a TM M. Show that the problem of determining whether $C \rightsquigarrow_M C'$ is unsolvable.

9.26. Show that the set of all $\psi(M)$ for TMs M that accept all inputs that are palindromes (possibly along with others) is undecidable.

9.27. Show that the set of all TMs that halt when started with the blank tape is computably enumerable. Is this decidable?

9.28. Let $f(n)$ be the maximum number of tape squares examined by any n-state TM with tape alphabet $\{0, 1, ƀ\}$ that halts when started with a blank tape. Show that $f : \mathbb{N} \to \mathbb{N}$ is not a computable function.

9.29. Is it true that any problem whose domain is finite is decidable?

9.30. Let M be a Turing machine with input alphabet $\{a, b\}$. Is the problem "$\mathcal{L}(M)$ contains two different strings of the same length" solvable?

9.31. Given a TM M and a string w, determine whether the following problems are solvable:
(a) Does M ever write ƀ on its tape on input w?
(b) Does M ever write a nonblank symbol on its tape on input w?
(c) Given a symbol σ, does M ever write σ on its tape on input w?

9.32. Show that the problem of determining whether the language accepted by a given TM equals $\{ww^R : w \in \{a, b\}^*\}$ is unsolvable.

9.33. Are the following problems solvable?
(a) Given a TM M, is $\mathcal{L}(M) = (\mathcal{L}(M))^R$?
(b) Given a TM M are there infinitely many TMs equivalent to M?
(c) Given a CFG G with the set of terminals as Σ, is $\mathcal{L}(G) = \Sigma^* - \{\varepsilon\}$?
(d) Given an LBA M, does M accept a string of odd length?

9.34. Are the following problems about a given pair of TMs M and N solvable?
(a) Is $\mathcal{L}(M) \subseteq \mathcal{L}(N)$?
(b) Is $\mathcal{L}(M) = \overline{\mathcal{L}(N)}$?
(c) Does there exist a string that both M and N accept, that is, is $\mathcal{L}(M) \cap \mathcal{L}(N) \neq \emptyset$?
(d) Is $\mathcal{L}(M) \cap \mathcal{L}(N)$ finite?
(e) Is $\mathcal{L}(M) \cap \mathcal{L}(N)$ decidable?
(f) Is $\mathcal{L}(M) \cap \mathcal{L}(N)$ computably enumerable?

9.35. Is the problem "Given a TM M, and two states p, q of M, whether there is a configuration with state component p that yields a configuration with state component q?" solvable? If we fix the state p, then?

9.36. Show that the halting problem for TMs remains unsolvable when restricted to machines with fixed but small number of states. Note that if arbitrarily large number of states are allowed, then the universal machine would do the job.

9.37. Let G, G' be unrestricted grammars. Determine whether the following problems are solvable or not:
(a) Is $(\mathcal{L}(G))^R$ computably enumerable?
(b) Is $\mathcal{L}(G) = (\mathcal{L}(G))^R$?
(c) Is $\mathcal{L}(G) \cap \mathcal{L}(G') = \emptyset$?
(d) Is $\mathcal{L}(G) = (\mathcal{L}(G))^*$?
(e) Let G' be a fixed unrestricted grammar such that $\mathcal{L}(G') \neq \emptyset$. Now, is $\mathcal{L}(G) \cap \mathcal{L}(G') = \emptyset$?

9.38. Are the following problems solvable?
(a) Does a TM M take at least $\ell(\psi(M))$ steps on some input?
(b) Does a TM M take at least $\ell(\psi(M))$ steps on all inputs?
(c) Does a given TM take at least 2009^{2009} moves on the input a^{2009}?
(d) Does a given TM reenters its initial state on some input?

9.39. Let $L = \{\psi(M) : M$ is a TM that accepts w^R whenever M accepts $w\}$. Show that L is undecidable. Is L computably enumerable?

9.40. We say that a TM M *uses k cells* on input w if, there is a configuration $qu\underline{\sigma}v$ of M such that $s\underline{b\!\!\!/}w \rightsquigarrow qu\underline{\sigma}v$ and $\ell(u\sigma v) \geq k$. Which of the following problems are (is) solvable:
(a) Given a TM M, a string w, and a number k, does M use k cells on input w?
(b) Let $f : \mathbb{N} \to \mathbb{N}$ be a computable total function. Given a TM M and a string w, does M use $f(\ell(w))$ cells on input w?
(c) Given a TM M and a string w, does there exist a k such that M uses k cells on input w?

9.4 About Finite Automata

We address questions about existence of algorithms for deciding or solving certain natural problems about regular grammars and finite automata. We start with the acceptance problem for DFAs:

> Does there exist an algorithm to determine whether any given DFA accepts a given string?

Yes; because you can have a total TM that simulates any given DFA. Moreover, this simulation can be written as an algorithm. Let us use the same notation $\psi(D)$ for a binary string encoding of a DFA D. The corresponding language for the acceptance problem for DFAs is:

$$L_{\text{ADFA}} = \{\psi(D)\#\,\psi(u) : \text{ the DFA } D \text{ accepts the string } u\}.$$

We then show that

Theorem 9.7. L_{ADFA} *is decidable.*

Proof. Construct a TM M that upon input as a binary string, checks whether it is in the form $\psi(D)\#\,\psi(u)$ for a DFA D and a string u. If not, then M enters its rejecting state $\hbar$; otherwise, it simulates D on u. If D comes to a final state after consuming the whole string u, M enters its accepting state h, otherwise, M enters its rejecting state $\hbar$. Clearly, M decides L_{ADFA}. □

Theorem 9.8. *The following problems about regular expressions, regular grammars, and finite automata are solvable:*

1. *Given an NFA N and a string u, does N accept u?*
2. *Given a regular expression E and a string u, is $u \in \mathcal{L}(E)$?*
3. *Given a regular grammar G and a string u, is $u \in \mathcal{L}(G)$?*
4. *Given a DFA D, is $\mathcal{L}(D) = \emptyset$?*
5. *Given a DFA D with input alphabet Σ, is $\mathcal{L}(D) = \Sigma^*$?*
6. *Given two DFAs, do they accept the same language?*
7. *Problems similar to (4–6) for NFAs, regular expressions, and regular grammars.*

Proof. 1. We have discussed an algorithm to convert any NFA to an equivalent DFA. Combine this algorithm with the algorithm given in the proof of Theorem 9.7.

2. Combine the algorithm of constructing an NFA from a regular expression, which accepts the language of the expression, with that of part 1.

3. There is an algorithm to construct an NFA from a regular grammar that accepts the same language as the grammar generates. Use this along with the algorithm in part 1.

4. The corresponding language is $L_{\text{EDFA}} = \{\psi(D) : D \text{ is a DFA and } \mathcal{L}(D) = \emptyset\}$.

We construct a TM M that decides L_{EDFA} as a subset of $\{0, 1\}^*$. M works as in the following:

> On input $u \in \{0, 1\}^*$, M first checks whether u is in the form $\psi(D)$ for some DFA D. If not, M enters its rejecting state $\hbar$. Otherwise, M
>
> (a) Marks the start state of D.
> (b) Repeats the following step until no new states are marked:
> Mark any state that has a transition coming into it from an already marked state.
> (c) If no final state is marked, then it enters the accepting state h, else it enters the rejecting state $\hbar$.

Now it is easy to see that M decides L_{EDFA}.

5. There is an algorithm to construct a DFA D' from a given DFA D such that $\mathcal{L}(D') = \overline{\mathcal{L}}(D)$. Now use part 4.

6. Revise the closure properties of regular languages. There is an algorithm to construct a DFA D from given DFAs D_1 and D_2 such that

$$\mathcal{L}(D) = (\mathcal{L}(D_1) \cap \overline{\mathcal{L}}(D_2)) \cup (\overline{\mathcal{L}}(D_1) \cap \mathcal{L}(D_2)).$$

Now, the problem is reduced to decide whether $\mathcal{L}(D) = \emptyset$. Use part 4.

7. As algorithms exist that construct NFAs, regular grammars, and regular expressions from a given DFA so that the language accepted or generated by them is also accepted by the DFA, the statement holds. □

Exercise 9.6. Show that the TM M constructed in the proof of the fourth statement of Theorem 9.8 decides L_{EDFA}.

Most of the interesting problems about regular languages are decidable. The paradox is that if you discover an unsolvable problem about finite automata, it will be really interesting!

An interesting problem concerning regular languages and TMs is that, given a TM, can it be determined whether it accepts a regular language or not. For a particular given TM, the answer may be "yes." But the question is whether there exists an algorithm to do it uniformly for all TMs. The language for the problem is:

$$L_{\text{RTM}} = \{\psi(M) : M \text{ is a TM and } \mathcal{L}(M) \text{ is regular}\}.$$

You have already seen in Theorem 9.6 that L_{RTM} is not decidable. Therefore, there is no such algorithm. You may ask whether a similar result holds if PDAs are taken instead of TMs, or when regular languages are replaced by context-free languages.

Problems for Section 9.4

9.41. Show that the problem of determining whether two given DFAs accept the same language is solvable. Give an intuitive explanation why for DFAs this problem is solvable, but for TMs, it is not.

9.42. Suppose L, L' are regular languages on Σ specified in terms of regular expressions and that w is any given string in Σ^*. Show that the following problems are solvable:
(a) Does $w \in L$?
(b) Is L empty?
(c) Is L finite?
(d) Is L infinite?
(e) Is $L = \Sigma^*$?
(f) Is $L = L'$?
(g) Does $w \in L - L'$?
(h) Is $L \subseteq L'$?
(i) Does $\varepsilon \in L$?
(j) Is $L = L'/L$?
(k) Is L palindromic, i.e., is $L = L^R$?
(l) Does $w^R \in L$?
(m) Given another regular language L'', is $L' = LL''$?
(n) Is $L = L^*$?
(o) Does L contain a string v such that w is a substring of v?
(p) Does L contain a string of even length?
(q) Does L contain infinite number of even-length strings?

9.43. Consider the problem of testing whether a DFA and a regular expression are equivalent. Express this problem as a language and then show that the language is decidable.

9.44. Show that the following languages are decidable:
(a) $\{\psi(D) : D$ is a DFA that does not accept any string $w \in \{0, 1\}^*$ with $\#_1(w)$ odd$\}$.
(b) $\{\psi(D) : D$ is a DFA that accepts some string of the form ww^R for $w \in \{0, 1\}^*\}$.

9.5 About PDA

We ask a simple question about context-free grammars (CFGs). Given a CFG G and a string u, can it be determined whether u is at all generated by G? If I actually give you such a G and a u, then you would start deriving u in G based on your intuition, and then possibly, by luck, derive it. If you fail in many attempts, then you may like to proceed towards a proof that u cannot be generated in G. But my question is different. Can you have an algorithm to determine whether any such G generates any such u?

You may think in terms of a PDA. Construct an equivalent PDA for G, for which there is an algorithm. If this PDA happens to be a DPDA, then run the DPDA on u. Depending upon whether the DPDA accepts u, the answer to the original question will be "yes" or "no." This is fairly algorithmic. But if the PDA is not a DPDA, then there might be more than one computation with the same input u, and it looks too awkward to follow through all those computations. Nonetheless, there are only a finite number of such computations, and an algorithm can find them all by brute force approach. Thus, the problem "whether $u \in \mathcal{L}(G)$ " is solvable.

Another way is to use a Chomsky normal form for G. Recall that in a Chomsky normal form CFG, each production is of the form $A \mapsto BC$ or $A \mapsto \sigma$ for nonterminals A, B, C, and terminal σ. Each application of a production either produces a terminal or increases the length of the (to be) derived string. Then you try all derivations of length $2\ell(u) - 1$ in the Chomsky normal form. If u is at all derived in any of these derivations, then answer is "yes," else, "no." Thus you have proved the following result:

Theorem 9.9. *The membership problem for CFGs (and for PDAs), that is, given an arbitrary CFG G and a string u, whether $u \in \mathcal{L}(G)$, is solvable.*

Notice that Theorem 9.9 is a restatement of the fact that the set of context-free languages form a subset of the set of decidable languages (over any alphabet Σ).

Similarly, the question whether a given CFG generates any string at all ($\mathcal{L}(G) \neq \emptyset$) is algorithmically solvable. How? Just revisit the proof of Pumping Lemma for CFLs. There we had chosen a magic number $n = k^m$, where m is the maximum number of nonterminals in G and k is the maximum length of α, when $X \mapsto \alpha$ is any production in G. It follows from Pumping Lemma that corresponding to each string of $\mathcal{L}(G)$ having length at least n, there exists a string of length at most n, which is also in $\mathcal{L}(G)$. That means, if G generates any string at all, then it must generate a string of length at most n. Now make a list of all strings of length at most n over the terminal alphabet, and apply Theorem 9.9. You have proved the following result:

Theorem 9.10. *The emptiness problem for CFGs (and for PDAs), that is, whether an arbitrary CFG G generates any string at all, is solvable.*

Proof. We have already proved it. Here is another proof similar to the marking scheme we have adopted for DFAs.

> First, mark all the terminal symbols of G. Next, repeatedly mark each nonterminal A if there is a production $A \mapsto \alpha$ and if all symbols of α have already been marked. Quit this loop when there are no more changes in the marking.

After the end of this loop, you will see that either A has been marked or some nonterminal symbols occurring in α have not been marked. By induction on the number of marking steps, it follows that if A is marked by this procedure, then there is a string u of terminals such that $A \Rightarrow u$. Induction on the length of the derivation $A \Rightarrow u$ also shows that if $A \Rightarrow u$ for some string u of terminals, then A has been marked. Therefore, some string w is in $\mathcal{L}(G)$ iff the start symbol S of G has been marked. This solves the problem whether $\mathcal{L}(G) = \emptyset$. □

Exercise 9.7. Complete the proof of Theorem 9.10 by carrying out the two inductive proofs.

Recall that context-free languages are not closed under complementation. Theorem 9.10 will thus be useless in deciding "whether a CFG generates all strings over its terminal alphabet" is a solvable problem or an unsolvable problem. Let us fix the terminal alphabet as Σ. The *All-CFG problem* can be posed as:

For an arbitrary CFG G, whether $\mathcal{L}(G) = \Sigma^*$?

We need to do some ground work to answer this question. This concerns computations of Turing machines. Let $M = (Q, \Sigma, \Gamma, \delta, s, h, \hbar)$ be a TM. Recall that a configuration of M is a snap-shot of a TM in action. It can be written as a string $uq\sigma v$, where M is currently in the state q, is scanning the symbol σ, to the left of which is the string u, and to the right of which is the string v. A computation of M is a sequence of configurations, where a configuration yields in one step the next configuration in the sequence. When a computation starts with an initial configuration and ends with an accepting configuration, we say that it is a *valid computation*.

A valid computation history is a valid computation written in a slightly twisted way. We use a new symbol, say, #, assumed not to have been used in M, to separate the configurations in a computation. A **valid computation history** of M is a string in either of the forms:

1. $\# w_1 \# w_2^R \# w_3 \# w_4^R \# \;\cdots\; w_{2m}^R \#$ OR
2. $\# w_1 \# w_2^R \# w_3 \# w_4^R \# \;\cdots\; w_{2m}^R \# w_{2m+1} \#$.

such that

(a) each w_i is a configuration of M,
(b) w_1 is an initial configuration of M,
(c) the last configuration (i.e., w_{2m} in the first case and w_{2m+1} in the second case) is an accepting configuration of M, and
(d) $w_i \overset{1}{\rightsquigarrow}_M w_{i+1}$, for each i.

As you know, the superscript R denotes the reversal of a string. As a shorthand, in this section, we will use the symbol $\rightsquigarrow$ for the one-step yield in M, that is, for $\overset{1}{\rightsquigarrow}_M$.

We start with a simple question: Can a TM be simulated by a PDA? (Consider generalized PDAs as mentioned in (6.4)) Silly question, because there are computably enumerable languages that are not context-free. But something related can be done; see the following example.

Example 9.1. Let M be a TM, whose transitions include $\sigma(q, a) = (p, b)$, $\delta(p, b) = (r, \not{L})$, and $\delta(r, b) = (t, \not{R})$. Suppose further that M has a configuration $abqab$ at some instant during its computation. Assume that there is a PDA P, which has $baqba$ on its stack, during its computation. The configuration of the TM is on the stack of P written bottom to top.

The next configuration of M can be reflected upon the stack of P. Surely, for this, P must somehow follow the working out of M. Since $\delta(q, a) = (p, b)$, we have $abqab \rightsquigarrow abpbb$. Thus following M, P should have on its stack $bbpba$ after a move. P does it by following the rule:

If $\delta(q, a) = (p, b)$ and vaq is on its stack, then replace vaq by vbp.

Here, v is any string not having a state component. Now the stack contains $bbpba$; read from top to bottom, as usual. The next move of M follows the transition $\delta(p, b) = (r, \not{L})$ after which, M has the configuration $brbba$. This is followed by P using the rule:

If $\delta(p, b) = (r, \not{L})$, and vbp is on the stack, then replace vbp by vrb.

Now M has the configuration $brbba$ and the transition $\delta(r, b) = (t, \not{R})$ brings it to the configuration $bbtba$. P follows it by using the rule:

If $\delta(r, b) = (t, \not{R})$, and vrb is on the stack, then replace vrb by vbt.

Thus configurations of a TM can be reproduced by a PDA on its stack, but in a reversed manner. □

Notice that the input alphabet and the stack alphabet of the PDA includes all state symbols, all tape symbols of the Turing machine, and the special symbol #. We also agree to use the same symbols on the stack as the input symbols.

Lemma 9.1. *The language* $\{u\#v^R : u \rightsquigarrow v\}$ *is a CFL.*

Proof. We construct a generalized PDA P for accepting the language, by following the idea of Example 9.1. Given a string $u\#v^R$ on its tape, P pushes u to the stack symbol-by-symbol and stops operating on its tape once it reads the # symbol. This means that u^R is stored on the stack, read from top to bottom. For example, if $u = aqb$, then P pushes a first, then q, and finally b. Next, P follows the transitions of M, which are hard-wired into the finite controls of P. It operates with the stack alone, keeping its reading head on the # symbol of the input. It does so by using the following rules:

Let Q be the set of states and Γ be the tape alphabet of M. Denote by σ, τ symbols from Γ, by x, a string from Γ^*, and by p, q, states from Q.

1. If $\delta(q, \sigma) = (p, \tau)$ and $x\sigma q$ is on the stack, then replace $x\sigma q$ by $x\tau p$.
2. If $\delta(q, \sigma) = (p, \not{L})$ and $x\sigma q$ is on the stack, then replace $x\sigma q$ by $xp\sigma$.
3. If $\delta(q, \sigma) = (p, \not{R})$ and $xq\sigma$ is on the stack, then replace $xq\sigma$ by $x\sigma p$.

Now, whenever $u \rightsquigarrow w$, the string w has been obtained from u by using a transition in either of the forms (1–3). In that case, the stack of P has been changed from u^R to w^R. The reading head is still on #. Here, operation on the stack is over, the reading head moves to the next square on the right and then P operates as follows:

4. If the symbol that P is scanning is the same as the top most symbol on the stack, then P's reading head goes to the next square on the right popping the matched symbol off the stack, else, P halts abruptly.

Now, if $u \# v^R$ is on the tape and v is on the stack, then only P would read the entire input and come to the stage where it has an empty stack. We just declare all states of P as final states. Thus P accepts $u \# v^R$ and only such strings where $u \rightsquigarrow v$. □

A similar result holds for valid computation histories when one of the yields is violated.

Lemma 9.2. *The language* $\{\# w_1 \# w_2^R \# \ \cdots \ w_n \# : w_i \not\rightsquigarrow w_{i+1}$ *for some* $i\}$ *is a CFL.*

Proof. We construct a PDA P for accepting this language. Let all states of P be final states. P nondeterministically selects some w_i that is preceded by some number of $\#$ s. This is the place where an error in the yield might occur. P pushes a new symbol, say, $\#$, to the stack initially. P then reads w_i and stores y in the stack from bottom to top such that $w_i \rightsquigarrow y$, as in the PDA constructed in the proof of Lemma 9.1. Thus, the rightmost symbol of y is on the top of the stack.

After reading $\#$ from the input, P compares the succeeding string w_{i+1} with y^R on the stack, read from top to bottom now. While comparing, it goes on popping off a matched symbol. If discrepancy occurs at some symbol (it is not #), then P scans the remaining symbols and empties the stack, accepting the string. If a perfect match between w_{i+1} and y^R occurs, then P finds $\#$ on the stack, in which case, it halts abruptly. Thus P accepts the input of the required form. □

On the contrary, the set of valid computation histories is not, in general, a CFL. We will not indulge in such details. We rather show that the All-CFG problem is unsolvable.

Theorem 9.11. *The problem that given an arbitrary CFG G with terminal alphabet* Σ*, whether* $\mathcal{L}(G) = \Sigma^*$ *is unsolvable.*

Proof. We reduce the acceptance problem of Turing machines to the All-CFG problem by employing the valid computation histories of a TM M.

Let M be a Turing machine and w be any input to it. A valid computation history $\# w_1 \# w_2^R \# \ \cdots \ \#$ of M is called a valid computation history of M on w whenever w_1 is an initial configuration of M with input w. Denote by V, the set of all valid computation histories of M on w. The set $V \subseteq \Delta^*$, for some alphabet Δ that has all the symbols of the state-set and the tape-alphabet of M along with the special symbols $\not L$, $\not R$, h, $\hbar$, and $\#$. (If you have used encodings of configurations w_i's, then Δ can be taken as $\{0, 1, \#\}$.) We write $\overline{V}$ for the complement of V in Δ^*. Now,

$$\overline{V} = V_1 \cup V_2 \cup V_3 \cup V_4,$$

where each V_i is determined by the ith property as given below.

1. u is not of the form $\# w_1 \# w_2^R \# \ \cdots \ \#$ for configurations w_j of M. (Otherwise, suppose u is of the form $\# w_1 \# w_2^R \# \ \cdots \ \# w' \#$, with u having $n+1$ #'s and, where $w' = w_n$ if n is odd, and $w' = w_n^R$ if n is even.)
2. w_1 is not the initial configuration of M on input w.
3. w' is not an accepting configuration of M.
4. $w_j \not\rightsquigarrow w_{j+1}$ for some j.

For example, $V_4 = \{\# w_1 \# w_2^R \# \cdots \# w' \# \in \Delta^* : w_j \not\vdash w_{j+1} \text{ for some } j\}$.

The languages V_1, V_2, V_3 are regular (Show it.) and by Lemma 9.2, V_4 is context-free. Hence $\overline{V}$ is a CFL with terminal alphabet Δ. Let G be a CFG that generates $\overline{V}$. Now, M does not accept w iff there is no valid computation history of M on w iff $\overline{V} = \emptyset$ iff $V = \Delta^*$ iff $\mathcal{L}(G) = \Delta^*$.

This completes the reduction and the proof. □

Exercise 9.8. Use lemma 9.2 to show that the intersection problem for CFGs, that is, for arbitrary CFGs G, G', whether $\mathcal{L}(G) \cap \mathcal{L}(G') = \emptyset$, is unsolvable.

Following similar lines, many more undecidable results about CFGs can be proved. we mention some of them here. Let G, G_1, G_2 be arbitrary CFGs and R be an arbitrary regular language. The following problems are unsolvable:

(a) Is $\mathcal{L}(G_1) = \mathcal{L}(G_2)$?
(b) Is $\mathcal{L}(G_1) \subseteq \mathcal{L}(G_2)$?
(c) Is $\overline{\mathcal{L}}(G)$ a CFL?
(d) Is $\mathcal{L}(G_1) \cap \mathcal{L}(G_2)$ a CFL?
(e) Is $\mathcal{L}(G) = R$?
(f) Is $R \subseteq \mathcal{L}(G)$?
(g) Is $\mathcal{L}(G)$ regular?
(h) Is G ambiguous?
(i) Is $\mathcal{L}(G)$ ambiguous?
(j) Does $\mathcal{L}(G)$ contain a palindrome?

Similar results hold for PDAs in place of CFGs. We mention some similar results for linear bounded automata (LBA).

(k) Acceptance problem for LBAs is solvable, that is, there is an algorithm for testing whether an arbitrary LBA accepts a given input string.
(l) The emptiness problem for LBAs is unsolvable, that is, there is no algorithm for deciding whether an arbitrary LBA accepts any string at all.

Problems for Section 9.5

9.45. Show that the finiteness problem for CFGs, that is, whether an arbitrary CFG generates only a finite number of strings, is solvable.

9.46. Either give an algorithm or show that no algorithm exists for solving each of the following decision problems:
(a) Is there a nonterminal of a CFG that comes up repeatedly in some derivation?
(b) Is the language of a CFG infinite?
(c) Is the language of a CFG regular?
(d) Given a regular language L and a CFG G, is $L \subseteq \mathcal{L}(G)$?
(e) Does a CFG generate a string of length less than some given number?
(f) Do a CFG and a regular grammar generate some common string?

(g) Is a CFG ambiguous?
(h) Is a CFL inherently ambiguous?
(i) Is intersection of two CFLs empty?
(j) Given two CFGs G and G', is $\mathcal{L}(G) \subseteq \mathcal{L}(G')$?
(k) Given two CFGs G and G', is $\mathcal{L}(G) = \mathcal{L}(G')$?
(l) Does a CFG generate every string over its alphabet?
(m) Does a CFG generate any string at all?
(n) Does a CFG ever generate a palindrome?
(o) Is $\mathcal{L}(G)$ finite for a given CFG G?
(p) Is a given context-sensitive language a CFL?
(q) The halting problem for PDAs: does there exist an algorithm to decide whether an arbitrary PDA will halt on a given arbitrary input w?
(r) The halting problem for the DPDAs.
(s) Given a PDA to find an equivalent PDA with fewest states.
(t) Given a PDA with just one state, whether it generates all strings.

9.47. Formulate and determine solvability of problems about PDAs analogous to those for CFGs stated in the preceding problem.

9.48. A suitable language for equivalence of two CFGs is $L = \{\psi(G_1)\#\psi(G_2) : \mathcal{L}(G_1) = \mathcal{L}(G_2)\}$, where $\psi(\cdot)$ is some binary encoding (Fix it.) of a CFG. You have seen that L is undecidable. Show that L is *co-computably enumerable,* that is, $\{0, 1\}^* - L$ is computably enumerable.

9.49. Show that the following problems are unsolvable:
(a) Is a given CFL a DCFL?
(b) Whether the intersection of two given CFLs is a CFL?
(c) Is the complement of a given CFL a CFL?

9.50. Which one of the following problems is solvable and which is not? Justify.
(a) Given a regular language L and a CFL L', is $L \subseteq L'$?
(b) Given a regular language L and a CFL L', is $L' \subseteq L$?

9.6 Post's Correspondence Problem

Here we discuss an interesting problem, a puzzle with dominoes. Each individual domino has an upper track and a lower track (imagine a small card board having two tracks); each track contains a string of symbols. The puzzle starts with a given finite collection of such dominoes. Assume that there are unlimited supplies of each kind of dominoes. (You can get xeroxed and use their copies, as many as you need.) The task is to place some (finite number) of the copies in such a way that the strings on the upper and lower tracks become exactly equal. Such an arrangement is called a match.

Example 9.2. Do the dominoes $\left[\frac{b}{bab}\right], \left[\frac{ba}{aa}\right], \left[\frac{abb}{bb}\right]$ have a match?

Solution. It is easy to make a match here. One such match is

$$\left[\frac{b}{bab}\right]\left[\frac{abb}{bb}\right]\left[\frac{ba}{aa}\right]\left[\frac{abb}{bb}\right],$$

where the concatenated string on the upper track matches with that on the lower track, concatenated as they are placed. The matched string here is $babbbaabb$. □

Exercise 9.9. Find a match with the dominoes $\left[\frac{aab}{a}\right], \left[\frac{ab}{abb}\right], \left[\frac{ab}{bab}\right], \left[\frac{ba}{aab}\right]$ or show that there is no match possible.

It looks easy to play this game. But our interest is in getting an algorithmic solution. You are now asked to write a program (an algorithm) that would solve all such domino games, and not just the one given in Example 9.2.

Remember that the same dominoes can be repeated as often as you require. That gives you more freedom, and in turn, makes it difficult to play the game. Post's correspondence problem is to find a match, if possible, for all such games. Let us formalize a bit.

Let Σ be an alphabet having at least two symbols. A **Post's correspondence system** P over Σ, or **PCS**, for short, is a finite ordered set of ordered pairs of strings $(u_1, v_1), \ldots, (u_n, v_n)$ from Σ^+. We write such an ordered set with n elements as the n-tuple $((u_1, v_1), \ldots, (u_n, v_n))$. A **match** in P is a sequence of indices $i_1, i_2, \ldots, i_k$, not necessarily distinct, such that $u_{i_1} u_{i_2} \cdots u_{i_k} = v_{i_1} v_{i_2} \cdots v_{i_k}$. **Post's Correspondence Problem**, or **PCP**, for short, is the problem of determining whether an arbitrary PCS has a match. An **instance of PCP** is a particular PCS where we seek a match.

Example 9.3. The PCS of Example 9.2 is the triple $P = ((u_1, v_1), (u_2, v_2), (u_3, v_3))$, where $u_1 = a, v_1 = aba, u_2 = ab, v_2 = bb, u_3 = baa, v_3 = aa$. The match found out there was the sequence of indices 1, 3, 2, 3. □

For an algorithmic solution of PCP, the suitable language (an encoding) is

$$L_{PCP} = \{\psi(P) : P \text{ is a PCS with a match}\}.$$

I leave the details of how you encode a PCS as binary strings: $L_{PCP} \subseteq \{0, 1\}^*$. All that you require is a TM that can find out the individual u_i's, v_i's, and their subscripts i, etc. from the encoding $\psi(P)$.

We will be interested in a particular type of PCP, called a **modified PCP**, which imposes certain restrictions on a match. Let $P = ((u_1, v_1), \ldots, (u_n, v_n))$ be a PCS. A **modified match** in P is a sequence of indices $1, i_1, i_2, \ldots, i_k$, such that $u_1 u_{i_1} u_{i_2} \cdots u_{i_k} = v_1 v_{i_1} v_{i_2} \cdots v_{i_k}$. That is, a modified match must start with the first domino. The modified PCP, or MPCP for short, is the PCP where we seek a modified match.

We try to reduce the acceptance problem of TMs to MPCP by using the technique of valid computations. In case of PDAs, we required a twisted way of writing the valid computations due to the presence of a stack. Here, we use the valid computations themselves.

Let $M = (Q, \Sigma, \Gamma, \delta, s, h, \hbar)$ be a Turing machine. As earlier, we write a configuration (q, u, σ, v) as the string $uq\sigma v$. We also use the symbol $\leadsto$ for the one-step yield relation $\overset{1}{\leadsto}_M$. A **valid computation of** M **on input** w is a string $\# w_1 \# w_2 \# \cdots w_n \#$ such that

1. each w_i is a configuration of M,
2. w_1 is an initial configuration of M with input string w,
3. w_n is an accepting configuration, and
4. $w_i \leadsto w_{i+1}$ for each i, $1 \le i \le n-1$.

As earlier, you may use encodings of configurations, though it is easier to proceed with the w_i's. We build a PCS P by mimicking a valid computation of M on w, and then show that M accepts w iff P has a modified match.

Suppose the valid computation of M on w is $\# w_1 \# w_2 \# \cdots w_n \#$. The initial configuration w_1 looks like $s\not{b}w$. Construction of P is achieved in seven stages depending upon the possible next-configurations of M. The seven stages are as follows.

Stage 1: Take $(\#, \# w_1 \#)$ as the first domino in P.

As computation of M begins with this configuration, simulation of M on w in P must start with such a domino. The next configuration in the valid computation depends upon the transition function δ. Accordingly, we construct the relevant dominoes in the next three stages.

Stage 2: If $\delta(q, \sigma) = (r, \tau)$, then add $(q\sigma, r\tau)$ to P.

Stage 3: If $\delta(q, \sigma) = (r, \not{L})$, then add $(\gamma q\sigma, r\gamma\sigma)$ to P, for each $\gamma \in \Gamma$.

Stage 4: If $\delta(q, \sigma) = (r, \not{R})$, then add $(q\sigma, \sigma r)$ to P.

Stages 5–7 fix the loose ends, so that a modified match may be obtained.

Stage 5: Add (γ, γ) to P, for each $\gamma \in \Gamma \cup \{\#\}$.

Stage 6: Add $(\gamma h, h)$ and $(h\gamma, h)$ to P, for each $\gamma \in \Gamma$.

Stage 7: Add $(h\#\#, \#)$ to P.

Example 9.4. Let $M = (\{s, q, p, r, t, k\}, \{a, b, c\}, \{a, b, c, \not{b}\}, s, \delta, h, \hbar)$ be a TM with transitions $\delta(s, \not{b}) = (s, \not{R})$, $\delta(r, b) = (h, b)$, $\delta(s, a) = (q, b)$, $\delta(q, b) = (r, \not{R})$, $\delta(r, c) = (p, a)$, $\delta(p, a) = (t, \not{R})$, $\delta(t, a) = (k, b)$, $\delta(k, b) = (s, \not{L})$. Give it the input $acaa$. Its computation (towards acceptance) proceeds as follows:

$$s\not{b}\,acaa \leadsto sacaa \leadsto qbcaa \leadsto brcaa \leadsto bpaaa \leadsto bataa \leadsto bakba$$
$$\leadsto bsaba \leadsto bqbba \leadsto bbrba \leadsto bbhba.$$

Accordingly, by Stage 1, we put $(\#, \# s\not{b}\, acaa\#)$ into P.
In Stage 2, add (rb, hb), (sa, qb), (rc, pa), (ta, tb).
In Stage 3, add (akb, sab), (bkb, sbb), (ckb, scb), $(\not{b}\, kb, s\not{b}\, b)$.
In Stage 4, add to P the dominoes $(s\not{b}, \not{b}\, s)$, (qb, br), (pa, at).
In Stage 5, add (a, a), (b, b), (c, c), $(\not{b}, \not{b})$, $(\#, \#)$.
In Stage 6, add (ah, h), (bh, h), (ch, h), $(\not{b}\, h, h)$, (ha, h), (hb, h), (hc, h), $(h\not{b}, h)$.
In Stage 7, add to P the domino $(h\#\#, \#)$.

Now, a modified match corresponding to the valid computation of M on input $acaa$ starts with the first domino $(\#, \#s\text{ƀ}\,acaa\#)$. Written as it appears on a domino:

$$\begin{array}{l}\# \\ \#s\text{ƀ}\,acaa\#.\end{array}$$

To match the strings on the lower and upper tracks (partially), we use Stage 4 and then Stage 5 dominoes bringing the game to:

$$\begin{array}{l}\#\ s\text{ƀ}\,acaa \\ \#s\text{ƀ}\,acaa\#\ \text{ƀ}\,sacaa.\end{array}$$

Up to this, we have simulated the TM M for one step of its computation. Once more apply Stage 5 dominoes $(\#, \#)$ and $(\text{ƀ}, \text{ƀ})$ (as separators) and follow up the transition $\delta(s, a) = (q, b)$ as in the second step of the computation, with the Stage 4 dominoes. Thus we extend the match, using Stage 5 dominoes, to:

$$\begin{array}{l}\#\ s\text{ƀ}\,acaa\#\ \text{ƀ}\,sacaa \\ \#s\text{ƀ}\,acaa\#\ \text{ƀ}\,sacaa\,\#\,\text{ƀ}\,qbcaa.\end{array}$$

The matching thus proceeds. Notice that the configurations occurring in order, in a valid computation of M on w, do occur in the partial match on the lower string one after another, with an extra ƀ after the $\#$. In this scheme of matching, we are reconstructing the valid computation of M on w. Let me write the already matched string as $-\,matched\,-$. Then the partial match now looks like:

$$\begin{array}{l}-\,matched\,- \\ -\,matched\,-\ \text{ƀ}\,bbhba\#.\end{array}$$

We use the Stage 5 dominoes $(\text{ƀ}, \text{ƀ})$ and (b, b), Stage 6 domino (bh, b), and Stage 5 dominoes (b, b), (a, a), $(\#, \#)$, in that order. The partial match now looks like:

$$\begin{array}{l}-\,matched\,-\ \text{ƀ}\,bbhba\# \\ -\,matched\,-\ \text{ƀ}\,bbhba\,\#\,\text{ƀ}\,bhba\#.\end{array}$$

See that one symbol from the unmatched string on the lower track has been reduced by the use of a Stage 6 domino. This process is continued and we arrive at the partial match

$$\begin{array}{l}-\,matched\,- \\ -\,matched\,-\ \text{ƀ}\,h\#.\end{array}$$

Using Stage 6 domino $(\text{ƀ}\,h, \text{ƀ})$ and then Stage 5 domino $(\#, \#)$, we have

$$\begin{array}{l}-\,matched\,- \\ -\,matched\,-\ h\#.\end{array}$$

Finally, we use Stage 7 domino $(h\#\#, \#)$ to obtain a match. □

Exercise 9.10. Prove by induction on the length of the valid computations of M on w that P has a modified match if there is such a valid computation of M on w.

Notice that a modified match in P starts with the first domino as given in Stage 1, where the second (lower) string outnumbers the single symbol $\#$ on the first (upper) string. And the only dominoes where the first string outnumbers the second are given in Stages 6 and 7. Thus in such a match, Stage 6 and/or Stage 7 dominoes are bound to occur. That means, the accepting state h must occur in the computation of M on w. Hence the statement in Exercise 9.10 is an "iff" statement. With this you have proved the following result:

Lemma 9.3. *There is an algorithm to construct a PCS P from any given TM M and a given string w so that M accepts the input w iff P has a modified match.*

Lemma 9.3 reduces the acceptance problem of TMs to MPCP. To complete the argument, we require a reduction of MPCP to PCP. Recall that such a reduction would mean utilizing an algorithm to decide "whether an arbitrary PCS has a match" for determining "whether an arbitrary PCS has a modified match." We give such a reduction in the proof of our main theorem below.

Theorem 9.12. *PCP is unsolvable. Moreover, the subproblem of PCP over an alphabet containing at least two symbols is unsolvable.*

Proof. Suppose PCP is solvable. That is, we have an algorithm, say, A that given an arbitrary PCS with at least two symbols in its alphabet, reports correctly whether this PCS has a match or not. We first show how to use this algorithm A for solving MPCP.

Let $P = ((u_1, v_1), \ldots, (u_n, v_n))$ be a PCS over an alphabet Σ containing at least two symbols. We give an algorithm, say, B to construct a PCS $\hat{P}$ corresponding to the PCS P.

Let $*$ and $\star$ be two new symbols not in Σ. For any string $u = \sigma_1\sigma_2\cdots\sigma_m$ with each $\sigma_i \in \Sigma$, define

$$*u = *\sigma_1 * \sigma_2 * \cdots * \sigma_m,$$

$$u* = \sigma_1 * \sigma_2 * \cdots * \sigma_m *,$$

$$*u* = *\sigma_1 * \sigma_2 * \cdots * \sigma_m * .$$

B constructs the new PCS $\hat{P}$, corresponding to P, by taking

$$\hat{P} = ((*u_1, *v_1*), (*u_2, v_2*), (*u_3, v_3*), \ldots, (*u_n, v_n*), (*\star, \star)).$$

For solving MPCP, suppose that the PCS P is given. we apply the algorithm B on P to obtain the PCS $\hat{P}$. Then, we use the algorithm A to determine whether $\hat{P}$ has a match or not. If there is a match in $\hat{P}$, then it must start with its first domino $(*u_1, *v_1*)$, as this is the only domino whose first symbols in the first component (top track) and the second component (bottom track) are same. That is, each match in $\hat{P}$ is a modified match. Conversely, each modified match in $\hat{P}$ is, trivially, a match. Thus, A determines whether there is a modified match in $\hat{P}$.

Moreover, each modified match in $\hat{P}$ gives rise to a modified match in P. Such a match in P is obtained by just deleting all occurrences of $*$ and $\star$ from a match in $\hat{P}$. Thus it is decidable whether P has a modified match or not. That is, MPCP is solvable. This result, of course, depends on the solvability of PCP, as we have assumed here.

To complete the reduction process, let M be a TM and w be an input to it. Using Lemma 9.3, construct the PCS P. M accepts w iff P has a modified match. This reduces the acceptance problem of TMs to MPCP whenever the dominoes of the PCS use at least two symbols. That is, if MPCP is solvable, then the acceptance problem for TMs is also solvable.

But the acceptance problem for TMs is not solvable. Hence, MPCP is not solvable. Therefore, the subproblem of PCP where dominoes use at least two symbols is unsolvable. It follows that PCP, in general, is also unsolvable. □

What is the catch in "at least two symbols?" If only one symbol is used in the dominoes, then write m_i for the number of symbols in the upper track of the ith domino, and n_i for that in the lower track. If for some k, $m_k = n_k$, then a match consists of a single copy of the kth domino. If $m_i > n_i$ for each i, then there is no match. Similarly, if $m_i < n_i$ for each i, then there is no match. Else, suppose $m_j > n_j$ and $m_k < n_k$ for some j, k. Then take $n_k - m_k$ copies of jth domino and $m_j - n_j$ copies of the kth domino for a match.

Problems for Section 9.6

9.51. Show that the PCS $((ab, abb), (b, ba), (b, bb))$ has no match.

9.52. Does the PCS $((a, ba), (bba, aaa), (aab, ba))$ have a match?

9.53. Find three matches in the PCS $((a, aaa), (ab, b)\,(abaaa, ab))$.

9.54. Let P be the PCS $((001, 01), (0011, 111), (11, 111), (101, 010))$. Does P have a match? Does P have a modified match?

9.55. Consider two PCSs $P = ((aa, ab), (a, bb), (bba, a), (b, a), (a, ab))$ and $P' = ((a, abb), (ab, ba), (ab, b), (aba, ba))$. Which one has a match and which one has not? Prove your claim.

9.56. Describe how a PCS P is encoded as the binary string $\psi(P)$.

9.57. If Σ is a singleton, see where do the reductions in the proof of Theorem 9.12 (possibly also in Lemma 9.3) go wrong.

9.58. A *weak match* in a PCS is a match (relaxed) where the concatenated string w can be obtained by concatenating the n first components and also it can be obtained by the n second components, for some n, but the pairs used in the two cases need not be the same pairs. Show that the problem of determining whether a PCS has a weak match is solvable.

9.59. Let $G = (\{S, A, B, C\}, \{a, b, c\}, S, \{S \mapsto aABb|Bbb,\ Bb \mapsto c,\ AC \mapsto aac\})$ be an unrestricted grammar. Corresponding to G construct a PCS with dominoes $(FS\#, F)$, (a, a), (b, b), (c, c), (A, A), (B, B), (C, C), (S, S), $(E, \#aaacE)$, $(aABb, S)$, (Bbb, S), (C, Bb), (aac, AC), and $(\#, \#)$. Show that the string $aaac \in \mathcal{L}(G)$, and corresponding to the derivation of $aaac$, there is a modified match in the PCS.

9.60. Let $G = (N, \Sigma, S, R)$ be an unrestricted grammar, and let $w \in \Sigma^*$. Let E, F, and # be three new symbols not in $N \cup \Sigma$. Corresponding to G and this string w, construct a PCS with dominoes $(FS\#, F)$, $(E, \#wE)$, $(\#, \#)$, and (σ, σ), for each $\sigma \in \Sigma$, (X, X) for each $X \in N$, and (y, x) for each production rule $x \mapsto y \in R$. Show that the PCS has a modified match iff $w \in \mathcal{L}(G)$. Conclude that MPCP is unsolvable.

9.61. Show that the following modifications to the PCP are unsolvable:
(a) Does there exist a first-at-last match to a given PCS, where a first-at-last match is a match whose last domino is the first domino of the PCS? In this terminology, a modified match is a first-at-first match.
(b) Does there exist a first-two-at-first match to a given PCS, where a first-two-at-first match is a match whose first two dominoes are the first two of the PCS, kept in the same order?

9.7 About Logical Theories

From a historical perspective, the theory of computation originated from a question about logical theories. Alan Turing first proved that validity of first order logic is, in general, undecidable. The question is a very natural and a very general one. You can see it easily enough if you view any branch of mathematics as an example of a logical theory, as would a reductionist do. For example, the theory of groups has four axioms, say, A_1, A_2, A_3, A_4, the closure axiom, the associativity axiom, axiom of the existence of an identity element, and axiom of the existence of inverse of any element, respectively. If A is a theorem of group theory, then the statement "if A_1, A_2, A_3, and A_4, then A" holds in first order logic, or as it is customarily called in the first order logic, this sentence is valid.

That is how, if there is an algorithm to solve all instances of validity problem, then that would decide for us whether any suggested theorem is indeed a theorem, in any branch of mathematics. Logical theories can be presented by first collecting all symbols that might be used in such a theory, and then specifying what the symbols mean. We plan to do that here in an informal and brief manner, and then proceed towards the issue of algorithmic solvability of validity in such theories.

The simplest of logical theories that keeps the general mathematical flavor alive is the propositional logic. A **propositional language** has an alphabet that includes the punctuation marks), (; the connectives $\neg$, $\wedge$, $\vee$, $\rightarrow$, $\leftrightarrow$; and the atomic propositions $p_0, p_1, p_2, \ldots, p_n$, for some $n \geq 0$. The connectives represent the English conjuncts: not, and, or, if...then, if and only if, respectively. Then we declare each atomic proposition as a proposition, and lay out the rule:

> if x, y are propositions, then $\neg x$, $(x \wedge y)$, $(x \vee y)$, $(x \rightarrow y)$, $(x \leftrightarrow y)$ are propositions.

Our propositional language consists of all **propositions** that are generated this only way.

Next, we define truth and falsity in such a logical language by interpreting them as true or false. An **interpretation** t is taken as an assignment of values 0 or 1 to the atomic propositions, say, 0 for falsity, and 1 for truth. The interpretation t is defined for all propositions by employing the rules: $t(\neg x) = 1$ iff $t(x) = 0$, $t(x \wedge y) = 1$ iff $t(x) = 1 = t(y)$, $t(x \vee y) = 0$ iff $t(x) = 0 = t(y)$, $t(x \to y) = 0$ iff $t(x) = 1$ and $t(y) = 0$, and $t(x \leftrightarrow y) = 1$ iff $t(x) = t(y)$. These rules are commonly presented as truth tables.

Then we say that a proposition is **valid** if all its interpretations assign it to 1. For example, the proposition $(p_5 \vee \neg p_5)$ has only two possible interpretations, namely, t_1, t_2, where $t_1(p_5) = 1$ and $t_2(p_5) = 0$. Now, $t_1(p_5 \vee \neg p_5) = 1$ and also $t_2(p_5 \vee \neg p_5) = 1$. Therefore, $(p_5 \vee \neg p_5)$ is valid.

Exercise 9.11. A proposition x is equivalent to y, written $x \equiv y$, iff for each interpretation t, $t(x) = t(y)$. Show that $x \equiv y$ iff $x \leftrightarrow y$ is valid.

If a propositional language has n atomic propositions, there are only 2^n number of interpretations. Thus given any proposition in a propositional language, you can just construct all possible interpretations and then go for evaluating the proposition to check whether in each case, the proposition is evaluated to 1 or not. That is, the **validity problem** in propositional languages is solvable. The set of valid propositions in a given propositional language is called a **propositional theory**. Solvability of the validity problem of a propositional language is also expressed as telling that a propositional theory is decidable.

However, propositional languages lack expressive power. Most mathematical theories require the quantifiers. For example, the existence of an identity element in a group requires the existential quantifier. These requirements are met in a first order language.

A **first order language** has an alphabet consisting of the punctuation marks), (, ','; the connectives $\neg, \wedge, \vee, \to, \leftrightarrow$; the quantifiers $\forall, \exists$; atomic propositions $p_0, p_1, p_2, \ldots, p_n$; the individual constants $c_0, c_1, \ldots, c_k$; the function symbols $f_0, f_1, \ldots, f_l$; predicates (relation symbols) $R_0, R_1, \ldots, R_m$; and variables (in fact, variable symbols) $x_0, x_1, x_2, \ldots$, where $n, k, l, m \in \mathbb{N}$. In every first order language, we also have the equality relation, denoted by the same $=$. The function symbols come with their *arities,* that is, in any first order language, we must first specify which function symbol can take how many arguments; this is its arity. Similarly, predicates or the relation symbols come with their arities. The equality relation is binary; it has arity 2. Notice that our predicates here can be binary, ternary, etc. unlike the properties (predicates) used in Rice's theorem, which were only unary. The list of variables is infinite, though only a finite number of them are used in any context. The formulas of a first order language are defined in two stages.

In the first stage, we define **terms** by declaring that individual constants and variables are terms. Then we use function symbols to generate more terms by using the rule:

> if f_i is a function symbol of arity j and $t_1, \ldots, t_j$ are terms, then $f_i(t_1, \ldots, t_j)$ is a term. Moreover, all terms are generated this way.

In the second stage, we define the **atomic formulas** by declaring that each atomic proposition is an atomic formula. Further, if $t_1, \ldots, t_j$ are terms and R_i is a predicate of arity j, then $R_i(t_1, \ldots, t_j)$ is an atomic formula. Similarly, for any terms s, t, the expression $(s = t)$ is also taken as an atomic formula. Notice that we agree to use infix notation for the equality relation. Each atomic formula is declared as a **formula**, and then other formulas are built up from atomic formulas by employing the following rule:

> if X, Y are formulas and x_i is a variable, $\neg X$, $(X \wedge Y)$, $(X \vee Y)$, $(X \rightarrow Y)$, $(X \leftrightarrow Y)$, $\forall x_i X$, and $\exists x_i X$ are formulas. All formulas are built up this way.

For example, $(\forall x_1 \exists x_2 R_2(x_1, x_2) \rightarrow \exists x_2 \forall x_1 R_3(f_1(c_1)))$ is a formula provided R_2 is binary, R_3 is unary, and f_1 is unary.

An occurrence of a variable in a formula may be free or bound. For example, in the formula $\forall x_1 \exists x_2 R_2(x_1, x_2)$, all occurrences of variables x_1 and x_2 are bound; they are bound by the quantifiers using them. While in the formula $\forall x_1 \exists x_3 R_2(x_1, x_2)$, the occurrence of x_2 (its only occurrence) is free, all occurrences of x_1, x_3 are bound. You can try to formally define the bound and free, occurrences of variables.

A **sentence** is a formula where each occurrence of each variable is bound. When you translate English sentences to a first order language, you always get a sentence; that is why these are so called. For example, let us translate the sentence

> Between squares of any two rational numbers, there is a rational number.

You may first build up vocabulary. Suppose you write $R_1(x)$ for "x is a rational number," $R_2(x, y, z)$ for "y is between x and z," and $f_1(u)$ for "square of u." Then you translate the sentence first to

> If y and z are rational numbers, then there is a rational number x between $f_1(y)$ and $f_1(z)$.

Next, you use the quantifiers properly, reading $\forall$ as "for all," and $\exists$ as "there exists," so that the translation looks like:

$$\forall x_1 \forall x_3 ((R_1(x_1) \wedge R_1(x_3)) \rightarrow \exists x_2 R_2(f_1(x_1), x_2, f_1(x_3))).$$

See that this first order formula is a sentence.

To complete the description of a first order language, we must specify how sentences are interpreted as true or false. An **interpretation** of a sentence starts with a nonempty set (a known concrete set), called the *domain* of the interpretation. The interpretation then identifies the individual constants as elements of the domain, function symbols as concrete functions on the domain, preserving their arity, predicates as relations on the domain, again preserving arity, and atomic propositions as concrete propositions on the domain. The quantifier $\forall$ is interpreted as "for each ... in the domain," and the quantifier $\exists$ as "there exists ... in the domain such that," and the connectives $\neg$, $\wedge$, $\vee$, $\rightarrow$, $\leftrightarrow$ as "not," "and," "or," "if ... then," "if and only if," respectively. Moreover, the variables are treated as variables whose values range over the domain.

If I is an interpretation with domain D, then a predicate R_i of arity j is interpreted by I as a relation over D of arity j; we write this relation as R_i^I. Similarly, the interpreted function symbol f_i is written as f_i^I and the interpreted individual constants c_i are written as c_i^I. The equality relation is interpreted as the equality relation in the domain D, which is written as $=$ instead of $=^I$.

An interpretation interprets the first order sentence as a concrete sentence over the domain D. If the sentence on the domain happens to be true, we say that the interpretation I is a **model** of the first order sentence.

Example 9.5. Let X be the sentence $\forall x_1 \exists x_2 R_1(x_1, x_2)$. Let us have an interpretation I with domain $\mathbb{N}$, the set of all natural numbers, where R_1^I is the relation "$\leq$." Notice that R_1 is a binary predicate and so is the relation "$\leq$." Then the sentence X is interpreted as

For each $x_1 \in \mathbb{N}$, there exists an $x_2 \in \mathbb{N}$ such that $x_1 \leq x_2$.

This is a sentence that speaks of natural numbers. It happens to be true. Thus I is a model of X.

Let J be the interpretation with domain as $\mathbb{N}$, where R_1^J is the relation of "$>$." Now J interprets X as

For each $x_1 \in \mathbb{N}$, there exists an $x_2 \in \mathbb{N}$ such that $x_1 > x_2$.

This sentence in $\mathbb{N}$ is false as 0 is greater than no number in $\mathbb{N}$. Thus, J is not a model of X. □

Example 9.6. Let Y be the sentence

$$\forall x_1((R_1(x_1) \rightarrow \neg R_2(f_1(x_1), c_1)) \wedge \exists x_2(R_1(x_2) \wedge R_3(x_2, x_1))).$$

Let I be the interpretation with domain as $\mathbb{N}$, where R_1^I is the unary relation of primeness, R_2^I is the equality relation "$=$," c_1^I is the element 0, R_3^I is the binary relation "$>$," and f_1^I is the successor function. Then I interprets the sentence Y as

For each natural number x_1, [(if x_1 is prime, then successor of x_1 is not equal to 0) and (there is a natural number x_2 such that x_2 is prime and $x_2 > x_1$)].

This happens to hold in $\mathbb{N}$, and thus I is a model of Y. □

A first order sentence is called **valid** if each interpretation of the sentence is a model. That is, a valid sentence holds (is true) in each of its interpretations.

For example, $\neg\forall x_1 R_1(x_1) \rightarrow \exists x_1 \neg R_1(x_1)$ is a valid sentence, while the sentence $\exists x_1 \forall x_2 R_2(x_1, x_2)$ is not valid. To see the invalidity of the latter sentence, interpret R_1 as "$\geq$" in $\mathbb{N}$. The sentence is interpreted as "There is a maximum natural number," which is false in $\mathbb{N}$. Thus, this interpretation is not a model of the sentence, and the sentence is not valid.

The **validity problem** for a first order language is expressed as:

Given an arbitrary first order sentence, is it valid?

Remember, you are not just asked to determine whether the given first order sentence has a model or not. To solve the problem, you are supposed to construct an algorithm (a TM or a program) that must determine the validity of each and every first order sentence.

We connect validity problem to Post's correspondence problem. Instead of writing P as an n-tuple, and then writing a match as a sequence of indices, we will write P as a set and use the informal and intuitive notion of a match in concatenating the dominoes in P.

Let $P = \{(u_1, v_1), (u_2, v_2), \ldots, (u_n, v_n)\}$ be a PCS over $\{0, 1\}$. Then each $u_i, v_i \in \{0, 1\}^*$, a binary string. Our goal is to construct a first-order language from P, with certain properties so that reduction might be possible. We choose an individual constant c_0, two unary function symbols f_0, f_1, and a binary predicate R_1. We think of c_0 as the empty string, f_0 as the function that concatenates its argument with a 0, f_1 as a function symbol that concatenates its argument with a 1. Using these nomenclature, we can write any binary string as compositions of f_0, f_1 evaluated on c_0.

For example, the binary string 0 is thought of as $\varepsilon 0$, which then is written as $f_0(c_0)$. Similarly, $f_1(f_0(c_0))$ represents the binary string (read the composition backward) 01. In general, the binary string $b_1 b_2 \cdots b_m$ of bits is represented by the term $f_{b_m}(f_{b_{m-1}}(\cdots (f_{b_2}(f_{b_1}(c_0))) \cdots))$, which we again abbreviate to $f_{b_1 b_2 \cdots b_m}(c_0)$ for better readability.

This nomenclature is like translating a given argument in English to a first order language by building an appropriate vocabulary. Here, we are translating the PCS P into a first-order language. The predicate R_1 below represents intuitively the initially available dominoes, and then how the game of concatenating the dominoes is played through successive moves.

We first express the fact that (u_i, v_i) are the available dominoes. Next, we say that if we have a domino (x, y) and the domino (u_i, v_i), then we can have a concatenated domino (an extended domino as) (xu_i, yv_i). A match in this game is then an extended domino with the same first and second components. The binary predicate R_1 is, in fact, this extended domino. Now you can read through the formal construction of a suitable formula, a sentence, as in the following:

Construct the sentence X, which is $((X_1 \wedge X_3) \rightarrow X_4)$, where

$$X_1 = R_1(f_{u_1}(c_0), f_{v_1}(c_0)) \wedge \cdots \wedge R_1(f_{u_n}(c_0), f_{v_n}(c_0)),$$

$$X_2 = R_1(f_{u_1}(x_1), f_{v_1}(x_2)) \wedge \cdots \wedge R_1(f_{u_n}(x_1), f_{v_n}(x_2)),$$

$$X_3 = \forall x_1 \forall x_2 (R_1(x_1, x_2) \rightarrow X_2), \text{ and}$$

$$X_4 = \exists x_3 R_1(x_3, x_3).$$

Our goal is to show that the sentence X is valid iff P has a match. We break up the proof into two parts.

Lemma 9.4. *If X is valid, then P has a match.*

Proof. This part is a bit trivial. If you have translated an English sentence to first-order, and the first-order sentence is valid, then in particular, the English sentence should be true. We give a similar argument.

As X is valid, each of its interpretations is a model. We construct an interpretation I using the intended meanings of the symbols. We take the domain as the set $\{0, 1\}^*$, the set of all binary strings, take $c_0^I = \varepsilon$, the empty string. We define the functions $f_0^I, f_1^I : \{0, 1\}^* \to \{0, 1\}^*$ by $f_0^I(x) = x\,0$ and $f_1^I(x) = x\,1$. Finally, we take R_1^I as the binary relation on $\{0, 1\}^*$ defined by

$$R_1^I = \{(x, y) \in \{0, 1\}^* \times \{0, 1\}^* : \text{there is a sequence of indices } i_1, \ldots, i_k \text{ such that } x = u_{i_1} u_{i_2} \cdots u_{i_k} \text{ and } y = v_{i_1} v_{i_2} \cdots v_{i_k}\}, \text{ where } (u_{i_j}, v_{i_j}) \in P.$$

In the model I, X_1 corresponds to the sentence

The pair $(u_i, v_i) \in R_1^I$, for each i, $1 \le i \le n$.

This holds as the sequence of indices here has only one term, namely, i. In I, X_3 reads as:

If the pair $(x, y) \in R_1^I$, then $(xu_i, yv_i) \in R_1^I$ for each i, $1 \le i \le n$.

This sentence also holds in I as the sequence of indices $i_1, \ldots, i_k$ for the pair (x, y) can be extended to the sequence $i_1, \ldots, i_k, i$.

Thus, I is a model of $X_1 \wedge X_3$. As X is valid, I is a model of $X_1 \wedge X_3 \to X_4$. That is, I is also a model of X_4. This means, the sentence that corresponds to X_4 holds in the model I, that is, there is $(w, w) \in R_1^I$ for some $w \in \{0, 1\}^*$. By the very definition of R_1^I, we have got a sequence $i_1, \ldots, i_k$ such that $w = u_{i_1} u_{i_2} \cdots u_{i_k} = v_{i_1} v_{i_2} \cdots v_{i_k}$. Therefore, P has a match. □

Lemma 9.5. *If P has a match, then any interpretation of X is a model of X.*

Proof. Assume that P has a match. We have a sequence of indices $i_1, \ldots, i_k$ such that $u_{i_1} \cdots u_{i_k} = v_{i_1} \cdots v_{i_k} = w$, say. Let I be any interpretation of X, with the nonempty domain as D. Then $c_0^I \in D$, $f_0^I, f_1^I : D \to D$, and $R_1^I \subseteq D \times D$. As P has only binary strings on its dominoes, we somehow assign these binary strings to elements of D. To this end, define a map $\phi : \{0, 1\}^* \to D$ recursively by

$$\phi(\varepsilon) = c_0^I, \ \phi(x0) = f_0^I(\phi(x)), \text{ and } \phi(x1) = f_1^I(\phi(x)).$$

If $b_1 b_2 \cdots b_j$ is a binary string with $b_i \in \{0, 1\}$, then

$$\phi(b_1 \cdots b_j) = f_{b_j}^I(f_{b_{j-1}}^I(\ \cdots\ f_{b_2}^I(f_{b_1}^I(c_0^I))\ \cdots)).$$

The last expression is again abbreviated to $f_{b_1 b_2 \cdots b_j}^I(c_0^I)$. Thus, with the abbreviation at work, we have $f_s(c_0^I) = \phi(s)$ for any binary string s. For example, $\phi(011) = f_1^I(f_1^I(f_0^I(c_0^I)))$. We are simply coding backward the elements of D, so to speak.

To show that I is a model of X, we assume that I is a model of X_1, I is a model of X_3, and then prove that I is a model of X_4. As I is a model of X_1, for each i, $1 \le i \le n$, $(f_{u_i}^I(c_0^I), f_{v_i}^I(c_0^I)) \in R_1^I$, that is,

$$(\phi(u_i), \phi(v_i)) \in R_1^I, \text{ for each } i, \ 1 \le i \le n.$$

Similarly, I is a model of X_3 implies that

$$\text{If } (\phi(s), \phi(t)) \in R_1^I, \text{ then } (\phi(su_i), \phi(tv_i)) \in R_1^I.$$

Starting with $(\phi(u_1), \phi(v_1)) \in R_1^I$, and repeatedly using the above statement, we obtain:

$$(\phi(u_{i_1}u_{i_2} \cdots u_{i_k}), \phi(v_{i_1}v_{i_2} \cdots v_{i_k})) \in R_1^I.$$

That is, $\phi(w, w) \in R_1^I$. I is a model of X_4. This completes the proof. □

Now, suppose we have an algorithm A such that given any sentence Y, A decides whether Y is valid or not. Let P be any given PCS over the alphabet $\{0, 1\}$. Construct the sentence X as in Lemma 9.4. Because of the Lemmas 9.4 and 9.5, A decides whether P has a match or not. However, PCP is not solvable. Therefore, there cannot exist such an algorithm A. You have thus proved the following statement:

Theorem 9.13. *The validity problem for first-order languages is unsolvable.*

This theorem is often expressed as "Validity of first order logic is undecidable." In fact, our proof tells something more. It says that in any first-order language with at least one individual constant, two unary function symbols, and one binary predicate, the validity problem is unsolvable. This statement can be strengthened by requiring the presence of only one binary predicate. However, validity of *monadic first-order logic* is decidable. That is, if we have a first-order language with only a finite number of unary predicates, and a finite number of individual constants, but no function symbols, then the validity problem there would be solvable. This logic is the so-called Aristotelian logic, and Aristotle had already given us an algorithm to determine the validity of an arbitrary sentence in his logic.

Problems for Section 9.7

9.62. Show that validity in monadic first order logic is decidable. You do not *have to* use the algorithm of Aristotle for showing this.

9.63. Let $Th(\mathbb{N}, +)$ denote the additive theory of natural numbers, that is, the set of all true sentences in natural numbers having only addition and no multiplication. Is the sentence $\exists x \forall y (x + y = y)$ in $Th(\mathbb{N}, +)$? Is $\exists x \forall y (x + y = x)$ in $Th(\mathbb{N}, +)$?

9.64. Prove that the theory $Th(\mathbb{N}, +)$ is decidable.

9.65. Denote by $Th(\mathbb{N}, <)$ as the set of all true sentences in natural numbers where we have only the relation of "less than." Show that $Th(\mathbb{N}, <)$ is a decidable theory.

9.66. For $m > 1$, let $\mathbb{Z}_m = \{0, 1, 2, \ldots, m - 1\}$. Define $+, \times$ on $\mathbb{Z}_m$ by the usual addition and multiplication but compounded modulo m. For example, if $m = 7$, then $4 \times 3 = 12 \mod 7 = 5$. Denote by $\mathbb{F}_m = (\mathbb{Z}_m, +, \times)$. Show that the theory $Th(\mathbb{F}_m)$, the set of all true sentences in $\mathbb{F}_m$, is decidable for each $m > 1$.

9.8 Other Interesting Problems

Theorem 9.13 says that the set of valid sentences of first order logic is not decidable. However, this set is computably enumerable, as we have proof procedures in first-order logic that do prove all (eventually all) valid sentences. This is essentially the contents of *Gödels's completeness proof* of a Hilbert-style axiomatic system for first-order logic. The undecidability of valid sentences of first-order logic also implies that we cannot hope to write a program that would either give a proof, whenever possible, of a mathematical conjecture, or would report to us that the conjecture is false. In this section, we will discuss (not prove) some of the undecidability results about other branches of mathematics.

To begin with, the theory of natural numbers with addition as the only operation, commonly written as $(\mathbb{N}, +)$, has been shown to be a decidable theory. This means that there is an algorithm which, given any sentence in this theory, decides whether the sentence is true or false. Similarly, the field of reals $(\mathbb{R}, +, \times)$ is also a decidable theory. However, the theory of natural numbers with addition and multiplication $(\mathbb{N}, +, \times)$ is not decidable. Had it been decidable, there would not have been unproved conjectures such as Goldbach's conjecture or the twin-primes conjecture.

There are also nontrivial problems that had been shown to be solvable. For example, let $p(x)$ be a polynomial in one variable, having rational coefficients. The predicate "$p(x)$ has a zero in the closed interval $[a, b]$ for given rationals a, b" is decidable. This means that there is an algorithm (Use Sturm's theorem.) that decides whether an arbitrary polynomial with rational coefficients has a zero between two given rational numbers. Moreover, there is an algorithm to compute the number of real zeros of such a polynomial between two rationals.

A related problem concerns Diophantine equations. Let $p(x_1, \ldots, x_n)$ be a polynomial in n variables having integer coefficients. The equation

$$p(x_1, \ldots, x_n) = 0$$

for which an *integer solution* is sought is called a *Diophantine equation*. We seek an algorithm that would determine whether an arbitrary Diophantine equation has a solution. This is the celebrated *Hilbert's tenth problem*. It has been shown that there is no such algorithm; Hilbert's tenth problem is unsolvable. In fact, something more has been shown. To expose the ideas, we need to improve our vocabulary a bit.

A *Diophantine predicate* is a unary predicate $M(x)$ expressed in the form

$$\exists x_1 \exists x_2 \cdots \exists x_n (p(x, x_1, x_2, \ldots, x_n) = 0), \text{ where } x \text{ ranges over integers.}$$

For example, the predicate "the integer x is a perfect cube" is Diophantine as this predicate can be expressed as $\exists x_1 (x - x_1^3) = 0$, where $p(x, x_1) = x - x_1^3$ is a polynomial with integer coefficients.

Let R be a unary predicate. We say that $R(x)$ is computably enumerable if the set $\{x : x \text{ is an integer and } R(x)\}$ is a computably enumerable set, that is, if all those integers satisfying the property R form a computably enumerable set. Matiyasevich proved that any predicate $R(x)$, where x ranges over integers, is computably enumerable iff it is Diophantine.

With Matiyasevich's theorem, you can get an easy reduction of acceptance problem of TMs to Hilbert's tenth problem, thereby showing the undecidability of the latter. Again, an easy but surprising corollary of Matiyasevich's theorem is: "there exists a polynomial with integer coefficients such that all its positive values are prime numbers, and this way all prime numbers can be obtained." It is so because the set of prime numbers is a computably enumerable subset of the set of all positive integers.

Unsolvability of a mathematical problem guarantees that there is no uniform effective procedure for each instance of the problem. It does not mean that a given instance of such a problem cannot be solved. For example, acceptance problem for TMs is unsolvable. But given a particular TM, it might be possible by simulating the machine on a given input, to determine whether it accepts the input or not. In the same vein, there is no effective procedure to determine whether any arbitrary sentence in $(\mathbb{N}, +, \times)$ is true or false. But we have proved many theorems and disproved many conjectures in this theory.

The unsolvability results prove the nonexistence of an effective procedure that may be able to tackle all the cases of a problem at a time. It does not say that we cannot prove any suggested theorem (provable sentence) of mathematics. Moreover, unsolvability results are silent about whether all true statements can be proved at all, say, case-by-case if not uniformly. For example, we have proof procedures in first-order logic that can prove all theorems eventually; for, that is enumerability involving completeness of proof procedures.

We say that a *problem is semi-decidable* when the corresponding language is computably enumerable. Moreover, this happens if an algorithm exists that answers "yes" in all yes-cases, and either answers "no" or does not answer anything in all no-cases. In the case of a first-order theory, there exist many proof systems, such as Hilbert-style axiomatic systems, natural deduction systems, and analytic tableaux. Soundness of such a proof system says that each provable first-order sentence is valid. Completeness of such a system guarantees that each valid first-order sentence has a proof.

These proof systems serve as algorithms in enumerating all valid sentences; they answer "yes" in all yes-cases. However, it can happen that each one fails in reporting a "no" answer when a given sentence is not valid. Therefore, validity problem for first-order logic is semi-decidable. Further, there exists gap between truth and provability in arithmetic. This is the celebrated incompleteness theorem of Gödel; see Problems 9.120 and 9.128 in the Additional Problems to this chapter.

Problems for Section 9.8

9.67. Let M be a TM and w be a string. Construct a formula $\phi_{M,w}(x)$ in the language of $Th(\mathbb{N}, +, \times)$ that contains a single free (unquantified) variable x so that the sentence $\exists x \phi_{M,w}(x)$ is true iff M accepts w.

9.68. Give a map reduction of the acceptance problem for TMs to $Th(\mathbb{N}, +, \times)$ by constructing the formula $\phi_{M,w}(x)$ from $\psi(M)\# \psi(w)$ using the preceding problem. Conclude that the theory $Th(\mathbb{N}, +, \times)$ is undecidable.

9.69. Let $P(x_1, x_2, \ldots, x_n)$ be an n-ary predicate of natural numbers. Define its characteristic function χ_P by $\chi_P(x_1, x_2, \ldots, x_n) = 1$ if $P(x_1, x_2, \ldots, x_n)$ holds, else, $\chi_P(x_1, x_2, \ldots, x_n) = 0$. We say that P is a *decidable predicate* if χ_P is a computable function, otherwise, we say that P is undecidable. Similarly, we say that P is a *semi-decidable predicate* if its partial characteristic function ξ_P defined by "$\xi_P(x_1, x_2, \ldots, x_n) = 1$ when $P(x_1, x_2, \ldots, x_n)$ holds, and $\xi_P(x_1, x_2, \ldots, x_n)$ is otherwise undefined" is computable. Prove the following:
(a) P is semi-decidable iff "there is a computable function g such that $(x_1, x_2, \ldots, x_n)$ is in the domain of g iff $P(x_1, x_2, \ldots, x_n)$ holds."
(b) P is semi-decidable iff "there is a decidable $(n+1)$-ary predicate R such that $P(x_1, x_2, \ldots, x_n)$ holds iff $\exists y R(x_1, x_2, \ldots, x_n, y)$ is true."
(c) If the $(n+1)$-ary predicate $Q(x_1, x_2, \ldots, x_n, y)$ is semi-decidable, then so is the n-ary predicate $\exists y Q(x_1, x_2, \ldots, x_n, y)$.
(d) If the $(n+m)$-ary predicate $Q(x_1, x_2, \ldots, x_n, y_1, y_2, \ldots, y_m)$ is semi-decidable, then so is the n-ary predicate $\exists y_1 \exists y_2 \cdots \exists y_m Q(x_1, x_2, \ldots, x_n, y_1, y_2, \ldots, y_m)$.
(e) "Provability" in first order logic is semi-decidable.
(f) Let $p(x_1, \ldots, x_m, y_1, \ldots, y_n)$ be a polynomial in $m+n$ variables with integer coefficients, then, $\exists y_1 \cdots \exists y_n (p(x_1, \ldots, x_m, y_1, \ldots, y_n) = 0)$ is a semi-decidable predicate. [Here, the variables range over integers.]
(g) A predicate P is decidable iff both P and $\neg P$ are semi-decidable.

9.70. Let $A \subseteq \mathbb{N}$. Prove that the following are equivalent:
(a) A is computably enumerable.
(b) The predicate $x \in A$ is semi-decidable.
(c) A is the domain of a unary computable partial function.
(d) There is a decidable predicate $R(x, y)$ such that $x \in A$ iff $\exists y R(x, y)$ is true.
(e) $A = \emptyset$ or A is the range of a unary computable total function.
(f) A is the range of a computable partial function.

9.71. Show that if the (numerical) predicate $P(x, y_1, \ldots, y_n)$ is semi-decidable, then the set $\{x : \exists y_1 \cdots \exists y_n P(x, y_1, \ldots, y_n)\} \subseteq \mathbb{N}$ is computably enumerable. Is the converse true?

9.72. Show that the following problems about C programs are unsolvable:
(a) Whether a given program loops forever on some input.
(b) Whether a given program ever produces an output.
(c) Whether two given programs produce the same output on all inputs.

9.9 Summary and Additional Problems

Our approach in this chapter has been greedy. We first proved Rice's theorem and then used it to show that many interesting problems about Turing machines are unsolvable. The reduction technique has been quite helpful in arriving at the unsolvability results. We then discussed solvability in the domains of finite automata and PDAs. As many unsolvable problems have been shown to be so by a reduction of Post's correspondence problem (PCP), we have discussed the unsolvability of PCP.

This has been demonstrated by the proof of unsolvability of validity problem in first order logic by reducing PCP to it. We then mentioned many interesting problems that have been shown to be unsolvable.

Reducibility relations have been discussed by Post; see [105, 111, 123]. Rice's theorem was proved in [109, 110]. The idea of valid computation histories was called as T-predicates by Kleene [64, 65]. Undecidable properties of CFLs are from [8, 11, 18, 33, 40, 52]. PCP has been introduced in [106].

To give a bit of history, the origin of Theory of Computation dates back to 1900. It came in the form of a problem:

> Find an algorithm to determine whether a given polynomial with integer coefficients has an integral root.

posed by David Hilbert at the International Congress of Mathematicians held at Paris. This was the tenth of 23 problems posed by Hilbert. As he rightly apprehended, the problems engaged the mathematicians around the globe throughout the last century. Some of them have not been solved yet.

To almost everyone present the problems appeared meaningful enough. As a byproduct, the discipline of Computer Science was born. However, it took around 35 years to make the tenth problem meaningful. For, only around 1935, the concept of an algorithm could be crystallized. It took another 35 years for solving Hilbert's tenth problem. In 1970, Matiyasevich proved that there cannot exist such an algorithm.

Sturm's theorem, which is used to show that calculation of zeros of a polynomial with rational coefficients can be solved algorithmically, is from [21]. Matiyasevich's theorem is from [83, 26]. For details regarding this see [25, 80]. For a deeper connection between logic, automata theory, and formal language theory, see [129].

In some of the problems below, I ask you to deal with oracle TMs; they were introduced by Turing [132]. See [125] also for oracle Turing machines. The arithmetic hierarchy and the analytic hierarchy can be found in [65, 66, 111, 119, 123]. The Post's tag system is from [89]. For recursion theorem, see [66].

Turing's proof of *Gödel's incompleteness theorem* is from Turing [131]. Gödel's proof can be found in [43, 44]. A simplified proof of the same along with Gödel's completeness theorem and the unprovability of consistency of arithmetic can be found in [1, 120]. You can also see [1, 120, 121] for various proof systems for first order logic.

Additional Problems for Chapter 9

9.73. Show that the halting problem for deterministic LBAs is decidable.

9.74. Let $\Sigma = \{0, 1\}$. A binary relation R on Σ^* is called *decidable* if the language $\{v2w : (v, w) \in R\}$ is decidable. Show that a language $L \subseteq \Sigma^*$ is decidable iff there exists a decidable binary relation R on Σ^* such that $L = \{v \in \Sigma^* : (v, w) \in R \text{ for some } w \in \Sigma^*\}$.

9.75. Show that the language $\{\psi(D) : D$ is a DFA that accepts w^R whenever it accepts $w\}$ is decidable. Express the last statement as the solvability of a problem about DFAs.

9.76. Suppose L and L' are regular languages over Σ. Show that the following problems are solvable:
(a) Is $shuffle(L, L') = L$?
(b) Is $tail(L) = L$?

9.77. Give algorithms to decide the following:
(a) For a given CFG G, does $\mathcal{L}(G)$ contain at least 2009 strings?
(b) Given a CFG and a symbol A, does there exist a derivation yielding a string (of terminals and possibly, nonterminals) in which the first symbol is A?

9.78. Let L be a CFL. Show that L^+ is computably enumerable.

9.79. A *useless state* in a PDA is a state that is never entered during any computation. Show that the problem of determining whether an arbitrary PDA has a useless state is solvable. Express the problem as a language.

9.80. Show that both the conditions spelled out below in detail are necessary for proving that L_P is undecidable in *Rice's theorem:*
(a) For any TMs M, M' with $\mathcal{L}(M) = \mathcal{L}(M')$, if $\psi(M) \in L_P$, then $\psi(M') \in L_P$.
(b) There are TMs M, M' such that $\psi(M) \in L_P$ but $\psi(M') \notin L_P$.

9.81. Are the following problems about an arbitrary TM M semi-decidable?
(a) Does $\mathcal{L}(M)$ contain at least two strings?
(b) Is $\mathcal{L}(M)$ finite?
(c) Is $\mathcal{L}(M)$ regular?
(d) Is $\mathcal{L}(M)$ context-free?
(e) Is $\mathcal{L}(M)$ context sensitive?
(f) Is $(\mathcal{L}(M))^R = \mathcal{L}(M)$?

9.82. For $L, L' \subseteq \Sigma^*$, define $L'/L = \{w \in \Sigma^* : vw \in L'$ for some $v \in L\}$.
(a) Show that if L, L' are computably enumerable, then so is L'/L.
(b) Show that each computably enumerable language over Σ can be written as L'/L for some CFLs L and L'.

9.83. Show that the language $\{\psi(M)\#\psi(M')\#0^{k+1} : M, M'$ are TMs where $\mathcal{L}(M) \cap \mathcal{L}(M')$ contains at least k strings$\}$ is computably enumerable but not decidable.

9.84. Let A be a computably enumerable set of descriptions of (ψ of) total TMs. Prove that there is a decidable language that is not decided by any TM listed in A.

9.85. Prove that for any two languages A and B, there exists a language C such that both A and B are Turing reducible to C.

9.86. Let $P(\cdot)$ be a property of *RE* such that the set of all TMs whose languages satisfy the property P is computably enumerable. If $A \subseteq B$ are languages such that $P(A)$ holds, then show that $P(B)$ holds.

9.87. Let R be the set of all TMs that accept regular languages. Show that neither R nor its complement in the set of all TMs is computably enumerable.

9.88. Let A be the set of all TMs that halt on all inputs. Show that neither A nor its complement is computably enumerable.

9.89. Show the following:
(a) There does not exist a computably enumerable list of total TMs such that each decidable language is represented by at least one TM in the list.
(b) There exists a computably enumerable list of TMs such that each decidable language is represented by at least one TM in the list and each TM in the list accepts some decidable language.

9.90. Let A be a computably enumerable set of descriptions of (ψ of) TMs. Prove that there is a decidable language B consisting of descriptions of TMs such that every TM in A has an equivalent TM having its description in B and vice versa.

9.91. Is the problem of determining whether for three given TMs M_1, M_2, M_3, $\mathcal{L}(M_3) = \mathcal{L}(M_1)\mathcal{L}(M_2)$ solvable?

9.92. A *ray automaton* consists of an infinite number of DFAs $D_0, D_1, \ldots,$ arranged in a line. They all have the same set of states Q, the same start state s, and the same transition function δ except D_0, which has a different transition function, as it has no left neighbor. They all start simultaneously in the state s and compute synchronously. In each step, D_i enters a new state, depending on its own current state and the current states of its two neighbors, following its own transitions. The ray automaton *halts* when D_0 enters its final state (only one such is available). There is no input alphabet. Formally define ray automata. Prove that the halting problem for ray automata is undecidable.

9.93. Show that for an arbitrary unrestricted grammar G, the problem of deciding whether $\mathcal{L}(G) = (\mathcal{L}(G))^*$ is unsolvable, by reducing the acceptance problem to this problem.

9.94. Prove that the emptiness problem for LBAs, that is, whether a given LBA accepts the empty language, is unsolvable.

9.95. Let G be a regular grammar and G' be any grammar. Is the problem $\mathcal{L}(G) = \mathcal{L}(G')$ solvable when G' is
(a) regular?
(b) context-free?
(c) linear?
(d) context-sensitive?
(e) unrestricted?

9.96. Is the problem "Given a CFG G, whether $\mathcal{L}(G) = \mathcal{L}(G)\mathcal{L}(G)$?" solvable?

9.97. Show that, determining whether the union of two DCFLs is a DCFL, is an unsolvable problem.

9.98. Prove that it is undecidable whether a given LBA halts on every input. Interpret the result for context-sensitive grammars.

9.99. Prove that the finiteness problem for TMs reduces to that of LBAs. Use this to show that the problem of determining whether a given LBA accepts a regular language is unsolvable.

9.100. Determine whether the following problems are solvable:
(a) Does a given TM ever move its read–write head more than 29 times during its computation on input a^{2009}?
(b) Does a given TM ever write more than 29 nonblank symbols on input a^{2009}?
(c) Does M ever move its read–write head to the left while computing on input w?
(d) Does M enter each of its states while computing on input w?
(e) Does a TM M started on an input w ever scan any tape square more than once?

9.101. Show that a nontrivial property P of RE is computably enumerable iff P satisfies the three conditions of monotonicity, compactness, and finite enumerability.

9.102. Let M be an arbitrary TM. Show that the problem of determining whether there exists a TM of shorter description than M accepting the same language $\mathcal{L}(M)$ is unsolvable. [See Problem 10.146.]

9.103. Consider only TMs with a left end (a dead end) beyond which the read–write head can never move. Note that these machines can simulate the standard ones. Restrict these TMs by requiring that they can never overwrite the input string; they can write on the blank squares to the right of the input. These restricted machines accept only the regular languages (prove). Show that given such a restricted TM, it is impossible to give an algorithm for constructing an equivalent DFA.

9.104. A nondeterministic two-head finite automaton or a *two-head NFA* is an NFA with two heads that can read but not write, and can move only left to right. The automaton starts in its initial state and with its two heads on the leftmost square of the tape. A typical transition is a quadruple of the form (p, q, a, b), where p, q are states and a, b are either symbols or the empty string ε. It means that being in state p, if the NFA reads a with its first head and b with its second head, then it enters state q. The NFA accepts an input string by moving both the heads off the right end of the string while by entering a final state. Show the following:
(a) "Given a two-head NFA N and an input string w, whether N accepts w" is solvable.
(b) "Given a two-head NFA N, whether it accepts some string" is unsolvable.

9.105. Informally describe a multitape TM that enumerates the set of all n such that M_n accepts w_n, where we have the sequence of all TMs and the sequence of all inputs written as M_i's and w_i's. Can a TM enumerate all these n in numerical order?

9.106. A language C *separates* two disjoint languages A and B if $A \subseteq C$ and $B \subseteq \overline{C}$. Show that if A, B are disjoint and co-computably enumerable languages, then there is a decidable language C that separates them. Recall that L is co-computably enumerable iff $\overline{L}$ is computably enumerable.

9.107. Unsolvability of most problems about CFLs can be seen via PCP instead of using valid computation histories. Consider the PCS $P = \{(u_1, v_1), \ldots, (u_n, v_n)\}$ over an alphabet Σ. Write $B = \{u_1, \ldots, u_n\}$ and $C = \{v_1, \ldots, v_n\}$. Take new symbols $a_1, a_2, \ldots, a_n$; these are called the index symbols. Define two CFGs G_B, G_C having the only nonterminals as B, C, respectively, and the productions as in the following:

$$B \mapsto u_1Ba_1|u_2Ba_2|\cdots|u_nBaN, \; B \mapsto u_1a_1|u_2a_2|\cdots|u_na_n;$$
$$C \mapsto v_1Ca_1|v_2Ca_2|\cdots|v_nCa_n, \; C \mapsto v_1a_1|v_2a_2|\cdots|v_na_n.$$

Let $L_B = \mathcal{L}(G_B)$ and $L_C = \mathcal{L}(G_C)$. The languages $\overline{L}_B$ and $\overline{L}_C$ are the complements of L_B and of L_C in the set $(\Sigma \cup \{a_1, \ldots, a_n\})^*$. Finally, we construct another CFG G_{BC} having nonterminals S, B, C with S as the start symbol, and the productions as $S \mapsto B|C$ along with all productions of G_B, and all of G_C. Attempt the following:

(a) Prove that the CFG G_{BC} is ambiguous iff the PCS P has a match.
(b) $\overline{L}_B$ and $\overline{L}_C$ are CFLs.
(c) Taking $\mathcal{L}(G_1) = L_B$ and $\mathcal{L}(G_2) = L_C$, show that the problem of determining whether for arbitrary CFGs G_1, G_2, $\mathcal{L}(G_1) \cap \mathcal{L}(G_2) = \emptyset$ is unsolvable.
(d) Taking $\mathcal{L}(G_1) = \overline{L}_B \cup \overline{L}_C$ and $\mathcal{L}(G_2) = (\Sigma \cup \{a_1, \ldots, a_n\})^*$, show that it is unsolvable whether the language of a given CFG G_1 is regular.
(e) Show that the problem of determining for an arbitrary CFG G and a regular expression E, whether $\mathcal{L}(E) \subseteq \mathcal{L}(G)$ is unsolvable.
(f) Show that it is unsolvable whether a given CFG is ambiguous or not.
(g) Taking $\mathcal{L}(G_1) = (\Sigma \cup \{a_1, \ldots, a_n\})^*$ and $\mathcal{L}(G_2) = \overline{L}_B \cup \overline{L}_C$, show that it is unsolvable whether for arbitrary CFGs G_1, G_2, $\mathcal{L}(G_1) \subseteq \mathcal{L}(G_2)$.
(h) Prove that $\overline{L}_B \cup \overline{L}_C$ is regular iff it equals $(\Sigma \cup \{a_1, \ldots, a_n\})^*$ iff the PCS P has a match.
(i) Show that it is unsolvable whether or not a CFG generates a regular language.
(j) Prove that it is unsolvable whether the complement of a CFL is also a CFL.

9.108. Show that the set of all $\psi(G)$ for CFGs G that generates at least one palindrome is undecidable.

9.109. Prove that there exist two languages neither of which is Turing reducible to the other.

9.110. A *Thue system* is a finite set of two-elements sets of strings over some alphabet Σ such as $T = \{\{u_1, v_1\}, \ldots, \{u_n, v_n\}\}$. Let $x, y \in \Sigma^*$ be strings. We say that x can be obtained from y in T if for some k with $1 \leq k \leq n$, there are $z_1, z_2 \in \Sigma^*$ such that either $x = z_1u_kz_2$ and $y = z_1v_kz_2$ happen or $x = z_1v_kz_2$ and $y = z_1u_kz_2$ happen. Notice that if x can be obtained from y, then y can also be obtained from x. We then extend this relation to an equivalence relation by taking its reflexive and transitive closure; that is, we say that x is equivalent to y in T if there is a finite sequence of strings $x_1, x_2, \ldots, x_m$ such that x can be obtained from x_1, each x_i can be obtained from x_{i+1}, and finally, x_m can be obtained from y. The *word problem for Thue systems* is the problem of determining whether given a Thue system T and given two strings x, y, the string x is equivalent to y in T or not. Show that the word problem for Thue systems is unsolvable.

9.111. A *weak–strong match* in a PCS is a match where the concatenated string w can be obtained by concatenating the n first-components and also it can be obtained by the n second-components, the pairs used in the two cases are required to be the same pairs but need not be in the same order. That is, we allow the n second-components to be permuted before concatenation. Is the problem of determining whether a PCS has a weak–strong match solvable?

9.112. A *Post's tag system* is a set of pairs of strings over an alphabet Σ with a designated *start* string. If (u, v) is a pair of strings and w is any string in Σ^*, we say that $uw \rightsquigarrow wv$; this defines a *move* in the system. Show that it is unsolvable, given a Post's tag system and a string x, whether $x \rightsquigarrow \varepsilon$ in zero or more moves.

9.113. There is a two-dimensional version of PCP, called the *tiling problem.* We have an infinite number of tiles, each one square unit with which we have to tile the first quadrant. The only restrictions are that a special tile is to be put at the lower left corner; it is called the *origin tile*; certain other tiles can touch each other horizontally and certain others can touch vertically. Tiles may not be rotated as we please. Thus the infinite number of tiles in a tiling system comprise an infinite supply from each kind of these finite types of tiles. The problem is to determine whether there is an algorithm to tile the first quadrant given a tiling system. A formal version of a tiling system is as follows. A tiling system is a quadruple (T, t_0, H, V), where T is a finite set of tiles (in fact, the types; having infinite supply from each type), $t_0 \in T$, and $H, V \subseteq T \times T$. A tiling by such a system is a function $f : \mathbb{N} \times \mathbb{N} \to T$ satisfying

$f(0, 0) = t_0$, and

$(f(m, n), f(m+1, n)) \in H,\ (f(m, n), f(m, n+1)) \in V$, for all $m, n \in \mathbb{N}$.

Show that the problem "given a tiling system, whether there exists a tiling by that system" is unsolvable.

9.114. Suppose we think of the tiles as being determined by the colors of their four edges, and that two similarly colored edges are allowed to touch each other. Further suppose that we are allowed to rotate the tiles and turn them over. Show that any nonempty set of tiles, with or without a given origin tile, can be used to tile the first quadrant.

9.115. Formulate precisely and then prove the following extension of *Rice's theorem:* every nontrivial property of pairs of computably enumerable sets is undecidable.

9.116. The set of propositions is defined over a countably infinite set of symbols $\{\neg, \wedge, \vee, \to, \leftrightarrow,), (\} \cup \{p_0, p_1, \ldots\}$, using the rules that each p_i is a proposition, and if p_i, p_j are propositions, then so are $\neg p_i$, and $(p_i \beta p_j)$, where $\beta \in \{\wedge, \vee, \to, \leftrightarrow\}$. It defines a CFG over an *infinite alphabet*. Using the prefix lemma (Lemma 2.1), show that this CFG is unambiguous.

9.117. Each atomic proposition p_i in the last problem can be rewritten as p followed by $i+1$ number of 0's. That is, the alphabet $\{\neg, \wedge, \vee, \to, \leftrightarrow,), (\} \cup \{p_0, p_1, \ldots\}$ can be generated as a language over $\Sigma = \{\neg, \wedge, \vee, \to, \leftrightarrow,), (, p, 0\}$. Finally, the set of propositions can be generated as a language over the alphabet Σ. Construct a CFG with terminals in Σ that generates the set of all propositions.

9.118. Construct a CFG for generating the set of all first order formulas.

9.119. The numbers 34 and 10 are the ASCII codes for the double quote and the newline, respectively. Understand, without executing, what the following C program does:

```
char *s=" char *s=%c%s%c;
% cmain() {printf(s,34,s,34,10,10);}%c";
main() {printf(s,34,s,34,10,10);}
```

9.120. *Turing's proof of Gödel's Incompleteness Theorem:* Let $T(\mathbb{N}, +, \times)$ denote the set of all statements in Peano's arithmetic ($\mathbb{N}$ with addition and multiplication) which have proofs, the theorems of $\mathbb{N}$. Let $Th(\mathbb{N}, +, \times)$ denote the set of all true statements of Peano's arithmetic, *the first order theory of* $\mathbb{N}$. Show that

(a) $T(\mathbb{N}, +, \times)$ is computably enumerable.

(b) $Th(\mathbb{N}, +, \times)$ is not computably enumerable.

(c) There exists at least one true statement in Peano's arithmetic which cannot be proved.

9.121. Show that there exists a computable total function from $\mathbb{N}$ to $\mathbb{N}$ that cannot be proved to be so in Peano's arithmetic.

9.122. Describe two TMs M and N such that when started on any input w, the TM M outputs $\psi(N)$ and the TM N outputs $\psi(M)$.

9.123. Here I ask you to prove a fixed-point theorem, called the *recursion theorem.* It is as follows. Use the notation S_x for the TM whose encoding, $\psi(S_x)$, is the binary string x. Also for any TM T, let $T(x)$ denote the contents of the tape of T when T has halted on input x; if T does not halt on x, then let $T(x)$ remain undefined. Further, suppose $f : \{0, 1\}^* \rightarrow \{0, 1\}^*$ be a computable total function. Then there exists a string $u \in \{0, 1\}^*$ such that $\mathcal{L}(S_u) = \mathcal{L}(M_{f(u)})$. Prove it by showing the following:

(a) Let f be computed by the TM M. Let N be the TM that on input x computes a description (encoding) of a TM K, where K on input y does the following:

K constructs M_x, simulates M_x on input x, if it halts, then it simulates M on $S_x(x)$, it interprets the result of the computation, that is, it identifies $M(S_x(x))$ as the description of a TM, and then simulates that TM on the original input y, accepting or rejecting as that machine accepts or rejects, respectively.

Show that N is a total TM and that $\mathcal{L}(S_{N(x)}) = \mathcal{L}(S_{f(S_x(x))})$.

(b) Let w be a description of the machine N, that is, $N = S_w$. Show that $u = N(w)$ satisfies $\mathcal{L}(S_u) = \mathcal{L}(M_{f(u)})$. [Such a u is called a *fixed point* of f.]

9.124. Give a short proof of Rice's theorem using the recursion theorem.

9.125. A TM M is called a *minimal* TM if, among all TMs accepting the same language $\mathcal{L}(M)$, M has the fewest number of states. Prove that there does not exist an infinite computably enumerable set of minimal TMs. Formulated another way, write $MIN_{TM} = \{\psi(M) : M \text{ is a minimal TM}\}$. Let $A \subseteq MIN_{TM}$ be any infinite set. Show that A is not computably enumerable.

9.126. Let $f : \{0, 1\}^* \to \{0, 1\}^*$ be a computable function. Prove that there is a TM M such that $f(\psi(M)) = \psi(M')$ for some TM M' equivalent to M. (The TM M is a *fixed point* of f.)

9.127. In the preceding problem, suppose f is the function that interchanges h and $\hbar$ of the machine encoded as binary strings, which are in the form $\psi(N)$ for some TM N. What would be a fixed point of f? Give an example of such a fixed point.

9.128. To construct the sentence "This sentence is not provable." in the theory $Th(\mathbb{N}, +, \times)$, we use the following *facts:*

Fact 1: Let M be a TM and w be a string. A formula $\phi_{M,w}(x)$ in the language of $Th(\mathbb{N}, +, \times)$ that contains a single free (unquantified) variable x can be algorithmically constructed so that the sentence $\exists x \phi_{M,w}(x)$ is true iff M accepts w.

Fact 2: There exists an algorithm A that checks whether a suggested proof of a sentence in $Th(\mathbb{N}, +, \times)$ is indeed a proof or not.

Now, Construct a TM M that operates as follows:

> On any input, M obtains its own description $\psi(M)$, via Recursion theorem. It then constructs the sentence $X = \neg\exists x \phi_{M,0}(x)$ using the above mentioned facts. Next, it runs algorithm A on X. Finally, if A reports success, then it accepts; if A halts but reports failure, then it rejects.

Show that

(a) The sentence X is true iff M does not accept 0.

(b) M cannot find a proof of X.

(c) *Gödel's Incompleteness Theorem*: X is true but not provable.

9.129. *Matiyasevich's Theorem* states that all computably enumerable sets are Diophantine. That is, if $A \subseteq \mathbb{N}$ is computably enumerable, then there is a polynomial $p(x, y_1, \ldots, y_n)$ in $n + 1$ variables and with integer coefficients such that "$x \in A$ iff $\exists y_1 \cdots \exists y_n(p(x, y_1, \ldots, y_n) = 0)$" holds.

Use Matiyasevich's theorem to show that a set $A \subseteq \mathbb{N}$ is computably enumerable iff A is the set of all nonnegative values taken by some polynomial $p(x_1, \ldots, x_n)$ with integer coefficients; for values of $x_1, \ldots, x_n$ taken from $\mathbb{N}$.

9.130. Prove that there is a computably enumerable set $A \subseteq \mathbb{N}$ such that $\overline{A}$ is infinite but $\overline{A}$ contains no infinite computably enumerable set. Ironically, such subsets of $\mathbb{N}$ are called *simple sets.*

9.131. An *oracle* for a language L is an external device that can report whether any given string is or is not in L. For instance, total TMs serve as oracles for decidable languages. An *oracle TM* is a TM that has an additional capability of querying an oracle. For example, an oracle TM that decides the acceptance problem can be used to decide the emptiness problem (how?). Show the following:

(a) If L is Turing reducible to L', then an oracle TM that decides L' can be used to decide L. (What about the converse?)

(b) An oracle TM decides more languages than decidable languages.

(c) There are languages that cannot be decided by oracle TMs.

9.132. *Relative Computation*: We may say an oracle TM is a TM that has, in addition to its read–write tape, another read-only tape with a special string written on it. The string is called the *oracle*. The machine can move on the oracle tape (its read-only tape) by left or right movements, but no writing is allowed. We usually think of an oracle as a specification of a set of strings. If the oracle is an infinite string over $\{0, 1\}$, then we regard it as the characteristic function of a set $C \subseteq \mathbb{N}$, where the nth bit of the oracle is 1 iff $n \in C$. A standard TM is simply an oracle TM with the null oracle ε. Let $A, B \subseteq \Sigma^*$. We say that A is *computably enumerable in* B if there is an oracle TM M with oracles in B such that $A = \mathcal{L}(M)$. In addition, if M is a total TM, then we write $A \leq_T B$ and say that A *is decidable in* B. (Can you identify this as Turing reduction?) Fix the alphabet as $\{0, 1\}$, and use the encoding $\psi(\cdot)$. Define the following classes of sets:

$\Sigma_1^0 = \{$ computably enumerable sets$\}$,
$\Sigma_{n+1}^0 = \{$ sets computably enumerable in some $B \in \Sigma_n^0\}$,
$\Delta_1^0 = \{$ decidable sets$\}$,
$\Delta_{n+1}^0 = \{$ sets decidable in some $B \in \Sigma_n^0, \}$
$\Pi_n^0 = \{$ complements of sets in $\Sigma_n^0.\}$

These classes comprise the so-called *Arithmetic hierarchy.* Prove the following:
(a) A set A is in Σ_n^0 iff there exists a decidable $(n + 1)$-ary predicate P such that $A = \{x : \exists x_1 \forall x_2 \exists x_3 \cdots Qx_n P(x, x_1, \ldots, y_n)\}$, where Q is $\forall$ if n is even, and Q is $\exists$, if n is odd.
(b) A set A is in Π_n^0 iff there exists a decidable $(n + 1)$-ary predicate P such that $A = \{x : \forall x_1 \exists x_2 \forall x_3 \cdots Qx_n P(x, x_1, \ldots, x_n)\}$, where Q is $\exists$ if n is even, and Q is $\forall$, if n is odd.
(c) $\Delta_n^0 = \Sigma_n^0 \cap \Pi_n^0$.

9.133. Using the previous problem, show the following:
(a) $L_H = \{\psi(M)\# \psi(w) : \exists y(M \text{ halts on } w \text{ in } y \text{ steps})\}$ is in Σ_1^0 but not in Π_1^0.
(b) The complement of L_H is in Π_1^0 but not in Σ_1^0.
(c) *Empty* $= \{\psi(M) : \mathcal{L}(M) = \emptyset\}$ is in Π_1^0.
(d) *Total* $= \{\psi(M) : M \text{ is a total TM}\}$ is in Π_2^0.
(e) *Finite* $= \{\psi(M) : \mathcal{L}(M) \text{ is finite}\}$ is in Σ_2^0.
(f) *CoFinite* $= \{\psi(M) : \overline{\mathcal{L}}(M) \text{ is finite}\}$ is in Σ_3^0.

9.134. Prove that each level of the arithmetic hierarchy is strictly contained in the next. That is, $\Sigma_n^0 \cup \Pi_n^0 \subseteq \Delta_{n+1}^0$ but $\Sigma_n^0 \cup \Pi_n^0 \neq \Delta_{n+1}^0$.

9.135. From the previous problem it follows that Σ_1^0 and Π_1^0 are not comparable. Prove that Σ_n^0 and Π_n^0 are not comparable for any $n \in \mathbb{Z}_+$, with respect to set inclusion.

9.136. A set A is called *RE-hard* if every computably enumerable set is map reducible to A. A set B is called *RE-complete* if B is both computably enumerable and

RE-hard. For example, the acceptance language $L_{\mathrm{A}} = \{\psi(M)\#\,\psi(w) : M \text{ accepts } w\}$ is RE-complete. Let $\mathcal{C}$ be a class of sets. We say that a set A is $\leq_m$-hard for $\mathcal{C}$ if each set in $\mathcal{C}$ is map reducible to A. Similarly, we say that a set B is $\leq_m$-complete for $\mathcal{C}$ if B is in $\mathcal{C}$ and B is $\leq_m$-hard for $\mathcal{C}$. Prove the following:

(a) If $A \leq_m B$ and $B \in \Sigma_n^0$, then $A \in \Sigma_n^0$, for each $n \in \mathbb{Z}_+$.
(b) If $A \leq_m B$ and $B \in \Delta_n^0$, then $A \in \Delta_n^0$, for each $n \in \mathbb{Z}_+$.
(c) If B is $\leq_m$-complete for Σ_n^0, then B is neither in Π_n^0 nor in Δ_n^0 nor in Σ_{n-1}^0.
(d) Both L_A and L_H are $\leq_m$-complete for Σ_1^0.
(e) *Empty* is $\leq_m$-complete for Π_1^0.
(f) *Total* is $\leq_m$-complete for Π_2^0.
(g) *Finite* is $\leq_m$-complete for Σ_2^0.
(h) *CoFinite* is $\leq_m$-complete for Σ_3^0.
(i) None of these problems is contained in any class lower in the hierarchy or Turing reduces to any problem complete for any class lower in the hierarchy.

Remark: It seems that the superscript 0 in the above hierarchies is unnecessary. But it is not, as the hierarchies do not stop at the arithmetic hierarchy. The entire arithmetic hierarchy consists of the first level of the so-called *analytic hierarchy.* A remarkable theorem by S. C. Kleene says that the sets of natural numbers definable by one universal *second order* quantifier (where we quantify over the relations and functions and not only over the elements of a set) are exactly the sets definable by first-order induction. The next in the analytic hierarchy are the sets Σ_n^1, Π_n^1, and Δ_n^1. Moreover, these classes also have natural *complete* problems.

10 Computational Complexity

10.1 Introduction

Now you know that certain problems can be solved by algorithms and certain others cannot be. In discussing the issue of algorithmic solvability, you have used the Church–Turing thesis, which asks you to believe in the dictum that algorithms are nothing but total Turing machines (TTM) that use potentially infinite tapes, an ideal which we will probably not be able to realize. Even if we realize this ideal, there is a possibility that a TM may take an impracticably long time for giving an answer. This can happen even if the machine operates too fast.

For example, suppose you want to visit 50 tourist destinations spread out all over the earth. You know the cost of traveling from each destination to the other. Depending upon which place to visit first, and which one next, you have now 50! possibilities from which you want to select the one that is the cheapest. The number of possibilities to be explored is too large, $50! > 100^{25}$. If computing the cost for one itinerary visiting all 50 destinations takes a billionth of a second (too fast indeed), then it will require no less than 10^{25} human life times to determine the cheapest itinerary. Thus, algorithmic solvability alone does not suffice; we are interested in practical solvability.

But what does it mean to say that a problem is practically solvable? Do we have to run an algorithm for each instance of a solvable problem and then determine its practical solvability? This is the most useless proposal, in fact, an impossible job. The simple reason is that there are, in all probability, an infinite number of instances of a problem.

Again, how do we measure time in this case? Measuring real time is useless. As technology progresses, computing time for the basic operations reduce, and how do we ascertain that our estimate is still applicable? Moreover, time is not the only factor in discussing efficiency or practicality of algorithms. You may be interested in the working space an algorithm might demand to solve a problem. The issue of practicality might also depend on the particular computational model we choose.

We fix Turing machines as our computational model. We will measure time in terms of the number of steps a TM takes in getting a solution. Of course, we must

A. Singh, *Elements of Computation Theory*, Texts in Computer Science,

decide whether our TM uses a single tape or multiple tapes, and whether the TMs are deterministic or nondeterministic.

And what about inputs? The same TM may arrive at a solution in different number of steps of computation working on different inputs. To make the matter simple, we will discuss the time taken by a TM as a function of the *lengths of inputs* rather than in absolute terms, or as functions of inputs themselves. However, for different inputs of same length, a TM may use different number of computational steps. Thus, this scheme of measuring time as a function of the lengths of input might fail. In that case, we consider the maximum time taken by a TM as a function of the lengths of inputs.

We could also decide to take the minimum time or the average time instead of the maximum time. But minimum time corresponds to those inputs with respect to which the TM behaves at its best, and if our input is not of that type, we may not get a solution in time; it would be unrealistically optimistic. The minimum time analysis will only give an opportunistic lower bound on the time required by a TM for performing a job. Just as the minimum time corresponds to the best case, maximum time corresponds to the worst case; it is pessimistic.

Average time may be a better alternative, but it is too vague. As we cannot run the algorithm on each of the possibly infinite inputs, we require a predefined distribution on the set of all inputs (relevant to the problem) in order to assess efficiency of an algorithm in the average case. Also, which average we would take: arithmetic mean, geometric mean, or any (but which) weighted average? Moreover, all these averages depend on the presumed distribution on the space of inputs. But then which distribution is the appropriate one?

On the other hand, the pessimistic method of analyzing the worst case would give us a guarantee that a particular algorithm would never take more time than what is estimated. It will give an upper bound for all the cases at once, with a possibility that an algorithm might perform better than expected. We agree to play safe by sticking to the worst case analysis. That is, we will take the maximum time taken by a TM as a function of the lengths of inputs.

10.2 Rate of Growth of Functions

Look at Fig. 10.1. Here, we assume that σ is an input symbol while τ is not. How much time the TM M takes for reversing a string? If you run the machine on the input ab, a string of length 2, the computation of M proceeds as follows:

$$\underline{\not b}ab \rightsquigarrow \underline{a}b \rightsquigarrow \underline{\tau}b \rightsquigarrow \underline{\not b}\tau b \rightsquigarrow \underline{\not b}\not b\tau b \rightsquigarrow \underline{a}\not b\tau b \rightsquigarrow a\not b\underline{\tau}b \rightsquigarrow a\not b\underline{a}b \rightsquigarrow a\not b a\underline{b} \rightsquigarrow a\not b a\underline{\tau}$$
$$\rightsquigarrow a\underline{\not b}a\tau \rightsquigarrow \underline{\not b}a\not b a\tau \rightsquigarrow \underline{b}a\not b a\tau \rightsquigarrow ba\not b a\underline{\tau} \rightsquigarrow ba\not b a\underline{b} \rightsquigarrow ba\not b ab\underline{\not b} \rightsquigarrow ba\not b a\underline{b}$$
$$\rightsquigarrow ba\not b a\underline{\not b} \rightsquigarrow ba\not b\underline{a} \rightsquigarrow ba\not b\underline{\not b} \rightsquigarrow ba\underline{\not b} \rightsquigarrow \underline{\not b}ba.$$

Our interest is in finding out for an input of length n, how many steps of computation is performed by M. For each input symbol σ, the machine M moves its read–write head to the right once, writes τ, performs $L_{\not b}$ twice, and then writes σ. The first $L_{\not b}$ uses m movements if the mth symbol is currently being scanned. (This was not shown correctly in the above trace of computation of M on input ab.)

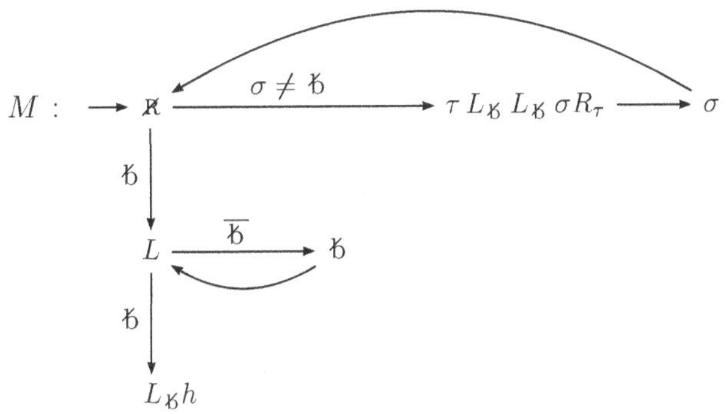

Fig. 10.1. String reversal.

Next L_{b} takes the same number of movements again. The simulation of R_τ involves $2m + 1$ movements, including the initial b̸ which is in the middle now. Thus, if the mth symbol is currently being scanned, then this is processed in altogether $1+1+m+m+1+2m+1+1 = 4m+4$ steps. Here, we assume that M takes the same amount of time in moving its read–write head one square to the left or right, and also writing or erasing a symbol on the scanned square.

After the reversal of the string has been written down to the left of the input string, M erases the input and comes back to the b̸, which is now to the left of the reversed string. For an input of length n, we see that after the reversal of the string, $2n + 1$ squares on the tape are occupied. M comes to the middle blank after erasing the input. This involves n left movements and n number of writing b̸. Next, it comes to the left of the reversed string in $n + 1$ left movements, totaling to $3n + 1$ units of time. That is, the total time taken by M is

$$\sum_{m=0}^{n}(4m + 4) + 3n + 1 = \{4(n(n + 1)/2\} + 4n + 3n + 1 = 2n^2 + 9n + 1.$$

We would thus say that the time complexity of M in reversing a string is $2n^2 + 9n + 1$. On the other hand, M uses n cells (tape squares) that contain the input, the n cells to the left of the b̸ symbol preceding the input, and then it halts on the left square to the left of the output. Altogether M requires $2n + 2$ cells on the tape. Omitting the input (since input can be given using a different input device in modern computers), M has altogether used $n + 2$ cells. We would then say that the space complexity of M in reversing a string is $n + 2$.

The important question is whether any TM, and not only this M, would require $2n + 2$ cells and $2n^2 + 9n + 2$ units of time (steps) in reversing a string of length n. We may modify the machine M in such a way that for strings of length less than a hundred, say, the new machine takes less time than $2n^2 + 9n + 1$ time. For example, we just forcefully drive the machine to a state after it finds out that the string has length less than 100. Then it just reproduces the reversal of the string, which it has already *memorized.* This can always be done as some of us just remember solutions of some problems without understanding a bit of it. Here the new TM behaves the same way in reproducing the reversal of a string by referring to a table. But we cannot do that for all strings. In that case, for the modified TM, the time complexity $2n^2 + 9n + 1$ still holds when $n \geq 100$. There is nothing sacrosanct about the number 100; it can be replaced by any other number.

The point is that the time complexity still holds for strings whose length is bigger than that finite number, that is, for *large* n. Moreover, for large n, the quantity $2n^2 + 9n + 1$ is less than or equal to $3n^2$. For example, you can see that for each $n > 10$, $2n^2 + 9n + 1 < 3n^2$. However, $2n^2 + 9n + 1$ is not less than cn, for large n, whatever constant c we may take. Notice that "large" n just means that all those n larger than some m. Moreover, for large inputs, it does not matter much whether the time taken is $3n^2$ or n^2.

Imagine a scenario where a TM handles inputs of a certain type very efficiently and inputs of certain other types, not so efficiently. To fix the idea, suppose the first type of inputs of length n are tackled by the machine in time $2n^2$, and a typical input of the second type is tackled in n^3 time. What would you conclude, the machine computes an arbitrary input of length n in time $2n^2$ or in time n^3? As we have agreed to consider the *worst case complexity*, we would say that the time complexity of the machine is n^3.

In what follows, we will be considering a very special class of problems, called the decision problems. These are the problems, as you now know, that ask for a "Yes" or a "No" answer. For example, reversing a string is not a decision problem, but whether a given string is the reverse of another is a decision problem. We consider decision problems due to their obvious importance, but it is not the case that other kinds of problems are not important. It is rather a matter of convenience; the fact is that decision problems can be converted easily to languages. The problem then would be a question of membership. Further, we will be concerned with calculating the complexity of a problem or a language rather than how much time a particular machine takes for solving that problem or for deciding that language. The crux is to go across the Turing machines deciding the same language.

Suppose a language is decided by a TM in time n^3 and by another TM in time n^4. This means that the first machine uses a maximum of n^3 steps to decide a string of length n, whether it is in the given language or not, whereas the second TM takes n^4 time at the most. The crucial strings that force the machines to go on a computation of maximum lengths may not be same for both the TMs. Our concern is what will be the time complexity of the language: n^3 or n^4? Notice that the worst case scenario does not ask us to go for n^4. The "worst case" applies to various strings of the same length when a particular TM is fixed. Here, we are discussing complexity across the machines. It is like, we have two algorithms to solve the same problem, and obviously, we will choose the best among them, if we are really interested in solving the problem. (Pretend that the inefficient algorithm has not been discovered at all.) We will stick to n^3 rather than n^4. The philosophy is that we analyze the worst case performance of the best available algorithm. Try to see how these philosophical choices are reflected in accepting the order notation of Landau described below.

Let $f, g : \mathbb{N} \to \mathbb{N}$. The function $f \in O(g)$ iff there exist constants $c > 0$ and $m \in \mathbb{N}$ such that for each $n \geq m$, $f(n) \leq cg(n)$. The set $O(g)$ denotes the set of functions of **growth rate** g. We allow imprecision and write $f \in O(g)$ as $f(n) = O(g(n))$ and read it as $f(n)$ is **big-oh of** $g(n)$.

For example, $5n+6 = O(n)$ means that the function $f : \mathbb{N} \to \mathbb{N}$ defined by $f(n) = 5n+6$ for each $n \in \mathbb{N}$ has, at the most, an order of growth as the function $g : \mathbb{N} \to \mathbb{N}$ defined by $g(n) = n$, for each $n \in \mathbb{N}$. It is so, because $5n + 6 \leq 6n$ for each $n \geq 6$; here, the constant c is 6 and m is 6.

Similarly, $5n + 6 = O(n^2)$ as $5n + 6 \leq 11n^2$ for $n \geq 1$, and also as $5n + 6 \leq n^2$ for $n \geq 6$.

However, $n^2 \neq O(5n + 6)$, as whatever c may be, we can always find an m so that for each $n \geq m$, $n^2 > 5n + 6$. For example, choose $c = 100$ and $m = 1100$. Now, if $n \geq 1100$, then $n^2 > 100(5n + 6)$. Though $n = O(5n + 6)$ holds trivially.

You must remember that $O(1)$ is the class of functions that are bounded above by some constant. The expression $f(n) = O(1)$ means that there is a constant c such that for each $n \in \mathbb{N}$, $f(n) \leq c$.

Writing $\log n$ for $\log_2 n$, we see that $\log n = O(n)$ as $\log n \leq n$ for each $n \geq 1$. And $(\log n)^2 = O(n)$ as $n \leq 2^{\sqrt{n}}$ for each $n \geq 4$.

Exercise 10.1. Determine whether $f(n) = O(g(n))$ and/or $g(n) = O(f(n))$, where
(a) $f(n) = n^3$ and $g(n) = 9n^2 + 5n - 1$.
(b) $f(n) = \log n$ and $g(n) = 1000$.
(c) $f(n) = n \log n$ and $g(n) = (\log n)^3$.
(d) $f(n) = n^{15}$ and $g(n) = (1.5)^n$.

We use the little-oh notation of Landau occasionally. Let $f, g : \mathbb{N} \to \mathbb{N}$ be functions. $f \in o(g)$ iff $\lim_{n\to\infty}[f(n)/g(n)] = 0$. Here, we assume that $g(n) \neq 0$ for large n, that is, for all values of n larger than some natural number m, so that the ratio becomes meaningful. The class $o(g)$ is the set of functions whose **growth rate is asymptotically less than** g. We will again use the less precise notation $f(n) = o(g(n))$ instead of $f \in o(g)$, and read it as $f(n)$ is **little-oh of** $g(n)$. For example, $5n + 6 = o(n^2)$ as $(5n + 6)/n^2 = 5/n + 6/n^2 \to 0$ as $n \to \infty$.

Exercise 10.2. Determine whether $f(n) = o(g(n))$ and/or $g(n) = o(f(n))$, where the functions $f(n)$, $g(n)$ are as in Exercise 10.1.

Other order symbols for denoting rates of growth are **big-Ω**, **little-ω**, and **big-Θ**. Let $f, g : \mathbb{N} \to \mathbb{N}$. We say that $f(n) = \Omega(g(n))$ iff there exist constants $c > 0$ and $m \in \mathbb{N}$, such that for each $n \geq m$, $f(n) \geq cg(n)$. $f(n) = \omega(g(n))$ iff $\lim_{n\to\infty}[f(n)/g(n)] = \infty$. And $f(n) = \Theta(g(n))$ iff $f(n) = O(g(n))$ and $g(n) = O(f(n))$.

These comparisons of numerical functions for large arguments are informally referred to as asymptotic comparisons. You would have, by now, guessed that there are functions that are not asymptotically comparable to each other. For example, the functions f, g defined by $f(n) = n$ and $g(n) = \lceil n^{(1+\sin n)} \rceil$ are not asymptotically comparable. Can you show it?

As $2n^2+9n+1 = O(n^2)$, we see that the TM M reverses a string in $O(n^2)$ time. We make further definitions that would help us in talking about the complexity issues. We make it a convention to write $f(n), g(n), s(n), t(n)$ for functions $f, g, s, t : \mathbb{N} \to \mathbb{N}$.

Problems for Section 10.2

10.1. In each of the following cases, determine whether $f(n) = O(g(n))$, $g(n) = O(f(n))$, $f(n) = o(g(n)$ and/or $g(n) = o(f(n))$:
(a) $f(n) = n^2 + 2n + 7$ and $g(n) = n^5$.
(b) $f(n) = \log n$ and $g(n) = 2009$.
(c) $f(n) = n \log n$ and $g(n) = (\log n)^3$.
(d) $f(n) = n \log n$ and $g(n) = n^2$.
(e) $f(n) = n$ and $g(n) = \sqrt{n}$.
(f) $f(n) = n^2$ and $g(n) = n(\log n)^2$.
(g) $f(n) = 3^n$ and $g(n) = 2^{cn}$, for some $c > 1$.
(h) $f(n) = 2^n$ and $g(n) = n!$.

10.2. Answer *true* or *false* to each of the following:
(a) $n^2 = o(2009\, n^2)$.
(b) $2009\, n^2 = o(n^{\sqrt{5}})$.
(c) $2009 \times 2^n = o(3^n)$.
(d) $2009 = o(1/n)$.
(e) $2009 = o(n)$.
(f) $n = o(\log n)$.
(g) $1/n = o(\log n)$.

10.3. Are the following true? Justify your answers.
(a) If $f(n) = a_0 + a_1 n + \cdots + a_k n^k$ and $g(n) = b_0 + b_1 n + \cdots + b_{k+m} n^{k+m}$, where a_i, b_j are any real numbers $a_k, b_{k+m} > 0$, $k \geq 0$, and $m \geq 1$, then $f(n) = O(g(n))$ but $g(n) \neq O(f(n))$.
(b) If $m, k \geq 1$, then $(\log n)^k = O(n^m)$ but $n^m \neq O((\log n)^k)$.
(c) $n^{O(1)}$ is the set of all functions that are bounded above by some polynomial.
(d) $f(n)^{O(1)}$ is the set of all functions that are polynomial in $f(n)$. Functions in $(\log n)^{O(1)}$ are called *polylogarithmic bounded functions*.

10.4. Show that if $f(n) = O(n^2)$, then $(f(n))^2 = O(n^4)$. Generalize this result to one involving the powers of n and the big-O notation.

10.5. Suppose for solving a problem we have two algorithms of which one runs in $f(n)$ time and the other runs in $g(n)$ time, where $f(n) = O(g(n))$ but $g(n) \neq O(f(n))$. Is it possible that the algorithm whose running time is $g(n)$ is preferable for values of n up to 2009, but after that the other algorithm is better? If so, propose a new algorithm that would be better than both of the earlier algorithms.

10.6. Consider the following two functions:

$$f(n) = n^2, \text{ if } n \text{ is odd; and } f(n) = n^3, \text{ if } n \text{ is even.}$$
$$g(n) = n^2, \text{ if } n \text{ is prime; and } g(n) = n^3, \text{ if } n \text{ is not a prime.}$$

Which of the following statements are true, and which are false?
(a) $f(n) = O(n^2)$.
(b) $f(n) = O(n^3)$.

(c) $g(n) = O(n^2)$.
(d) $g(n) = O(n^3)$.
(e) $f(n) = O(g(n))$.
(f) $g(n) = O(f(n))$.
(g) $n^3 = O(g(n))$.
(h) $n^2 = O(g(n))$.
(i) $n^3 = O(f(n))$.
(j) $n^2 = O(f(n))$.

10.7. Show that the function $n^{\log n}$ grows faster than any polynomial and also that it grows slower than 2^{cn} for any $c > 0$.

10.8. Define functions $f, g : \mathbb{N} \to \mathbb{N}$ by $f(n) = n$ and $g(n) = \lceil n^{(1+\sin n)} \rceil$. Show that neither $f(n) = O(g(n))$ nor $g(n) = O(f(n))$.

10.9. Consider the five types of notation such as $f(n) = O(g(n))$, $f(n) = \Omega(g(n))$, $f(n) = \Theta(g(n))$, $f(n) = o(g(n))$, and $f(n) = \omega(g(n))$ for comparing growth rates of functions. Which implications do hold between them? For example, if $f(n) = o(g(n))$, then does it mean that $f(n) = \omega(g(n))$? List and prove as many implications as possible.

10.3 Complexity Classes

Let M be a total deterministic Turing machine. The **space complexity** of M is $s(n)$ if $s(n)$ is the maximum number of squares M scans on any input of length n on all the tapes taken together, leaving aside the space used for the input. If M has a single tape, it is used for both reading and writing, and $s(n)$ is calculated leaving the cells used for the input. If M has many tapes, then it has one read-only tape, where we give inputs, and the other tapes are used for reading and writing. In this case, $s(n)$ is calculated by taking together used cells on all the tapes except the read-only tape. We do not take into account the space taken by the input. We only calculate the working space.

For a *nondeterministic* TM M, the space complexity is defined to be $s(n)$, where $s(n)$ is the maximum number of squares M scans on any branch of its computation for any input of length n. Here also we do not count the input string. In both the cases, we also say that M **runs in space** $s(n)$. Sometimes, to emphasize that M is an NTM, we would say that M runs on n-space $s(n)$.

The **time complexity** of M is $t(n)$, where $t(n)$ is the maximum number of steps that M takes on any input of length n. For time complexity, the steps taken for reading the input is taken into consideration, unlike the space complexity. Time complexity is also called the **running time** of M, and we say that M **runs in time** $t(n)$, and that M is a $t(n)$ **time Turing machine**.

If M is *nondeterministic*, then the time complexity or running time of M is $t(n)$, the maximum number of steps that M takes on any branch of its computation on any input of length n. In such a case, we also say that $t(n)$ runs in time $t(n)$ and that M is a $t(n)$ time NTM. Again, to be specific, for an NTM M, we would say that M runs on n-time $t(n)$. When NTMs are used to decide languages, we will use the terms *nondeterministic space complexity* and *nondeterministic time complexity* for space and time complexities, respectively. The corresponding space and time complexity classes are defined as follows:

$$DS(s(n)) = \{L : L \text{ is a language decided by an } O(s(n)) \text{ space multitape DTM}\},$$

$$NS(s(n)) = \{L : L \text{ is a language decided by an } O(s(n)) \text{ space multitape NTM}\},$$

$$DT(t(n)) = \{L : L \text{ is a language decided by an } O(t(n)) \text{ time multitape DTM}\},$$

$$NT(t(n)) = \{L : L \text{ is a language decided by an } O(t(n)) \text{ time multitape NTM}\}.$$

The class DS is also called *DSPACE*, and similarly DT, NS, and NT are called, respectively, as *DTIME*, *NSPACE*, and *NTIME*. If a language is decided by a deterministic TM in $5n^2+3n-2$ space, then it is in the class $DS(n^2)$ as $5n^2+3n-2 = O(n^2)$. However, the language is also in $O(n^{m+2})$ space as $5n^2 + 3n - 2 = O(n^{m+2})$ for each $m \in \mathbb{N}$. Possibly, the language is not in $O(n)$ space, but it cannot be so asserted in the absence of any further information. For example, if it so happens that we construct in a clever way another DTM that decides the same language in $O(n)$ space (linear space), then we have this sharpened result. This does not contradict the earlier statement that the language is in $O(n^2)$ space; it only improves that. Similar remark holds for time complexity also.

Sometimes space complexity is defined in terms of the total space requirements instead of the "working space" as we have defined. That way, minimum space requirement is linear, as the input itself consumes (linear) space. For time complexity, we have taken into account the input tape also. For all practical purposes, it is at least linear; this is so whenever the input is completely read.

Notice that there are trivial cases though; for example, the halting machine which never reads anything, but simply halts. The following examples will clarify the notions.

Example 10.1. Which complexity classes does a regular language belong to?

Solution. Let $D = (Q, \Sigma, \delta, s, F)$ be a DFA that accepts the given regular language L. We assume that $ⱨ \notin \Sigma$ and the symbols $S, h, \hbar \notin Q$ and that δ is a total function. We construct a DTM $M = (Q', \Sigma, \Sigma \cup \{ⱨ\}, \delta', S, h, \hbar)$, where $Q' = Q \cup \{S, h, \hbar\}$ and δ' is constructed from δ as in the following:

Initially, have the transition $\delta'(S, ⱨ) = (s, R)$ in M. Next, corresponding to each transition $\delta(q, \sigma) = p$ of D, add a transition $\delta'(q, \sigma) = (p, R)$ to M. Finally, corresponding to each final state $r \in F$ of D, add the transition $\delta'(r, \sigma) = (h, \sigma)$ and corresponding to each nonfinal state $t \in Q - F$, add a transition $\delta'(t, \sigma) = (\hbar, \sigma)$, for each $\sigma \in \Sigma$.

This means that M initially transfers control to the initial state of D, then simulates D, and finally, it enters the accepting state h when D has entered a final state, otherwise M enters the rejecting state $\hbar$. It is clear that M decides L. In deciding L, it has used $n + 1$ squares on the tape, but out of that the input itself occupies n squares. Thus, it has used only one additional square, the initial ⱨ. That is, $L \in DS(1)$. Notice that had we defined space complexity counting all squares including the input, the space complexity would have been $(n + 1)$ instead of 1.

Further, M decides L in $n + 2$ steps of computation including the initial right movement. Hence, $L \in DT(n + 2) = DT(O(n))$. Thus a regular language has space complexity $O(1)$ and time complexity $O(n)$. □

Example 10.2. What are the space and time complexities of the language $L = \{ucu^R : u \in \{0, 1\}^*\}$?

Solution. We design a two-tape DTM M for deciding L. Input is given on the first tape as usual. M first copies the string u from ucu^R to the second tape, symbol by symbol. At the end of this step, it has ucu^R on the first tape, scanning the symbol c, and u on the second tape, scanning the last symbol of u.

Next, it goes right leaving the c on the first tape and starts comparing the strings on the two tapes, symbol by symbol; the head on the first tape moving to the right, while the head on the second tape moving to the left. If a successful matching is obtained, then M enters its accepting halt state h. If at any stage, there is a failure, then M enters its rejecting halt state $\hbar$. For example, if there is no c on the input string, then M would have encountered a ƀ (before a c) at the end of the input string; in which case, M would enter the state $\hbar$. Similarly, if one of the heads do not find a matching symbol with the other, then M would enter the state $\hbar$.

In this process, we see that M has used only m squares on the second tape, when the input has length $2m + 1$. And this happens when it accepts the string. In the case of nonacceptance, the input may not be symmetric about c, or at the worst, there may not be any c at all in the input.

This is the worst case of an input for M. In such a case, M would have finished copying the input to the second tape. Of course, it goes to the rejecting state $\hbar$ after that. Thus, the maximum space required for M to work (leaving the input) is n, where n is the length of the input string. $L \in DS(n)$. Or that $L \in DS(O(n))$.

Similarly, each symbol can be written down in four steps, one for reading (first head), one for writing (second head), two for moving to the next square (both the heads). Matching of a symbol is done in two steps (only moving of heads). Thus the whole input requires $4n+2n$ plus some constant number for handling the initial steps, and the final steps. That is, $L \in DT(O(n))$. □

In Example 10.2, if we use a single-tape DTM, then the complexity may increase. For example, consider a DTM that matches the first symbol with the last symbol of the input and erases them. It continues till it encounters a c, or a 0, or a 1, or, a ƀ, and then decides accordingly. Then the DTM would not require any extra space than that is occupied by the input string, where it also writes symbols.

Similarly, it uses an $O(n^2)$ steps of computation, at the worst. This is so, because it has to go back and forth for matching and erasing the matched symbols. Thus, we may assert that $L \in DS(n)$ and $L \in DT(O(n^2))$ when a single tape machine is used.

Exercise 10.3. Find exactly how much time the DTM with a single tape (described above) takes in deciding the language L of Example 10.2.

Example 10.3. Let $L = \{\psi(M) : M \text{ is an NFA and } \mathcal{L}(M) \neq \emptyset\}$, where $\psi(M)$ is the binary encoding of M. What is the nondeterministic time complexity of L?

Solution. We design an NTM N for deciding L. Given $\psi(M)$ as an input, N first marks the initial state of M. Next, it selects nondeterministically, an input symbol and changes the marking by simulating one step of M. This step is repeated 2^k times if M has k states. Finally, N *accepts* $\psi(M)$ if ever a final state of M has been marked during the repeated execution of the second stage, else, N *rejects* the input.

Essentially, N chooses a string of length 2^k or less and simulates M on it. Because of the Pumping Lemma for regular languages, it is enough to decide strings of length 2^k or less because, if N accepts any string at all, then it must accept one of length no more than 2^k. Hence N decides L.

The space used by N are for marking various symbols in the input string, and that requires an $O(n)$ space, where n is the length of the input string $\psi(M)$. If the input string is not of the form $\psi(M)$, for any NFA M, then of course, N rejects the input, requiring no extra space at all. Thus, $L \in NS(O(n))$. However, time required for running N is exponential, as at any stage, it uses an $O(2^k)$ steps at the most; and in general, k can be linear in n. Thus, $L \in NT(O(2^n))$.

Can you show that $L \in DS(O(n^2))$ and also $L \in DT(O(2^n))$? □

Exercise 10.4. Show the following containment relations:
(a) $DS(s(n)) \subseteq NS(s(n))$.
(b) $DT(t(n)) \subseteq NT(t(n))$.
(c) $DT(f(n)) \subseteq DS(f(n))$.
(d) $NT(s(n)) \subseteq NS(f(n))$.

Clever designing of algorithms can save space and time. For example, consider the problem of determining whether a given string is at all generated by a given CFG. In general, there can be an exponential number of derivations to try. However, there is a cubic time algorithm to solve this problem; it is known as CYK algorithm. The algorithm uses a technique called *dynamic programing.* For each substring v of u, it derives the set of all nonterminals that would possibly generate v, and then proceeds inductively on the length of v. See Problem 10.88.

The language $\{0^n 1^n : n \in \mathbb{N}\}$ as a subset of $\{0, 1\}^*$ can be decided in $O(n^2)$ time, as all that one has to do is go on deleting a 0 from the left and a 1 from the right iteratively. However, there are $O(n \log n)$ time algorithms that decide the same language. For example, here is one:

ALGORITHM $0n1n$

1. If a 0 is found to the right of a 1, then *reject.*
2. Repeat the following until tape is empty:
   a. If total number of 0's and 1's is odd, *reject.*
   b. Erase every other 0 starting with the first 0, and then erase every other 1 starting with the first 1.
3. If tape is empty, then *accept*, else *reject.*

Exercise 10.5. Show that Algorithm $0n1n$ decides the language $\{0^n 1^n : n \in \mathbb{N}\}$ in $O(n \log n)$ time.

Notice that if we show that a certain decision problem has time complexity $t(n)$, it will remain so for all future time to come. How sharp is this estimate depends on the currently available algorithm. Tomorrow we may get a sharper result proving that the same decision problem has time complexity $\tau(n)$, where $\tau(n) = o(t(n))$. That will not invalidate our result of today; that will only sharpen it.

Problems for Section 10.3

10.10. Given a set of n numbers $x_1, x_2, \ldots, x_n$ (unsorted) and a key number x, it is required to determine if the set contains x. Describe an algorithm to fulfil the requirement. How much time does your algorithm take? Express the time taken in big-O notation.

10.11. Given a set of n numbers $x_1, x_2, \ldots, x_n$ (unsorted), you are asked to find if any of the numbers has a duplicate in the set. Suggest an algorithm to do the job, and express the time taken by the algorithm in order notation. Also see if an implementation of the algorithm in a Turing machine requires more time.

10.12. Repeat the preceding problem for determining if the set contains any triplicates.

10.13. Find an $O(n)$ time algorithm for determining membership in $\{ww : w \in \{a, b\}^*\}$ using a two-tape TM. What is the best you can expect for a standard TM?

10.14. Try to show that determining membership in $\{a^n b^n : n \in \mathbb{N}\}$ is in $DT(\Theta(n^2))$. What if we use a two-tape DTM? Can we sharpen the result to $DT(\Theta(n))$?

10.15. Show that
(a) $\{ww : w \in \{a, b\}^*\} \in NT(n)$.
(b) $\{www : w \in \{a, b\}^+\} \in DT(n)$.
(c) $\{ww^R w : w \in \{a, b\}^+\} \in DT(n)$.

10.16. Each regular language is in $DT(n)$. Is the converse true?

10.17. Determine in each of the following cases whether one of the given pair of complexity classes is properly contained in the other:
(a) $DS(n^2)$ and $DS(f(n))$, where $f(n) = n$ for even n, and $f(n) = n^2$ for odd n.
(b) $DS(n)$ and $DS((\lceil \log n \rceil)^n)$.
(c) $NS(2^n)$ and $NS(3^n)$.
(d) $DT(2^n)$ and $DT(3^n)$.

10.18. Let G be an undirected connected graph with m vertices and n edges, with each edge being assigned a positive integer as its weight. A spanning tree in G is a cycleless subgraph of G having $n - 1$ edges. A minimum spanning tree is a spanning tree with the sum of weights of edges minimum among all spanning trees. *Kruskal's algorithm* constructs a minimum spanning tree in G by first ordering the edges according to their weights, ties are broken arbitrarily. Next, it selects an edge with minimum weight. It adds an edge to the already selected edges by selecting an edge with minimum weight from among the unselected ones, provided it does not form a cycle with the selected ones. The process terminates when $n - 1$ edges have been selected. How much time does Krushkal's algorithm take in constructing a minimum spanning tree? Express the time taken in big-O notation.

10.4 Space Complexity

Recall that $DS(s(n))$ denotes the class of languages, or of the decision problems that are represented by these languages, that are decided by multitape total deterministic Turing machines with a read-only tape where the input is given, using a maximum of $s(n)$ number of squares on the other tapes taken together, on any input of length n. Similar is the class $NS(s(n))$, where the machines are nondeterministic and $s(n)$ is the maximum number of working squares used in any branch of computation of the NTM. We begin with the question I asked you in the solution of Example 10.3 :

$$\text{Is } L = \{\psi(M) : M \text{ is an NFA and } \mathcal{L}(M) \neq \emptyset\} \in DS(O(n^2))?$$

As $L \in NS(O(n))$, answer to this question is in the affirmative due to the following theorem.

Theorem 10.1. (Savitch) *Let* $s : \mathbb{N} \to \mathbb{N}$ *be such that* $s(n) \geq \log n$ *for each* $n \geq 1$. *Then* $NS(s(n)) \subseteq DS((s(n))^2)$.

Proof. Let M be a two-tape NTM with a separate read-only tape on which inputs are given, and a working tape. For an input w to M define a *configuration of M on w* as a quintuple (q, σ, u, τ, v), where q is the current state of M, σ is the currently scanned symbol on the input (on the read-only tape), τ is the currently scanned symbol on the working tape, u is the string to the left of τ, and v is the string to the right of τ on the working tape. The input w, as it is, is not a part of the configuration of M on w; it is on the read-only tape. The configurations on w reflect only the working tape, the current state, and the currently scanned symbol.

Suppose M decides a language L in space $s(n)$. We construct a DTM D which tests whether one of the configurations of M on w can yield another configuration of M on w within a specified number of steps. Notice that by solving this problem, where C_1 is the initial configuration of M on w, C_2 is the final configuration of M on w, and t is the maximum number of steps that M uses, we can determine whether M accepts the input w or not. The following algorithm, named as, $yields$, when implemented as a DTM just does this.

ALGORITHM $yields(C_1, C_2, t)$

1. If $t = 1$, then check whether $C_1 = C_2$ or whether C_1 yields C_2 in one step using the transitions of M. If one of these two tests succeeds, then output *accept*, else, output *reject*.
2. If $t > 1$, then for each configuration C_n of M on w, using space $s(n)$,
   (a) run $yields(C_1, C_m, \lceil t/2 \rceil)$;
   (b) run $yields(C_m, C_2, \lceil t/2 \rceil)$.
3. If both 2(a) and 2(b) output *accept*, then output *accept*, else output *reject*.

The algorithm *yields* recursively searches for an intermediate configuration C_m of M on w to confirm that C_1 yields C_2. It saves space significantly by reusing the space for the sub-problems. However, it requires some more space for saving the recursive stack for future backtracking. Each level of the recursion uses an $O(s(n))$ space to store a configuration of M on w; depth of the recursion is $\log t$, where t is the maximum number of steps M takes on any branch of its computation on w. As $t = 2^{O(s(n))}$, $\log t = O(s(n))$ is the depth of the recursion in *yields*.

The DTM D uses *yields* as a subroutine. On input w, it outputs the result of $yields(C_1, C_2, 2^{ds(n)})$, where d is such a number that M takes decision on a string of length n in $2^{ds(n)}$ steps or less. Notice that there exists one such d as $L \in DS(s(n))$. Clearly, D decides the language L as M does so. Moreover, each level of recursion in *yields* uses an $O(s(n))$ space to store a configuration of M on w and depth of the recursion is $\log(ds(n)) \leq ds(n)$. D uses $O(d(s(n))^2) = O((s(n))^2)$ space, that is, $L \in DS((s(n))^2)$. □

Savitch's theorem uses a multitape DTM to show that a language that is decided in $O(s(n))$ n-space (nondeterministic space) can be decided in $O((s(n))^2)$ space (deterministic space). It says nothing about how big this constant in the $O((s(n))^2)$ is. However, this constant can be made as small as you like by suitably decoding the input alphabet of D.

Theorem 10.2. (Tape Compression) *Let $L \in DS(s(n))$. Then for any constant $c > 0$, there exists a DTM D such that D decides L having space complexity $c \cdot s(n)$.*

As mentioned earlier, the idea is to regard strings as inputs to a given machine as symbols for D. Suppose we have a machine M having input alphabet $\{a, b\}$. We take the alphabet $\{1, 2, 3, 4, 5, 6\}$, where the strings a, b, aa, ab, ba, bb are encoded as $1, 2, 3, 4, 5, 6$, respectively. The new machine works with the symbols $1, 2, 3, 4, 5, 6$ instead. Then space requirement is halved compared to that in M. With this intuition, we leave the Tape Coversion Theorem unproved; see Sect. 10.10 for references.

If a language L is decided by an NTM in space $p(n)$, where p is a polynomial in n, Savitch's theorem guarantees that there is a DTM that decides L in space $(p(n))^2$, which is again a polynomial in n. This means that the class of languages that are decidable in **nondeterministic polynomial space** coincides with the class of languages that are decidable in **deterministic polynomial space**. This is written, in symbols, as $NPS = DPS$, where

$$NPS = \cup_{k \in \mathbb{N}} NS(n^k), \quad DPS = \cup_{k \in \mathbb{N}} DS(n^k).$$

Savitch's theorem requires that the space complexity $s(n)$ of a language be at least $\log n$. For large n, $(\log n)^2 \neq O(\log n)$, but $\log n \leq (\log n)^2$. Hence, there is no guarantee that the complexity classes $NS(\log n)$ and $DS(\log n)$ are equal. We make a shorthand

$$LS = DS(\log n), \quad NLS = NS(\log n).$$

In other words, LS consists of languages that are decidable in **deterministic logarithmic space** and NLS is the class of languages that are decidable in **nondeterministic logarithmic space**. LS and NLS are also written as LOGSPACE and NLOGSPACE, respectively. As each deterministic machine is also nondeterministic, $LS \subseteq NLS$. Whether $LS = NLS$ is an open problem. However, there are such languages in NLS that if any one of them is shown to be in LS, then NLS will be equal to LS. Such languages are called NLS-complete languages.

Let Σ be an alphabet. A function $f : \Sigma^* \to \Sigma^*$ is called a **log space computable function** if f is computed by a DTM with a read-only tape, a write-only tape, and a read–write working tape that uses a maximum of $O(\log n)$ squares on the working tape for any input of length n. A language $L_1 \subseteq \Sigma^*$ is **log space reducible** to a language $L_2 \subseteq \Sigma^*$, written $L_1 \leq_L L_2$, if there exists a log space computable function $f : \Sigma^* \to \Sigma^*$ such that for each $w \in \Sigma^*$, we have $w \in L_1$ iff $f(w) \in L_2$. Such a function is called a **log space reduction** of L_1 to L_2. A language $L' \subseteq \Sigma^*$ is called ***NLS*-complete** if $L' \in NLS$ and each $L \in NLS$ is log space reducible to C.

A log space reduction is a map reduction that uses a log space computable function. We show that if a language is log space reducible to another language known to be in LS, then the former language is also in LS; this would prove the following result:

Theorem 10.3. *Let L be an NLS-complete language. If $L \in LS$, then $NLS = LS$.*

Proof. Let L be an NLS-complete language and $L \in LS$. Let $L' \in NLS$. We show that $L' \in LS$.

As L is NLS-complete and $L' \in LS$, there is a log space computable function $f : \Sigma^* \to \Sigma^*$ and that $L' = \{f(w) : w \in L\}$. Let M be the DTM that computes f in space $\log n$. as $L \in LS$, we also have a DTM M_L that decides L in space $\log n$. The combined machine $M_L \to M$ decides L', but may not be in space $\log n$ due to the reason that length of $f(w)$ may exceed $O(\log n)$. We modify $M_L \to M$ to obtain another machine M', which operates just like $M_L \to M$ but working on the individual symbols of $f(w)$ separately, or as and when required, rather than $f(w)$ at a time.

In this modification, the machine M_L computes individual symbols of $f(w)$ as requested by M'. The simulation keeps track of where on the input the reading head of M' was while operating with $f(w)$. The modified combination M' takes more time than $M_L \to M$, but it requires no more space than that is used by M. As M uses only an $O(\log n)$ space, so does L' since only one square from M_L is marked and used for M to act on it. Therefore, $L' \in LS$. □

To mention an example of an NLS-complete language, consider a directed graph G. The problem is to determine whether, given any two vertices v_1 and v_2 of G, there is a directed path from v_1 to v_2 in G. The appropriate language for this problem uses an encoding of vertices and edges of G. Suppose $\psi(v)$ and $\psi(G)$ denote the encodings of the vertices v and the graph G (including edges) into binary strings. Construct the language

$$\text{PATH} = \{\psi(G)\#\,\psi(u)\#\,\psi(v) : G \text{ is a directed graph in which there is a path from the vertex } u \text{ to the vertex } v\}.$$

The language PATH is a language over $\{0, 1, \#\}$. It can be shown that PATH is an NLS-complete language; see Sect. 10.10 for references. It is yet unknown whether $\text{PATH} \in LS$ or not.

Define ***CoNLS*** as the class of languages whose complements are in *NLS*. It can be shown that the complement of PATH is in *NLS*, that is, PATH $\in$ *CoNLS*. Because of *NLS*-completeness of PATH, it follows that $CoNLS = NLS$. It is also known that *CoNLS* is a proper subclass of *DPS* though it is not yet known whether LS is a proper subclass of *NLS*. The results in this regard may be summarized as

$$LS \subseteq NLS = CoNLS \subsetneq DPS = NPS.$$

Problems for Section 10.4

10.19. Design a DTM that accepts the language $\{wcw^R : w \in \{a, b\}^*\}$ in $O(n)$ space.

10.20. Design a DTM that accepts the language $\{a^n b^n c^n : n \in \mathbb{N}\}$ using an $O(n)$ space. Can the algorithm be improved to accept the same language in $O(\log n)$ space?

10.21. Prove: Suppose the language L is accepted by an NTM with space complexity $s(n)$. Then for any $c > 0$, L is accepted by an NTM with space complexity $c \cdot s(n)$.

10.22. Show that $\{0^n 1^n : n \in \mathbb{N}\}$ is log space reducible to $\{a^n b^n : n \in \mathbb{N}\}$.

10.23. Using NL-completeness of PATH, show that $NL \subseteq P$.

10.24. Prove the following.
(a) If $f(n) = O(g(n))$, then $DS(f(n)) \subseteq DS(g(n))$.
(b) If $f(n) = O(g(n))$, then $NS(f(n)) \subseteq NS(g(n))$.
(c) $DS(s(n)) \subseteq DS(2^{O(s(n))})$.

10.25. Which of the following are true, which false, and which ones you think cannot be determined to be true or false on the basis of the material discussed so far?
(a) $NS(\log n) \subseteq DS((\log n)^2)$.
(b) $NS(n) \neq DS(n^2 \log n)$.
(c) $NS(n^2) \subseteq DS(n^4)$.
(d) $NS(2^n) \subseteq DS(4^n)$.
(e) $DS(n^6) \subseteq DS(n^8)$.

10.26. Suppose that f and g are functions that can be computed in logarithmic space. Show that the composition $f \circ g$ can be computed in logarithmic space. Deduce the transitivity of log space reduction, that is, the composition of two log space reductions is a log space reduction.

10.27. Suppose L is log space reducible to L' and $k \geq 1$. Prove the following:
(a) If $L' \in DS((\log n)^k)$, then $L \in DS((\log n)^k)$.
(b) If $L' \in NS((\log n)^k)$, then $L \in NS((\log n)^k)$.

10.28. Is the language of properly balanced parentheses in LS?

10.29. Assume that there is an *NLS*-complete language L such that $\overline{L} \in NLS$. Show that $CoNLS = NLS$.

10.30. Does there exist an *NLS*-complete language whose complement is in *NLS*?

10.31. Explain how reductions can be viewed as subroutine calls.

10.5 Time Complexity

Analogous to space complexity, we have defined time complexity by using multitape Turing machines. Depending on whether we use a deterministic or a nondeterministic machine, the time complexity classes $DT(t(n))$ and $NT(t(n))$ have been defined. You must remember that in case of NTMs, the failures are not charged. Let me clarify.

Suppose L is a language decided by a multitape NTM M. Suppose further that on an input of length n, M goes on a computation of length at most $t(n)$ and reports that the string is in L or not in L. The point is, there can be many other computations of length more than $t(n)$ for deciding membership of the input in L; or even there can be many computations of M that do not decide any thing within $t(n)$ steps. As, for each string of length n, there exists a successful computation of length at most $t(n)$ that reports whether the string is in L or is not in L, we say that L is decided by M in time $t(n)$. Recall that for a function $t : \mathbb{N} \to \mathbb{N}$, the **time complexity classes** $DT(t(n))$ and $NT(t(n))$ are defined as

$$DT(t(n)) = \{L : L \text{ is a language decided by some } O(t(n)) \text{ time multitape DTM}\},$$

$$NT(t(n)) = \{L : L \text{ is a language decided by some } O(t(n)) \text{ time multitape NTM}\}.$$

The time complexity class $NT(t(n))$ involves the decision process of an NTM. How does an NTM decide a language? First of all, we require the NTM M to be total. That means, if Σ is the input alphabet of M, then for each $w \in \Sigma^*$, we have two possibilities. The first possibility is that there is *some* computation of M on input w, which leads to the accepting halt state h, in which case, M accepts w, and $w \in L$. The second possibility is that *each* computation of M on input w leads to the rejecting state $\hbar$, in which case, M rejects w and $w \notin L$.

Thus, the decision problem for L is not the same as that of its complement, $\overline{L}$. This happens in case of a DTM; but not necessarily for an NTM. Why?

Well, how do you resolve this metaproblem for a DTM? Suppose a DTM decides a language $L \subseteq \Sigma^*$. If $w \in L$, then the DTM goes eventually to its accepting state h or to its rejecting state $\hbar$; it is a single computation that ends in either of these states. So, to decide $\overline{L}$, you just reverse the roles of the states h and $\hbar$. But in case of an NTM M, reversing roles of h and $\hbar$ does not decide $\overline{L}$.

It so happens as there can be a string $w \in \Sigma^*$ on which there is a computation of M leading to h, and yet there is another computation that leads to $\hbar$. In this case, $w \in L$. Now, reversing the roles of h and $\hbar$ shows that the new *NTM* accepts w as earlier; it does not reject w as intended. Therefore, when we say that an NTM M decides a language, it *does not really decide* in the same sense as a DTM.

How does it affect $NT(t(n))$? Suppose $L \in NT(t(n))$, where L is a language over Σ. Then there exists an NTM, say, M that decides this L in time $t(n)$. Had it been a DTM, given any string $w \in \Sigma^*$, M would reach in $t(\ell(w))$ steps or less, one of the halt states h or $\hbar$. In the NTM M, a halt state is still reached within $t(\ell(w))$ steps; further, if $w \in L$, then there exists at least one computation path in the computation tree of M on input w, which ends in the state h, and if $w \in \overline{L}$, then all the branches in the computation tree end in the state $\hbar$. In both the cases, the time that M takes to reach one of the halt states is measured vertically; it is the depth of the computation tree of M on input w.

Therefore, to determine whether M has reached h, we scan only one successful computation and charge it; we count the number of steps in such a branch. To determine whether M has reached $\hbar$ in every branch, we scan each of the computation branches but charge only the branch having maximum length, and not the sum total of all lengths of the branches. This is the anomaly in the notions DT and NT. However, once you understand this point, we will allow the imprecise phrase *an* NTM *decide a language*, bowing to the traditions.

If D is a DTM that decides a language L, we will say that D is a *deterministic decider for* L. Similarly, M is a *nondeterministic decider for* L would mean that M is an NTM that decides L.

Do multitape or singletape machines affect the time complexity classes greatly? In answering this question we assume that the time complexity of each of our languages is at least linear, that is, $t(n) \geq n$ for each $n \in \mathbb{N}$. This is a reasonable assumption because if a TM reads its input completely, it would take at least this much time. In a way, we are not considering the trivial machines that make a decision without reading the input completely.

Theorem 10.4. *Let $t : \mathbb{N} \to \mathbb{N}$ be a function with $t(n) \geq n$ for each $n \in \mathbb{N}$. Let M be a multitape DTM that decides a language L in time $O(t(n))$. Then there exists a single-tape DTM that decides L in time $O((t(n))^2)$.*

Proof. We use a construction similar to that in simulating a multitape machine by a single tape machine. Let M be a k-tape DTM. The single tape DTM D starts with writing the contents of all the k tapes of M to its (single) tape, leaving a ƀ in between. For each single movement of a head of M, the contents of the appropriate string on the tape of D is shifted one square (to the left or right as necessary) and then the movement action is taken.

Simulation of M by D uses a shift operation that may take an $O(t(n))$ time at the most corresponding to each step of M. There are altogether a maximum of $O(t(n))$ steps of M. Thus, D may take a maximum of $O(t(n)) \times O(t(n))$ time in simulating M. D uses an additional time of $O(t(n))$ for writing all the contents of the k tapes of M on its own tape in proper format. Hence, this simulation uses $O(t(n)) \times O(t(n)) + O(t(n)) = O((t(n))^2)$ time at the most. □

Exercise 10.6. In the proof of Theorem 10.4, where did we use the assumption that $t(n) \geq n$ for each $n \in \mathbb{N}$?

The same simulation proves an analogous result for nondeterministic machines.

Theorem 10.5. *Let $t : \mathbb{N} \to \mathbb{N}$ be a function with $t(n) \geq n$ for each $n \in \mathbb{N}$. Then, for each multitape NTM that decides a language in time $t(n)$, there exists a single tape NTM that decides the same language in time $O((t(n))^2)$.*

Contrary to Savitch's theorem, we do not have an efficient simulation of nondeterministic deciders by a deterministic decider with regard to time complexity. The simulations take an exponential increase in time complexity as the following theorem shows. Of course, the theorem does not assert that we cannot have a more efficient simulation; the fact is that we do not have one yet.

Theorem 10.6. *Let $t : \mathbb{N} \to \mathbb{N}$ be a function with $t(n) \geq n$ for each $n \in \mathbb{N}$. Let M be a single-tape NTM that decides a language L in time $t(n)$. Then there exists a DTM that decides L in time $2^{O(t(n))}$.*

Proof. Earlier we have described how a three-tape DTM could simulate an NTM. The same simulation would serve our purpose. Notice that the simulation uses a brute force approach; it simulates the branches of computations of the NTM breadth-first way.

Let M be a single tape NTM that decides a language L in time $t(n)$ on an input of length n. Each branch of computation of M has depth at most $t(n)$. As M is nondeterministic, for a state–symbol pair, there can be many choices of the next-move; nonetheless, they are finite in number. Suppose c is the maximum number of such choices of next-moves for each state–symbol pair in the transition relation of M. That means, each node in the computation tree of M on a given input of length n has at the most c children. Thus the computation tree has a maximum of $c^{t(n)}$ number of leaves.

The DTM D limits its search to the point when a success is met, that is, the relevant portion of the tree, looked at breadth-first way, has depth at most $t(n)$. The number of nodes in this portion of the computation tree of M cannot exceed $t(n)\, c^{t(n)}$ nodes. Hence, the running time of D is $O(t(n)c^{t(n)})$. However, D is a three-tape DTM. By Theorem 10.5, there is a DTM that decides L in $O((t(n)c^{t(n)})^2)$ time, which is same as $2^{O(t(n))}$. □

Exercise 10.7. Show that if $t : \mathbb{N} \to \mathbb{N}$ satisfies $t(n) \geq n$ for each $n \in \mathbb{N}$, and $c > 1$ is any real number, then $O((t(n)c^{t(n)})^2) = 2^{O(t(n))}$.

Because of the absence of any speedup result regarding the simulation of a nondeterministic decider, by a deterministic decider, we may expect a gap between *DPT* and *NPT*. This is unlike the scenario in space complexity, where Savitch's theorem does the trick. But what are *DPT* and *NPT*?

DPT is the class of languages that are decidable in polynomial time by single tape DTMs. Similarly, *NPT* consists of all languages that are decidable in polynomial time by single tape NTMs. From Theorem 10.5, it does not matter whether we use multitape or single tape machines for deciding languages as long as our interest lies in bigger classes such as *DPT* or *NPT*.

We make further shorthands and call *DPT* as *P* and *NPT* as *NP*. Along with ***P***, ***NP***, we also define ***EXP***, the class of languages decidable in exponential time.

$P = DPT = \cup_{k \in \mathbb{N}} DT(n^k)$

= The set of languages decidable by multitape DTMs in time $t(n)$, where $t(n)$ is a polynomial in n, the length of the input string.

$NP = NPT = \cup_{k \in \mathbb{N}} NT(n^k)$

= The set of languages decidable by multitape NTMs in time $t(n)$, where $t(n)$ is a polynomial in n, the length of the input string.

$EXP = EXPTIME = DT(2^{n^{O(1)}}) = \cup_{k>0} DT(2^{n^k})$

= The set of languages decidable by multitape DTMs in an exponential time in n, the length of the input string.

You can define *NEXP* in a similar way; see Problem 10.127.

It is now clear that $P \subseteq NP \subseteq EXP$. The first inclusion holds as each DTM is also an NTM, and the last inclusion, due to Theorem 10.6. We will have a useful characterization of *NP* in the next section.

At this stage you may revisit the algorithms we have discussed in the context of regular languages and context-free languages; see Sects. 3.2–3.5, 4.5, 5.6, and 6.3. The following statements summarize their time complexity.

Theorem 10.7. *The following algorithms run in polynomial time:*

1. *Construction of a regular grammar G_D from a DFA D so that $\mathcal{L}(G_D) = \mathcal{L}(D)$. (Lemma 3.3)*
2. *Construction of an NFA N_G from a regular grammar G so that $\mathcal{L}(N_G) = \mathcal{L}(G)$. (Lemma 3.4)*
3. *Construction of an NFA N_E from a regular expression E so that $\mathcal{L}(N_E) = \mathcal{L}(E)$. (Lemma 3.5)*
4. *Construction of a DFA D' with minimum number of states from a DFA D so that $\mathcal{L}(D') = \mathcal{L}(D)$. (State Minimization Algorithm in Theorem 4.9)*
5. *An algorithm to decide whether two DFAs accept the same language. (For example, using (4) twice.)*
6. *An algorithm to decide whether a given string is in the language represented by a regular expression.*
7. *An algorithm to decide whether a given string is at all generated by a regular grammar.*
8. *An algorithm to decide whether a given string is at all accepted by a DFA.*
9. *An algorithm to decide whether a given string is at all accepted by an NFA.*
10. *Construction of a CFG G' in Chomsky Normal Form from a CFG G so that $\mathcal{L}(G') = \mathcal{L}(G) - \{\varepsilon\}$. (Theorem 5.1)*
11. *Construction of a CFG G' in Greibach Normal Form from a CFG G so that $\mathcal{L}(G') = \mathcal{L}(G) - \{\varepsilon\}$. (Theorem 5.2)*
12. *Construction of a PDA P_G from a CFG G so that $\mathcal{L}(P_G) = \mathcal{L}(G)$. (Lemmas 6.2–6.3).*
13. *Construction of a CFG G_P from a PDA P so that $\mathcal{L}(G_P) = \mathcal{L}(P)$. (Lemmas 6.4–6.5).*

14. An algorithm to decide whether a given string is at all generated by a given CFG. (CYK Algorithm, not proved in this book; see Problem 10.88 in Sect. 10.10.)
15. An algorithm to decide whether a given string is at all accepted by a given PDA.

Theorem 10.8. *The following algorithms run on exponential time:*

1. *Construction of a DFA D_N from an NFA N so that $\mathcal{L}(D_N) = \mathcal{L}(N)$. (Lemma 3.2)*
2. *Construction of regular expression E_N from an NFA N so that $\mathcal{L}(E_N) = \mathcal{L}(N)$. (State Elimination Algorithm in Lemma 3.6)*
3. *An algorithm for deciding whether two NFAs accept the same language. (For example, using (1) and Theorem 10.7(4))*
4. *An algorithm for deciding whether two regular grammars (or two regular expressions) generate the same language.*

In view of the above results we do not yet have any answers to the following questions:

1. Does there exist a polynomial time algorithm to solve any of the problems in Theorem 10.8 above?
2. Does there exist a polynomial time algorithm to construct an NFA with least number of states from a given NFA so that they accept the same language?
3. Does there exist a polynomial time algorithm to construct a DFA with least number of states from a given NFA so that they accept the same language?

Problems for Section 10.5

10.32. Prove the following.
(a) $DT(t(n)) = DT(c \cdot t(n))$ for any constant $c > 0$.
(b) If $n = o(g(n))$ and $f(n) = O(g(n))$, then $DT(f(n)) \subseteq DT(g(n))$.
(c) If $n = o(g(n))$ and $f(n) = O(g(n))$, then $NT(f(n)) \subseteq NT(g(n))$.

10.33. Is it true that $DS(s(n)) \subseteq DT(2^{O(s(n))})$?

10.34. Show that
(a) If $L \in DT(O(n^2))$, then $L \in DT(f(n))$, where $f(n) = \max(n^2, 2n + 20)$.
(b) If $L \in DT(cn)$ for some $c > 2$, then $L \in DT(\lceil 2.1\, n + 20 \rceil)$.
(c) If $L \in DT(f(n))$, $L' \in DT(g(n))$, then $L \cup L' \in DT(f(n) + g(n) + O(n))$.

10.35. Construct a list of TMs that run in linear time such that every language decidable in linear time is accepted by some TM on the list. Build a TM that diagonalizes over the list. Then conclude that there exists a decidable language that is not decidable in linear time.

10.36. Show that there is no computable total function $f(n)$ such that every decidable language is in $DT(f(n))$.

10.37. What are the space and time complexities of the language $\{a^n b^n : n \in \mathbb{N}\}$? Improve your algorithm to get a best space-complexity bound. Improve your first algorithm again to obtain the best time-complexity bound. Can the algorithm be improved so that the best bounds for both of space- and time-complexity are achieved simultaneously?

10.38. Show that there exist polynomial time algorithms for the following:
(a) Constructing a PDA equivalent to a given CFG.
(b) Constructing a CFG equivalent to a given PDA.
(c) Constructing a CFG in CNF equivalent to a given CFG.
(d) Converting a PDA accepting by final states to one accepting by empty stack.
(e) Converting a PDA accepting by empty stack to one accepting by final states.

10.39. Can you find a polynomial time algorithm for converting a CFG to one in GNF? If yes, give such an algorithm and show why is it polynomial. If no, give reasons.

10.40. Show that whether a CFG generates any string at all (emptiness check) can be checked in linear time.

10.41. Show that the problem "whether a symbol is reachable in a CFG" can be checked in liner time.

10.42. Show that the problem "whether a given nonterminal in a CFG is nullable (derives ε)" can be checked in linear time.

10.43. Give linear time algorithms for determining the following:
(a) Which symbols appear in some derived string of terminals or nonterminals?
(b) Which nonterminals are nullable?

10.44. Call a nonterminal symbol A of a CFG a *productive nonterminal* if a string of terminals can be derived from A. Give a polynomial time algorithm for finding all productive nonterminals of a given CFG.

10.45. Give an algorithm for deciding whether the language of a given CFG is infinite. Can you find a polynomial time algorithm?

10.46. Show that the *problem of relatively prime*, that is, "Given two integers $m, n > 1$, whether m, n are relatively prime", is in P.

10.47. Show that $\{0^n : n \text{ is a prime number}\}$ is in P.

10.48. Show that the problem of "determining whether a given $n \in \mathbb{N}$ in binary is a prime number" is in $DT(O(2^n))$.

10.49. Prove that the *graph connectedness problem,*: that is, "Is a given undirected graph connected?," is in P.

10.50. Show that the problem of "determining whether a given undirected graph has a triangle" is in P. This is called the *triangle problem* in graphs.

10.51. Suppose there is a 2^n time bounded reduction of L to L' and that $L' \in DT(2^n)$. What do you conclude about L? Justify.

10.52. Which of the following are true, which false, and which ones you think cannot be determined to be true or false on the basis of the material discussed so far?
(a) $DT(2^n) \neq DT(n^2 2^n)$.
(b) $NT(n) \subseteq DS(n^2)$.
(c) $NT(n^6) \subseteq DS(n^8)$.
(d) $EXP \subseteq DT(2^{n^{O(1)}}) = \cup_{k>0} DT(2^{n^k})$.

10.6 The Class *NP*

How does a nondeterministic decider work, in general? If a problem is to be solved, an NTM might nondeterministically choose a supposed solution and then tries to see whether it is indeed a solution. If the chosen one is not a solution, then it chooses or generates yet another, until it cannot generate a new one, and then it reports that the problem has no solution. If it succeeds, then we would try to calculate the time taken by the NTM for arriving at a decision, in the worst case. We neglect the cases of failures of the NTM in course of guessing a solution and verifying it.

In an NTM, nonacceptance of a string is never charged, whereas acceptance of a string is charged only for a successful guess, and then we calculate possible maximum time. This means that if a nondeterministic decider decides a language in time $t(n)$, then a worst case for accepting some string of length n has taken $t(n)$ time. Because of this reason, *NP* is often described as the class of languages in which membership can be verified in a polynomial time. If we consider *NP* as the class of problems (decision problems, in fact), then any suggested solution to such a problem can be verified in a polynomial time. To see this version of *NP*, we define a verifier of a language.

A total DTM (an algorithm) M is called a **verifier** for the language

$$L = \{w : M \text{ accepts } \psi(w)\#\, \psi(c) \text{ for some string } c\}.$$

A **polynomial time verifier** is a verifier that runs on a polynomial time in the length of w. A string such as c in the ordered pair (w, c) is called the **certificate** or a *proof of membership*. We say that L is a **polynomial time verifiable language** if there is a polynomial time verifier for L.

Informally, the string $\psi(w)\#\, \psi(c)$ is replaced by the ordered pair (w, c) as the former is simply an encoding of the latter. Though, in a verifier, time spent on manipulating the certificate c is not explicitly charged, it has to be accessed and used. Thus the length of c must be bounded by a polynomial in the length of w. We will illustrate this with an example from propositional logic.

Recall that a propositional language is built upon a finite number of symbols, called atomic propositions, $\{p_0, p_1, \ldots, p_n\}$ using the connectives $\neg, \wedge, \vee, \rightarrow, \leftrightarrow$. When a proposition uses the only connectives $\neg$, $\vee$, and $\wedge$, we say that it is a *boolean formula*. In a propositional language, a **literal** is defined as any atomic proposition or the negation of it. For example, $p_0, p_5, \neg p_1, \neg p_0, \neg p_{10}$ are literals. A **disjunctive clause** is a disjunction ($\vee$'s) of literals. For example, $p_0 \vee \neg p_7$, $p_5 \vee p_{11} \vee \neg p_5 \vee \neg p_3$ are disjunctive clauses. A **conjunctive normal form**, also called a **cnf**, is a conjunction ($\wedge$'s) of disjunctive clauses. (Do not confuse with a CNF, which is an acronym for Chomsky Normal Form.) For example,

$$(p_2 \vee \neg p_3 \vee p_5) \wedge (p_1 \vee p_2 \vee p_{10} \vee \neg p_5) \wedge (p_0 \vee p_3 \vee p_6 \vee p_8)$$

is a cnf. It is easy to see that every proposition is logically equivalent to one in cnf. For example, the proposition $(p_2 \leftrightarrow p_3) \leftrightarrow p_4$ is equivalent to the cnf $(p_2 \vee p_4) \wedge (\neg p_3 \vee p_4) \wedge (\neg p_2 \vee p_3 \vee \neg p_4)$. Here, we say that two propositions x and y are equivalent ($x \equiv y$) iff whatever interpretation (an assignment of values of 0 and 1 to the atomic propositions) i you take, $i(x) = i(y)$. Please go through Sect. 9.7 if you have forgotten how connectives are interpreted. It can be easily shown that $x \to y \equiv \neg x \vee y$, $x \leftrightarrow y \equiv (\neg x \vee y) \wedge (x \vee \neg y)$, $\neg(x \wedge y) \equiv \neg x \vee \neg y$, $\neg(x \vee y) \equiv \neg x \wedge \neg y$ and $\neg\neg x \equiv x$. Moreover, the connectives $\wedge$ and $\vee$ distribute over each other, that is, $(x \wedge y) \vee z \equiv (x \vee z) \wedge (y \vee z)$ and $(x \vee y) \wedge z \equiv (x \wedge z) \vee (y \wedge z)$. The connectives $\wedge$ and $\vee$ are associative and commutative. Using these equivalences, any proposition can be brought to an equivalent cnf. For example,

$$
\begin{aligned}
&(p_2 \to p_3) \leftrightarrow p_4 \\
&\equiv (\neg p_2 \vee p_3) \leftrightarrow p_4 \\
&\equiv (\neg(\neg p_2 \vee p_3) \vee p_4) \wedge ((\neg p_2 \vee p_3) \vee \neg p_4) \\
&\equiv ((\neg\neg p_2 \wedge \neg p_3) \vee p_4) \wedge (\neg p_2 \vee p_3 \vee \neg p_4) \\
&\equiv ((p_2 \wedge \neg p_3) \vee p_4) \wedge (\neg p_2 \vee p_3 \vee \neg p_4) \\
&\equiv ((p_2 \vee p_4) \wedge (\neg p_3 \vee p_4)) \wedge (\neg p_2 \vee p_3 \vee \neg p_4) \\
&\equiv (p_2 \vee p_4) \wedge (\neg p_3 \vee p_4) \wedge (\neg p_2 \vee p_3 \vee \neg p_4).
\end{aligned}
$$

A proposition is called **satisfiable** if it has a model, that is, if there is an interpretation that assigns the cnf to 1. Of course, we follow the rules that $i(x \wedge y) = 1$ iff $i(x) = i(y) = 1$; $i(x \vee y) = 0$ iff $i(x) = i(y) = 0$; and $i(\neg x) = 1$ iff $i(x) = 0$.

For example, the cnf $(p_2 \vee p_4)$ is satisfiable as the interpretation i with $i(p_2) = i(p_4) = 1$ evaluates $(p_2 \vee p_4)$ to 1. Similarly, $(\neg p_3 \vee p_4)$ is also satisfiable as the interpretation i with $i(p_3) = 0$, $i(p_4) = 1$ satisfies it (assigns it to 1). Even the interpretation i with $i(p_3) = i(p_4) = 1$ satisfies $(\neg p_3 \vee p_4)$. You can see that the disjunctive clause $(\neg p_2 \vee p_3 \vee \neg p_4)$ is also satisfiable. But this does not make the cnf $x = (p_2 \vee p_4) \wedge (\neg p_3 \vee p_4) \wedge (\neg p_2 \vee p_3 \vee \neg p_4)$ satisfiable. To show that it is satisfiable, we must have at least one interpretation j that satisfies all the disjunctive clauses in it simultaneously. This also happens here, as the interpretation j with $j(p_2) = 0$, $j(p_3) = 0$, $j(p_4) = 1$ satisfies all these disjunctive clauses. Hence the cnf x is satisfiable.

Moreover, there is a way to construct a cnf from a given boolean formula in polynomial time so that the boolean formula is satisfiable iff the constructed cnf is satisfiable; see Problem 10.68 at the end of this section. If you are mystified, do not worry; we will be concerned with cnfs only, forgetting how we have arrived at that. The satisfiability problem for cnfs is defined below.

The **satisfiability problem**, abbreviated to **SAT**, is:

> Does there exist an algorithm which, given any cnf, decides whether the cnf is satisfiable?

As a language, SAT $= \{\psi(x) : x$ is a satisfiable cnf$\}$. Informally, it is the set of all satisfiable boolean formulas in cnf. Answer to the question that SAT raises is "yes," as we can simply construct all possible interpretations appropriate to the cnf by assigning 0's and 1's to the atomic propositions occurring in the cnf, and then compute the value assigned to the given cnf under each interpretation.

That is, if $x = x_1 \wedge x_2 \wedge \cdots \wedge x_m$ is a cnf and that the atomic propositions occurring in x are $p_1, p_2, \ldots, p_n$, then we construct $k = 2^n$ interpretations $i_1, i_2, \ldots, i_k$ by assigning 0's and 1's to the p_js. Then, we evaluate the cnf x under each interpretation i_j systematically. When one of them satisfies x, we stop and report that x is satisfiable, else, all the k interpretations evaluate x to 0, and thus x is unsatisfiable.

In this procedure, how much time is spent on deciding satisfiability of x? Well, how much time do we give for evaluating x under a given interpretation? In the worst case, an interpretation i must evaluate each disjunctive clause x_l. Such an evaluation looks for at least one literal in x_l whose value under i is 1. Search for such a literal is linear in the number of literals, that is, in the worst case, it will scan all the literals, one after another.

Thus evaluation of x under an interpretation takes a polynomial time in the length of x. As there are 2^n interpretations to try, it may take at the most an exponential time for a DTM to determine satisfiability of a cnf.

What if an NTM is used to decide the satisfiability problem? Let us use a three-tape NTM, on the first tape of which is given the cnf x as input. On the second tape, the NTM nondeterministically generates an interpretation i, and on the third tape it outputs the value of $i(x)$ after evaluating it. If the output value on the third tape happens to be 1, then it stops and accepts the input. Otherwise, it rejects.

Notice that owing to nondeterminism, the phrase "otherwise, reject" is heavily loaded. It does not mean that if, after one generation of an interpretation, the interpretation does not satisfy x, then it goes to the reject state. It means that if x is at all satisfiable, then at least once such generation of an interpretation will succeed in satisfying x.

As computing $i(x)$ takes only a polynomial time and the unsuccessful attempts are not charged, the NTM decides satisfiability of the cnf x in a polynomial time. Thus, we have proved the following result:

Theorem 10.9. SAT $\in$ *NP*.

The verification stage in the above NTM is deterministic, and to accomplish the job, the NTM has acted as a verifier. The certificate c was the generated model. The above technique of solving SAT shows a close connection between membership in *NP* and existence of a polynomial time verifier.

Theorem 10.10. *Let L be a language. $L \in NP$ iff L is polynomial time verifiable.*

Proof. If $L \in NP$, we have an NTM M that decides L in polynomial time. Take the certificate c as a string representing a nondeterministic choice that M might take for a string w. Notice that once this choice of a branch of computation is fixed, M behaves deterministically. The verifier V uses the pair (w, c) and simulates M on w with the nondeterministic choice specified by c. If this branch of computation of M accepts w, then V goes to the accept state, else V enters the reject state. Clearly, V is a polynomial time verifier for L as M takes a polynomial time to decide L.

Conversely, suppose V is a polynomial time verifier for L. Without loss of generality, assume that V, as a Turing machine, runs in time at most n^k on any input of length n. We construct an NTM M, which, on an input w of length n, chooses a certificate c of length at most n^k nondeterministically. Then, M simulates V on the pair (w, c) and accepts w if V accepts (w, c), else M rejects w. Clearly, M decides L, as V does so, in time n^k. □

Theorem 10.10 formalizes the intuition that *NP* is the class of languages (decision problems) for which a suggested solution can be verified to determine whether it is indeed a solution or not in polynomial time.

Related to the class *NP* is the class ***CoNP***, where $L \in CoNP$ iff its complement $\overline{L} \in NP$. Closure of *NP* with respect to complementation leads to this class; see the problems below.

Problems for Section 10.6

10.53. Suppose L' is log space reducible to L. Prove that if $L \in P$, then $L' \in P$.

10.54. Prove that P is closed under the operations of union, intersection, complementation, concatenation, reversal, inverse homomorphism, and Kleene star.

10.55. Prove that *NP* is closed under union, intersection, concatenation, reversal, inverse homomorphism, and Kleene star.

10.56. It is yet an open problem whether *NP* is closed under complementation, that is, whether $NP = CoNP$? Are the following true? Justify.
(a) If $NP \neq CoNP$, then $P \neq NP$.
(b) If $P \neq NP$, then $NP \neq CoNP$.
(c) *NP* is closed under complementation iff the complement of some *NP*-complete problem is in *NP*.

10.57. It is known that $P \subseteq NP \subseteq DPS$ and $P \subseteq CoNP \subseteq DPS$. Which of the following are known and which are yet unknown?
(a) $P \neq \emptyset$.
(b) $NP - P \neq \emptyset$.
(c) $CoNP - P \neq \emptyset$.
(d) $(NP \cap CoNP) - P \neq \emptyset$.
(e) $DPS - (NP \cup CoNP) \neq \emptyset$.

10.58. Prove that $NP \subseteq EXP$.

10.59. Show that SUBSUM, the *subset sum problem*, that is, "Given a set A of numbers $m_1, m_2, \ldots, m_k$ and a target number t, does there exist a subset B of A such that the sum of all numbers in B is t?" is in *NP*.

10.7 *NP*-Completeness

Once we know that a certain problem is in P or in *NP*, we would be interested in knowing the status of other related problems. In such cases, often the tool of problem reduction comes handy. Suppose L is known to be in P and we have a reduction of another language L' to L. In which case, can we say that L' is also in P? To give a meaningful answer, the process of reduction should not take more than a polynomial time. Such reductions are called polynomial time reductions.

Let Σ be an alphabet. A function $f : \Sigma^* \to \Sigma^*$ is called a **polynomial time computable function** if there exists a polynomial time DTM that, upon input w, gives output $f(w)$ for each $w \in \Sigma^*$. Let L, L' be languages over Σ. L is **polynomial time reducible** to L', written $L \leq_P L'$, if there exists a polynomial time computable function $f : \Sigma^* \to \Sigma^*$ such that for each $w \in \Sigma^*$, we have $w \in L$ iff $f(w) \in L'$. Such a function f is called a **polynomial time reduction** of L to L'. A polynomial time reduction is also called a *Karp reduction*.

As you have rightly thought, a polynomial time reduction is a map reduction; moreover, it must also be computable in a polynomial time. It should now be clear that a polynomial time reduction preserves membership in P, but in which direction?

Theorem 10.11. *Let L, L' be languages over an alphabet Σ. If $L \leq_P L'$ and $L' \in P$, then $L \in P$.*

Proof. You need not read this proof; it is easy to work it out. As $L' \in P$, there is a polynomial time DTM M_2 that decides whether a string $w \in \Sigma^*$ of length m is in L' or not in time $O(m^k)$ for some $k \in \mathbb{N}$. As $L \leq_P L'$, there is a function $f : \Sigma^* \to \Sigma^*$ and a DTM M_1 such that M_1 outputs $f(u)$ for a given $u \in \Sigma^*$ of length m in $O(m^r)$ time for some $r \in \mathbb{N}$. Moreover, $u \in L$ iff $f(u) \in L'$.

Now, the M_1 in the combined DTM $M_1 \to M_2$, upon input $u \in \Sigma^*$ of length n, computes $f(u)$ in time $O(n^r)$ time. Then the M_2 in the combined machine decides whether $f(u) \in L'$ (or equivalently, $u \in L$) in time $O((n^r))^k$. Thus, $M_1 \to M_2$ decides $u \in L$ in a polynomial time. □

If we think of P as the class of decision problems that are easily solvable as many computer scientists think, then polynomial time reduction of L to L' says that L is as easy as L'. Looked another way, we may say that L' is as hard as L. What about the problems in *NP*? If L is polynomial time reducible to L' and $L' \in NP$, then the polynomial time verifier for L' combined with the polynomial time reduction (the machine that computes it) would provide a polynomial time verifier for L. Hence Theorem 10.11 can be read correctly replacing P by *NP* throughout. That is,

Theorem 10.12. *Let L, L' be languages over an alphabet Σ. If $L \leq_P L'$ and $L' \in NP$, then $L \in NP$.*

A little thought shows that even P can be replaced by EXP, the same way. In fact, any class of decision problems that possibly take more time than a polynomial time remains so under a polynomial time reduction. An important class of problems, called *NP*-hard problems, are obtained by employing this idea. This class contains all those decision problems that are at least as hard to solve as any typical *NP*-problem.

A language L is called ***NP*-hard** if every language $L' \in NP$ is polynomial time reducible to L. A language L is called ***NP*-complete** if $L \in NP$ and L is *NP*-hard.

NP-hardness looks like a big constraint, because to show that a language is *NP*-hard, one must have a polynomial time reduction of each and every language in *NP* to this language. Nonetheless, there are *NP*-hard languages (or *NP*-hard problems). The first such problem to be discovered was SAT.

The problems in *NP* can differ wildly from each other. The fact that they are in *NP* guarantees that each of those is decided in polynomial time by an NTM. The key idea in the construction of a polynomial time reduction of any problem in *NP* to SAT is to encrypt the work-out of a polynomial time NTM as a cnf. This encryption is analogous to the symbolization of an argument in English to propositional logic.

Let M be an NTM with a single tape having $Q = \{q_0, q_1, \ldots, q_m\}$ as the set of states and the tape alphabet $\Gamma = \{\gamma_1, \gamma_2, \ldots, \gamma_p\}$, where γ_1 is the symbol ƀ. We take q_0 as the initial state, q_1 as the accepting state h, and q_2 as the rejecting state $\hbar$. The transition relation Δ is a finite subset of $(Q \times \Gamma) \times (Q \times (\Gamma \cup \{Ⱡ, Ɍ\}))$. Let w be an input to M with length n. Assume that M is a polynomial time decider that decides the language $L \subseteq \Sigma^*$, where Σ is some subset of $\Gamma - \{\gamma_1\}$. There is a polynomial $f : \mathbb{N} \to \mathbb{N}$ such that each $w \in \Sigma^*$ of length n is decided by M ($w \in L$ when M accepts w and $w \notin L$ if M rejects w) in time $f(n)$.

The string w is given as an input to M the standard way, the read–write head of M scanning the ƀ just to the left of w. Number this square containing the ƀ as 0, to the left of it as $-1, -2, -3 \cdots$, and to the right of it as $1, 2, 3, \cdots$ in succession. As M decides w in time $f(n)$, only the tape squares numbered $-f(n)$ to $f(n)$ might, in the worst case, be used by M during its computation on w. Without loss of generality, assume that M accepts w in time $f(n)$ exactly, as otherwise, when M accepts w in less than $f(n)$ steps, we can let it work till $f(n)$th step in a trivial manner such as changing its states, or going left and right, etc.

Exercise 10.8. Let M be an NTM that accepts a string $w \in L$ of length n in less than $f(n)$ steps, where L is any language. Show that there is an NTM N that accepts any string $w \in L$ of length n in exactly $f(n)$ steps.

For $-f(n) \leq i \leq f(n)$, $1 \leq j \leq p$, $0 \leq k \leq f(n)$, and $0 \leq r \leq m$, we introduce the following atomic propositions with their associated meanings for the working out of M :

x_{ij}^k : The ith tape square contains γ_j at kth step of computation of M.

y_{ik} : The read–write head of M is scanning the ith tape square at kth step of computation of M.

z_{rk} : M is in state q_r at kth step of its computation.

For example, the following things happen initially:

x^0_{01} : The 0th tape square contains γ_1 (= ƀ) at 0th step of computation of M.

y_{00} : The read–write head of M is scanning the 0th tape square at 0th step of computation of M.

z_{00} : M is in state q_0 at 0th step of its computation.

The initial configuration of M on input $w = \gamma_{i1}, \gamma_{i2} \cdots \gamma_{in}$ with initial state q_0, when the read–write head is scanning the 0th square containing the symbol γ_1 (=ƀ) is described as follows:

$$A: \quad x^0_{(-f(n))1} \wedge x^0_{(-f(n)+1)1} \wedge \cdots \wedge x^0_{(-1)1} \wedge z_{00} \wedge y_{00} \wedge x^0_{01} \wedge x^0_{1i_1}$$
$$\wedge x^0_{2i_2} \wedge \cdots \wedge x^0_{ni_n} \wedge x^0_{(n+1)1} \wedge \cdots \wedge x^0_{f(n)1}$$

This also says that each square beyond the nth one contains γ_1.

$$B_k: \quad \left[\bigvee_{-f(n) \le i \le f(n)} y_{ik} \right] \wedge \left[\bigwedge_{-f(n) \le i \ne l \le f(n)} (\neg y_{ik} \vee \neg y_{lk}) \right],$$

$$B: \quad B_1 \wedge B_2 \wedge \cdots \wedge B_{f(n)}$$

As $\neg y_{ik} \vee \neg y_{jk} \equiv y_{ik} \to \neg y_{jk} \equiv y_{jk} \to \neg y_{ik}$, the cnf B_k says that at the kth step of computation of M, only one tape square is scanned. The cnf B says that at each step of computation of M, only one tape square is scanned.

$$C_{ik}: \quad \bigwedge_{1 \le j \ne l \le p} (\neg x^k_{ij} \vee \neg x^k_{il}), \qquad C: \quad \bigwedge_{-f(n) \le i \le f(n)} \ \bigwedge_{1 \le k \le f(n)} C_{ik}$$

As $\neg x^k_{ij} \vee \neg x^k_{il} \equiv x^k_{ij} \to \neg x^k_{il} \equiv x^k_{il} \to \neg x^k_{ij}$, C_{ik} says that at kth step of computation of M, the ith square contains only one symbol from Γ. Then, C asserts that at each step of computation of M, each square contains only one symbol from Γ.

$$D_k: \quad \bigwedge_{0 \le r \ne l \le m} (\neg z_{rk} \vee \neg z_{lk}), \qquad D: \quad \bigwedge_{0 \le k \le f(n)} D_k$$

The cnf D_k says that at kth step of computation, M stays at only one state. Thus, D means that at any step of computation, M stays at only one state.

Next, using our vocabulary we translate each transition of M into a clause. There are three possible types of transitions, such as symbol writing, left moving, or right moving. We take them in turn.

For a transition of the form $t = (q_\alpha, \gamma_\beta, q_\theta, \gamma_\tau) \in \Delta$, we take the cnf:

$$S_{\alpha\beta\theta\tau}: \quad (\neg x_{i\beta}^{k} \vee \neg y_{ik} \vee \neg z_{\alpha k} \vee z_{\theta(k+1)}) \wedge (\neg x_{i\beta}^{k} \vee \neg y_{ik} \vee \neg z_{\alpha k} \vee y_{i(k+1)}) \wedge (\neg x_{i\beta}^{k} \vee \neg y_{ik} \vee \neg z_{\alpha k} \vee x_{i\tau}^{k+1}).$$

This says that if M is in state q_α scanning the symbol γ_β on the ith square at kth step of computation, then it goes to state q_θ, writing γ_τ on the ith square at $(k+1)$th step of its computation. This is so as the cnf $S_{\alpha\beta\theta\tau}$ is equivalent to $(z_{\alpha k} \wedge y_{ik} \wedge x_{i\beta}^{k}) \to (z_{\theta(k+1)} \wedge y_{i(k+1)} \wedge x_{i\tau}^{k+1})$.

For transitions of the form $t = (q_\alpha, \gamma_\beta, q_\theta, \not{L}) \in \Delta$, we take the cnf

$$L_{\alpha\beta\theta}: \quad (\neg x_{i\beta}^{k} \vee \neg y_{ik} \vee \neg z_{\alpha k} \vee z_{\theta(k+1)}) \wedge (\neg x_{i\beta}^{k} \vee \neg y_{ik} \vee \neg z_{\alpha k} \vee y_{(i-1)(k+1)}) \wedge (\neg x_{i\beta}^{k} \vee \neg y_{ik} \vee \neg z_{\alpha k} \vee x_{(i-1)\tau}^{k+1}).$$

The cnf $L_{\alpha\beta\theta}$ is equivalent to $(z_{\alpha k} \wedge y_{ik} \wedge x_{i\beta}^{k}) \to (z_{\theta(k+1)} \wedge y_{(i-1)(k+1)} \wedge x_{(i-1)\tau}^{k+1})$. It thus says that if M is in state q_α scanning the symbol γ_β on the ith square at kth step of computation, then it goes to state q_θ, moves to the $(i-1)$th square at $(k+1)$th step of its computation.

The transitions $t = (q_\alpha, \gamma_\beta, q_\theta, \not{R}) \in \Delta$ of the third type are translated to

$$R_{\alpha\beta\theta}: \quad (\neg x_{i\beta}^{k} \vee \neg y_{ik} \vee \neg z_{\alpha k} \vee z_{\theta(k+1)}) \wedge (\neg x_{i\beta}^{k} \vee \neg y_{ik} \vee \neg z_{\alpha k} \vee y_{(i+1)(k+1)}) \wedge (\neg x_{i\beta}^{k} \vee \neg y_{ik} \vee \neg z_{\alpha k} \vee x_{(i+1)\tau}^{k+1}).$$

This says that if M is in state q_α scanning the symbol γ_β on the ith square at kth step of computation, then it goes to state q_θ, moves to the $(i+1)$th square at $(k+1)$th step of its computation. Notice that the proposition that captures this description is $(z_{\alpha k} \wedge y_{ik} \wedge x_{i\beta}^{k}) \to (z_{\theta(k+1)} \wedge y_{(i+1)(k+1)} \wedge x_{(i+1)\tau}^{k+1})$, which is equivalent to the cnf $R_{\alpha\beta\theta}$.

Thus the transitions in Δ and the computation of M following these transitions for correct updating of states, position of the read–write head, and the currently scanned square are encapsulated by the cnf

$$E: \quad \bigwedge_{-f(n)\le i\le f(n)} \; \bigwedge_{1\le j\le p} \; \bigwedge_{0\le k\le f(n)} \; \bigwedge_{t\in\Delta} E_{ijk}^{t}.$$

The proposition E_{ijk}^{t} is read as either $S_{\alpha\beta\theta\tau}$ or $L_{\alpha\beta\theta}$ or $R_{\alpha\beta\theta}$, according to the form of the transitions $t \in \Delta$. The cnf E says that correct updating of the states, scanned symbols, and the positions of the read–write head is achieved by using the transitions in M.

Finally, M accepts each string at $f(n)$th step of its computation. We translate this statement as follows:

$$F: \quad z_{1f(n)}.$$

The subscript 1 in $z_{1f(n)}$ refers to the state q_1, which is the accepting halt state. Notice that this statement takes a simpler form due to our assumption that M halts at exactly $f(n)$th step.

Theorem 10.13. (Cook–Levin) SAT *is NP-complete.*

Proof. By Theorem 10.9, SAT $\in NP$. To see that SAT is NP-hard, we use the above construction. Our aim is to give a polynomial time reduction of any language in NP to SAT. Any language in NP is decided by a polynomial time verifier. Thus, membership problem for any such language is same as the acceptance problem of any such verifier. It is enough to see how such verifiers can be reduced to SAT instances.

Let M be a polynomial time verifier for a language $L \in NP$. Suppose $w \in \Sigma^*$ is given as an input to the NTM M with a certificate c. The certificate supplies the choice of a branch of computation for this w. By following this branch, M decides whether $w \in L$ or not. Now, M accepts w iff the cnf $X = A \wedge B \wedge C \wedge D \wedge E \wedge F$ is satisfiable. Notice that the certificate c correctly constructs the cnf E. However, we must show that the construction above takes a polynomial time. To this end, we count the time taken by each of the six cnfs.

The number of atomic propositions, that is, the x_{ij}^k, y_{ik}, z_{rk}, is altogether of the order of $O((f(n))^2)$. Thus A contains at the most an $O((f(n))^2)$ literals. B_k has similarly an $O(f(n))$ literals; so B has at most $O((f(n))^2)$ literals. Similarly, C has $O((f(n))^2)$, D has $O(f(n))$, E has $O((f(n))^2)$, and F has $O(f(n))$ literals, at the most. Thus X contains at most an $O((f(n))^2)$ literals. As an $O((f(n))^2)$ number of letters can be encoded as an $O(\log f(n))$ digited binary number, the length of the cnf X is of the order $O((\log f(n) \cdot f(n))^2)$, which is (bounded above by) of the order $O((f(n))^4)$, a polynomial in n, the length of w. This completes the proof. □

Cook–Levin theorem shows that the notion of NP-completeness is not vacuous; there is at least one NP-complete language. Moreover, because of transitivity of polynomial reductions, we now have a technique to prove existence of other NP-complete languages. See the following theorem.

Theorem 10.14. *Let L and L' be languages. If L is NP-hard and $L \le_P L'$, then L' is NP-hard.*

Proof. Suppose L is NP-hard. Let $\hat{L} \in NP$. Then there is a polynomial time reduction f from $\hat{L}$ to L. As $L \le_P L'$, there is a function g, which is a polynomial time reduction of L to L'. Then the composition $g \circ f$ is a polynomial time reduction of $\hat{L}$ to L'. □

In addition to the hypothesis of Theorem 10.14, if the language L' is in NP, then NP-completeness of L is passed to L'.

Theorem 10.15. *Let L be an NP-complete language. Let $L' \in NP$ be such that $L \le_P L'$. Then L' is NP-complete.*

We will discuss some more examples of NP-complete problems in the next section. In almost all those cases, we will use the technique of polynomial time reduction, starting from SAT. However, what is the fun in knowing these difficult but next-to-easy problems? The reason is the following result.

Theorem 10.16. *Let L be an NP-complete language. If $L \in P$, then $P = NP$.*

Proof. Let $\hat{L} \in NP$. As L is NP-complete, there is a polynomial time reduction f from $\hat{L}$ to L. That is, f is computed by a polynomial time DTM, say, M_f. As $L \in P$, there is a polynomial time decider, say, M which decides L. Then the combined machine $M_f \to M$ decides $\hat{L}$ in polynomial time. Therefore, $\hat{L} \in P$. Hence, $NP \subseteq P$. As $P \subseteq NP$, we have $P = NP$. □

I repeat. If you can show that any one of the NP-complete problems is in P, then all NP-problems will have to be in P. Constructively, if you can find a polynomial time algorithm for deciding any one of the NP-complete problems, then that algorithm can be used to solve each and every problem in NP. For example, if SAT $\in P$, then $P = NP$. As all approaches to prove SAT $\in P$ have failed, people believe that probably $P \neq NP$. However, to show this, one must prove that SAT $\notin P$ or that some NP-problem is not in P.

Problems for Section 10.7

10.60. Suppose that f and g are functions that can be computed in time $\cup_{i=0}^{\infty} DT(n^i)$. Show that the composition $f \circ g$ can be computed in time n^k for some $k \in N$. Deduce the transitivity of polynomial time reduction.

10.61. Show that if every NP-hard language is DPS-hard, then $DPS = NP$.

10.62. Let L be any nonempty proper subset of Σ^* for an alphabet Σ. Show that if $P = NP$ and $L \in P$, then L is NP-complete.

10.63. Let L_1, L_2, L_3 be three languages over some alphabet Σ that does not contain the symbol #. Assume that $L_1 \in P$, L_2 is NP-complete, and $L_3 \notin NP$. Answer in each case below (a–e) whether the given language is
(i) in P,
(ii) in NP but not NP-complete,
(iii) NP-complete,
(iv) not in NP, or
(v) none of (i-iv):

(a) $L_i \cup L_j$, for $1 \leq i, j \leq 3$.
(b) $L_i \cap L_j$, for $1 \leq i, j \leq 3$.
(c) $L_i L_j$, for $1 \leq i, j \leq 3$.
(d) $L_i \# L_j$, for $1 \leq i, j \leq 3$.
(e) $\Sigma^* - L_i$, for $1 \leq i \leq 3$.

10.64. Let $L = \{\psi(M)\#\psi(w) : M \text{ accepts } w \text{ after at most } 2^{\ell(w)} \text{ steps}\}$. Mimic the undecidability of the acceptance problem to prove that $L \notin P$.

10.65. Show that the *Euler cycle problem*, that is, "Given an undirected graph G, does there exist a cycle in G that uses each edge exactly once?" is in P. What about Euler cycle problem in directed graphs?

10.66. The *isomorphism problem* in undirected graphs asks for the existence of an isomorphism between two given undirected graphs. Show that the isomorphism problem of graphs is in *NP*. Is it *NP*-complete?

10.8 Some *NP*-Complete Problems

The order notation characterization of complexity classes led us to the following containments:

$$LS \subseteq NLS = CoNLS \subseteq P \subseteq NP \subseteq DPS = NPS \subseteq EXP \subseteq EXPSPACE.$$

It is generally believed that all the containments above are proper containments, though only the following have been known to be so:

$$P \subsetneqq EXP, \; NLS \subsetneqq DPS, \; DPS \subsetneqq EXPSPACE.$$

You are invited to try them and make your mark in the mathematical world by proving any of the other containments to be proper or by showing that some of them are not so.

It is also believed that *NP* and *CoNP* are different. Notice that when L is decided by an NTM in polynomial time, it does not mean that $\overline{L}$ is decided in polynomial time. This is due to the anomaly discussed in Sect. 10.5. However, you can easily see that P is closed under complementation, that is, $P = Co\text{-}P$.

To give an evidence for the conjecture that *NP* and *CoNP* are different, consider the problem **UNSAT**:

Given a cnf x, to determine whether x is unsatisfiable.

Though SAT is in *NP*, it seems that UNSAT is not; at least the method of proof for the former fails in the latter case. To see this, suppose x is a given cnf. Design an NTM M that generates an interpretation i and evaluates x under it. If $i(x) = 0$, then we cannot assert that x is unsatisfiable. It becomes necessary to generate all the interpretations and check in each case whether x is evaluated to 0 or not. On the other hand, if for one i, we see that $i(x) = 1$, then we conclude that x is not unsatisfiable. That means a string (cnf, here), which is not in the language (corresponding to UNSAT), is accepted in polynomial n-time, while a string in that language is accepted in an exponential n-time (also exponential time).

You should remember the warning that this argument is not a proof of $NP \neq CoNP$. It is only an evidence that supports the belief that possibly $NP \neq CoNP$, and this signals the asymmetry in the decision process of an NTM. The evidence is a week evidence; despite the fact that we have not been able to find even a single *NP*-complete problem whose complement is in *NP*.

Exercise 10.9. Prove that $NP = CoNP$ iff there exists an *NP*-complete problem whose complement is in *NP*.

There are many interesting problems that can be seen naturally to be in *CoNP*. A problem related to UNSAT is **VAL**, or sometimes written as **TAUT**:

Given a proposition, whether it is valid or not.

VAL asks whether each interpretation of the given proposition is its model or not. It is in *CoNP* because the problem UNSAT is in *CoNP*. The connection between VAL and UNSAT is that a proposition E is valid iff its negation $\neg E$ is unsatisfiable.

You should not be misguided here by the fact that validity of a cnf can be checked quite easily. (How?) It is because conversion of a proposition to a cnf that preserves validity is not a polynomial process unlike the conversion that preserves satisfiability; see the problems following this section. What about the validity problem when we have propositions in dnf, that is, when the proposition is given as a disjunction of conjunctive clauses? Is it in *NP* or in *CoNP*?

It is generally argued that P represents the easy or tractable problems. That is, if a decision problem or its corresponding language is known to be in P, then the problem can be solved for all practical purposes. It is thought that spending a polynomial time is well within our reach, though polynomials of degree four or more become practically unmanageable for large inputs.

For example determining whether an integer is prime or not has been proved to be solvable in time $O(n^{11})$. However, the cryptographic systems those keep their secrets under the risk of being broken by primality testing have not yet been broken. Because of such reasons, many believe that P is too huge a complexity class to represent truly tractable problems.

On the other hand, the next best in our containment of complexity classes is *NP*, the decision problems, which have polynomial time verifiers. It seems that most of the interesting problems are in *NP*, and are, in fact, *NP*-complete. It would be a nice piece of magic to see that discovering a solution to a problem coincide with verifying a suggested solution (in polynomial time).

Human experience shows that the jobs of discovering a solution and verifying a suggested solution are a way apart in terms of the time they would take. It is thus generally believed that $P \neq NP$. Because of its aesthetic appeal, it stands out among all other open problems concerning the left-out containments of the complexity classes.

We have seen that to settle this problem, it would be enough to show whether one (just one) of the *NP*-complete problems is in P or not. We do not know whether the solution will be so straight forward to solve $P \stackrel{?}{=} NP$ problem. The *NP*-complete problems represent hard problems next to polynomial time solvable problems, and due to our ignorance of $P \stackrel{?}{=} NP$, they are, in fact, at the boundary of tractable and intractable problems.

In this section, we discuss some other interesting *NP*-complete problems.

We begin with a SAT related problem. A k**cnf** (Read it as k cnf.) is a cnf where each disjunctive clause has at most k literals. For example, $(\neg p_1 \vee p_2) \wedge p_3 \wedge \neg p_5$ is a 2cnf, whereas $(p_1 \vee \neg p_0) \wedge \neg p_3 \wedge (p_4 \vee p_6 \vee p_5)$ is a 3cnf. The problem k**SAT** is,

Given a $k$cnf, is it satisfiable?

It is the subproblem of SAT where each cnf has clauses with at most k literals. First, we see that 1SAT $\in P$; how? A 1SAT instance is a 1cnf that looks like $q_1 \wedge q_2 \wedge \cdots \wedge q_m$, where each q_i is a literal. To see whether it is satisfiable, you just assign 1 (i.e., true) to each literal, and the 1cnf is satisfiable. This procedure may fail when some q_i is negation of some q_j. In that case, the 1cnf is unsatisfiable. This heuristic is directly implemented in the following algorithm:

ALGORITHM 1SAT(i, m)

1. Take a 1cnf $q_i \wedge \cdots \wedge q_m$ as an input.
2. If any of $q_{i+1}, \ldots, q_m$ is the negation of q_i, then stop and report that the 1cnf is unsatisfiable. Else, delete q_i from the 1cnf and run Algorithm 1SAT$(i + 1, m)$.
3. Report that the 1cnf is satisfiable.

Exercise 10.10. Show that 1SAT$(1, m)$ correctly solves 1SAT. Also give an iterative algorithm to solve 1SAT.

In the worst case, the Algorithm 1SAT$(1, m)$ will go through m number of recursive steps dealing with $m+(m-1)+\cdots+1$ steps. Thus, 1SAT solves 1SAT in $O(m^2)$ steps where m is the number of literals in the given 1cnf; 1SAT $\in P$.

What about 2SAT? You can now try an algorithm for checking satisfiability of a 2cnf employing the same idea as in Algorithm 1SAT$(1, m)$. Of course, it needs some modifications.

Exercise 10.11. Try a modification of 1SAT$(1, m)$ for solving 2SAT.

Suppose there is a single literal that is a clause in the given 2cnf. If its negation is also a clause, then the 2cnf is unsatisfiable. So, this should be our first check. If you get a clause, which is a single literal, but the negation of the literal is not a clause, then in order that the cnf is satisfiable, this literal must be assigned to 1.

This amounts to omitting the literal from the cnf, as conjunction of the remaining clauses of the cnf (omitting this single-literal clause, which has been assigned to 1) is satisfiable iff the original 2cnf is satisfiable. Also, you can delete all those clauses wherever this literal occurs, as all those clauses are also evaluated to 1.

Thus, the updated 2cnf is now smaller and it contains less number of atomic propositions than the original. This is written below as the Stage 1 of our procedure for 2SAT.

On the other hand, when each clause in the 2cnf is a two-literals clause, we try to get a model of the 2cnf by assigning a chosen literal to 0, 1, alternately. Suppose we assign a literal (Call it the chosen literal.) in a clause to 0, to begin with. Then, this particular clause can be satisfied, provided the other literal in it is satisfied.

Notice that if the clause that contains this chosen literal has another literal, then that other literal must also be satisfied. Further, if any clause contains the negation of this chosen literal, then that clause is automatically satisfied as the negation of the chosen literal is assigned 1.

It may quite well happen that the 2cnf has no model where the chosen literal is assigned to 0. In that case, we will not succeed in satisfying the 2cnf by following the method described in the above paragraph. We then go back and assign the chosen literal as 1. In this case, the clauses that contain this chosen literal are satisfied; and any clause that contains the negation of this chosen literal is satisfied if (only) the other literal in it is satisfied.

We use these observations to describe a procedure for solving 2SAT. Let x be a given 2cnf. We make it a point to write both the clauses $p \vee q$ and $q \vee p$ as $p \vee q$, taking commutativity of $\vee$ into consideration. The finer details of the procedure are revealed in three stages as given below.

ALGORITHM 2SAT

Stage 1 : Scan the clauses in x for a single-literal clause. If x contains no single-literal clause, then perform Stage 2. Otherwise, let p be a single-literal clause of x. Scan further for the single-literal clause $\neg p$. If $\neg p$ is also a clause of x, then report that x is unsatisfiable. If $\neg p$ is not a clause of x, then scan x from the beginning. Drop each clause that contains p and drop $\neg p$ from each clause that contains $\neg p$. That is, update x by deleting all clauses of the form $p \vee q$ and by replacing each clause of the form $\neg p \vee q$ by q.

Repeat Stage 1 as long as possible. If the result is empty, that is, every clause has been deleted in the process, then report that x is satisfiable. Otherwise, the result is a 2cnf A, each clause of which has exactly two literals. Then perform Stage 2.

Stage 2 : Take the first literal in the first clause of A; say p, in the first clause $p \vee q$ of A. Scan A from the beginning. Drop the literal p from each clause that contains p, and drop each clause that contains $\neg p$, from A. That is, update A by replacing each clause of the form $p \vee q$ by q, and by deleting each clause of the form $\neg p \vee q$. Call the updated 2cnf as B. Execute Stage 1 repeatedly on the updated 2cnf. This will result in one of the following three cases.

(a) Reporting that B is unsatisfiable
(b) An empty cnf
(c) A cnf C having only two-literals clauses

In the case (a), execute Stage 3 as given below. In the case (b), report that x is satisfiable. In the case (c), repeat Stage 2 on C.

Stage 3 : Go back to the 2cnf A. Let $p \vee q$ be the first clause of A. Scan each clause of A. Update A to D by dropping each clause of the form $p \vee r$. Scan D for an occurrence of the literal $\neg p$. If $\neg p$ does not occur in D, then execute Stage 2 on D. Otherwise, update D to E by dropping the literal $\neg p$ from each clause of the form $\neg p \vee r$. Now, E has at least one single-literal clause. Execute Stage 1 on E repeatedly. This will halt resulting in either of the following:

(a) Reporting that E is unsatisfiable
(b) An empty cnf
(c) A cnf F having only two-literals clauses

In the case (a), report that x is unsatisfiable. In the case (b), report that x is satisfiable. In the case (c), execute Stage 2 on F.

Exercise 10.12. Show that ALGORITHM 2SAT correctly solves 2SAT. The catch is to see that in Stage 2, the 2cnf A is satisfiable if the 2cnf C is satisfiable.

Stage 1 of the procedure eliminates at least one atomic proposition, and Stages 2 and 3 together eliminate one. Moreover, unsatisfiability of the 2cnf x is reported while executing Stage 1. The worst case scenario corresponds to the case when all atomic propositions are eliminated one by one in Stage 2 followed by an execution of Stage 3, and when finally, x is found to be satisfiable. Initially, the 2cnf is scanned for checking for a possible application of Stage 1, anyway.

Suppose n is the length of the 2cnf. Initial scanning of the 2cnf for a single literal clause takes an $O(n)$ time. Repeated execution of Stage 1 can take, at the most, an $O(n^2)$ time. Thus, executing Stage 3 for a literal, which turns out to be an unsuccessful attempt, takes an $O(n^2)$ time. It is followed by an execution of Stage 2, which again takes an $O(n^2)$ time. So, finally a literal is eliminated in $O(n^2)$ time. At the worst, each literal is eliminated; thus the maximum time amounts to an $O(n^3)$. Therefore, 2SAT $\in P$.

We have thus proved the following:

Theorem 10.17. *1*SAT $\in P$ *and 2*SAT $\in P$.

There is another procedure to solve 2SAT; see Problem 10.72 at the end of this section. In this respect, 3SAT is very different from 1SAT and 2SAT.

Theorem 10.18. 3SAT *is NP-complete.*

Proof. We use the technique of problem reduction. Let $L \in NP$. As there is already a polynomial time reduction of L to SAT, due to Theorems 10.15 and 10.16, it is enough to show that SAT $\leq_P$ 3SAT. To this end, consider a cnf x. We construct a 3cnf y corresponding to this x such that x is satisfiable iff y is satisfiable. For this construction, we consider any disjunctive clause C having more than three literals, and then go for construction of a 3cnf C' corresponding to C, so that C is satisfiable iff C' is. And finally, we just take conjunctions of all these 3cnfs C' to obtain y.

Consider a clause $p \vee q \vee r \vee s$ having four literals. We introduce a new symbol z and take $z \leftrightarrow (r \vee s)$ to be valid. Then

$$\begin{aligned} & p \vee q \vee r \vee s \\ & \equiv ((p \vee q) \vee z) \wedge (z \leftrightarrow (r \vee s)) \\ & \equiv (p \vee q \vee z) \wedge (\neg z \vee r \vee s) \wedge (z \vee \neg r) \wedge (z \vee \neg s). \end{aligned}$$

By induction, each clause can now be written as a cnf. However, there is a more efficient way. We need only satisfiability to be preserved, not necessarily equivalence.

Let $C = p_1 \vee p_2 \vee \cdots \vee p_m$ be a clause, where $m > 3$. Construct

$$\begin{aligned} C' = {} & (p_1 \vee p_2 \vee z_1) \wedge (p_3 \vee \neg z_1 \vee z_2) \wedge (p_4 \vee \neg z_2 \vee z_3) \\ & \wedge \cdots \wedge (p_{m-2} \vee \neg z_{m-3} \vee z_{m-2}) \wedge (p_{m-1} \vee p_m \vee \neg z_{m-2}), \end{aligned}$$

where $z_1, \ldots, z_{m-2}$ are new propositional variables. We first show that C is satisfiable iff the cnf C' is satisfiable. To show this, take an interpretation i such that $i(C') = 1$. If $i(C) = 0$, then $i(p_j) = 0$ for each j with $1 \leq j \leq m$. Looking at individual clauses of C', we find that $i(z_1) = 1$, then $i(z_2) = 1$. Continuing further, we see that $i(z_3) = i(z_4) = \cdots i(z_{m-2}) = 1$. But then i of the last clause $(p_{m-1} \vee p_m \vee \neg z_{m-2})$ in C' is 0. This contradicts $i(C') = 1$. Therefore, $i(C) = 1$.

Conversely, suppose that $i(C) = 1$. As C' contains new literals (the z_i's), we construct an extension of i, which we write as i again taking care of these z_i's where $i(p_j)$'s remain the same. As $i(C) = 1$, not all p_j's are 0 under i. Let k be the smallest index such that $i(p_k) = 1$. Then, we set

$$i(z_1) = i(z_2) = \cdots = i(z_{k-1}) = 1, \text{ and } i(z_k) = i(z_{k+1}) = \cdots = i(z_{m-2}) = 0.$$

It is easy to verify that under this (extended) interpretation i, each of the clauses in C' is evaluated to 1. Therefore, $i(C') = 1$ as desired.

The reduction of SAT to 3SAT first replaces each clause C having more than three literals in any cnf by the corresponding 3cnf C', thus obtaining a 3cnf y for the cnf x. Notice that if C has m occurrences of literals, then C' has at the most $3m$ occurrences of literals. (Exactly how many?) Hence the length of the corresponding 3SAT instance y is at the most an $O(n)$ for a cnf of length n. Thus, the reduction is a polynomial time reduction. Therefore, SAT $\leq_P$ 3SAT, and as SAT is *NP*-complete, so is 3SAT. □

Exercise 10.13. Show that the extended interpretation as defined in the proof of Theorem 10.18 does satisfy the 3cnf C', if the original interpretation satisfies C.

Analogous to 3SAT, many other problems related to SAT have been shown to be *NP*-complete. Some of them are

MAXSAT: Given a cnf x and a positive integer k, does there exist an interpretation that satisfies at least k of the clauses in x?

MAX2SAT: Given a cnf x, each clause of which has at most two literals, and a positive integer k, does there exist an interpretation that satisfies at least k of the clauses in x?

Surprisingly, MAX2SAT is *NP*-complete though 2SAT is in P.

Next, we take up an interesting problem from graph theory. A ***k*-clique** in an undirected graph is a subgraph with k vertices each of which is connected to the other by an edge. It is a subgraph of the original graph with k vertices which is, by itself, a complete graph. The problem **CLIQUE** is:

> Given an undirected graph and an integer $k \geq 2$, does there exist a k-clique in the graph?

For the problem CLIQUE, the corresponding language is:

$$\text{CLIQUE} = \{(G, k) : \text{the undirected graph } G \text{ has a } k\text{-clique}\}.$$

Formally, the ordered pair (G, k) is replaced by its encoding $\psi(G)\#\,\psi(k)$; we choose to work with the informal notation. In the following, we show that 3SAT reduces to CLIQUE in polynomial time. The reduction introduces vertices corresponding to atomic propositions and triples of vertices to clauses. Such small components used in reductions are sometimes called *gadgets*.

Theorem 10.19. CLIQUE *is NP-complete.*

Proof. To show that CLIQUE $\in NP$, we construct a nondeterministic algorithm N for accepting the language CLIQUE as described below:

> Given the ordered pair (G, k) as an input, N nondeterministically selects a subset V of k vertices from G. It then joins an edge between each pair of these vertices. Next, it checks whether each constructed edge is an edge in G or not. If the answer is affirmative, then N accepts (G, k), else it rejects (G, k).

Constructing edges between the k selected vertices from G requires an $O(k^2)$ time. Checking whether each such constructed edge is really an edge in G takes a linear time with respect to the number of edges in G. Hence, N is a polynomial time nondeterministic decider for CLIQUE. Therefore, CLIQUE $\in NP$.

Towards a polynomial time reduction of 3SAT to CLIQUE, we give an algorithm to construct an undirected graph G corresponding to a given 3cnf x having k disjunctive clauses. As a clause of the form p can be rewritten as $p \vee p \vee p$, and one of the form $p \vee q$ as $p \vee p \vee q$, assume that each clause in x has exactly three literals. That is, let

$$x = (p_1 \vee q_1 \vee r_1) \wedge (p_2 \vee q_2 \vee r_2) \wedge \cdots \wedge (p_k \vee q_k \vee r_k),$$

where p_i, q_i, r_i are literals, not necessarily distinct. We construct the graph G having $3k$ vertices with labels as

$$p_1, q_1, r_1, p_2, q_2, r_2, \ldots, p_k, q_k, r_k,$$

respectively. The labels are taken from their occurrences in the clauses directly. We treat the $3k$ vertices into k groups, each group having three vertices labeled as p_i, q_i, r_i. The edges are constructed by joining each vertex to the other satisfying the following restrictions:

1. There is no edge between vertices of equal index, that is, no edge between p_i, q_i; between q_i, r_i; and between r_i, p_i. That means, there is no edge between the vertices in the same group. There can be only edges from vertices in one group to vertices in another group.
2. There is no edge between two vertices (belonging to different groups) if their labels are complementary. For example, if the literal p_1 equals p and the literal q_2 is $\neg p$, then even though the vertices labeled p_1 and q_2 are members of different groups, there cannot be an edge between them.

For an illustration, see Example 10.4 below. Then, come back to the proof.

We must show that x is satisfiable iff G has a k-clique. To this end, let f be an interpretation such that $f(x) = 1$. Then in every clause, at least one literal is assigned to 1 by f. We select the corresponding vertex from G. If more than one literal from a clause is assigned to 1 by f, we choose one of them arbitrarily. These vertices form a clique in G as there is an edge from one to the other, by our construction.

Conversely, suppose G has a k-clique. This clique does not contain vertices labeled with literals from the same group, that is, from the same clause of x, due to the first restriction. Moreover, by the second restriction, no pair of vertices with labels as complementary literals are connected by an edge. Thus, assigning all nodes in the clique to 1 defines an interpretation of x, which also satisfies x.

It is easy to see that the construction of a graph from a 3cnf as described above can be carried out in a polynomial time in the length of the 3cnf. This completes the proof. □

Example 10.4. Let $x = (p_1 \vee p_2 \vee p_2) \wedge (\neg p_2 \vee \neg p_2 \vee p_3) \wedge (\neg p_1 \vee \neg p_3 \vee p_4)$ be a cnf having exactly three literals in each of its clauses. Construct the graph corresponding to x by following the idea in the proof of Theorem 10.19.

Solution. The graph G corresponding to x has nine vertices; name them as $1, 2, \ldots 9$. The labels of these vertices are, respectively, $p_1, p_2, p_2, \neg p_2, \neg p_2, p_3, \neg p_1, \neg p_3$, and p_4. The first group of vertices are 1,2,3; the second group consists of vertices 4,5,6; and the vertices in the third group are 7,8,9. The first restriction says that there is no edge between 1 and 2; between 2 and 3; between 3 and 1; between 4 and 5; between 5 and 6; between 6 and 4; between 7 and 8; between 8 and 9; and between 9 and 7. The second restriction says that there is no edge between 2 and 4 as 2 is labeled as p_2 and 4 is labeled as $\neg p_2$. Similarly, there is no edge between 3 and 4; between 2 and 5; between 3 and 5; between 1 and 7; and between 6 and 8. Thus the graph has edges connecting 1 to 4,5,6,8,9; 2 to 6,7,8,9; 3 to 6,7,8,9; 4 to 1,7,8,9; 5 to 1,7,8,9; 6 to 1,2,3,7,9; 7 to 2,3,4,5,6; 8 to 1,2,3,4,5; and vertex 9 to 1,2,3,4,5,6. See Fig. 10.2. □

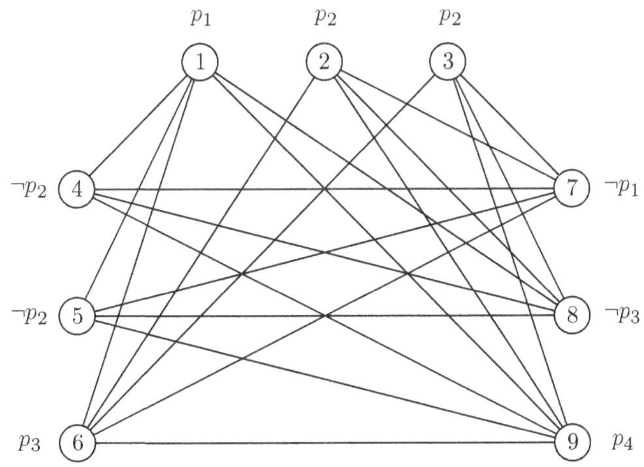

Fig. 10.2. Graph for Example 10.4.

Exercise 10.14. Show that the reduction of 3SAT to CLIQUE as given in the proof of Theorem 10.19 is a polynomial time reduction.

In addition to CLIQUE, many more graph theoretic problems have been shown to be *NP*-complete. Some of them are:

IS: In an undirected graph G, an independent set is a subset of the vertices such that there is no edge between any pair of vertices in this subset. The *independent subset problem* asks for the existence of an independent subset of size greater than a given integer $k > 1$.

VC: In an undirected graph G, a vertex cover is a subset V of vertices of G, where each edge of G is incident to some vertex in V. The *vertex cover problem* asks for a vertex cover of size no more than a given integer $k > 1$.

HP: A Hamiltonian path in an undirected graph G is a path that goes through every vertex exactly once. The *Hamiltonian path problem* asks for the existence of a Hamiltonian path joining two given vertices u, v in G.

The *Hamiltonian cycle problem* asks for the existence of a cycle (a path ending at the starting vertex) that goes through each vertex exactly once in a given undirected graph.

CN: Given an undirected graph G and an integer $k \geq 1$, can G be colored with k colors so that no adjacent vertices are of the same color? The acronym CN stands for the *chromatic number problem*; it is also called the *vertex colouring problem*.

Some other interesting *NP*-complete problems are:

EC: Given a finite set A and a family F of subsets of A, an exact cover is a subfamily S of F such that the sets in S are pairwise disjoint and have A as their union. The *exact cover problem* asks for the existence of an exact cover, given the set A and the family F.

SUBSUM: Given a finite multiset A of natural numbers, and a target number t, all represented in binary, the *subset sum problem* asks for the existence of a sub-multiset B of A such that the numbers in B add up to t.

KNAPSACK: For a finite set A, two functions $s, p : A \to \mathbb{Z}_+$, and two positive integers S and P, a knapsack is a subset B of A such that $\sum_{x \in B} s(x) \leq S$ and $\sum_{x \in B} p(x) \geq P$. The *knapsack problem* asks for the existence of a knapsack B given A, s, p, S, and P.

PARTITION: Given a finite set A of natural numbers, all represented in binary, the *partition problem* asks for the existence of a subset B of A such that the numbers in B add up to the same as in its complement $A - B$, that is, $\sum_{a_i \in B} a_i = \sum_{a_i \notin B} a_i$.

TSP: Given a set $\{c_1, c_2, \ldots, c_m\}$ of $m \geq 2$ cities and an $m \times m$ matrix of natural numbers d_{ij} of distances between cities c_i and c_j (thus $d_{ii} = 0$ and $d_{ij} = d_{ji} > 0$ for $i \neq j$), the *traveling salesperson problem* asks for the shortest tour of the cities, that is, a bijection π on $\{1, 2, \ldots, m\}$, $\pi(i)$ being the city visited as ith in the tour, so that the cost $c(\pi) = d_{\pi(1)\pi(2)} + d_{\pi(2)\pi(3)} + \cdots + d_{\pi(m)\pi(1)}$ of the tour is a minimum.

Problems for Section 10.8

10.67. Any proposition can be converted to a boolean formula by eliminating the connectives $\rightarrow$ and $\leftrightarrow$, using the laws of implication and the biconditional: $A \rightarrow B \equiv (\neg A \vee B)$ and $A \leftrightarrow B \equiv (A \wedge B) \vee (\neg A \wedge \neg B)$. Show that this conversion preserves equivalence. Is this conversion a polynomial process?

10.68. Any boolean formula can be converted to an equivalent cnf. In general, this conversion is an exponential process. However, there exists a polynomial time conversion of boolean formulas that preserves satisfiability. This process of polynomial time *cnf conversion* justifies our starting point with a cnf for defining SAT. To see this, prove the following:

(a) Let $X = C_1 \wedge C_2 \wedge \cdots \wedge C_m$ and $Y = D_1 \wedge D_2 \wedge \cdots \wedge D_n$ be two cnfs, where the C_i, D_j are disjunctive clauses. Let x be a propositional variable occurring neither in X nor in Y. Construct the cnf:

$$Z = (C_1 \vee x) \wedge (C_2 \vee x) \wedge \cdots \wedge (C_m \vee x) \wedge (D_1 \vee \neg x) \wedge (D_2 \vee \neg x) \cdots \wedge (D_n \vee \neg x).$$

Suppose f is a model of X. Extend f to the interpretation g by assigning x to 0 (0 for falsity and 1 for truth). Then, g is a model of Z.

(b) Let X, Y, Z be as in (a). Each model of $X \vee Y$ can be extended to a model of Z. Conversely, each model of Z is a model of $X \vee Y$.

(c) Given a boolean formula, move the $\neg$ sign to the variables by using the laws of De Morgan and the double negation:

$$\neg(A \vee B) \equiv \neg A \wedge \neg B, \ \neg(A \wedge B) \equiv \neg A \vee \neg B, \ \neg\neg A \equiv A.$$

It takes a time polynomial in the length of the given boolean formula.

(d) Define the conversion procedure inductively by assuming that the given expression is in either of the forms $A \wedge B$ or $A \vee B$, where A, B are in cnf. Note that for an actual construction, you have to identify the innermost propositions that are not in cnf, and then build it up. In the first case, the expression is a cnf. In the second case, use the construction in (a) by taking A, B as X, Y for obtaining the expression Z. If E is a proposition and F is the cnf obtained by this construction, then E is satisfiable iff F is satisfiable.

(e) The construction of F from E as described here takes an $O(n^2)$ time, where n is the length of the expression E.

10.69. Does there exist a validity preserving polynomial time conversion of any boolean formula to a dnf?

10.70. Define *CoNP*-completeness. Show that VAL is *CoNP*-complete.

10.71. To prove $P = NP$, is it enough to show that VAL$\in P$? Why?

10.72. An alternative procedure to solve 2SAT is based on the observation that each clause $p \vee q$ is equivalent to both $\neg p \rightarrow q$ and $\neg q \rightarrow p$. Given a 2cnf x having n atomic propositions $p_1, \cdots, p_n$, construct a directed graph G_x by taking $2n$ vertices $p_1, \neg p_1, \cdots, p_n, \neg p_n$. Corresponding to each clause of the form $p \vee q$, join an edge from the vertex $\neg p$ to q, and join another edge from the vertex $\neg q$ to p. First, show that x is unsatisfiable iff there is a vertex p in G_x such that there is a path from

p to $\neg p$, and also there is a path from $\neg p$ to p. Next, show that determining whether such a vertex p exists in G_x takes polynomial time.

10.73. Consider the language PATH= $\{\psi(G)\#\,\psi(u)\#\,\psi(v) : G$ is an undirected graph in which there is a path from the vertex u to the vertex $v\}$. Show the following:
(a) The space complexity of PATH is $O((\log n)^2)$.
(b) PATH is in $NS(\log n)$.
(c) A DTM that enumerates all possible paths in G for deciding PATH has time complexity $O(2^n)$, where n is the number of vertices in G.

10.74. Can you improve the time complexity in Problem 10.73(c)?

10.75. Let G be a graph with four vertices: v_1, v_2, v_3, v_4. For $1 \le i, j \le 4$, let x_{ij} denote a propositional variable representing "there is an edge between the vertices v_i and v_j." An interpretation then represents a graph with these four vertices. We can then express any property about the graphs with four vertices involving edges by propositions involving these variables. Write expressions for the following properties:
(a) G is a connected graph.
(b) G has a Hamiltonian cycle.
(c) G contains a clique of size 3.
(d) G contains at least one isolated vertex.

10.76. The problem **4TA** is: Given a proposition, does it have at least four models? (*Models* are the satisfying truth assignments.) Show that 4TA is NP-complete.

10.77. Using a complementary graph of an undirected graph as the one with the same vertex set but having all and only those edges which are not in the original graph, show that the problem of determining an independent set can be reduced to the problem of determining whether there is a clique. With $\wp(k)$ as the binary representation of k, define

$L = \{\psi(G)\#\,\wp(k)$: the undirected graph G has an independent set of size at least $k\}$.
$L' = \{\psi(G)\#\,\wp(k)$: the undirected graph G has a clique of size at least $k\}$.
Give a polynomial time and log space reduction of L to L'.

10.78. Here is the description of an algorithm to solve the partition problem: Let k be the sum of all integers in the set $A = \{a_1, \ldots, a_n\}$ divided by 2. If k is not an integer, then there is no B. Otherwise, for each i, $0 \le i \le n$, define the sets: $B(i) = \{b \le k : b$ is the sum of numbers in some subset of $\{a_1, \ldots, a_i\}\}$. If $B(n)$ is known, then testing whether $k \in B(n)$ solves the problem. $B(n)$ is computed as follows:

```
Initialize B(0) := {0};
for i = 1 to n, do
  B(i) := B(i − 1);
  for j = a_i to k, do
    if j − a_i ∈ B(i − 1) then B(i) := B(i) ∪ {j}
  od
od
```

Prove that this algorithm solves the partition problem correctly in time $O(nk)$. Had the numbers in A been given in unary notation, this algorithm could have solved the partition problem in a polynomial time. Argue why this is not a polynomial time algorithm.

10.79. Define appropriate certificates for PARTITION and CLIQUE.

10.80. The *longest path problem* is: Given an undirected graph G and a positive integer k, does G contain a simple path of length at least k? Show that there is a polynomial time and log space reduction of the Hamiltonian path problem to the longest path problem.

10.9 Dealing with *NP*-Complete Problems

Read the TSP as formulated in the last section once again. It asks for a shortest tour and not mere existence of a shortest tour. Existence of a shortest tour is anyway guaranteed, be it unique or not. Associated with this problem are two others, where we might ask for the cost of a shortest tour, or we ask for a tour whose cost is no more than a given positive integer k. Such problems are called *optimization problems*. An optimization problem always comes with a cost function. For example, MAXSAT is an optimization problem where the cost is the number of clauses that are satisfiable, and we want this cost to be maximized. That is, given an instance of the problem, a set of disjunctive clauses, we want a subset of maximum size such that all clauses in the subset can be satisfied by some interpretation simultaneously. In general, there are three kinds of problems associated with an optimization problem. They are the following:

1. *Construction Problem*: Given an instance of the optimization problem, construct an optimum solution.
2. *Evaluation Problem*: Given an instance of the optimization problem, find the cost of an optimum solution.
3. *Decision Problem*: Given an instance of the optimization problem and an integer k, determine whether there exists a solution whose cost is
   less than or equal to k if the problem is a minimization problem, or,
   more than or equal to k if the problem is a maximization problem.

The optimization problem CLIQUE has three versions such as

CLIQUEC: Given an undirected graph G, find a clique of maximum size in G.

CLIQUEE: Given an undirected graph G, find the maximum size of a clique in G.

CLIQUED: Given an undirected graph G and an integer $k \geq 2$, does there exist a k-clique in G?

The traveling salesperson problem formulated in the last section as TSP is infact, **TSPC**, the construction version of TSP. The other two versions are:

TSPE: Given a set $\{c_1, c_2, \ldots, c_m\}$ of $m \geq 2$ cities and an $m \times m$ matrix of natural numbers d_{ij} of distances between cities c_i and c_j (thus $d_{ii} = 0$ and $d_{ij} = d_{ji} > 0$ for $i \neq j$), determine the minimum cost $c(\pi) = d_{\pi(1)\pi(2)} + d_{\pi(2)\pi(3)} + \cdots + d_{\pi(m)\pi(1)}$ among all possible tours π, where π is a bijection on $\{1, 2, \ldots, m\}$.

TSPD: Given a set $\{c_1, c_2, \ldots, c_m\}$ of $m \geq 2$ cities, an $m \times m$ matrix of natural numbers d_{ij} of distances between cities c_i and c_j (thus $d_{ii} = 0$ and $d_{ij} = d_{ji} > 0$ for $i \neq j$), and an integer $k \geq 2$, does there exist a bijection π on $\{1, 2, \ldots, m\}$ so that the cost $c(\pi) = d_{\pi(1)\pi(2)} + d_{\pi(2)\pi(3)} + \cdots + d_{\pi(m)\pi(1)} \leq k$.

Well, we have now three versions of the same optimization problem. Are they computationally equivalent? Is it that if one of them is in some complexity class, the other is necessarily in the same class? Can we decide these questions, at least, for the problems in *NP*?

To make the things simpler, let us constrain our optimization problems to be *NP*-like. An *NP-like Optimization problem* is an optimization problem satisfying all of the following conditions:

Each instance of the problem is recognizable in polynomial time.
The cost of a solution can be computed in polynomial time.
A suggested solution to a given instance can be verified to be a solution or not, in polynomial time.

Suppose *NPO* is an *NP*-like optimization problem. Let us write the construction, evaluation, and the decision version of *NPO* as *NPOC, NPOE,* and *NPOD*, respectively. If *NPOC* has a solution, the cost of such a solution can be computed in polynomial time. This would solve *NPOE* with an extra overhead of a polynomial time only. Similarly, if we have solved *NPOE*, then given any instance of *NPO*, we know how to get the cost of an optimum solution. Now, a mere comparison of this cost with k solves the decision problem *NPOD*.

Conversely, assume, without loss of generality, that *NPO* is a minimization problem. Being a meaningful minimization problem, we have an upper bound C for the cost of any optimum solution. For example, in TSP, the cost of an arbitrary tour provides an upper bound for the minimum cost. Now, between 0 and C, we can use binary search for k and solve the evaluation problem *NPOE*. It is like finding out the number between 0 and 100 that your friend has thought by asking him questions such as "Is it less than 50?"; if yes, "Is it less than 25?", and so on. This takes only $\log k$ time solving the *NPOD*.

Writing an arrow $\longrightarrow$ for the phrase "is no harder to solve than," we can summarize our discussion as follows.

$$NPOD \longrightarrow NPOE \longrightarrow NPOC, \quad NPOE \longrightarrow NPOD.$$

Here, "no harder to solve than," in fact, means "modulo polynomial time." For example, the first arrow above says that if we have a solution for *NPOD*, then with perhaps an additional polynomial time, we will be able to solve *NPOE*.

It is an open problem whether *NPOC* $\longrightarrow$ *NPOE*, that is, whether there exists an optimization problem whose construction version is strictly harder to solve than the evaluation version. A partial answer to this question is available. It says that if *NPOD* is *NP*-complete, then *NPOC* $\longrightarrow$ *NPOE*. As a consequence, we see that if *NPOD* is *NP*-complete, then so are *NPOE* and *NPOC*. In the summary section to this chapter, you will find pointers to references for similar results, and also for an almost complete list of interesting optimization problems.

So you see, there are many interesting problems that are *NP*-hard. For these problems we do not yet have a polynomial time algorithm. However, the problems are of such practical importance that we need to solve them; question is how? The obvious choice is to use an algorithm any way, even if it takes more than a polynomial time. The risk is that we may not get a solution during our life time!

One common approach is to look for special subclasses of such a problem; look for patterns, etc. For example, if a graph happens to be a tree, then most *NP*-complete problems in this special case can be solved in polynomial time. The vertex cover problem for perfect graphs is in *P*. Depending on your intuition on the problem at hand, you may look for special graphs such as bipartite graphs or regular graphs, etc.

Another approach is to use a near optimal algorithm. This approach suits best for the optimization problems such as MAXSAT or TSP, where all that we require is an approximate solution to the problem, and the time to get such an approximate solution is comparably affordable. When an *NP*-complete problem admits of such an approximation, we would focus on a polynomial time approximate algorithm. In view of this, the problems can be seen to be of three types.

The first kind consists of those problems that do not admit of a polynomial time algorithm for an approximate solution.

The second type has those problems that admit of approximate solutions but the difference between an approximate solution and the optimum solution is always bigger than some quantity. Instead of the difference between an optimal solution and an approximate solution, one often considers the ratio. This approach is pursued for optimization problems, where some sort of cost is involved. In such a case, we say that an *approximate algorithm* (for a problem) is *k-optimal* if it always finds an approximate solution that does not produce a cost more than k times that of the optimal solution.

And the problems of the third type admit of approximate solutions with the difference between an approximate solution and an optimum solution being controllable. It is the third type of problems for which this approach of computing an approximate solution is truly applicable.

Another approach for *NP*-complete problems is to follow *good algorithmic practices* such as branch-and-bound, backtracking, and divide-and-conquer for constructing algorithms that may be exponential in the worst case, but work well in many practical instances. For example, Davis–Putnam algorithm for SAT uses backtracking, and there are many branch-and-bound algorithms for solving CLIQUE, divide-and-conquer algorithms for TSP.

Especially for optimization problems, a common approach is to use the so-called *local improvements*. In this strategy, we obtain an approximate solution by following another method, or just guess it. Next, we change some values of some of the parameters slightly. If the new solution is an improvement over the old, then we adopt it and try to improve further by changing the same parameter. Else, we use the old approximation, and change values of another parameter, and repeat the process. For example, the sophisticated and the best known heuristic algorithm of Lin–Kernighan for TSP uses local improvement.

A twist on local improvements is the so-called method of *simulated annealing*. Here, we allow occasional changes in the opposite direction towards an optimal solution, as it often happens in the physics of cooling solids. The *genetic algorithms, neural networks*, and some other related methods follow this approach. It has been found that for many practical problems, these biology-motivated heuristic algorithms often perform better than other algorithms. In contrast, these methods are exponential, in the worst case, and that to, they give only an approximate solution. But they work! And why do they work is perhaps worth pursuing.

Problems for Section 10.9

10.81. Formulate the construction problem, the evaluation problem, and the decision problem corresponding to MAXSAT.

10.82. Formulate an optimization problem version of the independent set problem. Next, formulate the three versions ISC, ISE, and ISD.

10.83. The *dominating set problem* is stated as: Given a directed graph G and an integer k, does there exist a subset S of k vertices from G such that for each vertex u of G not in S, there is a vertex $v \in S$ so that G has an edge from v to u? This problem is NP-complete. Show that there is a polynomial time algorithm for solving the dominating set problem when G is a tree.

10.10 Summary and Additional Problems

In this chapter, we have tried to measure the performance of algorithms (total TMs) in solving problems with respect to how much space or time they require. To be on the safe side, we considered the worst-case performance only. Moreover, our concern was further restricted to analyzing the performance with respect to the length of an input string. We have introduced the order notations and then classified the solvable problems according to the resources such as space and time that they need.

In discussing space complexity, we have seen that certain problems can be solved in logarithmic space and certain others cannot be. Though the problem whether $LS = NLS$ remains open, we argued that if any one of certain problems can be shown to be in LS, then $LS = NLS$ would be proved. Such problems were named as NLS-complete problems.

Similar to space complexity, we have come across the NP-complete problems in discussing time complexity. If we would be able to show any one of the NP-complete problems to be in P, then NP would become equal to P. The analogy between space complexity and time complexity breaks down when we consider the complementary problems. We have seen that $CoNLS = NLS$, but we do not know whether $CoNP = NP$. We had shown that the satisfiability of boolean formulas is an NP-complete problem, and then mentioned many other such problems. Similar to map reductions in solvability, log space reductions and polynomial time reductions were the main tools we used in proving NLS-hardness and NP-hardness.

Though Hopcroft and Ullman [58] attribute the original idea of the CYK algorithm to J. Cocke, the algorithm first appeared in Kasami [63] and independently, in Younger [135]. Improvements over the CYK algorithm for context-free recognition can be found in [28, 133].

For subrecursive concepts about hardness, completeness, and arithmetic hierarchy, see [23, 62, 76, 128]. The texts [20, 36, 53, 72, 92, 98] provide good introduction to the theory of complexity of algorithms along with proofs that many interesting problems are NP-complete. Many ways of tackling NP-complete problems are the major themes addressed in [55, 99, 108]. You may like to see [5] and the references there for approximate algorithms of optimization problems. You may also like to see the forthcoming book by S. Arora and B. Barak on Computational Complexity, a draft of which is available on the net.

In some of the exercises, you will find mention of information theoretic complexity theory. This exciting notion resulted from the works of Shannon and Chaitin; see [13] and the references therein. This also includes Chaitin's theorem. For Parikh's sentence mentioned in Problem 10.146 below, see [101]. The fact that primality testing can be done in polynomial time is in [3]. We have not discussed an important issue, the complexity of boolean circuits. You can see [46, 122] for an introduction to the topic.

Additional Problems for Chapter 10

10.84. Show that any computation that takes $O(T(n))$ time on a single-tape off-line TM can be performed on a standard TM in time $O(T(n))$.

10.85. Show that any computation that can be performed on a standard TM in time $O(T(n))$ can also be performed in time $O(T(n))$ on a TM with a left-end on its tape.

10.86. Write $\psi(G)$ for an encoding of a CFG as a binary string, where we also denote by $\psi(X)$ the binary encoding of any symbol X involved in G. Define two languages as in the following:

$L = \{\psi(G_1)\#\,\psi(G_2) : G_1, G_2$ are CFGs with $\mathcal{L}(G_1) = \mathcal{L}(G_2)\}$.

$L' = \{\psi(G)\#\,\psi(A)\#\,\psi(B) : G$ is a CFG and A, B are nonterminals of G, and the terminal strings derived from A and B are the same$\}$.

Answer the following with justification:
(a) Is L' polynomial time reducible to L?
(b) Is L polynomial time reducible to L'?
(c) What about the NP-completeness of L and that of L'?

10.87. Prove that the *state minimization algorithm* for DFAs as given below runs on polynomial time. (Is it the same as we have described in Sect. 4.5?)

State Minimization Algorithm:

On input as a DFA $M = (Q, \Sigma, \delta, s, F)$, do the following:
  Remove all states from Q that are not reachable from s.
  Construct the undirected graph G as follows:
  Take Q as the set of vertices of G.
  Join each state in F to each state in $Q - F$.
  Add additional edges as follows:
    Repeat until no new edges are added to G:
      For each distinct pair of states $p, q \in M$ and each $\sigma \in \Sigma$,
      if there is an edge joining $\delta(p, \sigma)$ and $\delta(q, \sigma)$,
        then join p and q by an edge.
  For each state $q \in Q$, form the set
    $[q] = \{r \in Q$: no edge joins q and r in $G\}$.
    Take $Q' = \{[q] : q \in Q\}$, $s' = [s]$, $F' = \{[q] : q \in F\}$, and
    $\delta'([q], \sigma) = [\delta(q, \sigma)]$, for each $q \in Q$ and $\sigma \in \Sigma$.
  Output the DFA $M' = (Q', \Sigma, \delta', s', F')$.

10.88. The CYK *algorithm* for checking membership of a string in a CFL depends upon the following considerations:
You have a CFG G in CNF. You also have a string of terminals $w = a_1 a_2 \cdots a_n$. Define substrings $w_{i,j} = a_i \cdots a_j$ and subsets $N_{i,j} = \{A : A$ is a nonterminal with $A \Rightarrow w_{i,j}\}$. Show that
(a) $A \in N_{i,i}$ iff $A \mapsto a$ is a production of G. Thus, $N_{i,i}$ can be computed by looking at w and the productions of G.
(b) For $j > i$, $A \Rightarrow w_{i,j}$ iff there is a production $A \mapsto BC$, and $B \Rightarrow w_{i,k}$, $C \Rightarrow w_{k+1,j}$, for some k with $i \le k < j$.
(c) $N_{i,j} = \cup_{i \le k < j}\{A : A \mapsto BC$, for some $B \in N_{i,k}$, $C \in N_{k+1,j}\}$.
(d) The subsets $N_{1,1}, \ldots, N_{n,n}, N_{1,2}, \ldots, N_{n-1,n}, N_{1,3}, \ldots, N_{n-2,n}, \ldots, N_{1,n}$ can thus be computed.
(e) $w \in \mathcal{L}(G)$ iff $S \in N_{1,n}$.
Write the CYK algorithm. It is so named after the three mathematicians who devised it; J. Cocke, D.H. Younger, and T. Kasami. It is also called *CKY algorithm*.

10.89. Apply CYK algorithm to determine whether the string *aabbb* is generated by the CFG with productions $S \mapsto AB$, $A \mapsto a|BB$, $B \mapsto b|AB$. What are the intermediary subsets $N_{i,j}$ for $1 \le i \le j \le 5$?

10.90. Show that the CYK algorithm takes an $O(n^3)$ time, where $n = \ell(w)$.

10.91. Show how the CYK algorithm can be converted to a parsing method by keeping track of how the subsets $N_{i,j}$ are computed. Then, find a parsing of *aab* using the productions $S \mapsto AB$, $A \mapsto a|BB$, $B \mapsto b|AB$.

10.92. Modify the CYK algorithm to output the number of distinct parse trees for the given input, rather than just reporting membership in the language.

10.93. A nondeterministic algorithm for determining membership in a CFL, given its CFG, proceeds by guessing which sequence of productions is applied in the derivation of the string. Show that if the CFG contains no ε-production, then this is an $O(n)$ time nondeterministic algorithm. What happens if the CFG contains an ε-production?

10.94. It is obvious now that each regular language is in $DT(n)$; each CFL is in $DT(n^3)$, and that each CFL is in $NT(n)$. Show that if L is a context-sensitive language, then there is an m such that $L \in DT(n^m)$. Does it follow that there is m such that each context-sensitive language is in $DT(n^m)$?

10.95. Show that there are languages that are not in $NT(O(2^{n^k}))$, for any $k \in \mathbb{N}$. Is any such language decidable?

10.96. A disjunctive clause is called a *Horn clause* if at most one propositional variable in it is unnegated. Show that SAT restricted to horn clauses can be solved in polynomial time.

10.97. A *Horn clause* can be defined as a proposition of the form $p_1 \wedge p_2 \wedge \cdots p_n \to \bot$, where $\bot$ denotes falsity, and $p_1, \ldots, p_n$ are propositional variables. Show that the emptiness problem for CFLs, that is, deciding whether for a given CFG G, $\mathcal{L}(G) = \emptyset$, reduces to the satisfiability problem for finite conjunctions of Horn clauses, and vice versa. From this reduction, what do you conclude about the emptiness problem of CFLs?

10.98. Given an undirected graph G, show that the problem of "determining whether G is a tree" is in $DS(\log n)$.

10.99. Prove that $NLS = CoNLS$.

10.100. Give an example of an NLS-complete context-free language.

10.101. Show that the problem of determining whether two regular expressions represent the same language is in DPS.

10.102. An undirected graph is *bipartite* if its vertices can be partitioned into two sets so that each edge joins some vertex in one set to some vertex in the other. Show that the problem of "determining whether a given undirected graph is bipartite" is in NLS.

10.103. A directed graph is *strongly connected* if from any vertex to any other vertex there is a directed path in the graph. Show that the problem of determining whether a directed graph is strongly connected is NLS-complete.

10.104. For a cnf with c clauses and m propositional variables, show that an NFA with number of states of the order $O(cm)$ can be constructed in polynomial time that accepts all nonsatisfying interpretations, represented as binary strings of length m. Conclude that minimization of NFAs cannot be realized in polynomial time unless $P = NP$.

10.105. Using diagonalization, show that the following languages are NP-complete:
(a) $\{\psi(M)\#\,\psi(w)\#\,0^k : M$ is an NTM that accepts w in k or fewer steps$\}$.
(b) $\{\psi(M)\#\,\psi(w) : M$ is an NTM that accepts w in $3\ell(w)$ or fewer steps$\}$.

10.106. Show that the language $L = \{\psi(M)\#\,0^k : M = (Q, \Sigma, \Gamma, \delta, s, h, \hbar)$ is a DTM and there is a $w \in \Sigma^*$ such that for every $q \in Q$, there is an $m \leq k$ for which $s\underline{b}\,w \rightsquigarrow_M qu\underline{\sigma}v$ in m steps, for some $u, v \in \Sigma^*, \sigma \in \Sigma\}$ is NP-complete.

10.107. A homomorphism h is called a *nonerasing homomorphism* if it maps no symbol to ε. Attempt to answer whether the following classes of languages are closed under nonerasing homomorphisms, and under general homomorphisms:
(a) P.
(b) NP.
(c) The class of decidable languages.
(d) The class of computably enumerable languages.

10.108. Show that P is closed under nonerasing homomorphisms iff $P = NP$.

10.109. Show the following under the assumption that $P = NP$:
(a) There exists a polynomial time algorithm that constructs a model of a given boolean formula, provided that the formula is satisfiable.
(b) There exists a polynomial time algorithm for finding a largest clique in a given undirected graph.
(c) Any integer can be factored into primes in polynomial time.

10.110. In 3COLOR problem, we are given an undirected graph, and are asked to determine whether its vertices can be colored with three colors such that no two adjacent vertices have the same color. Show that 3COLOR is NP-complete.

10.111. Let USUBSUM be the subset sum problem when the numbers in the set A are given in unary notation instead of binary. Show that USUBSUM is in P.

10.112. Define the *directed Hamiltonian path problem* as "Given a directed graph, does there exist a directed Hamiltonian path in the graph?". Show that there is a log space reduction of Hamiltonian path problem to the directed Hamiltonian path problem.

10.113. Define the *fixed vertices Hamiltonian path problem* as "Given an undirected graph and two vertices in it, does there exist a Hamiltonian path joining the two vertices?". Show that there is a polynomial time reduction of the Hamiltonian path problem to the fixed vertices Hamiltonian path problem.

10.114. A *polynomial Turing reduction* from L to L' is a two-tape TM M with three distinguished states q_1, q_2, q_d, where the yield relation with configurations having state components q_1, q_2 is defined the usual way. But, for q_d, it is defined differently. From state q_d, M does not change anything on its input tapes; it just goes to state q_1 or to q_2 according as the string on its second tape is in L or not. This is considered one move of M. Further M is assumed to decide L' in a polynomial time. Prove the following:

(a) If there is a polynomial Turing reduction from L_1 to L_2, and one from L_2 to L_3, then there is a polynomial Turing reduction from L_1 to L_3.
(b) Suppose that there is a polynomial Turing reduction from L_1 to L_2. If $L_2 \in P$, then $L_1 \in P$.
(c) There is a polynomial Turing reduction from the Hamiltonian cycle problem to the Hamiltonian path problem.

10.115. Let ModExp be the problem of determining whether $a^b \equiv c \mod p$, given the positive integers a, b, c, p in binary. Show that ModExp is in P.

10.116. Let G be an undirected graph. A simple path in G is a path where no vertex is repeated. Write $\rho(k)$ for the binary representation of a natural number k. Define the following two problems:

$$\text{MPATH} = \{\psi(G)\#\,\psi(s)\#\,\psi(t)\#\,\rho(k) : G \text{ contains a path of length at most } k \text{ from } s \text{ to } t\}$$

$$\text{LPATH} = \{\psi(G)\#\,\psi(s)\#\,\psi(t)\#\,\rho(k) : G \text{ contains a path of length at least } k \text{ from } s \text{ to } t\}.$$

Show that MPath is in P and LPath is NP-complete.

10.117. Recall that a permutation π on the set $\{1, 2, \ldots, n\}$ is a bijection on the set. For any $k \in \mathbb{N}$, π^k is the permutation obtained by composing π with itself k times; with π^0 as the identity map. The *problem of permutation power* is to determine whether $\pi^k = \tau$, given the permutations π, τ, and the binary representation of $k \in \mathbb{N}$. Show that the problem of permutation power is in P.

10.118. Let $f(n)$ be a computable total function. Prove the following *hierarchy theorems*:

(a) There is a decidable language L not in $DS(f(n))$.
(b) There is a decidable language L not in $DT(f(n))$.
(c) There is a computable total function $g(n)$ such that $DS(f(n)) \subsetneqq DS(g(n))$.
(d) There is a computable total function $g(n)$ such that $DT(f(n)) \subsetneqq DT(g(n))$.
(e) If $f(n) \geq \log n$ and L is accepted by an $f(n)$ space bounded TM, then L is accepted by an $f(n)$ space bounded total TM.
(f) Suppose there is a TM that uses $g(n)$ squares for some input of length n, whatever be this n. If $f(n) \geq \log n$, $g(n) \geq \log n$, and $\lim_{n\to\infty}(f(n)/g(n)) = 0$, then there is a language in $DS(g(n))$ but not in $DS(f(n))$.
(g) Suppose there is a TM that takes exactly $g(n)$ time on each input of length n, whatever be this n. If $\lim_{n\to\infty}[f(n)\log(f(n))/g(n)] = 0$, then there is a language in $DT(f(n))$ but not in $DT(g(n))$.
(h) If $L \in DT(f(n))$, then $L \in DS(f(n))$.
(i) Suppose there is a TM that takes $f(n)$ time on each input of length n. Then $DT(f(n)\log f(n)) \subseteq DS(f(n))$.
(j) If $L \in DS(f(n))$ and $f(n) \geq \log n$, then there is a constant $c > 0$ (possibly depending on L) such that $L \in DT(c^{f(n)})$.

(k) If $L \in NT(f(n))$, then there is a constant $c > 0$ (possibly depending on L) such that $L \in DT(c^{f(n)})$.
(l) Suppose there are TMs M, M' such that on any input of length n, M uses exactly $g(n)$ squares and M' uses exactly $h(n)$ squares. If $f(n) \geq n$ and $h(n) \geq n$, then $DS(g(n)) \subseteq DS(h(n))$ implies $DS(g(f(n))) \subseteq DS(h(f(n)))$.
(m) Suppose there are TMs M, M' such that on any input of length n, M uses exactly $g(n)$ squares and M' uses exactly $h(n)$ squares. If $f(n) \geq n$ and $h(n) \geq n$, then $NS(g(n)) \subseteq NS(h(n))$ implies $NS(g(f(n))) \subseteq NS(h(f(n)))$.
(n) Suppose there are TMs M, M' such that on any input of length n, M runs exactly for $g(n)$ time and M' runs exactly for $h(n)$ time. If $f(n) \geq n$ and $h(n) \geq n$, then $DT(g(n)) \subseteq DT(h(n))$ implies $DT(g(f(n))) \subseteq DT(h(f(n)))$.
(o) Suppose there are TMs M, M' such that on any input of length n, M runs exactly for $g(n)$ time and M' runs exactly for $h(n)$ time. If $f(n) \geq n$ and $h(n) \geq n$, then $NT(g(n)) \subseteq NT(h(n))$ implies $NT(g(f(n))) \subseteq NT(h(f(n)))$.
(p) Let $r \geq 0$ and $c > 0$. Then $NS(n^r) \subsetneqq NS(n^{r+c})$.

10.119. Let $s, t : \mathbb{N} \to \mathbb{N}$ be functions such that $s(n) \geq \log n$ and $t(n) \geq n \log n$. We say that $s(n)$ is *space constructable* if the function that maps 0^{n+1} to the binary representation of $s(n)$ is computable in $O(s(n))$ space. Similarly, $t(n)$ is *time constructable* if the function that maps 0^{n+1} to the binary representation of $t(n)$ is computable in $O(t(n))$ time. Prove the *hierarchy theorems* as summarized in the following:
(a) *Space hierarchy theorem*: Let $s(n)$ be a space constructable function. There is a language L such that L is decidable in $O(s(n))$ space but not in $o(s(n))$ space.
(b) *Time hierarchy theorem*: Let $t(n)$ be a time constructable function. There exists a language L such that L is decidable in $O(t(n))$ time but not in $o(t(n)/\log t(n))$ time.

10.120. Using the hierarchy theorems, show that
(a) $DT(2^n) \subsetneqq DT(n^2 2^n)$.
(b) $DT(2^n) \subsetneqq DT(n2^n)$.
(c) $DT(2^{2^n}) \subsetneqq DT(2^{2^n+n})$.
(d) Let $r > 0$ and $c \geq 1$. Show that $NS(r^n) \subsetneqq NS((r+c)^n)$.
(e) $NLS \subsetneqq DPS \subsetneqq EXPSPACE$. [Define $EXPSPACE$.]

10.121. *Borodin's gap theorem*: For each computable total function $f(n) \geq n$, there is a computable function $s(n)$ such that $DS(s(n)) = DS(g(s(n)))$. Along with this formulate and prove the other three gap theorems replacing DS by NS, DT, and NT.

10.122. Show that there is a computable function $f(n)$ such that

$$DS(f(n)) = NS(f(n)) = DT(f(n)) = NT(f(n)).$$

10.123. Let $f(n)$ be a computable total function. Show that there is a monotonically nondecreasing function $g(n)$ such that $g(n) \geq f(n)$, $g(n) \geq n^2$, and there is a TM that uses exactly $g(n)$ squares for each input of length n.

10.124. Prove *Blum's speed-up theorem*: Let $f(n)$ be a computable total function. There exists a decidable language L such that for any TM accepting L in space $s(n)$, there is a TM that accepts L in space $s'(n)$ so that $f(s'(n)) \le s(n)$ for all but a finite number of n's.

10.125. Prove the *Honesty theorem for space*: There exists a computable total function $f(n)$ such that for any space complexity class $\mathcal{C}$, there is a function $s(n)$ with $DS(s(n)) = \mathcal{C}$, and $s(n)$ is computable in $f(s(n))$ space.

10.126. The *bounded tiling problem* asks whether there exists a tiling of an $s \times s$ square, given a bounded tiling system with a positive integer s. A bounded tiling system prescribes the first row of tiles instead of the origin tile. It is formalized as follows. A bounded tiling system is a quintuple $B = (T, H, V, s, f_0)$, where $s \in \mathbb{Z}_+$ and $f_0 : \{0, 1, \ldots, s-1\} \to T$ is a function, and T is a finite set of tiles. The problem asks whether there exists a function $f : \{0, 1, \ldots, s-1\} \times \{0, 1, \ldots, s-1\} \to T$ such that

$$f(m, 0) = f_0(m) \text{ for all } m < s;$$
$$(f(m, n), f(m+1, n)) \in H \text{ for all } m < s-1, n < s;$$
$$(f(m, n), f(m, n+1)) \in V \text{ for all } m < s,\ n < s-1.$$

Such a tiling extends the function f_0 to f. Show that the bounded tiling problem is NP-complete.

10.127. Consider the *binary bounded tiling problem*, where in the bounded tiling problem we are not given the first row of tiles, but only an origin tile d_0, as in the original tiling problem. Further, the size s of the square to be tiled is given in binary. Prove the following:

(a) There is a reduction from the language $\{\psi(M)$: the TM M halts on the empty string within $2^{\ell(\psi(M))}$ steps$\}$ to the (language of) binary bounded tiling problem.
(b) The binary bounded tiling problem is not in P.
(c) Let $NEXP$ be the class of all languages decided by NTMs in time 2^{n^k} for some $k > 0$. Then the binary bounded tiling problem is in $NEXP$.
(d) All languages in $NEXP$ are polynomial time reducible to the binary bounded tiling problem. That is, the binary bounded tiling problem is $NEXP$-complete.

10.128. Show that MAX2SAT, that is, "Given a set of disjunctive clauses with at most two literals each, and an integer k, does there exist an interpretation that satisfies at least k of the clauses?" is NP-complete.

10.129. Recall that the growth rate of the function $n^{\log n}$ lies strictly between that of the polynomials and the exponentials. Suppose that there is an NP-complete problem that has a solution that takes an $O(n^{\log n})$ time by a DTM. What could you say about the running time of any problem in NP?

10.130. Describe complements of each of the problems below. Decide whether the problem is in NP or in $CoNP$. If the problem or its complement is NP-complete, then supply a proof.

(a) TRUESAT: Given a proposition that is satisfied by the interpretation that assigns each variable to 1, does there exist another model of the proposition?

(b) FALSESAT: Given a proposition that is falsified by the interpretation that assigns each variable to 0, does there exist another interpretation that falsifies the proposition?
(c) DOUBLESAT: Given a proposition, do there exist at least two interpretations that satisfy it?
(d) NEARVAL: Given a proposition, whether there is at most one interpretation that falsifies it?

10.131. Suppose that there is a bijection f on the set of n-bit integers such that $f(x)$ can be computed in polynomial time, while $f^{-1}(x)$ cannot be computed in polynomial time. Show that the language $\{(x, y) : f^{-1}(x) < y\}$ is in $(NP \cap CoNP) - P$.

10.132. A function $f : \mathbb{N} \to \mathbb{N}$ is called *subexponential* if for every $c > 0$, $f(n) = O(2^{n^c})$. Show that if some NP-complete language can be shown to be decided in subexponential time, then all NP-languages can be decided in subexponential time.

10.133. Define a problem or its corresponding language L to be *DPS-complete* if $L \in DPS$ and all languages in *DPS* are log space reducible to L. Let L be a *DPS*-complete problem. Prove the following:
(a) If $L \in P$, then $P = DPS$.
(b) If $L \in NP$, then $NP = DPS$.
What happens if we define *DPS*-completeness using polynomial time reduction instead of using log space reduction?

10.134. Prove that the following problems are *DPS*-complete:
(a) REGALL: Given a regular expression R over an alphabet Σ, is $\mathcal{L}(R) = \Sigma^*$?
(b) *Context sensitive recognition problem*: Given a context sensitive grammar G and a string w, whether $w \in \mathcal{L}(G)$?

10.135. Let $L_{\text{CFE}} = \{\psi(G) : G \text{ is a CFG and } \mathcal{L}(G) = \emptyset.\}$ Prove that every language in P is log space reducible to L_{CFE}. This says that the emptiness problem for CFGs is complete for P with respect to log space reduction. Note that we do not yet have a deterministic algorithm to recognize L_{CFE} in $(\log n)^k$ space.

10.136. Prove that the context-sensitive languages occupy the bottom of *DPS*. That is, if $L_{\text{CS}} = \{\psi(G)\#\,\psi(w) : G$ is a context-sensitive grammar and w is a string over the terminal alphabet of $G\}$, then show that L_{CS} is in $NS(O(n)) \cap DS(O(n^2))$.

10.137. *Graph reachability problem*: Given a directed graph with $v_1, v_2, \ldots, v_n$ as its vertices, does there exist a path from v_1 to v_n? Prove that every problem in $NS(O(n))$ is log space reducible to the graph reachability problem. Note that we do not yet have a deterministic algorithm to recognize the language of graph reachability problem in $\log n$ space.

10.138. Given an undirected graph G with two vertices s and t each of degree 1. There are two players called S and C, for Short, and Cut, respectively. Each selects a different vertex other than s, t, in each move alternately, which then belongs to the player for the rest of the game. S starts the game and wins if he is able to select a set of vertices that make up for a path between s and t. C wins if all the vertices have been selected but S could not win. The *Shannon Switching Game problem* is: Given an undirected graph G, can S win no matter what choices C makes? Show that this problem is *DPS*-complete.

10.139. Recall that propositions with only connectives as $\neg$, $\vee$, and $\wedge$ are called boolean formulas. An expression of the form $Q_1x_1Q_2x_2 \cdots Q_nx_nE$ is called a *quantified boolean formula* when E is a boolean formula with the propositional variables $x_1, \ldots, x_n$, and any Q_i is either $\forall$ or $\exists$. When Q_ix_i occurs in a quantified boolean formula F, we say that in F, the variable x_i has been quantified. Further, in a quantified boolean formula each variable is quantified exactly once. (In fact, we are considering a restricted class of quantified boolean formulas here.) The meanings of the quantifiers $\forall$ and $\exists$ are as follows. The expression $\forall x E$ means that E is evaluated to 1 (true) when all occurrences of x are assigned to 1, and also, when all occurrences of x are assigned to 0 (false). Similarly, the expression $\exists x E$ means that E is true when all occurrences of x are assigned to 1, or when all occurrences of x are assigned to 0. Thus the meaning of a quantified boolean formula is a statement about when the unquantified expression in it is true or not. We say that the quantified boolean formula is true or false, according as the statement that expresses its meaning is true or false. The *quantified boolean formula problem*, denoted by QBF, is: Given a quantified boolean formula is it true? Prove the following:
(a) QBF is in *DPS*.
(b) QBF is *DPS*-complete.

10.140. Prove that the following problems are *NP*-complete by using the suggested reductions:
(a) Reduce SAT to MAXSAT.
(b) Reduce SAT to IS, the independent set problem.
(c) Reduce 3SAT to DSAT, also called the *double-sat,* the problem of determining whether a given cnf has at least two models.
(d) Reduce 3SAT to CN, the chromatic number problem.
(e) Reduce 3SAT to SUBSUM, the subset sum problem.
(f) Reduce IS to CLIQUE.
(g) Reduce IS to VC, the vertex cover problem.
(h) Reduce VC to VCWT, the vertex cover problem for graphs without triangles.
(i) Reduce VCWT to CC, the *clique cover problem*: Given an undirected graph G and an integer $k \geq 1$, does there exist k cliques in G such that each vertex of G is in at least one of the k cliques?
(j) Reduce SAT to the problem of inequivalence of $*$-free regular expressions, which asks for determining whether two given $*$-free regular expressions represent two different languages.
(k) Reduce SAT to EC, the exact cover problem.
(l) Reduce EC to DHP, the Hamiltonian path problem in directed graphs.
(m) Reduce EC to KNAPSACK.
(n) Reduce KNAPSACK to PARTITION.
(o) Reduce DHP to HP, the Hamiltonian path problem in undirected graphs.

(p) Reduce HP to TSP.
(q) Reduce HP to ST, the *spanning tree problem*: Given an undirected graph G and a tree T with the same number of vertices, is T isomorphic to some spanning tree of G?

10.141. Prove that the Hamiltonian path problem (HP) is NP-complete even when restricted to planar graphs. Note that NP-completeness of HP also holds for four-connected graphs.

10.142. Consider the traveling salesperson problem, where the distances satisfy the triangle inequality, that is, $d_{ij} + d_{jk} \geq d_{ik}$, for all possible cities i, j, k. Give a polynomial time algorithm to find a tour whose cost is within twice the minimum cost.

10.143. Prove that if there exists a polynomial time algorithm to find a tour whose cost is within twice the minimum cost, in an arbitrary TSP, then $P = NP$.

10.144. Let $DP = \{L - L' : L, L' \in NP\}$. This is the first language class in the so-called called *difference hierarchy*. A DP-complete language is a language in DP to which every other language in DP is polynomial time reducible. Let $\wp(j)$ denote the binary representation of $j \in \mathbb{N}$. Prove that the following languages are DP-complete:
(a) $\{\psi(G)\#\ \wp(m)\#\ \psi(G')\#\ \wp(n) : G$ has an m-clique and G' has no n-clique$\}$.
(b) $\{\psi(G)\#\ \wp(k) :$ the largest clique in G is of size $k\}$.

10.145. Is it possible that the following questions are undecidable:
(a) $P \subseteq DS((\log n)^k)$ for some $k \geq 1$?
(b) $NS(O(n)) = DS(O(n))$?
(c) Is it that each context-sensitive language is deterministic context-sensitive?
(d) $P = NP$?

10.146. We have not discussed the information theoretic complexity. An elegant result in this context is *Chaitin's theorem*, which is often stated as "There is no algorithm to determine whether a proof of a theorem is the shortest." Here, I ask you to supply an informal proof of another version of Chaitin's theorem, that is, "There is no algorithm to determine whether an algorithm is the shortest among all algorithms that do the same job as the given one." (Compare it with Problem 9.125.) Here is a cute idea:

The set of all TMs is computably enumerable. Thus, there is an algorithm that enumerates all algorithms. Use $\psi(A)$ to write the binary representation of an algorithm A. A *cute algorithm* is an algorithm A such that among all algorithms doing the same job as A, $\ell(\psi(A))$ is the shortest. Suppose that there is an algorithm B that tests cuteness of all algorithms. That is, given $\psi(A)$ as an input to B, it determines whether A is cute or not. Define an algorithm C as in the following:

On input as a binary number n,
Enumerate all algorithms X such that $\ell(\psi(X)) > n$.
Run B on the enumerated algorithms until a cute algorithm D is found.
Run D.

(a) What happens if C is given the input $\ell(\psi(C)) + 1$?
(b) Conclude that there is no algorithm for testing cuteness.

Notice that the enumeration used in C is not done at a time as there are infinitely many algorithms. The enumeration gives one algorithm, the next steps in C are performed, and then the enumeration gives the next algorithm, and so on. You can try to give an informal proof of the first version of Chaitin's theorem using a theorem generator instead of an algorithm enumerator. However, this informal proof is analogous to proving Gödel's incompleteness theorem by using the sentence "This sentence is not provable," or using the sentence "This sentence has no short proof" to show that there exists a sentence in the system of natural numbers having a long proof but the fact that it is provable has a short proof (*Parikh's sentence*).

Answers and Hints to Selected Problems

Problems of Chapter 1

1.2 For each $a \in A$, there may not be one $b \in A$ such that aRb. For example, the relation $\{(b, b)\}$ on the set $\{a, b\}$.

1.4 $R \subseteq A^2$. So, $R \in 2^{A^2}$. Number of such R is 2^{n^2}. Reflexive R's are 2^{n^2-n} in number. And $2^{n(n+1)/2}$ symmetric R's.

1.5 $xR^{-1}y$ iff yRx. Hence they have the same equivalence classes.

1.6 Take $f(x)$ as the number of elements in x.

1.7 $B = \{[x] : x \in A\}$. Define $f : A \to B$ by $f(x) = [x]$.

1.9 $\cup\mathcal{A} = \emptyset$. Use De Morgan's law to get $\cap\mathcal{A} = A$.

1.12 $f(x) = f(y)$ implies $x = g(f(x)) = g(f(y)) = y$. Hence f is injective. Now, $f(g(y)) = y$ shows that $f(x) = y$ for $x = g(y)$. Hence f is surjective.

1.13 (d) $g \circ f$ injective implies f is injective; g need not be injective.
(e) $g \circ f$ surjective implies g is surjective; f need not be surjective.

1.15 Use Problem 1.12.

1.24 Suppose the collection of all sets is a set; denote it by S. Each element of 2^S is a set, and hence is in S. That is, $2^S \subseteq S$. Then $|2^S| \le |S|$. This contradicts Cantor's theorem.

1.28 Let P_n be the set of all polynomials of degree n with rational coefficients. Define $f : P_n \to \mathbb{Q}^{n+1}$ by $f(a_0 + a_1x + \cdots + a_nx^n) = (a_0, a_1, \ldots, a_n)$. This f is injective. Hence P_n is countable. The set of all polynomials with rational coefficients is $\cup_{n\in\mathbb{Z}_+} P_n$, and hence, is countable. Each such polynomial is in some P_n and has n roots. Therefore, all roots of such polynomials are countable in number.

1.33 $\cup_{r<1}\{x \in \mathbb{R} : a - r \le x \le a + r\} = \{x \in \mathbb{R} : a - 1 < x < a + 1\}$ and $\cap_{r>0}\{x \in \mathbb{R} : a - r < x < a + r\} = \{a\}$.

1.36 Yes.

1.39 Let the $n+2$ positive integers be $x_1 > x_2 > \ldots > x_{n+2}$. Consider the set $A = \{x_1 + x_k, x_1 - x_k : k = 2, 3, \ldots, n+2\}$. By the pigeon hole principle, at least two of them have the same remainder when divided by $2n$. If they are $y_i, y_j \in A$, then consider the difference $y_j - y_i$. This is in the form $\pm(x_i \pm x_j)$ and is divisible by $2n$.

1.41 Write the $n+1$ numbers in the form $x_k = 2^{q_k} y_k$, where y_k is odd. How many y_k's are there at the most?

1.42 For $x \in \mathbb{R}$, let $\lfloor x \rfloor$ denote the largest integer less than or equal to x. Consider the numbers $\sqrt{2} - \lfloor\sqrt{2}\rfloor$, $2\sqrt{2} - \lfloor 2\sqrt{2}\rfloor$, $\ldots$, $(n+1)\sqrt{2} - \lfloor(n+1)\sqrt{2}\rfloor$. All these are between 0 and 1. Divide the 0–1 interval into n subintervals each of length $1/n$. By pigeon hole principle, at least two of these numbers lie in the same subinterval. Their difference is of the form $m + n\sqrt{2}$ and it lies between 0 and $1/n$.

Problems of Chapter 2

Section 2.2

2.4 $bbbbaa$, $baaaaabaaaab \notin L^*$.

2.7 (d) Yes. (i) Yes. (j) No.

2.13 (c) No; e.g., $(L^*)^* = L^* \neq L$, in general. (d) Yes, $\{\varepsilon\}$. (e) Yes.

Section 2.3

2.17 (a) $a(aa)^*$. (b) $(aa)^* \cup (aaa)^* \cup (aaaaa)$. (c) $aa(aaa)^*$.

2.18 $a^*b(a \cup b)^*$.

2.19 (c) $(0 \cup 1)^*00(0 \cup 1)^*$. (d) $(1 \cup 01)^*(0 \cup \varepsilon)$. (e) $1^* \cup (1 \cup 011)^*01^*$.
(g) $00000^*(\varepsilon \cup 1 \cup 11 \cup 111)$. (i) $(\varepsilon \cup 0 \cup 00 \cup 0^3)1^* \cup 0^*1^41^* \cup (0 \cup 1)^*10(0 \cup 1)^*$.
(l) Split into three cases: $m \geq 1, n \geq 2$; $m \geq 2, n \geq 2$; and $m \geq 3, n = 1$.
(o) Enumerate all cases with $\ell(u) = 2$. (s) $(1^*01^*01^*)^* \cup 1^*$.

2.20 (a) $\{a, aa, aaa, \ldots, b, ab, aab, \ldots\}$. (b) $\{a, bb, aa, abb, ba, bbb, \ldots\}$.
(c) $\{u00v : u, v \in \{0,1\}^*\}$. (d) $\{a^{2m}b^{2n+1} : m, n \in \mathbb{N}\}$.
(e) All strings having no pair of consecutive zeros.

2.21 (f) Generate all permutations of the three symbols, giving six terms that can be $\cup$-ed together; e.g., one of these six is $(a \cup b \cup c)^*a(a \cup b \cup c)^*b(a \cup b \cup c)^*c(a \cup b \cup c)^*$.
(g) $((a \cup b \cup c)^3)^*$.

2.22 (c) Yes.

2.24 E has a starred sub-expression.

Section 2.4

2.29 (a) $S \mapsto abS|\varepsilon$. (b) $S \mapsto aaS|aa$. (c) $S \mapsto a|aS|bS$. (d) $S \mapsto bA$, $A \mapsto aA|bA|\varepsilon$.

(e) $S \mapsto aS|aA,\ A \mapsto bA|bB,\ B \mapsto c|cB$. (f) $S \mapsto aaS|A|aB,\ A \mapsto \varepsilon|bbA,\ B \mapsto c|cB$. (h) $S \mapsto A|B$. A generates the first language and B generates the second.

2.30 (d) See solution to Problem 2.21(f) in Sect. 2.3.
(e) $S \mapsto \varepsilon|aA|bA|cA,\ A \mapsto aB|bB|cB,\ B \mapsto aS|bS|cS$.

Section 2.5

2.34 The DFA should accept $\Sigma^* abc \Sigma^*$.

2.36 Initial state s, $F = \{t\}$, $\delta(s, a) = p$, $\delta(s, b) = u$, $\delta(p, a) = q$, $\delta(p, b) = p$, $\delta(q, a) = r$, $\delta(q, b) = p$, $\delta(t, a) = t$, $\delta(t, b) = r$, $\delta(u, a) = q_5$, $\delta(u, b) = u$.

2.37 (c) Initial state q_0, $F = \{q_8\}$, and Δ has the transitions:
$(q_0, a, q_1), (q_0, b, q_9), (q_i, b, q_{i+1})$, for $i = 1\,(1)\,7$; (q_i, a, q_9), for $i = 1\,(1)\,5$; $(q_6, a, q_6), (q_7, a, q_6), (q_6, b, q_7), (q_8, b, q_8), (q_8, a, q_6), (q_9, a, q_9), (q_9, b, q_9)$.
(e) Initial and final state q_0, and Δ has the transitions:
$(q_0, a, q_1), (q_0, b, q_1), (q_0, a, q_2), (q_0, b, q_2), (q_2, a, q_0), (q_2, b, q_0)$.
(h) Initial state A, $F = \{D, G, H\}$ and the transitions are:
$\delta(A, a) = D, \delta(A, b) = B, \delta(B, a) = E, \delta(B, b) = C, \delta(C, a) = F, \delta(C, b) = A$, $\delta(D, a) = G, \delta(D, b) = E, \delta(E, a) = H, \delta(E, b) = F, \delta(F, a) = I, \delta(F, b) = D$, $\delta(G, a) = A, \delta(G, b) = H, \delta(H, a) = B, \delta(H, b) = I, \delta(I, a) = C, \delta(I, b) = G$.

2.38 (a) Initial state q_0, $F = \{q_0, q_3\}$, and the transitions are:
$\delta(q_0, 0) = q_1, \delta(q_0, 1) = q_0, \delta(q_1, 0) = q_2, \delta(q_1, 1) = q_0, \delta(q_2, 0) = q_4$, $\delta(q_2, 1) = q_3, \delta(q_3, 0) = q_1, \delta(q_3, 1) = q_0, \delta(q_4, 0) = q_4, \delta(q_4, 1) = q_4$.

Section 2.6

2.46 (c) Initial state q_0, $F = \{q_0, q_1, q_2, q_3\}$, and the transitions are:
$\delta(q_i, a) = q_{i+1}, \delta(q_i, b) = q_i$, for $i = 0, 1, 2, 3$, and $\delta(q_4, a) = \delta(q_4, b) = q_4$.
(d) Initial state q_0, $F = \{q_2\}$, $\delta(q_0, a) = q_1$, $\delta(q_0, b) = q_3$, $\delta(q_1, a) = q_3$, $\delta(q_1, b) = q_2$, $\delta(q_2, a) = q_2$, $\delta(q_2, b) = q_2$, $\delta(q_3, a) = q_3$, $\delta(q_3, b) = q_3$.
(e) Initial state p, $F = \{p, q, r\}$, $\delta(p, a) = q$, $\delta(p, b) = p$, $\delta(q, a) = r$, $\delta(q, b) = p$, $\delta(r, a) = r$, $\delta(r, b) = t$, $\delta(t, a) = t$, $\delta(t, b) = t$.
(m) Initial state s, $F = \{r\}$, $\delta(s, a) = r$, $\delta(s, b) = p$, $\delta(p, a) = q$, $\delta(p, b) = s$, $\delta(q, a) = p$, $\delta(q, b) = r$, $\delta(r, a) = s$, $\delta(r, b) = q$.
(n) Initial and final state s, $\Delta = \{(s, a, r), (s, b, p), (p, a, q), (p, b, s), (q, a, p), (q, b, r), (r, a, s), (r, b, q)\}$.

2.47 Drive all final states to a single new final state by an epsilon transition. Old final states are now nonfinal. Not possible for DFAs, e.g., try for $\{\varepsilon, a\}$.

Additional Problems for Chapter 2

2.52 $(111 \cup 110 \cup 101)(0 \cup 1)^3 \cup 1(0 \cup 1)^6(0 \cup 1)^*$.

2.57 For an easy solution, use Pumping Lemma of Sect. 4.3.

2.62 (a) q to q labeled $(0, 0)/0, (0, 1)/1, (1, 0)/1$; q to p labeled $(1, 1)/0$, p to p labeled $(0, 1)/0, (1, 0)/0, (1, 1)/1$; p to q labeled $(0, 0)/1$.

2.71 The addition $0101 + 0110 = 1011$ is written as the string *abcd* of triples: $a = (0, 0, 1)$, $b = (1, 1, 0)$, $c = (0, 1, 1)$, $d = (1, 0, 1)$. Write 0110 below 0110 and then 1011 as the third line. Now read the columns to get a, b, c, d.

Problems of Chapter 3

Section 3.2

3.2 The set of all strings whose nth symbol from the end is 1. All the 2^n states are relevant.

3.7 Initial state p, $F = \{q, r\}$, $\Delta = \{(p, a, q), (p, \varepsilon, r), (r, a, r)\}$.

3.8 (b) Initial state p, $F = \{q, r, s, t\}$, $\delta(p, 0) = q$, $\delta(p, 1) = t$, $\delta(q, 0) = r$, $\delta(q, 1) = s$, $\delta(r, 0) = r$, $\delta(r, 1) = s$, $\delta(s, 0) = u$, $\delta(s, 1) = u$, $\delta(t, 0) = u$, $\delta(t, 1) = u$, $\delta(u, 1) = u$, $\delta(u, 0) = v$, $\delta(v, 0) = v$, $\delta(v, 1) = v$. Notice the renaming of states.
(c) Initial state p, $F = \{q\}$, $\delta(p, 0) = q$, $\delta(p, 1) = q$, $\delta(q, 1) = q$, $\delta(q, 0) = r$, $\delta(r, 0) = q$, $\delta(r, 1) = q$.

Section 3.3

3.11 Initial state p, $F = \{r\}$, $\delta(p, a) = q$, $\delta(q, a) = s$, $\delta(q, b) = r$, $\delta(s, b) = p$.

3.12 Initial state p, $F = \{r\}$, and transitions $\delta(p, 0) = \delta(p, 1) = \delta(q, 0) = q$, $\delta(q, 1) = \delta(r, 0) = \delta(r, 1) = r\}$.

3.14 (a) Construct a DFA and then the grammar. One such has productions: $S \mapsto aA|bB|\varepsilon$, $A \mapsto bC|aS$, $B \mapsto aC|bS$, $C \mapsto aB|bA$.

3.15 (a) Initial state p, $F = \{r\}$ and transitions (p, a, p), (p, b, q), (p, b, r), (q, a, q), (q, a, r), (q, b, p).
(b) Initial state s, $F = \{p\}$, $\Delta = \{(s, a, s), (s, b, s), (s, a, r), (r, b, q), (q, a, p)\}$.

3.16 Initial state q_0, $F = \{q_2\}$, $\delta(q_0, a) = q_1$, $\delta(q_1, b) = \delta(q_2, a) = \delta(q_2, b) = q_2$, $\delta(q_0, b) = \delta(q_1, a) = q_3$, $\delta(q_3, a) = \delta(q_3, b) = q_4$, $\delta(q_4, a) = \delta(q_4, b) = q_5$, $\delta(q_5, a) = q_5$, $\delta(q_5, b) = q_5$.

Section 3.4

3.18 Initial and final state p, $\Delta = \{(p, a, q), (q, b, r), (r, c, p), (r, \varepsilon, p)\}$. No.

3.20 (f) Initial state p, $F = \{r, v\}$, $\Delta = \{(p, \varepsilon, q), (p, \varepsilon, s), (q, a, r), (r, a, r), (s, a, t), (t, b, u), (u, a, u), (u, \varepsilon, v), (v, b, v)\}$.
(g) Initial state p, $F = \{q\}$, $\Delta = \{(p, a, p), (p, b, p), (p, b, q), (q, b, r), (q, a, s), (r, b, s), (s, \varepsilon, q)\}$.

Section 3.5

3.24 Initial state p, Final state r, $\Delta = \{(p, b, q), (q, b, q), (q, b, r), (p, a, r), (p, b, r)\}$.

3.28 (a) Remove r. (b) $a^*(a \cup b)ab(bb \cup ab \cup aa^*(a \cup b)ab)^*$.

3.29 (b) $(ab \cup (aa \cup b)(ba)^*bb^*)^*$.

Additional Problems for Chapter 3

3.32 Initial state p, $F = \{q\}$, $\delta = (p, a) = r$, $\delta(p, b) = q$, $\delta(q, a) = s$, $\delta(q, b) = p$, $\delta(r, a) = p$, $\delta(r, b) = s$, $\delta(s, a) = q$, $\delta(s, b) = r$.

3.39 From a machine with states in Q that accepts L, build another with states in $Q \times \{0, 1, 2\}$. The second component tells the number of errors met so far. Using non-determinism, guess the string $y \in L$ similar to the input x, and where the errors are.

Problems of Chapter 4

Section 4.2

4.3 As earlier, with $F = (Q_1 \times F_2) \cup (F_1 \times Q_2)$.

4.5 (c) Initial state q_0, $F = \{q_0, q_1, q_2, q_3, q_5\}$, $\delta(q_i, a) = q_{i+1})$, for $i = 0(1)4$, and $\delta(q_5, a) = q_5$.

4.12 (b) Closure under concatenation and reversal.

4.14 Take $L = \Sigma^*$. Conclude that any L' is regular.

Section 4.3

4.17 (a) Otherwise, every walk has a finite number of edges, and the accepted string is of length at most some fixed number. The language is finite.
(b) If not, then such a cycle can be used to generate a string of arbitrary length that can be accepted.

4.18 (f) Consider $\overline{L} \cap a^*b^*$. (g) $w = a^m b^n a^{2m} = xyz$. Take xy^3z.
(m) Consider its complement.

4.20 (n) No. (o) No. (p) No.

4.21 (b) Use $h(a) = h(b) = a$, $h(c) = c$. (d) No. (e) No. (i) No.

4.22 (a) Consider a nonregular language and its complement. (b) $L'' = L \cap (L')^R$
(c) Take a nonregular infinite language. It is an infinite union of singletons.

4.24 (a) One case is: $m = 0$, $k = 0$, $n > 6$. Get a regular expression.
(b) Choose $w = a^7 b^j a^j$. One of $m > 6$ or $k \leq n$ is violated after suitable pumping.

4.26 (b) $\#_u(E) = \#_d(E) = \#_l(E) = \#_r(E)$. (c) L is not regular. Prove it.

4.28 Converse of Theorem 4.6 does not hold. For example, consider the language $\{a^n b^n : n \in \mathbb{N}\}$.

Section 4.4

4.29 First, give a simple verbal description of the language of the DFA.

4.30 $D = (\{p, q, r\}, \{0, 1\}, \delta, p, \{s\})$ with $\delta(p, 0) = \delta(p, 1) = \delta(q, 0) = q$, $\delta(q, 1) = \delta(r, 0) = \delta(r, 1) = r$.

Section 4.5

4.33 Prove by contradiction.

4.37 (d) Initial state p, $F = \{p, q, r, t\}$, $\delta(p, a) = q$, $\delta(q, a) = r$, $\delta(r, a) = s$, $\delta(s, a) = \delta(s, b) = \delta(t, a) = t$.

4.38 Initial state p, $F = \{q\}$, $\delta(p, 0) = p$, $\delta(p, 1) = \delta(q, 0) = \delta(q, 1) = q$.

4.39 Prove it by contradiction.

Additional Problems for Chapter 4

4.42 $L \cap L'$ is finite. So $L' = ((L \cup L') \cap \overline{L}) \cup (L \cap L')$ is regular.

4.44 With $a = (0, 0, 1)$, $b = (0, 0, 1)$, $c = (1, 0, 0)$, $d = (0, 1, 0)$, $e = (1, 0, 1)$, $f = (0, 1, 0)$, the multiplication 001010×000101 is represented by the string $abcdef$. Consider $(2^n + 1) \times (2^n + 1)$ in binary.

4.46 (d) Use DFAs. (e) Yes.
(f) Mark all states q such that a path exists from the initial state to q, and then from q there is a path to a final state. Next, make each such q a final state.
(m) Compute all states that are reachable from the initial state of the DFA by taking a string of length 2. Add ε-transitions to all these states from a new initial state.
(v) From a DFA for L, duplicate all states by labels q_o, q_e. Corresponding to each transition from (p, a, q), have two transitions in the new NFA by adding (p_o, a, q_e) and (p_e, ε, q_o); make q_e initial if q is initial, make both q_o and q_e final states if q is a final state.

4.59 (a) No. Choose $w = (ab)^k aa(ba)^k$. (b) No.

4.65 (b) What happens if two cycles in the transition diagram of a DFA accepting the language share a vertex? And what if they don't?

Problems of Chapter 5

Section 5.2

5.10 Not easy to describe.

5.17 Yes.

5.8 The only possible derivations giving aab as prefix, ends with aba.

5.18 (e) $\{a^m b^n : n > m \geq 1\}$. (f) $\{a^m b^n a^n b^m : m, n \geq 1\}$.
(g) $\{a^m b^n : m \neq n, m \geq 1 \text{ or } n \geq 1\}$. (h) $\{w \in \{a, b\}^+ : aa$ does not occur in $w\}$.
(i) $\{a^m b^n : m \neq n\}$. (j) $\{ab(bbaa)^n bba(ba)^n : n \in \mathbb{N}\}$.
(k) $\{w \in \{a, b\}^* : \#_a(w) = \#_b(w), \#_a(x) \geq \#_b(x)$ for any prefix x of $w\}$.

5.19 (c) $S \mapsto Ab,\ A \mapsto aAb,\ A \mapsto \varepsilon$. (e) $S \mapsto \varepsilon|aSbb|aSbbb$.
(k) $S \mapsto \varepsilon|aA|aaA|aaaA,\ A \mapsto aAb|B,\ B \mapsto \varepsilon|Bb$. (m) $S \mapsto AaAaAaA,\ A \mapsto aA|bA|\varepsilon$.
(q) $S \mapsto aSbb|abb$. (r) $S \mapsto aSb|aA|bB|a|b,\ A \mapsto aA|a,\ B \mapsto bB|b$.
(s) $S \mapsto \varepsilon|SaSbSaS|SaSaSbS|SbSaSaS$. (u) $S \mapsto aAb|bBa,\ A \mapsto aAb|\varepsilon,\ B \mapsto bBa|\varepsilon$.
(w) $S \mapsto aSb|ab|aAb,\ A \mapsto bA|b$.

5.20 (b) $S \mapsto aS|A,\ A \mapsto bAc|bc$ (f) $S \mapsto \varepsilon|aSc|bAc,\ A \mapsto \varepsilon|bAc$. (g) Consider two cases: $k = m - n$ and $k = n - m$. The first case is solved by $S \mapsto \varepsilon|B|aSc,\ B \mapsto \varepsilon|aBa$. (j) $S \mapsto X|Y,\ X \mapsto AC,\ A \mapsto \varepsilon|aAb,\ C \mapsto \varepsilon|Cc,\ Y \mapsto BD,\ B \mapsto \varepsilon|aB,\ D \mapsto E|bDc,\ E \mapsto \varepsilon|Ec$. (k) $S \mapsto aA|aBc|Cc,\ A \mapsto bAc|bc,\ B \mapsto aBc|b,\ C \mapsto aCb|ab$.

5.21 $\{(ab)^n : n \geq 0\}$.

5.24 For $k = 2$, add $S' \mapsto SS$ to G.

5.28 $S \mapsto (S \cup S)|(SS)|S^*|\emptyset$.

Section 5.4

5.36 (a) Derive aab. (c) $S \mapsto \varepsilon|aS|aCbS,\ C \mapsto \varepsilon|aCbC$ disambiguates.

5.39 (b) $S \mapsto aA,\ A \mapsto b|aAB,\ B \mapsto b$.

5.40 (a) aab has two left-most derivations. (b) ab has two left-most derivations. An equivalent unambiguous CFG has productions $S \mapsto \varepsilon|A,\ A \mapsto ab|AA|aAb$.

5.42 From the DFA, get a regular grammar. Leaving aside rules of the form $A \mapsto \varepsilon$, it is a simple CFG. Next, the rules of the kind $A \mapsto \varepsilon$ never create any ambiguity.

5.45 (a) Yes. (b) Yes. (c) No. (d) No.

Section 5.5

5.46 $S \mapsto a|aC|Aa|Ba|AaC|ABa|BaC|ABaC,\ A \mapsto B|C|BC,\ B \mapsto b,\ C \mapsto D,\ D \mapsto c$.

5.47 $S \mapsto a|bc|bb|Aa,\ A \mapsto a|bb|bc,\ B \mapsto a|bb|bc$.

5.48 $S \mapsto aB|aBB,\ B \mapsto bb|bBb$.

5.50 (b) $S \mapsto A|B|AB,\ A \mapsto b|bA|bAA,\ B \mapsto a|aB|aBB$.

5.51 $S \mapsto S{+}A|A \times B|(S)|a|b|Ca|Cb|C0|C1,\ A \mapsto A \times B|(S)|a|b|Ca|Cb|C0|C1,\ B \mapsto (S)|a|b|Ca|Cb|C0|C1,\ C \mapsto |a|b|Ca|Cb|C0|C1$.

5.53 $S \mapsto CA,\ A \mapsto a,\ C \mapsto b$.

5.54 $S \mapsto b|bA,\ A \mapsto bb|bbA; \mathcal{L}(G) = a^+a.$

5.59 $S \mapsto aA,\ A \mapsto BB,\ B \mapsto \varepsilon|aBb$. Elimination of the ε-production results in $S \mapsto a|aA,\ A \mapsto B|BB,\ B \mapsto ab|aBb.$

Section 5.6

5.60 (a) $S \mapsto ASB|AB,\ A \mapsto b|bA|bAS,\ B \mapsto a|aa|A|aS|Sa|SaS.$
(b-c) $S \mapsto AB|ASB,\ A \mapsto b|bA|bAS,\ B \mapsto a|b|aa|bA|aS|Sa|bAS|SaS.$
(d) $S \mapsto AB|AE,\ A \mapsto b|CA|CF,\ B \mapsto a|b|CA|CF|DD|DS|SD|SG, C \mapsto b,$ $D \mapsto a, E \mapsto SB,\ F \mapsto AS,\ G \mapsto DS.$

5.63 (d) $S \mapsto CA,\ A \mapsto DX,\ C \mapsto XB,\ B \mapsto XY,\ D \mapsto CZ,\ X \mapsto a,\ Y \mapsto b,$ $Z \mapsto c.$
(f) $S \mapsto AB|XB|YY|ZA,\ A \mapsto WY,\ B \mapsto ZA|YY,\ Z \mapsto YY,\ W \mapsto XY, \mapsto a,$ $Y \mapsto b.$
(h) $S \mapsto AA_1|UB|a|SA|AS,\ S \mapsto AA_1|UB|a|SA|AS,\ A \mapsto b|AA_1|UB|a|AS|SA,$ $A_1 \mapsto SA,\ U \mapsto a,\ B \mapsto b.$

5.64 (b) Label of an interior node is a nonterminal.

5.65 Use a CNF. Then, for $A \mapsto a$, use $A \mapsto aXY,\ X \mapsto \varepsilon,\ Y \mapsto \varepsilon$ with new nonterminals X, Y.

5.66 (c) $S \mapsto bA|bBSB,\ A \mapsto b,\ B \mapsto a.$ (g) $S \mapsto bS|bA|bBS,\ A \mapsto a,\ B \mapsto b.$
(i) $S \mapsto aB|aBB|bAB,\ A \mapsto a|aB|bA,\ B \mapsto a.$

5.68 . Conjecture which strings are derived from A, B and use induction on the length of derivations.

Additional Problems for Chapter 5

5.70 Is it $\{a, b\}^*$?

5.71 Use induction to show that $S \Rightarrow x,\ A \Rightarrow y,\ B \Rightarrow z$ simultaneously, where $\#_a(x) = 2 \cdot \#_b(x),\ \#_a(y) = 2 \cdot \#_b(y) + 1,\ \#_a(z) = 2 \cdot \#_b(z) - 2.$

5.92 $S \mapsto aA,\ A \mapsto a$ generates the same language as $S \mapsto aa.$

5.97 Show by induction on the number of steps in a derivation that any string of terminals and nonterminals derived from the start symbol has odd length.

Problems of Chapter 6

Section 6.2

6.3 (b) $\{a\} \cup ab^+a$. (e) aa^*ba^*.

6.8 (a) Initial state p, $F = \{q\}$, $\Delta = \{(p, a, A, p, AA), (p, b, A, p, BA), (p, a, B, p, AB), (p, b, b, p, BB), (p, a, \varepsilon, p, A), (p, b, \varepsilon, p, B), (p, \varepsilon, A, q, A), (p, \varepsilon, b, q, B),$

$(q, a, A, q, \varepsilon), (q, b, B, q, \varepsilon)\}$. (c) Initial state p, $F = \{q\}$, $\Delta = \{(p, \varepsilon, \varepsilon, q, \varepsilon), (q, a, \varepsilon, q, AA), (q, a, A, q, AAA), (q, b, A, q, \varepsilon)\}$. (f) Use nondeterminism to generate one or two tokens by the transitions $(p, a, \varepsilon, q, A), (p, a, \varepsilon, q, AA)$, etc.

6.10 For the bottom-of-the- stack symbol, see ∇ in Sect. 6.6.

Section 6.3

6.12 (a) Initial state p, $F = \{r\}$, $\Delta = \{(p, \varepsilon, \varepsilon, q, S), (q, a, S, q, SA), (q, a, S, q, \varepsilon), (q, b, A, q, B), (q, b, B, q, \varepsilon), (q, \varepsilon, \varepsilon, r, \varepsilon)\}$.
(c) Initial and final state q, $\Delta = \{(q, \varepsilon, S, q, ASB), (q, \varepsilon, S, q, C), (q, \varepsilon, C, q, BCA), (q, \varepsilon, C, q, S), (q, \varepsilon, C, q, \varepsilon), (q, a, A, q, \varepsilon), (q, b, B, q, \varepsilon), \}$.
(e) Initial state p, $F = \{r\}$, $\Delta = \{(p, \varepsilon, \varepsilon, q, S), (q, \varepsilon, \varepsilon, r, \varepsilon), (q, a, S, q, A), (q, a, A, q, ABC), (q, a, A, q, \varepsilon), (q, b, A, q, B), (q, b, B, q, \varepsilon), (q, c, C, q, \varepsilon)\}$.
(f) Initial state p, $F = \{s\}$, $\Delta = \{(p, a, \varepsilon, q, \varepsilon), (q, a, \varepsilon, r, \varepsilon), (r, a, \varepsilon, r, AA), (r, a, A, r, AAA), (r, b, A, s, A), (s, b, A, s, \varepsilon)\}$.
(g) Initial state p, $F = \{q\}$, $\Delta = \{(p, \varepsilon, \varepsilon, q, S), (q, a, S, q, SSS), (q, a, S, q, B), (q, b, B, q, \varepsilon)\}$.

6.13 (d) What is the language?

6.14 (a) Initial state p, final state q, $D = \{(p, \varepsilon, \varepsilon, p, SA), (p, a, S, p, SB), (p, b, S, p, \varepsilon), (p, b, B, p, \varepsilon), (p, \varepsilon, A, q, \varepsilon)\}$.

6.15 Yes. From PDA construct a CFG; from this CFG construct a PDA.

Section 6.4

6.18 (g) Choose $a^n b^{2n} b^{2n} a^n$. (l) Choose $a^{n^2} b^n$.

Section 6.5

6.20 (a) Intersect it with $a^* b^* c^*$. (e) Closure under union. (g) Closure under union. (i) Closure under union.

6.21 (b) Write it as $\{a^m b^m\} \cap \bar{(}L)$ where $L = \{a^{2009} b^{2009}\}$. (c) Is a CFL (d) Is a CFL. (o) Closure under union. (s) Construct a PDA. (t) Choose a^{pq} where both p, q are primes bigger than n. (v) Yes. Show that $\{w \in \{a, b, c\}^* : \#_a(w) < \#_b(w)\}$ is a CFL by breaking into eight cases.

6.22 Replace each production $A \mapsto x$ in G by $A \mapsto x^R$. Then use induction on the lengths of all derived strings of terminals and nonterminals.

Section 6.6

6.25 (a) Initial and final states p, $\Delta = \{(p, a, \nabla, q, A\nabla), (q, a, A, q, AA), (q, b, A, r, \varepsilon), (r, b, A, r, \varepsilon), (r, b, \nabla, p, \nabla)\}$. (e) Initial state p, $F = \{r, s\}$, $\Delta = \{(p, a, \nabla, s, \nabla), (s, a, \nabla, q, AA), (q, a, A, q, AA), (q, b, A, r, \varepsilon), (r, b, A, r, \varepsilon), (r, b, \nabla, s, \nabla)\}$.
(g) Construct a PDA. How to replace a transition of the form $(p, \varepsilon, \nabla, \varepsilon, \nabla)$?
(k) Once c is encountered it matches the number of accumulated A's by popping off one a step.

6.28 What is its complement?

Additional Problems for Chapter 6

6.46 Mark the two middle blocks in $a^n b^n a^n b^n$.

6.47 (b) Consider $a^n b^n c^{(n+n!)}$.

6.51 You may think of a/b as acb.

6.52 (a) $S \mapsto aSA,\ A \mapsto bAD,\ D \mapsto bC,\ C \mapsto aBC$.

6.53 Use a two-standard form CFG and then get a PDA from it.

6.57 $\{a^m b^m c^n : m, n \in \mathbb{N}\} \cap \{a^n b^m c^m : m, n \in \mathbb{N}\}$ is not a CFL.

6.58 Can you modify a nonregular language that satisfies the Pumping Lemma for regular languages?

6.59 Use a PDA.

6.65 (i) Use induction on $\ell(w)$.

6.66 Use the previous problem.

6.68 Use the previous problem.

6.73 Use ultimate periodicity.

6.77 Use Parikh's Theorem and the previous problem.

6.80 (a) Use a PDA. (c) Is a CFL.

6.89 Use DPDAs.

Problems of Chapter 7

Section 7.2

7.1 Yes.

7.2 None is in $\mathcal{L}(G)$.

7.6 $S \mapsto aSBC|aBC,\ CB \mapsto BC,\ aB \mapsto ab,\ bB \mapsto bb,\ bC \mapsto bc,\ cC \mapsto cc$.

7.8 $\mathcal{L}(G) = \{ab^m c^n d : m, n \geq 1\}$, a regular language.

7.9 (c) $S \mapsto aaaA,\ A \mapsto aAb|\varepsilon$. (d) $S \mapsto AbA,\ A \mapsto aAb|\varepsilon,\ B \mapsto bB|\varepsilon$.

7.10 (a) $S \mapsto XYZ,\ XY \mapsto aXA,\ XY \mapsto bXB,\ Aa \mapsto aA,\ Ab \mapsto bA,\ Ba \mapsto aB,\ Bb \mapsto bB,\ aY \mapsto Ya,\ bY \mapsto Yb,\ XY \mapsto \varepsilon,\ Z \mapsto \varepsilon$.
(b) $S \mapsto aSb|aAb,\ A \mapsto bAa|ba$.
(c) $S \mapsto aB|bA,\ A \mapsto a|aS|bAA,\ B \mapsto b|bS|aBB$.

7.11 (a) $S \mapsto XAY, X \mapsto XD, DY \mapsto Y, DA \mapsto AAD, X \mapsto \varepsilon, Y \mapsto \varepsilon, A \mapsto a$.
(d) $S \mapsto A|Sc,\ A \mapsto ab|aAb$.

7.13 (a) $\emptyset$. (b) Only strings of length 2^n are generated. (c) $\{a^m b^n c^n : m \geq 0, n \geq 1\}$. (d) $\{ab^m c^n d : m, n \geq 1\}$. (e) $\{a^{n^2} : n \geq 1\}$. (f) $\{a^{n^2} : n \geq 1$. (g) $\{a^n b^n a^n : n \geq 1\}$. (h) $\{ww : w \in \{a, b\}^*$. (i) $\{a^n b^n c^n : n \geq 1\}$. (j) $\{a^{n+1} b^{n+k} : n \geq 1, k \geq -1\}$.

Section 7.3

7.15 $\delta(s, \text{ƀ}) = (s, \text{ƀ})$.

7.16 $\delta(s, \sigma) = (s, \text{Ɍ})$ if $\sigma \neq a$; $\delta(s, a) = (h, a)$.

Section 7.4

7.22 (b) Replace first a by c, last b by d, come back to c, go right, continue. (c) $\delta(s, \text{ƀ}) = (s, \text{Ɍ})$, $\delta(s, a) = \delta(q, a) = \delta(q, b) = (q, \text{Ɍ})$, $\delta(q, \text{ƀ}) = (h, \text{ƀ})$. (d) $\delta(s, \text{ƀ}) = (p, \text{Ɍ})$, $\delta(p, b) = (q, \text{Ɍ})$, $\delta(q, a) = (r, \text{Ɍ})$, $\delta(r, b) = (r, \text{Ɍ})$, $\delta(r, a) = (h, \text{Ɍ})$. (e) $\delta(s, \text{ƀ}) = (p, \text{Ɍ})$, $\delta(p, \text{ƀ}) = (h, \text{ƀ})$, $\delta(p, a) = (q, \text{Ɍ})$, $\delta(p, b) = (q, \text{Ɍ})$, $\delta(q, a) = \delta(q, b) = (p, \text{Ɍ})$, $\delta(q, \text{ƀ}) = (\hbar, \text{ƀ})$.

Section 7.5

7.28 Put left and right markers and move them cell by cell in each direction.

7.34 (a) 1. Go from left to right, crossing off every other a.
2. If in stage 1, the tape contained a single a, accept.
3. If in Stage 1, the tape contained odd (but >1) number of a s, reject.
4. Return to the original blank cell. Go to stage 1.

(b) Remember and delete the first symbol, search for c on the right, match and delete the next symbol. Remove the blank by shifting suitably. Go to the original blank cell, and start over again.

(d) 1. Scan from left to right and check if the string is in $a^*b^*c^*$; else reject.
2. Come back to the original blank cell.
3. Delete one a.
4. Replace one b by a d, delete a c; do this until all b are replaced. Change all d's to b's.
5. Continue to execute Stage 3–4 if there is another a to be deleted.
6. If all a's are deleted, check whether all c's have been deleted. If yes, accept, else, reject.

Section 7.6

7.37 Move all a from the beginning of the input s to another tape, and then match the contents of the two tapes.

7.38 Can it write β instead of ƀ ?

7.40 No. Represent each σ by a σ' (new). If needed, revert back to σ.

7.41 Replace $\delta((p, \{a, b\}) = (q, \text{Ɍ})$ by $\delta(p, \sigma) = (q, \text{Ɍ})$, for each $\sigma \notin \{a, b\}$.

Section 7.7

7.44 Generate n on another tape, and then compare the nth symbol of u and v. If not equal, accept; else, reject.

7.47 (c) Nondeterministically guess the middle of the input and then compare the left and the right strings. (h) For $n > 1$, generate nondeterministically a positive integer m and check whether n is a multiple of m.

7.49 The DTM may be difficult, while the NTM can guess where is this 0.

Additional Problems for Chapter 7

7.53 (b) $S \mapsto A|B$, $A \mapsto aaaA|a$, $B \mapsto aaaB|aa$.

7.54 (b) Even length strings not in the form ww^R are of two types: $uaww^Rbv$ or $ubww^Rav$ with $\ell(u) = \ell(v)$.

7.56 Corresponding to a production $u \mapsto v$ with $\ell(u) > \ell(v)$, insert $\ell(v) - \ell(u)$ number of new nonterminals on the right side and then take them to ε. For instance, $AB \mapsto C$ can be replaced by $AB \mapsto CD$, $D \mapsto \varepsilon$.

7.60 No. Use a start symbol whose right hand side is the starting string.

7.61 Make a disjoint copy of the symbols. Introduce rules to remove symbols from the left end, and for appending the same symbols on the right end.

7.63 Use another tape to keep track of the states of the standard TM.

7.64 Yes.

7.67 Use β instead.

7.75 (a) First identify $x_1 = \varepsilon$, $V_1 = \varepsilon$, $x_2 = \varepsilon$, $y_1 = a$, $W_1 = V_1$, $y_2 = b$. Next, identify $V_1 = ab$ leaving others as in the first. $\mathcal{L}(P) = \{a^n b^n : n \in \mathbb{N}\}$.
(b) Identities representing addition of natural numbers in unary.
(c) $(N \cup \Sigma, \{V_1, V_2\}, \{x\}, \{V_1 x V_2 \mapsto V_1 ax V_2 a | V_1 bx V_2 b | V_1 cx V_2 c\})$ with $N = \{x\}$ and $\Sigma = \{a, b, c\}$. (d) $\{a^{2^n} : n \geq 1\}$.

Problems of Chapter 8

Section 8.2

8.3 (a) Find the first blank, write a 0, find next blank, move left, remove two 0's.

8.8 Write a c on both the sides of the string; bring them closer symbol by symbol. Replace the middle cc by c.

8.9 (f) $\lceil x \rceil$ is the least integer greater than or equal to x. Thus, $\lceil n/2 \rceil = m$ or $m + 1$ according as $n = 2m$ or $2m + 1$.
(i) Euclidean algorithm is given by:
if $n = 0$ or $n = m$, then return m, else, return $\gcd(n, m \mod n)$.

8.10 How does $n + 1$ look like compared to n in binary notation?

8.19 L is the output language.

Section 8.3

8.24 Computably enumerable languages are not closed under complementation.

8.25 It computes the function f, where $f(u, v) = 1$ if u is a substring of v, else, $f(u, v) = 0$.

8.26 For example, test whether an integer is prime.

8.28 $\overline{L_1} = \cup_{i=2}^{k} L_i$, and finite union of computably enumerable languages is computably enumerable.

Section 8.5

8.35 Yes, as there are only finite number of splits of any string from L^+.

8.36 (e) Computably enumerable. (g) Not computably enumerable.
(k) Computably enumerable. (l) Not computably enumerable.

8.38 If L is computably enumerable, design an NTM M for $h(L)$ as follows. Suppose w is the input to M. On a second tape, M guesses some string v and then checks whether $h(x) = w$. Next, it simulates the TM for L on v. If v is accepted, then M accepts w.

For the decidable languages, consider the language L consisting of strings of the form $\psi(M)\#w@2^{n+1}$, where M has input alphabet $\{0, 1\}$, and $w \in \{0, 1\}^*$. The string is in L iff M accepts w after making at most n moves. L is decidable as we may simulate M on w for n moves and then decide whether or not to accept. Take the homomorphism h defined by $h(0) = 0, h(1) = 1, h(\#) = h(@) = h(2) = \varepsilon$. Apply h to L. We find that $h(L)$ is the acceptance language, which is not decidable.

Section 8.6

8.40 (a) $S \mapsto abc|aAbc,\ Ab \mapsto bA,\ Ac \mapsto Bbcc,\ bB \mapsto Bb,\ aB \mapsto aa|aaA$.
(b) Create $a^n Bc^n D$. The markers B and D assure the correct number of b's and d's created. Then, B travels to the right to meet D. Finally, we create one d and a return messenger that puts b in the right place. That is, $S \mapsto aAcD|aBcD,\ A \mapsto aAc|aBc,\ Bc \mapsto cB,\ Bb \mapsto bB,\ BD \mapsto Ed,\ cE \mapsto Ec,\ bE \mapsto Eb,\ aE \mapsto ab$.

8.41 Right-linear: $S \mapsto aaA,\ A \mapsto aA|B,\ B \mapsto bbbC,\ C \mapsto bC|\varepsilon$.
Left-linear: $S \mapsto Abbb,\ A \mapsto Ab|B,\ B \mapsto aaC,\ C \mapsto aA|\varepsilon$.

8.42 Yes. Prove it.

8.45 (b) Choose $a^n b^{2n} a^n$. (c) Choose $a^n b^{2n} a^n$. (e) Choose $({}^n a\,)^n + ({}^n a\,)^n$.

8.47 (b) Go on dividing n in a^n successively by $2, 3, 4, \ldots$ until the string is accepted or rejected. You may have to rewrite the quotient on the input squares and delete others.

(f) Use a three-tape machine. On the third tape, keep the current value of $\ell(w)$. On the second tape keep markers after each $\ell(w)$ squares. Compare the strings between the markers.

8.51 $\mathcal{L}(G)$ is linear.

8.52 $aab(ab)^*$.

8.53 Both generate $\{a^n b^n c^n : n \geq 1\}$.

8.54 You can find a suitable grammar. Alternatively, an LBA starts at opposite ends and stops when a match is found.

8.55 A nonregular CFG can generate a regular language.

Additional Problems for Chapter 8

8.61 Suppose M' accepts L'. Construct M as follows: Given input w, M changes its input to $1w$ and simulates M'. If M' accepts, then so does M. But then, $w \notin L$ due to the definition of L'. On the other hand, when M' does not accept $1w$, M does not accept w. That is, $\mathcal{L}(M) = \overline{L}$. Thus, L is decidable, a contradiction. Hence L' is not computably enumerable.

8.63 (a) k_f is the number of states of the TM M_f that on input 0^{n+1} halts with $0^{f(n)+1}$.
(b) If so, then $f(n) = \beta(2n)$ is also a total computable function. Use (a).

8.65 See that $SwS \Rightarrow XwZ \Rightarrow XZw^R \Rightarrow w^R$. Thus, it computes $f(w) = w^R$.

8.68 Replace productions $A \mapsto Bv$ by $A \mapsto v^R B$ and $A \mapsto v$ by $A \mapsto v^R$. Then use induction on the length of a derived string.

8.70 Replace $A \mapsto a_1 \cdots a_m B b_1 \cdots b_n$ by $A \mapsto a_1 X,\ X \mapsto a_2 \cdots a_m B b_1 \cdots b_n$.

8.72 Follow the line of proof as for computably enumerable languages and see that the conditions of an LBA are met.

8.73 Consider L^2, where $L = \{a^m b^m : m \in \mathbb{N}\}$.

8.74 Get a linear grammar for LL' from those of L and L'.

8.77 Yes.

8.83 For decidability, use enumeration in lexicographic ordering.

8.84 (a) $\{0^n 1^n : n \geq 1\}$. (b) $\max(m - n, 0)$.

8.88 A CFL is accepted by a PDA in which the number of symbols in its stack never exceeds the length of the input by more than one.

8.92 (a) Composition of the successor function $succ$ with itself.
(b) Recursively define from functions obtained by combining zr, id, and $succ$.
$plus(m, 0) = m$; $plus(m, n + 1) = succ(plus(m, n))$.
(c) $mult(m, 0) = zr_1(m)$; $mult(m, n + 1) = plus(m, mult(m, n))$.
(d) $exp(m, 0) = succ(zr_1(m))$; $exp(m, n + 1) = mult(m, exp(m, n))$.

8.93 $f = p \cdot g + (1 \sim p) \cdot h$.

8.94 (c) $digit(m, n, p) = div(rem(n, exp(p, m)), exp(p, (m \sim 1)))$.
(d) $sum_f(m, 0) = 0, sum_f(m, n + 1) = sum_f(m, n) + f(m, n + 1)$.

8.97 (a) $\log(m, n) = \mu\ p[greater\text{-}than\text{-}or\text{-}equal(exp(m + 2, p), n + 1)]$.

Problems of Chapter 9

Section 9.2

9.1 Modify the transitions of a given machine D in such way that q would behave like a halt state.

9.3 Construct D_w from D that writes w on the tape, and then simulates D.

9.4 How many n-state TMs are there which halt? From a machine M and a number m, construct D which always halts with one of two answers: M on the blank tape halts in at most m moves, or M on the blank tape takes more than m moves. Use this D and the TM that computes f to solve the blank-tape problem.

9.5 Modify the halting transitions of M so that it first writes σ and then halts.

9.7 (a) Decidable.

9.12 L_{A} is not map-reducible to $L_{\emptyset}$.

9.13 Yes.

9.16 $\overline{L}_{\text{A}}$ is not computably enumerable, while both L_A and $\overline{L}_A$ are undecidable.

Section 9.3

9.24 This shows that problems and their complements are essentially different. They are similar only when either, so each, is decidable.

9.25 Reduce the acceptance problem to this problem by modifying M in such a way that C' leads to a halted configuration in one step.

9.26 Rice's theorem.

9.27 No; a nontrivial property of RE.

9.29 Yes.

9.30 No.

9.31 (b) Yes. If M has n states, consider the first n moves of M. (c) Unsolvable.

9.34 (a) No. Choose $\mathcal{L}(N) = \emptyset$.

9.36 Show that each TM can be simulated by one having only two states other than h and $\hbar$; you may have to enlarge the tape alphabet.

9.37 (c) Take $\mathcal{L}(G') = \Sigma^*$. (d) Unsolvable. (e) Unsolvable.

9.40 (a) Solvable. (b) Solvable. (c) Unsolvable.

Section 9.4

9.42 (g) $L - L'$ is regular. (h) $L \cup L'$ is regular. Is $L' = L \cup L'$?
(k) Construct a DFA for L^R from a DFA for L. Then, check equality.
(p) If L over $\{a, b\}$ contains no even length string, then $L \cap (aa \cup ab \cup ba \cup bb)^* = \emptyset$.

Section 9.5

9.46 (a) Solvable. (b) Solvable. (c) Unsolvable. (d) Solvable. (e) Solvable.
(g) Unsolvable. (h) Unsolvable. (i) Unsolvable. (j) Unsolvable. (k) Unsolvable.
(l) Unsolvable. (m) Solvable. (n) Unsolvable. (o) Solvable. Use Pumping Lemma.
(p) Unsolvable. (s) Unsolvable. (t) Solvable. Such a PDA accepts all strings iff it accepts all strings of length 1.

Section 9.6

9.52 No.

9.54 Yes. No.

9.55 P has a match, P' does not have.

9.61 (a) Mimic the reduction of MPCP to PCP.

Section 9.7

9.64 Write a given sentence ϕ in prenex form, where all the quantifiers are in the beginning. Suppose there are k quantifiers. Give an algorithm that constructs a DFA D_i for each i from 0 to k. The DFA D_i recognizes the collection of all strings representing i-tuples of numbers that make the formula ϕ_i true, where ϕ_i is obtained from ϕ by chopping of the first i occurrences of quantifiers. Use D_i to construct D_{i-1}. Once A_0 is constructed, test whether A_0 accepts the empty string. If it does, ϕ is true.

Section 9.8

9.67 $\phi_{M,w}(x)$ states that x is a suitably encoded accepting computation history of M on w.

9.69 (b) This says that semi-decidability can be had by an unbounded search monitoring at each stage, a decidable property. (c) Use part (b).
(e) The predicate $Pr(p, \phi)$ that stands for "p is a proof of ϕ," is decidable. Encode p and ϕ as numbers. (f) The Diophantine predicate $p(\cdots) = 0$ is decidable, by Sturm's theorem.

9.72 (c) This can be interpreted in two ways.

Additional Problems for Chapter 9

9.81 (d) Given M for accepting $L_\emptyset$ construct an NTM M' that might either accept $\mathcal{L}(M) = \emptyset$ or $\mathcal{L}(M) = \{01\}$. In any case, make sure that if $\mathcal{L}(M) = \emptyset$, then $\mathcal{L}(M')^R = \mathcal{L}(M')$. The NTM M' first checks whether its input is 01, and rejects if not. Next, it guesses an input w for M. Finally, it simulates M on w. If M accepts, then M' accepts its own input, 01.

9.82 (b) Use valid computation histories.

9.86 Use the fact that every semi-decidable property of RE is monotone.

9.87 Take the property P, where $P(X)$ means X is regular. Use Problem 9.86 once with $A = \emptyset$, $B = \{0^n 1^n : n \in \mathbb{N}\}$, and second time with $A = \{0^n 1^n : n \in \mathbb{N}\}$, $B = \Sigma^*$.

9.89 (a) Use diagonalization directly. (b) Given M, let M' be a TM that on input w simulates M on all $v \leq w$, where $\leq$ is a computable linear order on Σ^*.

9.91 Show that the corresponding language is not computably enumerable.

9.93 Construct the TM D so that if M accepts w, then D accepts a^*; and if M does not accept w, then D accepts $\emptyset$.

9.94 Use valid computation histories.

9.96 No.

9.100 (d) Unsolvable.

9.104 (b) Reduce PCP to this problem.

9.105 No. If it were possible to enumerate all these n in numerical order, then the set of all (M_i, w_i) would have been decidable. But these n can be enumerated by a TM that simulates M_n on w_n in rounds. In the kth round, we try only a limited number of TMs for some limited number of steps, in such a way that eventually in all rounds taken together, we use all the TMs in the sequence on all the corresponding inputs.

9.107 (a) A match gives two derivations, one via productions of G_B and the other via that of G_C. Moreover, neither G_B nor G_C is ambiguous, hence conclude the converse.
(b) Construct PDAs for $\overline{L}_B$, $\overline{L}_C$. (c) $\mathcal{L}(G_1) \cap \mathcal{L}(G_2)$ is the set of matches in a suitable PCS.
(d) $\mathcal{L}(G_1)$ is missing those strings that represent matches for a suitable PCS. $\mathcal{L}(G_2)$ is missing nothing. Thus, their languages are the same iff the PCS has no match.
(f) Use part (a). (g) $\mathcal{L}(G_1) \subseteq \mathcal{L}(G_2)$ iff a suitable PCS has no match.
(i) If uv is a match in P, then $u \in \Sigma^*$, $v \in \{a_1, \ldots, a_n\}^*$ is the reversal of a corresponding string, and $uv \notin \overline{L}_B \cup \overline{L}_C$. Define the homomorphism h with $h(0) = u$, $h(1) = v$. Show that $\overline{L}_B \cup \overline{L}_C$ is not regular. Now, what about $h^{-1}(\overline{L}_B \cup \overline{L}_C)$?
(j) Show that $\overline{L}_{BC}$ is a CFL iff a suitable PCS has no match. You may need to use inverse homomorphisms and Ogden's lemma to force equality in the lengths of certain substrings.

9.108 Reduce PCP to this problem by constructing, from each instance of PCP, a grammar whose language contains a palindrome iff the PCP instance has a match.

9.110 Reduce the acceptance problem.

9.112 The proof follows similar lines as for PCP. Remember to modify the construction in such a way that when a TM M enters its accepting state, the current string can eventually be erased!

9.113 Reduce the acceptance problem to tiling.

9.116 Using induction on the length of y, show that if x, y are propositions and x is a prefix of y, then $x = y$.

9.117 $S \mapsto pA|\neg S|(S \wedge S)|(S \vee S)|(S \rightarrow S)|(S \leftrightarrow S),\ A \mapsto \varepsilon|A0.$

9.119 (a) Some TM can enumerate all provable statements.
(b) Reduce the acceptance problem to $Th(\mathbb{N})$; use valid computation histories.

9.121 Use diagonalization.

9.124 Let P be a nontrivial property of computably enumerable languages. If it were decidable for TMs M whether $\mathcal{L}(M) \in P$, then we could construct a computable map f having no fixed point.

9.126 Construct M by "On input w, obtain, via recursion theorem, $\psi(M)$. Compute $f(\psi(M))$ to get $\psi(M')$. Simulate M' on w." The crux is to show that $f(\psi(M)) = \psi(M')$ for some M'.

9.128 (b) If so, then M accepts 0, and then X would be false.

9.129 Consider the polynomial $p(x, y) = x - (x + 1)(q(x, y))^2$. Then $p(x, y)$ is nonnegative iff $q(x, y) = 0$.

9.131 (c) How many languages are decided by oracle TMs?

9.133 (c) *Empty* $= \{\psi(M) : \forall x \forall y (M$ does not accept x in y steps$)\}$. Next, $\forall x \forall y$ can be combined into one using the map that takes (i, j) to $i + (i + j)(i + j + 1)/2$.
(d) *Total* $= \{\psi(M) : \forall x \exists y (M$ halts on x in y steps$)\}$.
(e) *Finite* $= \{\psi(M) : \exists n \forall x \forall y (\ell(x) \leq n$ or M does not accept x in y steps$\}$. The two universal quantifiers can be combined into one.
(f) *CoFinite* $= \{\psi(M) : \exists n \forall x \exists y (\ell(x) \leq n$ or M accepts x in y steps$)\}$.

Problems of Chapter 10

Section 10.2

10.1 (a) $f(n) = O(g(n)),\ f(n) = o(g(n)),\ g(n) \neq O(f(n)),\ g(n) \neq o(f(n))$.
(h) $f(n) = o(g(n)),\ g(n) \neq O(f(n))$.

10.2 (a) False. (b) True. (c) True.

10.3 (a) True. (d) True.

Section 10.3

10.15 (a) Guess the middle of the string, copy the second part to second tape, then compare the strings.

10.16 Where is $\{a^n b^n : n \in \mathbb{N}\}$?

Section 10.4

10.20 Yes.

10.21 Use Tape compression.

10.22 The map reduction here uses an $O(1)$ space.

10.23 A TM that uses $f(n)$ space runs in time $2^{O(f(n))}$. Moreover, PATH is in P.

10.25 (a) True. (c) True. (d) True.

10.28 Yes.

Section 10.5

10.36 Use diagonalization. Consider $\Sigma = \{0, 1\}$. Assume that strings in Σ^+ have been arranged in lexicographic order, and that we have an enumeration of all TMs as $M_1, M_2, \ldots$. Next, define $L = \{w_i : M_i$ does not accept w_i in $f(\ell(w_i))$ or fewer steps$\}$. Show that L is decidable but is not in $DT(f(n))$.

10.38 (a) Linear. (b) $O(n^3)$. (c) $O(n^2)$. (d) Linear. (e) Linear.

10.40 An $O(n^2)$ algorithm for emptyness check is easy to construct. Modify it.

10.41 Use emptiness check.

10.42 Use emptiness check.

10.46 Use the Euclidean algorithm for computing the gcd, the greatest common divisor, of two numbers. It uses $\gcd(m, n) = \gcd(m, m \mod n)$.

10.48 This result has been improved recently. See [3].

10.49 An undirected graph is connected iff there is a path between every pair of vertices. A crude algorithm for checking connectedness is: mark some vertex; mark a vertex (another) if there is an edge between it and some marked vertex. Finally, check whether all vertices have been marked.

Section 10.6

10.54 For concatenation, suppose the input is of length n. For each i between 1 and $n-1$, test whether positions 1 through i holds a string in L_1 and positions $i+1$ through n holds a string in L_2. If so, accept. If the test fails for all i, reject. The running time of this test is at most n times the sum of the running times of the recognizers for L_1 and L_2.
For Kleene star, let $x \in L^*$, where $L \in P$. Consider all substrings of x; one by one with increasing length. Reviewing the CYK algorithm may help.

10.56 (a) True. (c) True.

10.58 Suppose L is decided by an NTM with time bound $p(n)$. Construct a DTM that decides L in time $c^{p(n)}$ for some constant $c > 1$. The DTM mimics the NTM carrying out all possible computations of length 1, then those of length 2, and so on.

10.59 Take the subset as the certificate.

Section 10.7

10.64 If $L \in P$, then the complement of $\{\psi(M) : M$ accepts $\psi(M)$ after at most $2^{\ell(\psi(M))}$ steps$\}$ is also in P.

10.65 Use Euler's characterization that an undirected graph G is Eulerian iff G is connected and each vertex of G has even degree. A similar result also holds for directed graphs.

Section 10.8

10.68 (b) Part (a) provides one case. For the other case, take f to be a model of Y.

10.75 (b) As there are only three Hamiltonian cycles in G. the formula is: $(x_{12} \wedge x_{23} \wedge x_{34} \wedge x_{14}) \vee (x_{13} \wedge x_{23} \wedge x_{24} \wedge x_{14}) \vee (x_{13} \wedge x_{34} \wedge x_{24} \wedge x_{12})$.

10.77 Show: the reduction can be carried out in $O(1)$ space, and in $O(n)$ time.

10.78 As numbers in A are given in binary, length of k is not a polynomial in n.

Additional Problems for Chapter 10

10.86 (a) Yes. Change the start symbols of G to A, and to B for obtaining the CFGs G_1 and G_2, respectively. This change provides a polynomial time reduction.
(b) Yes. (c) L, L' are not even decidable!

10.89 The string $aabbb$ is generated by the CFG. $N_{1,1} = \{A\}$, $N_{2,2} = \{A\}$, $N_{3,3} = \{B\}$, $N_{4,4} = \{B\}$, $N_{5,5} = \{B\}$, $N_{1,2} = \emptyset$, $N_{2,3} = \{S, B\}$, $N_{3,4} = \{A\}$, $N_{4,5} = \{A\}$, $N_{1,3} = \{S, B\}$, $N_{2,4} = \{A\}$, $N_{3,5} = \{S, B\}$, $N_{1,4} = \{A\}$, $N_{2,5} = \{S, B\}$, $N_{1,5} = \{S, B\}$.

10.90 It needs to compute $n(n+1)/2$ sets $N_{i,j}$. Each such set may take $O(n)$ time, at the most.

10.91 $S \in N_{1,3}$ because $S \mapsto AB$ with $A \in N_{1,1}$ and $B \in N_{2,3}$. $A \in N_{1,1}$ as $A \mapsto a$. $B \in N_{2,3}$ because $B \mapsto AB$ with $A \in N_{2,2}$ and $B \in N_{3,3}$. As $A \mapsto a$, $A \in N_{2,2}$. Finally, $B \in N_{3,3}$ because $B \mapsto b$. This gives the derivation: $S \Rightarrow AB \Rightarrow aB \Rightarrow aAB \Rightarrow aaB \Rightarrow aab$; and then the required parsing.

10.99 Show that the complement of PATH is in NLS.

10.107 (b) NP is closed under nonerasing homomorphisms.

10.108 For one of the implications, consider $L = \{\psi(x)2\psi(y) : x$ is a proposition and y is a model of $x\}$. Show that $L \in P$.

10.109 (c) *NP* is a class of languages and *factoring* is a function. Thus saying that factoring is in *NP* will not do.

10.110 Prove that 3COLOR is in *NP*. Next, construct a polynomial time reduction of 3COLOR to SAT

10.115 $b = 2^k$ is worth trying.

10.116 Reduce HP to LPATH

10.117 First, try k as a power of 2.

10.118 (c) Choose L not from $DS(f(n))$, and a TM accepting L whose running time is $h(n)$. Take $g(n) = \max(f(n), h(n))$.
(f) Use diagonalization. The infimum condition says that there is an input w of length n such that $(\log t) f(n) < g(n)$, where t is the number of tape symbols of any considered TM.
(o) There are positive integers m, k such that $r \leq m/k$ and $r + c \geq (m+1)/k$. Prove that $NS(n^{m/k}) \subsetneq NS(n^{(m+1)/k})$.

10.123 Use the space complexity of a TM that computes $f(n)$.

10.128 Consider $C = \{p, q, r, t, \neg p \vee \neg q, \neg q \vee \neg r, \neg r \vee \neg p, p \vee \neg t, q \vee \neg t, r \vee \neg t\}$. Show that if an interpretation is a model of $\{p, q, r\}$, then it can be extended to a model of at most seven of the clauses from C and no more. Moreover, there is exactly one model of $\{p, q, r\}$ that can be extended to at the most six of the clauses from C. Use this gadget to reduce 3SAT to MAX2SAT

10.129 There would be some constant c such that an *NP* problem could be solved in time $O(n^{c \log n})$.

10.130 (a) Reduce SAT to TRUESAT. Suppose we are given an expression E with variables $x_1, x_2, \ldots, x_n$. Convert E to E' as follows:
(i) First, test if E is true when all variables are true. If so, E is satisfiable; so convert it to a specific expression $x \vee y$ that we know is in TRUESAT.
(ii) Otherwise, let $E' = E \vee (x_1 \wedge x_2 \wedge \cdots x_n)$. It is a polynomial time reduction. E' is true when all variables are true.

If E is in SAT, then it is satisfied by some interpretation, which does not assign all variables to 1, because we tested such an interpretation and found E to be false. Thus, E' is in TRUESAT. Conversely, if E' is in TRUESAT, then as $x_1 \wedge x_2 \wedge \cdots \wedge x_n$ is true only when all of them are 1, E must be satisfiable.

10.131 Guess z, compute $f(z)$ deterministically in polynomial time, and test whether $f(z) = x$. When the guess of z is correct, we have $f^{-1}(x)$. Compare it with y, and accept the pair (x, y) if $z < y$. This shows that $L \in NP$. To show that $L \in CoNP$, we require the set of inputs that are not of the form (x, y), with $f^{-1}(x) < y$, is in *NP*. It is easy to check for ill-formed inputs. For the rest of the inputs, guess z, compute $f(z)$, test if $f(z) = x$, and then test if $z \geq y$. If both tests succeed, then $f^{-1}(x) \geq y$, so (x, y) is in $\overline{L}$.

10.134 (a) For any polynomially bounded TM M and an input w, construct, in polynomial time, a regular expression R that generates all strings that are not sequences of configurations of M leading to the acceptance of w.

10.139 (a) If $F = F_1 \vee F_2$, then (if F_1 is true, F is true, else, (F is as is F_2)). If $F = \exists x\, E$, then (Evaluate E by taking each x in it as 0. If result is 1, then F is true. Else, evaluate E by taking each x as 1. The result is the value of F.)
(b) For a DTM M that uses $p(n)$ space, and an input w of length n, construct a quantified boolean formula that is true iff M accepts w. You will have to consider translating the moves of M as quantified boolean formulas analogous to the construction in Cook's theorem.

10.146 (a) As C progresses, B finds a cute algorithm, say, D that satisfies $\ell(\psi(D)) > \ell(\psi(C))$. Next, D is run. Thus, C does exactly what D does, eventually. Then D is not cute; a contradiction.

References

1. Abramsky, S., Gabbay, D., Maibaum, T. (eds.): Handbook of Logic in Computer Science, Oxford University Press, London (1992–2000)
2. Ackermann, W.: Zum Hilbertschen aufbau der reellen zahlen. Math. Annalen. **99**, 118–133 (1928)
3. Agarwal, M., Kayal, N., Saxena, N.: PRIMES is in P. Ann. Math. **160**, 781–793 (2004)
4. Aho, A.V., Sethi, R., Ullman, J.D.: Compilers: Principles, Techniques, and Tools. Addison-Wesley, Reading, MA (1986)
5. Ausiello, G., Crescenzi, P., Protasi, M.: Approximate solution of NP optimization problems. Theor. Comp. Sci. **150**, 1–55 (1995)
6. Backhouse, R.C.: Closure Algorithms and the Star-height Problem of Regular Languages. Ph.D. Thesis, Imperial College, London (1975)
7. Backus, J.W.: The syntax and semantics of the proposed international algebraic language of the Zurich ACM-GAMM conference. In: Proc. Int. Conf. on Information Processing, pp. 125–132. UNESCO (1959)
8. Bar-Hillel, V., Perls, M., Shamir, E.: On formal properties of simple phrase structure grammars. Zeitschrift fur Phonetik, Sprachwissenschaft, und Kommunikationsforschung. **14**, 143–172 (1961)
9. Barendregt, H.P.: The Lambda Calculus. North-Holland, Amsterdam (1984)
10. Bird, M.: The equivalence problem for deterministic two-tape automata. J. Comp. Sys. Sci. **7**, 218–236 (1973)
11. Cantor, D.C.: On the ambiguity problem of backus systems. J. ACM. **9**, 477–479 (1962)
12. Cantor, G.: Über eine eigenschaft des inbegriffes aller reellen algebraischen zahlen. J. für die Reine und Angewandte Mathematik. **77**, 258–262 (1874). Reprinted in: Georg Cantor Gesammelte Abhandlungen, pp. 115–118. Springer-Verlag, Berlin (1932)
13. Chaitin, G.J.: Information-theoretic complexity. IEEE Trans. Inform. Theory. **IT-20**, 10–15 (1974)
14. Chomsky, N.: Three models for the description of languages. IRE Trans. Inform. Theory. **2–3**,113–124 (1956)
15. Chomsky, N.: On certain formal properties of grammars. Inform. Control. **2**, 137–167 (1959)

16. Chomsky, N.: Context-free grammars and pushdown storage. Tech. Rep., MIT Research Lab in Electronics. Cambridge, MA (1962)
17. Chomsky, N., Miller, G.A.: Finite state languages. Inform. Control. **1**, 91–112 (1958)
18. Chomsky, N., Schützenberger, M.P.: The algebraic theory of context-free languages. In: Braffort, P., Hirschberg, D. (eds.) Computer Programming and Formal Systems, pp. 118–161. North-Holland, Amsterdam (1963)
19. Church, A.: An unsolvable problem of elementary number theory. Amer. J. Math. **58**, 345–363 (1936)
20. Cobham, A.: The intrinsic computational difficulty of functions. In: Proc. 1964 Congress for Logic, Math. Phil. of Sci., pp. 24–30. North-Holland, Amsterdam (1964)
21. Cohn, P.M.: Algebra, Vol 2. Wiley, NY (1977)
22. Conway, J.H.: Regular Algebra and Finite Machines. Chapman and Hall, London (1971)
23. Cook, S.A.: The complexity of theorem proving procedures. In: Proc. Third Symp. Theory of Computing, pp. 151–158. Assoc. Comput. Mach., NY (1971)
24. Curry, H.B.: An analysis of logical substitutions. Am. J. Math. **51**, 363–384 (1929)
25. Cutland, N.: Computability: An Introduction to Recursive Function Theory. Cambridge University Press, NY (1980)
26. Davis, M.: Hilbert's tenth problem is unsolvable. Amer. Math. Monthly. **80**, 233–269 (1973)
27. Denning, P.J., Dennis, J.B., Qualitz, J.E.: Machines, Languages, and Computation. Prentice Hall, Englewood-Cliffs, NJ (1978)
28. Earley, J.: An efficient context-free parsing algorithm. Comm. of ACM. **13**, 94–102 (1970)
29. Ehrenfeucht, A., Parikh, R., Rozenberg, G.: Pumping lemmas and regular sets. SIAM J. Comput. **10**, 536–541(1981)
30. Evey, J.: Application of pushdown store machines. In: Proc. Fall Joint Computer Conf., pp. 215–227. AFIPS Press, Montvale, NJ (1963)
31. Fischer, P.C.: On computability by certain class of restricted Turing machines. In: Proc. Fourth Symp. Switching Circuit Theory and Logical Design, pp. 23–32 (1963)
32. Fischer, P.C.: Turing machines with restricted memory access. Inform. Control. **9**, 364–379 (1966)
33. Floyd, R.W.: On ambiguity in phrase-structure languages. Comm. ACM. **5**, 526–534 (1962)
34. Gallier, J., Hicks, A.: The Theory of Languages and Computation, web page: http://www.cis.upenn.edu/ jean/gbooks/tc.html. Cited 15 Nov 2008
35. Gamow, G.: One, Two, Three, . . . Infinity: Facts and Speculations of Science, The Viking Press, NY (1961)
36. Garey, M.R., Johnson, D.S.: Computers and Intractability: A Guide to the Theory of NP-completeness. W.H. Freeman, NY (1979)
37. Ginsburg, S.: Examples of abstract machines. IEEE Trans. Elec. Comp. **EC-11**, 132–135 (1962)

38. Ginsburg, S.: The Mathematical Theory of Context-free Languages. McGraw-Hill, NY (1966)
39. Ginsburg, S., Greibach, S.A.: Deterministic context-free languages. Inform. Control. **9**, 563–582 (1966)
40. Ginsburg, S., Rose, G.F.: Some recursively unsolvable problems in ALGOL-like languages. J. Assoc. Comput. Mach. **10**, 29–47 (1963)
41. Ginsburg, S., Rose, G.F.: Preservation by languages by transducers. Inform. Control. **9**, 153–176 (1966)
42. Ginsburg, S., Spanier, E.H.: Quotients of context-free languages. J. ACM. **10**, 487–492 (1963)
43. Gödel, K.: Über formal unentscheidbare Sätze der Principia Mathematica und verwandter Systeme I. Monatshefte für Mathematik und Physik. **38**, 173–198 (1931)
44. Gödel, K.: On undecidable propositions of formal mathematical systems. In: Davis, M. (ed.) The Undecidable, pp. 5–38. Raven Press, Hewitt, NY (1965)
45. Goldstine, J.: A simplified proof of Parikh's theorem. Discrete Math. **19**, 235–240 (1977)
46. Greenlaw, R., Hoover, H.J.: Fundamentals of the Theory of Computation: Principles and Practice. Morgan Kaufmann, NY (1998)
47. Greibach, S.A.: A new normal form theorem for context-free phrase structure grammars. J. Assoc. Comput. Mach. **12**, 42–52 (1965)
48. Gries, D., Schneider, F.B.: A Logical Approach to Discrete Math, Springer, NY (1993)
49. Haines, L.: Generation and Recognition of Formal Languages. Ph.D. thesis, MIT, Cambridge, MA (1965)
50. Halmos, P.R.: Naive Set Theory. Van Nostrand Reinhold Co., NY (1960)
51. Harrison, M.A.: Introduction to Formal Language Theory. Addison-Wesley, Reading, MA (1978)
52. Hartmanis, J., Hopcroft, J.E.: Structure of undecidable problems in automata theory. In: Proc. Ninth Symp. Switching and Automata Theory, pp. 327–333. IEEE (1968)
53. Hartmanis, J., Stearns, R.E.: On the computational complexity of algorithms. Trans. Am. Math. Soc. **117**, 285–306 (1965)
54. Herms, H.: Enumerability, Decidability, Computability. Springer-Verlag, NY (1969)
55. Hochbaum, D.: Approximation Algorithms for NP-hard Problems. PWS Pub., Boston (1996)
56. Hodges, A.: Alan Turing: The Enigma. Walker and Co., NY (2000)
57. Hopcroft, J.E.: An $n \log n$ algorithm for minimizing the states in a finite automaton. In: Kohavi, Z. (ed.) The Theory of Machines and Computation, pp. 189–196. Academic Press, NY (1971)
58. Hopcroft, J.E., Ullman, J.D.: Introduction to Automata Theory, Languages, and Computation. Addison-Wesley, Reading, MA (1979)
59. Hopcroft, J.E., Motwani, R., Ullman, J.D.: Introduction to Automata Theory, Languages, and Computation, 3rd ed. Pearson Education, NY (2007)
60. Huffman, D.A.: The synthesis of sequential switching circuits. J. Franklin Institute. **257**, 161–190 & 275–303 (1954)

61. Jaffe, J.: A necessary and sufficient pumping lemma for regular languages. SIGACT News. **10**, 48–49 (1978)
62. Karp, R.M.: Reducibility among combinatorial problems. In: Miller, R.E., Thatcher, J.W. (eds.) Complexity of Computer Computations, pp. 85–103. Plenum Press, NY (1972)
63. Kasami, T.: An efficient recognition and syntax algorithm for context-free languages. Tech. Rep. **AFCRL-65–758**, Air Force cambridge Research Lab., Redford, MA (1965)
64. Kleene, S.C.: General recursive functions of natural numbers. Math. Annalen. **112**, 727–742 (1936)
65. Kleene, S.C.: Recursive predicates and recursive quantifiers. Trans. Am. Math. Soc. **53**, 41–74 (1943)
66. Kleene, S.C.: Introduction to Metamathematics. D. Van Nostrand, Princeton, NJ (1974)
67. Kleene, S.C.: Representation of events in nerve nets and finite automata. In: Shannon, C.E., McCarthy, J. (eds.) Automata Studies, pp. 3–41. Princeton University Press, Princeton, NJ (1956)
68. Knuth, D.E.: On the translation of languages from left to right. Inform. Control. **8**, 607–639 (1965)
69. Knuth, D.E., Morris Jr., J.E., Pratt, V.R.: Fast pattern matching in strings. SIAM J. Comput. **6**, 323–350 (1976)
70. Kozen, D.C.: A completeness theorem for Kleene algebras and the algebra of regular events. Inform. Comput. **110**, 366–390 (1994)
71. Kozen, D.C.: Automata and Computability. Springer-Verlag, NY (1997)
72. Kozen, D.C.: Theory of Computation. Springer-Verlag, NY (2006)
73. Kuich, W.: The Kleene and Parikh theorem in complete semirings. In: Ottmann, T. (ed.) Proc. 14th Colloq. Aut. Lang. and Progr., Lect. Notes in Comp. Sci. **257**, pp. 212–225. EATCS, Springer-Verlag, NY (1987)
74. Kuroda, S.Y.: Classes of languages and linear bounded automata. Inform. Control. **7**, 207–223 (1964)
75. Landweber, P.S.: Three theorems on phrase structure grammars of type 1. Inform. Control. **6**, 131–136 (1963)
76. Levin, L.L.: Universal sorting problems. Problemi Predachi Informatsii. **9**, 265–266 (1973)
77. Lewis, P.M., Rosenkrantz, D.J., Stearns, R.E.: Compiler Design Theory. Addison-Wesley, Reading, MA (1976)
78. Lewis, H.R., Papadimitriou, C.H.: Elements of the Theory of Computation, 2nd ed. Pearson Education, NJ (1998)
79. Linz, P.: An Introduction to Formal Languages and Automata, 4th ed. Jones and Bartlett, MA (2006)
80. Manin, Y.I.: A Course in Mathematical Logic. Springer-Verlag, NY (1977)
81. Manna, J.: Mathematical Theory of Computation. McGraw-Hill, NY (1974)
82. Markov, A.A.: The Theory of Algorithms. Trudy Math. Steklov Inst. (1954). Eng. Trans.: National Science Foundation, Washington, DC (1961)
83. Matiyasevich, Y.: Enumerable sets are Diophantine. Dokl. Akad. Nauk SSSR. **191**, 279–282 (1970). Eng. Trans.: Soviet Math. Dokl. **11**, 354–357 (1970)

84. McCulloh, W.S., Pitts, W.: A logical calculus of the ideas immanent in nervous activity. Bull. Math. Biophy. **5**, 39–47 (1943)
85. McNaughton, R., Yamada, H.: Regular expressions and state graphs for automata. IEEE Trans. Elec. Computs. **9**, 39–47 (1960)
86. Mealy, G.H.: A method for synthesizing sequential circuits. Bell Sys. Tech. J. **34**, 1045–1079 (1955)
87. Meyer, A.R., Ritchie, D.M.: The complexity of loop programs. In: Proc. ACM Natl. Meeting, pp. 465–469 (1967)
88. Milner, R.: Operational and algebraic semantics of concurrent processes. In: van Leuwen, J. (ed.) Handbook of Theoretical Computer Science, Vol B, pp. 1201–1242. North-Holland, Amsterdam (1990)
89. Minsky, M.L.: Recursive unsolvability of Post's problem of tag and other topics in the theory of Turing machines. Ann. Math. **74**, 437–455 (1961)
90. Minsky, M.L.: Computation: Finite and Infinite Machines. Prentice Hall, Englewood-Cliffs, NJ (1967)
91. Moore, E.F.: Gadanken experiment on sequential machines. Automata Studies. 129–153 (1956)
92. Moret, M.E.: The Theory of Computation. Addison-Wesley, NY (1998)
93. Myhill, J.: Finite automata and the representation of events. In: Technical Note WADD, pp. 57–624. Wright Patterson AFB, Dayton, Ohio (1960)
94. Naur, P.: Revised report on the algorithmic language Algol 60. Comm. of the ACM. **6**, 1–7 (1963). Reprinted in: Rosen, S. (ed.) Programming Systems and Languages, pp. 79–118. McGraw-Hill, NY (1967)
95. Nerode, A.: Linear automaton transformations. Proc. Am. Math. Soc. **9**, 541–544 (1958)
96. Oettinger, A.G.: Automatic syntactic analysis and the pushdown store. In: Proc. Symp. on Appl. Math. Vol 12., pp. 104–129. American Math. Soc., Providence, RI (1961)
97. Ogden, W.G.: A helpful result for proving inherent ambiguity. Math. Sys. Theory **2**, 191–194 (1968)
98. Papadimitriou, C.H.: Computational Complexity. Addison-Wesley, Reading, MA (1994)
99. Papadimitriou, H., Steiglitz, K.: Combinatorial Optimization: Algorithms and Complexity, 2nd ed. Dover, NY (1997)
100. Parikh, R.: On context-free languages. J. Assoc. Comput. Mach. **13**, 570–581 (1966)
101. Parikh, R.J.: Existence and Feasibility in Arithmetic. J. Symbolic Logic **36**, 494–508 (1971).
102. Polya, G., Mathematics and Plausible Reasoning, Vol. I: Induction and Analogy in Mathematics. Princeton University Press, Priceton, NJ (1954)
103. Post, E.: Finite combinatory process – formulation I. J. Symb. Logic. **1**, 103–105 (1936)
104. Post, E.: Formal reductions of the general combinatorial decision problem. Am. J. Math. **65**, 197–215 (1943)
105. Post, E.: Recursively enumerable sets of positive natural numbers and their decision problems. Bull. Am. Math. Soc. **50**, 284–316 (1944)

106. Post, E.: A variant of recursively unsolvable problem. Bull. Am. Math. Soc. **52**, 264–268 (1946)
107. Rabin, M.O., Scott, D.S.: Finite automata and their decision problems. IBM J. Res. Develop. **3**, 115–125 (1959)
108. Reeves, C.R.: Modern Heuristic Techniques for Combinatorial Problems. John Wiely, NY (1993)
109. Rice, H.G.: Classes of recursively enumerable sets and their decision problems. Trans. Am. Math. Soc. **89**, 25–59 (1953)
110. Rice, H.G.: On completely recursively enumerable classes and their key arrays. J. Sym. Logic. **21**, 301–341 (1956)
111. Rogers Jr., H.: Theory of Recursive Functions and Effective Computability. McGraw-Hill, NY (1967)
112. Salomaa, A.: Two complete axiom systems for the algebra of regular events. J. Assoc. Comput. Mach. **13**, 158–169 (1966)
113. Salomaa, A.: Formal Languages. Academic Press, NY (1973)
114. Scheinberg, S.: Note on Boolean properties of context-free languages. Inform. Control. **3**, 372–375 (1960)
115. Schützenberger, M.P.: On context-free languages and pushdown automata. Inform. Control. **6**, 246–264 (1963)
116. Seiferas, J.I., McNaughton, R.: Regularity preserving relations. Theor. Comp. Sci. **2**, 147–154 (1976)
117. Shepherdson, J.C.: The reduction of two-way automata to one-way automata. IBM J. Res. Develop. **3**, 198–200 (1959)
118. Shepherdson, J.C., Sturgis, H.C.: Computability of recursive functions. J. Assoc. Comput. Mach. **10**, 217–255 (1963)
119. Shoenfield, J.R.: Degrees of Unsolvability. North-Holland, Amsterdam (1971)
120. Singh, A., Goswamy, C.: Fundamentals of Logic. Indian Council of Philosophical Research, New Delhi (1998)
121. Singh, A.: Logics for Computer Science. PHI, New Delhi (2003)
122. Sipser, M.: An Introduction to the Theory of Computation. PWS Pub. Co. NY (1997)
123. Soare, R.I.: Recursively Enumerable Sets and Degrees. Springer-Verlag, Berlin (1987)
124. Soare, R.I.: Computability and recursion. Bull. Symb. Logic. **2**, 284–321 (1996)
125. Soare, R.I.: Computability and incomputability. In: Proc. Third Conf. on Computability in Europe, CIE 2007, Siena, Italy, June 18–23, 2007. Lect. Notes in Comp. Sci. No. 4497, S.B. Cooper, B. Löwe, A. Sorbi (Eds.) Springer-Verlag, Berlin, Heidelberg (2007)
126. Stanat, D., Weiss, S.: A pumping theorem for regular languages. SIGACT News **14**, 36–37 (1982)
127. Stearns, R.E., Hartmanis, J.: Regularity preserving modifications of regular expressions. Inform. Control **6**, 55–69 (1963)
128. Stockmeyer, L.J.: The polynomial-time hierarchy. Theor. Comp. Sci. **3**, 1–22 (1976)
129. Thomas, W.: Languages, automata, and logic. In G. Rozenberg and A. Salomaa, eds., *Handbook of Formal Languages, volume III*, 389–455, Springer-Verlag, New York (1997)

130. Thompson, K.: Regular expression search algorithms. Comm. ACM **11**, 419–422 (1968)
131. Turing, A.M.: On computable numbers with an application to the Entscheidungsproblem. Proc. Lond. Math. Soc. **42**, 230–265 (1936), Erratum: Ibid. **43**, 544–546 (1937)
132. Turing, A.M.: Systems of logic based on ordinals. Proc. Lond. Math. Soc. **42**, 230–265 (1939)
133. Valiant, L.G.: General context-free recognition in less than cubic time. J. Comp. Sys. Sci. **10**, 308–315 (1975)
134. Weil, P.: Algebraic recognizability of languages. In: Lecture Notes in Computer Science No. 3153, Fiala, J., Koubek, V., Kratochvl, J. (eds.) Springer-Verlag, 149–175 (2004)
135. Younger, D.H.: Recognition and parsing of context-free languages in time n^3. Inform. Control. **10**, 189–208 (1967)

Index